P9-CSB-699

Springer Texts in Statistics

Advisors:
Stephen Fienberg   Ingram Olkin

Springer
New York
Berlin
Heidelberg
Barcelona
Budapest
Hong Kong
London
Milan
Paris
Tokyo

Springer Texts in Statistics

Alfred: Elements of Statistics for the Life and Social Sciences
Berger: An Introduction to Probability and Stochastic Processes
Blom: Probability and Statistics: Theory and Applications
Brockwell and Davis: An Introduction to Times Series and Forecasting
Chow and Teicher: Probability Theory: Independence, Interchangeability, Martingales, Second Edition
Christensen: Plane Answers to Complex Questions: The Theory of Linear Models, Second Edition
Christensen: Linear Models for Multivariate, Time Series, and Spatial Data
Christensen: Log-Linear Models
Creighton: A First Course in Probability Models and Statistical Inference
du Toit, Steyn and Stumpf: Graphical Exploratory Data Analysis
Edwards: Introduction to Graphical Modelling
Finkelstein and Levin: Statistics for Lawyers
Jobson: Applied Multivariate Data Analysis, Volume I: Regression and Experimental Design
Jobson: Applied Multivariate Data Analysis, Volume II: Categorical and Multivariate Methods
Kalbfleisch: Probability and Statistical Inference, Volume I: Probability, Second Edition
Kalbfleisch: Probability and Statistical Inference, Volume II: Statistical Inference, Second Edition
Karr: Probability
Keyfitz: Applied Mathematical Demography, Second Edition
Kiefer: Introduction to Statistical Inference
Kokoska and Nevison: Statistical Tables and Formulae
Lehmann: Testing Statistical Hypotheses, Second Edition
Lindman: Analysis of Variance in Experimental Design
Madansky: Prescriptions for Working Statisticians
McPherson: Statistics in Scientific Investigation: Its Basis, Application, and Interpretation
Mueller: Basic Principles of Structural Equation Modeling
Nguyen and Rogers: Fundamentals of Mathematical Statistics: Volume I: Probability for Statistics
Nguyen and Rogers: Fundamentals of Mathematical Statistics: Volume II: Statistical Inference
Noether: Introduction to Statistics: The Nonparametric Way
Peters: Counting for Something: Statistical Principles and Personalities
Pfeiffer: Probability for Applications
Pitman: Probability
Robert: The Bayesian Choice: A Decision-Theoretic Motivation

Continued at end of book

Jim Pitman

PROBABILITY

Springer

Jim Pitman
Department of Statistics
University of California
Berkeley, CA 94720

Mathematical Subject Classification (1992): 60–01

Library of Congress Cataloging-in-Publication Data
Pitman, Jim.
Probability / Jim Pitman.
p. cm. -- (Springer texts in statistics)
Includes bibliographical references and index.
ISBN 0-387-97974-3 (U.S.) -- ISBN 3-540-97974-3
1. Probabilities. I. Title. II. Series.
QA273.P493 1993
519.2--dc20 92-39051

Production managed by Karen Phillips; manufacturing supervised by Vincent Scelta.
Photocomposed pages prepared using the author's LaTeX file.
Printed and bound by R.R. Donnelley and Sons, Harrisonburg, VA.
Printed in the United States of America.

9 8 7 6 (Corrected sixth printing, 1997)

ISBN 0-387-97974-3 Springer-Verlag New York Berlin Heidelberg
ISBN 3-540-97974-3 Springer-Verlag Berlin Heidelberg New York SPIN 10567834

Preface

Preface to the Instructor

This is a text for a one-quarter or one-semester course in probability, aimed at students who have done a year of calculus. The book is organized so a student can learn the fundamental ideas of probability from the first three chapters without reliance on calculus. Later chapters develop these ideas further using calculus tools.

The book contains more than the usual number of examples worked out in detail. It is not possible to go through all these examples in class. Rather, I suggest that you deal quickly with the main points of theory, then spend class time on problems from the exercises, or your own favorite problems. The most valuable thing for students to learn from a course like this is how to pick up a probability problem in a new setting and relate it to the standard body of theory. The more they see this happen in class, and the more they do it themselves in exercises, the better.

The style of the text is deliberately informal. My experience is that students learn more from intuitive explanations, diagrams, and examples than they do from theorems and proofs. So the emphasis is on problem solving rather than theory.

Order of Topics. The basic rules of probability all appear in Chapter 1. Intuition for probabilities is developed using Venn and tree diagrams. Only finite additivity of probability is treated in this chapter. Discussion of countable additivity is postponed to Section 3.4. Emphasis in Chapter 1 is on the concept of a probability distribution and elementary applications of the addition and multiplication rules. Combinatorics appear via study of the binomial and hypergeometric distributions in Chapter 2. The concepts of mean and standard deviation appear in a preliminary form in this chapter, motivated by the normal approximation, without the notation of random variables. These concepts are then developed for discrete random variables in Chapter 3. The main object of the first three chapters is to get to the circle of ideas around the normal approximation for sums of independent random variables. This is achieved by Section 3.3. Sections 3.4 and 3.5 deal with the standard distributions on the non-negative integers. Conditional distributions and expectations, covariance and correlation for discrete distributions are postponed to Chapter 6, nearby treatment of the same concepts for continuous distributions. The discrete theory could be done right after Chapter 3, but it seems best to get as quickly as possible to continuous things. Chapters 4 and 5 treat continuous distributions assuming a calculus background. The main emphasis here is on how to do probability calculations rather than rigorous development of the theory. In particular, differential calculations are used freely from Section 4.1 on, with only occasional discussion of the limits involved.

Optional Sections. These are more demanding mathematically than the main stream of ideas.

Terminology. Notation and terms are standard, except that *outcome space* is used throughout instead of sample space. Elements of an outcome space are called *possible outcomes.*

Pace. The earlier chapters are easier than later ones. It is important to get quickly through Chapters 1 and 2 (no more than three weeks). Chapter 3 is more substantial and deserves more time. The end of Chapter 3 is the natural time for a midterm examination. This can be as early as the sixth week. Chapters 4, 5, and 6 take time, much of it spent teaching calculus.

Preface to the Student

Prerequisites. This book assumes some background of mathematics, in particular, calculus. A summary of what is taken for granted can be found in Appendices I to III. Look at these to see if you need to review this material, or perhaps take another mathematics course before this one.

How to read this book. To get most benefit from the text, work one section at a time. Start reading each section by skimming lightly over it. Pick out the main ideas, usually boxed, and see how some of the examples involve these ideas. Then you may already be able to do some of the first exercises at the end of the section, which you should try as soon as possible. Expect to go back and forth between the exercises and the section several times before mastering the material.

Exercises. Except perhaps for the first few exercises in a section, do not expect to be able to plug into a formula or follow exactly the same steps as an example in the text. Rather, expect some variation on the main theme, perhaps a combination with ideas of a previous section, a rearrangement of the formula, or a new setting of the same principles. Through working problems you gain an active understanding of the concepts. If you find a problem difficult, or can't see how to start, keep in mind that it will always be related to material of the section. Try re-reading the section with the problem in mind. Look for some similarity or connection to get started. Can you express the problem in a different way? Can you identify relevant variables? Could you draw a diagram? Could you solve a simpler problem? Could you break up the problem into simpler parts? Most of the problems will yield to this sort of approach once you have understood the basic ideas of the section. For more on problem-solving techniques, see the book *How to Solve It* by G. Polya (Princeton University Press).

Solutions. Brief solutions to most odd numbered exercises appear at the end of the book.

Chapter Summaries. These are at the end of every chapter.

Review Exercises. These come after the summaries at the end of every chapter. Try these exercises when reviewing for an examination. Many of these exercises combine material from previous chapters.

Distribution Summaries. These set out the properties of the most important distributions. Familiarity with these properties reduces the amount of calculation required in many exercises.

Examinations. Some midterm and final examinations from courses taught from this text are provided, with solutions a few pages later.

Acknowledgments

Thanks to many students and instructors who have read preliminary versions of this book and provided valuable feedback. In particular, David Aldous, Peter Bickel, Ed Chow, Steve Evans, Roman Fresnedo, David Freedman, Alberto Gandolfi, Hank Ibser, Barney Krebs, Bret Larget, Russ Lyons, Lucien Le Cam, Maryse Loranger, Deborah Nolan, David Pollard, Roger Purves, Joe Romano, Tom Salisbury, David Siegmund, Anne Sheehy, Philip Stark, and Ruth Williams made numerous corrections and suggestions. Thanks to Ani Adhikari, David Aldous, David Blackwell, David Brillinger, Lester Dubins, Persi Diaconis, Mihael Perman and Robin Pemantle for providing novel problems. Thanks to Ed Chow, Richard Cutler, Bret Larget, Kee Won Lee, and Arunas Rudvalis who helped with solutions to the problems. Thanks to Carol Block and Chris Colbert who typed an early draft. Special thanks to Ani Adhikari, who provided enormous assistance with all aspects of this book. The graphics are based on her library of mathematical graphics routines written in PostScript. The graphics were further developed by Jianqing Fan, Ed Chow, and Ofer Licht. Thanks to Ed Chow for organizing drafts of the book on the computer, and to Bret Larget and Ofer Licht for their assistance in final preparation of the manuscript.

Jim Pitman, January 1993

Contents

1 Introduction

This chapter introduces the basic concepts of probability theory. These are the notions of:

- *an outcome space*, or set of all possible outcomes of some kind;
- *events* represented mathematically as *subsets* of an outcome space; and
- *probability* as a function of these events or subsets.

The word "event" is used here for the kind of thing that has a probability, like getting a six when you roll a die, or getting five heads in a row when you toss a coin five times. The probability of an event is a measure of the likelihood or chance that the event occurs, on a scale from 0 to 1. Section 1.1 introduces these ideas in the simplest setting of equally likely outcomes. Section 1.2 treats two important interpretations of probability: approximation of long-run frequencies and subjective judgment of uncertainty. However probabilities are understood or interpreted, it is generally agreed that they must satisfy certain rules, in particular the basic *addition rule*. This rule is built in to the idea of a *probability distribution*, introduced in Section 1.3. The concepts of *conditional probability*, and *independence* appear in Section 1.4. These concepts are further developed in Section 1.5 on *Bayes' rule* and Section 1.6 on sequences of events.

1.1 Equally Likely Outcomes

Probability is an extension of the idea of a *proportion*, or ratio of a part to a whole. If there are 300 men and 700 women in a group, the proportion of men in the group is

$$\frac{300}{300 + 700} = 0.3 = 30\%$$

Suppose now that someone is picked at random from this population of men and women. For example, the choice could be made by drawing at random from a box of 1000 tickets, with different tickets corresponding to different people. It would then be said that

- the *probability* of choosing a woman is 70%;
- the *odds in favor* of choosing a woman are 7 to 3 (or 7/3 to 1); and
- the *odds against* choosing a woman are 3 to 7 (or 3/7 to 1).

So in thinking about someone picked at random from a population, a proportion in the population becomes a probability, and something like a sex ratio becomes an odds ratio.

There is an implicit assumption here: "*picked at random*" means everyone has the same chance of being chosen. In practice, for a draw at random from a box, this means the tickets are similar, and well mixed up before the draw. Intuitively, we say different tickets are *equally likely*, or that they have the *same chance*. In other words, the draw is *honest*, *fair*, or *unbiased*. In more mathematical language, the *probability* of each ticket is the same, namely, 1/1000 for an assumed total of 1000 tickets.

For the moment, take for granted this intuitive idea of equally likely outcomes. Represent the set of all possible outcomes of some situation or experiment by Ω (capital omega, the last letter in the Greek alphabet). For instance, Ω would be the set of 1000 people (or the 1000 corresponding tickets) in the previous example. Or $\Omega = \{head, tail\}$ for the result of tossing a coin, or $\Omega = \{1, 2, 3, 4, 5, 6\}$ for rolling an ordinary six-sided die. The set Ω is called the *outcome space*. Something that might or might not happen, depending on the outcome, is called an *event*. Examples of events are "person chosen at random is a woman", "coin lands heads", "die shows an even number". An event A is represented mathematically by a subset of the outcome space Ω. For the examples above, A would be the set of women in the population, the set comprising the single outcome $\{head\}$, and the set of even numbers $\{2, 4, 6\}$.

Let $\#(A)$ be the number of outcomes in A. Informally, this is the number of chances for A to occur, or the number of different ways A can happen. Assuming equally likely outcomes, the probability of A, denoted $P(A)$, is defined to be the corresponding proportion of outcomes. This would be 700/1000, 1/2, and 3/6 in the three examples.

Equally Likely Outcomes

If all outcomes in a finite set Ω are equally likely, the probability of A is the number of outcomes in A divided by the total number of outcomes:

$$P(A) = \frac{\#(A)}{\#(\Omega)}$$

Probabilities defined by this formula for equally likely outcomes are fractions between 0 and 1. The number 1 represents certainty: $P(\Omega) = 1$. The number 0 represents impossibility: $P(A) = 0$ if there is no way that A could happen. Then A corresponds to the empty set, or set with no elements, denoted $\emptyset$. So $P(\emptyset) = 0$. Intermediate probabilities may be understood as various degrees of certainty.

Example 1. **Picking a number between 1 and 100.**

Suppose there is a box of 100 tickets marked $1, 2, 3, \ldots, 100$. A ticket is drawn at random from the box. Here are some events, with their descriptions as subsets and their probabilities obtained by counting. All possible numbers are assumed equally likely.

Event	Subset of $\{1, 2, \ldots, 100\}$	Probability
the number drawn has one digit	$\{1, 2, \ldots, 9\}$	9%
the number drawn has two digits	$\{10, 11, \ldots, 99\}$	90%
the number drawn is less than or equal to the number k	$\{1, 2, \ldots, k\}$	k%
the number drawn is strictly greater than k	$\{k+1, \ldots, 100\}$	$(100 - k)$%
the sum of the digits in the number drawn is equal to 3	$\{3, 12, 21, 30\}$	4%

Example 2. **Rolling two dice.**

A fair die is rolled and the number on the top face is noted. Then another fair die is rolled, and the number on its top face is noted.

Problem 1. What is the probability that the sum of the two numbers showing is 5?

Solution. Think of each possible outcome as a pair of numbers. The first element of the pair is the first number rolled, and the second element is the second number rolled. The first number can be any integer between 1 and 6, and so can the second number. Here are all the possible ways the dice could roll:

$$\begin{array}{cccccc}
(1,1) & (1,2) & (1,3) & (1,4) & (1,5) & (1,6) \\
(2,1) & (2,2) & (2,3) & (2,4) & (2,5) & (2,6) \\
(3,1) & (3,2) & (3,3) & (3,4) & (3,5) & (3,6) \\
(4,1) & (4,2) & (4,3) & (4,4) & (4,5) & (4,6) \\
(5,1) & (5,2) & (5,3) & (5,4) & (5,5) & (5,6) \\
(6,1) & (6,2) & (6,3) & (6,4) & (6,5) & (6,6)
\end{array}$$

The collection of these 36 pairs forms the outcome space Ω. Assume these 36 outcomes are equally likely. The event "the sum of the two numbers showing is 5" is represented by the subset $\{(1,4),(4,1),(2,3),(3,2)\}$. Since this subset has 4 elements,

$$P(\text{sum of two numbers showing is } 5) = \frac{4}{36} = \frac{1}{9}$$

Problem 2. What is the probability that one of the dice shows 2, and the other shows 4?

Solution. The subset corresponding to this event is $\{(2,4),(4,2)\}$. So the required probability is $2/36 = 1/18$.

Problem 3. What is the probability that the second number rolled is greater than the first number?

Solution. Look at the pairs in the outcome space Ω above, to see that this event corresponds to the subset

$$\begin{array}{ccccc}
(1,2) & (1,3) & (1,4) & (1,5) & (1,6) \\
 & (2,3) & (2,4) & (2,5) & (2,6) \\
 & & (3,4) & (3,5) & (3,6) \\
 & & & (4,5) & (4,6) \\
 & & & & (5,6)
\end{array}$$

These are the pairs above the diagonal in Ω. There are 15 such pairs, so the probability that the second number rolled is greater than the first is $15/36$.

Problem 4. What is the probability that the second number rolled is less than the first number rolled?

Solution. The subset of Ω corresponding to this event is the set of pairs below the diagonal. There are just as many pairs below the diagonal as above. So the probability that the second number rolled is less than the first number is also $15/36$.

Example 3. **Rolling two n-sided dice.**

Repeat the above example for two rolls of a die with n faces numbered $1, 2, \ldots, n$, assuming $n \geq 4$.

Problem 1. Find the chance that the sum is 5.

Solution. Now there are n^2 possible pairs instead of $6^2 = 36$. But there are still just 4 possible pairs with sum 5. Hence

$$P(\text{sum is } 5) = 4/n^2$$

Problem 2. Find the chance that one roll is a 2, the other is a 4.

Solution. By the same argument $P(\text{a 2 and a 4}) = 2/n^2$.

Problem 3. Find the chance that the second number is greater than the first.

Solution. Now all pairs above the diagonal must be counted in an $n \times n$ matrix of pairs. There are no such pairs in the bottom row, 1 in the next, 2 in the next, and so on up to $(n-1)$ pairs in the top row, so the number of pairs above the diagonal is

$$\#(\text{above}) = 1 + 2 + 3 + \cdots + (n-1) = \frac{1}{2}n(n-1)$$

pairs altogether (see Appendix 2 on sums.) This gives

$$P(\text{second number is greater}) = \frac{\#(\text{above})}{\#(\text{total})} = \frac{\frac{1}{2}n(n-1)}{n^2} = \frac{1}{2}\left(1 - \frac{1}{n}\right)$$

Remark. Here is another way to find $\#(\text{above})$, which gives the formula for the sum of the first $n-1$ integers (used above) as a consequence. Since

$$\#(\text{below}) + \#(\text{above}) + \#(\text{diagonal}) = \#(\text{total}) = n^2$$

and $\#(\text{below}) = \#(\text{above})$ by symmetry, and $\#(\text{diagonal}) = n$,

$$\#(\text{above}) = (n^2 - n)/2 = \frac{1}{2}n(n-1)$$

Problem 4. Find the chance that the first number is bigger.

Solution. Same as above, by the symmetry used already.

Note. As $n \to \infty$,

$$P(\text{two numbers are equal}) = \frac{\#(\text{diagonal})}{\#(\text{total})} = \frac{n}{n^2} = \frac{1}{n} \to 0$$

hence

$$P(\text{second bigger}) = P(\text{first bigger}) = \frac{1}{2}\left(1 - \frac{1}{n}\right) \to \frac{1}{2}$$

Odds

In a setting of equally likely outcomes, odds in favor of A give the ratio of the number of ways that A happens to the number of ways that A does not happen. The same ratio is obtained using probabilities instead of numbers of ways. Odds against A give the inverse ratio. More generally, just about any ratio of chances or probabilities can be called an odds ratio.

Gamblers are concerned with another sort of odds, which must be distinguished from odds defined as a ratio of chances. These are the odds offered by a casino or bookmaker in a betting contract, called here ***payoff odds*** to make the distinction clear. If you place a $1 bet on an event A, and the payoff odds against A are 10 to 1, you stand to win $10 if A occurs, and lose your $1 if A does not occur. In a casino you first pay your $1. If A occurs you get back a total of $11. This is your winnings of $10 plus your $1 back. If A does not occur, the casino keeps your $1. The price of $1 is ***your stake***, the $10 is the ***casino's stake***, and the $11 is the ***total stake***.

The connection between payoff odds and chance odds is an ancient principle of gambling, understood long before mathematicians decided that probabilities were best represented as numbers between 0 and 1. Around 1584, a colorful gambler and scholar of the Italian Renaissance, named Cardano, wrote a book on games of chance. Considering the roll of a die, Cardano said,

> *I am as able to throw a 1, 3 or 5 as 2, 4 or 6. The wagers are therefore laid in accordance with this equality if the die is honest, and if not, they are made so much the larger or smaller in proportion to the departure from true equality.*

First there is the idea of equally likely outcomes, then a heuristic connecting payoff odds and chance odds:

The Fair Odds Rule

In a fair bet, the payoff odds equal the chance odds.

That is to say, in a fair bet on an event A, where you win if A occurs and the casino wins otherwise, the ratio of your stake to the casino's stake should be the ratio of probabilities $P(A)$ to $1 - P(A)$. Put another way, your stake should be proportion $P(A)$ of the total stake.

Example 4. **House percentage at roulette.**

A Nevada roulette wheel has equally spaced pockets around its circumference. The pockets are numbered 1 through 36, 0 and 00. The wheel is spun, a ball inside the wheel is released, and by the time the motion stops the ball has been caught in one of the pockets. A *play* is a bet that the ball will fall in one of a certain set of pockets, with the payoff odds as shown below in Figure 1.

FIGURE 1. Layout of a Nevada roulette table. Key to colors: 0 and 00 = Green, *unshaded numbers* = Red, *shaded numbers* = Black.

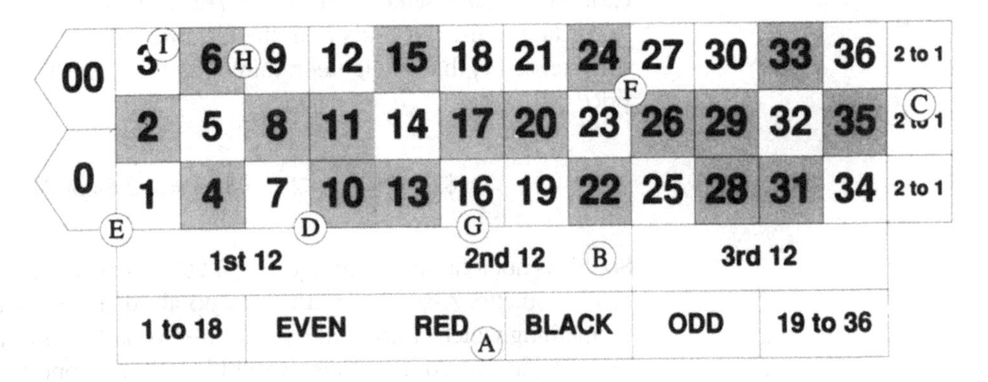

Each letter inside a circle, Ⓐ, Ⓑ, Ⓒ, etc., indicates a typical *play*.

Play	Set of winning numbers	Payoff odds
A. Even money play	Group of 18 numbers as marked in the box	1 to 1
B. Dozen play	12 numbers marked in the box	2 to 1
C. Column play	12 numbers in column (shown here as a row)	2 to 1
D. Line play	Six numbers above	5 to 1
E. House special	0, 00, 1, 2, 3	6 to 1
F. Quarter play	Four numbers in square	8 to 1
G. Street play	Three numbers above	11 to 1
H. Split play	Two adjoining numbers	17 to 1
I. Straight play	Single number	35 to 1

Over the long run, each number comes up about equally often. The obvious probabilistic assumption is that all 38 possible numbers are equally likely. To find the chance odds against a play, just take the ratio of losing numbers to winning numbers for that play. For example, the 38 numbers are divided into 18 numbers which are red, 18 numbers which are black, and two numbers are green (0 and 00). So the chance odds against red are 20 to 18, as are the chance odds against black. Put another way,

$$P(\text{red}) = P(\text{black}) = 18/38$$

The house offers bets on red at *even* odds, that is to say, payoff odds of 1 to 1. You can think about this in the following way. Suppose you stake \$1 on red. The casino then has a stake of \$1 on *not red*. The total stake in the game is $\$1 + \$1 = \$2$. According to the fair odds rule, the fair price to pay would be proportion $P(\text{red})$ of the total stake, that is, eighteen thirty-eighths of \$2. The \$1 you pay exceeds the fair price by

$$\$1 - \left(\frac{18}{38} \times \$2\right) = \$\frac{1}{19} = 5.26\% \text{ of } \$1 = 5.26 \text{ cents}$$

So this is not a fair bet. The figure of 5.26% is called the ***house percentage*** for bets on red at roulette. Assuming red comes up about 18/38 of the time over the long run (something which casinos take great care to ensure), this means that if you repeatedly bet a dollar on red, the house would be taking money from you at an average rate of 5.26 cents a game. If you bet a dollar 100 times in a row, you can expect to lose \$5.26. Of course you might lose more or less than this amount, depending on your luck. For example, there is a 26.5% chance that you will be ahead after 100 single-dollar bets on red. This chance is figured by assuming that every possible string of 100 numbers is equally likely to appear on your 100 plays, and finding the proportion of these strings with more than 50 reds. That is quite a hefty calculation, not to be undertaken until Chapter 2. But it gives you an idea of how far the method of equally likely outcomes can be pushed.

The argument just used to calculate the house percentage on red can be generalized to calculate the house percentage on any bet whatever. Consider a bet on A at payoff odds of r_{pay} to 1 against. If you stake \$1 on A, the house stakes $\$r_{\text{pay}}$, so the total at stake is $\$(r_{\text{pay}} + 1)$. According to the fair odds rule, the fair price to pay would be proportion $P(A)$ of the total stake, that is,

$$\$P(A)(r_{\text{pay}} + 1)$$

So out of your \$1 bet, the fraction taken by the house is $1 - P(A)(r_{\text{pay}} + 1)$. That is to say

$$\text{House Percentage} = [1 - P(A)(r_{\text{pay}} + 1)] \times 100\%$$

For example, in a straight play at roulette, that is, a bet on a single number, the chance odds are 37 to 1 against, corresponding to a probability of 1 in 38. But the payoff odds are only 35 to 1 against. So for a straight play, the house percentage is

$$[1 - \frac{1}{38}(35 + 1)] \times 100\% = 5.26\%$$

the same as for bets on red or black. For single numbers there is a neat way of checking this house percentage. Imagine there are 38 gamblers, each of whom bets on a different number. Then the house collects \$38 from each spin of the wheel. But one and only one of the gamblers wins each time. After each spin, the house pays off exactly \$36, the winning gambler's payoff of \$35 plus \$1 back. So the house collects $\$38 - \$36 = \$2$ for sure from every spin. If this cost of \$2 is thought of as shared equally among the 38 gamblers, the result is a cost of $\$2/38 = 5.26$ cents per gambler. This is the house percentage. Over the long run, the different numbers come up about equally often. So each player would end up paying about that amount per game.

Exercises 1.1

1. In a certain population of adults there are twice as many men as women. What is the proportion of men in the population:

a) as a fraction;

b) as a percent;

c) as a decimal?

Repeat for a population in which there are four men to every three women.

2. Suppose a word is picked at random from this sentence. Find:

a) the chance that the word has at least 4 letters;

b) the chance that the word contains at least 2 vowels (a, e, i, o, u);

c) the chance that the word contains at least 4 letters and at least 2 vowels.

3. Sampling with and without replacement.

Sampling with replacement:

A box contains tickets marked $1, 2, \ldots, n$. A ticket is drawn at random from the box. Then this ticket is replaced in the box and a second ticket is drawn at random. Find the probabilities of the following events:

a) the first ticket drawn is number 1 and the second ticket is number 2;

b) the numbers on the two tickets are consecutive integers, meaning the first number drawn is one less than the second number drawn.

c) the second number drawn is bigger than the first number drawn.

Sampling without replacement:

d) Repeat a) through c) assuming instead that the first ticket drawn is not replaced, so the second ticket drawn must be different from the first.

4. Suppose I bet on red at roulette and you bet on black, both bets on the same spin of the wheel.

   a) What is the probability that we both lose?

   b) What is the probability that at least one of us wins?

   c) What is the probability that at least one of us loses?

5. Suppose a deck of 52 cards is shuffled and the top two cards are dealt.

   a) How many ordered pairs of cards could possibly result as outcomes?

   Assuming each of these pairs has the same chance, calculate:

   b) the chance that the first card is an ace;

   c) the chance that the second card is an ace (explain your answer by a symmetry argument as well as by counting);

   d) the chance that both cards are aces;

   e) the chance of at least one ace among the two cards.

6. Repeat Exercise 5, supposing instead that after the first card is dealt, it is replaced, and shuffled into the deck before the second card is dealt.

7. Suppose two dice are rolled. Find the probabilities of the following events.

   a) the maximum of the two numbers rolled is less than or equal to 2;

   b) the maximum of the two numbers rolled is less than or equal to 3;

   c) the maximum of the two numbers rolled is exactly equal to 3.

   d) Repeat b) and c) for x instead of 3, for each x from 1 to 6.

   e) Denote by $P(x)$ the probability that the maximum number is exactly x. What should $P(1) + P(2) + P(3) + P(4) + P(5) + P(6)$ equal? Check this for your answers to d).

8. Repeat Exercise 7 for two rolls of a fair n-sided die for an arbitrary n instead of 6.

9. The chance odds against an event occurring are 10 to 1. What is the chance of the event? What if the odds were 5 to 1 against?

10. Calculate the chance of a win and the house percentage for each of the bets at roulette described below the layout in Figure 1.

11. Show that if the fair (chance) odds against an event are r_{fair} to 1, then in a bet at payoff odds of r_{pay} to 1 the house percentage is

$$\frac{r_{\text{fair}} - r_{\text{pay}}}{r_{\text{fair}} + 1} \times 100\%$$

1.2 Interpretations

James Bernoulli (1654 – 1705), one of the founders of probability theory, put it like this:

> *Probability is the degree of certainty, which is to the certainty as a part is to a whole.*

This conveys the right intuitive idea. And it points correctly to the rules of proportion as the mathematical basis for a theory of probability. But it leaves open the question of just how probabilities should be interpreted in applications.

This section considers two important interpretations of probability. First, the *frequency interpretation* in which probabilities are understood as mathematically convenient approximations to long-run relative frequencies. Second, the *subjective interpretation* in which a probability statement expresses the *opinion* of some individual regarding how certain an event is to occur. Which (if either) of these interpretations is "right" is something which philosophers, scientists, and statisticians have argued bitterly for centuries. And very intelligent people still disagree. So don't expect this to be resolved by the present discussion.

Frequencies

A *relative frequency* is a proportion measuring how often, or how frequently, something or other occurs in a sequence of observations. Think of some experiment or set of circumstances which can be repeated again and again, for example, tossing a coin, rolling a die, the birth of a child. Such a repeatable experiment may be called a *trial*. Let A be a possible result of such a trial: for example, the coin lands heads, the die shows a six, the child is a girl. If A happens m times in n trials, then m/n is the *relative frequency* of A in the n trials.

Example 1. Coin tossing.

Suppose a coin is tossed ten times, and the observed sequence of outcomes is

$$t, h, h, t, h, h, h, t, t, h,$$

where each t indicates a tail and each h a head. The successive relative frequencies of heads in one toss, two tosses, and so on up to ten tosses are then

$$\frac{0}{1}, \frac{1}{2}, \frac{2}{3}, \frac{2}{4}, \frac{3}{5}, \frac{4}{6}, \frac{5}{7}, \frac{5}{8}, \frac{5}{9}, \frac{6}{10},$$

as graphed in Figure 1. Figure 2 shows what usually happens if you plot a similar graph of relative frequencies for a much longer series of trials.

A general rule, illustrated in Figure 2, is that relative frequencies based on larger numbers of observations are less liable to fluctuation than those based on smaller

FIGURE 1. Relative frequencies in a series of 10 coin tosses.

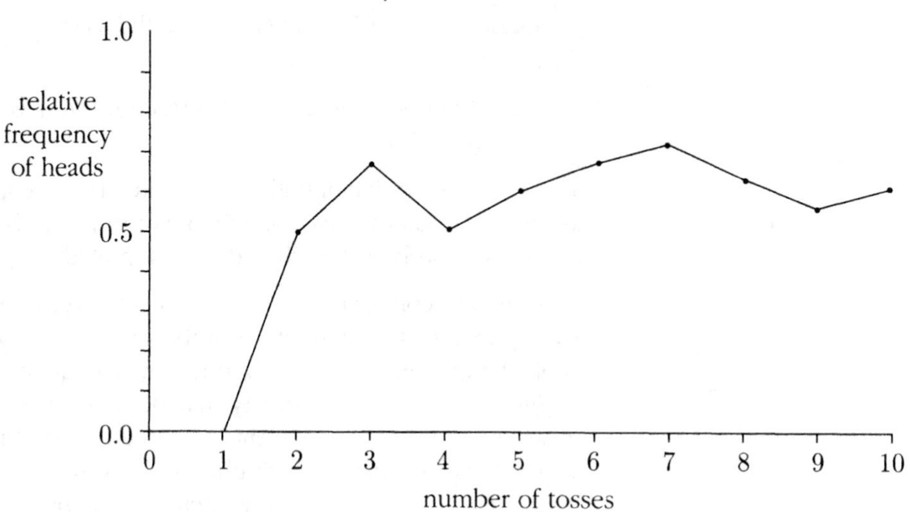

numbers. It is observed that almost regardless of the precise nature of the experimental trials in question, or what feature A of the trials is recorded, the relative frequency of A based on n trials tends to stabilize as n gets larger and larger, provided that the conditions of the trial are kept as constant as possible. This phenomenon is called the *statistical regularity* of relative frequencies, or the *empirical law of averages.*

In coin tossing, heads and tails usually come up about equally often over a long series of tosses. So the long-run relative frequency of heads is usually close to $1/2$. This is an empirical fact, closely linked to our intuitive idea that heads and tails are equally likely to come up on any particular toss. Logically, there is nothing to prevent the relative frequency of heads in a long series of tosses from being closer to, say, $1/4$, or $2/3$, than to $1/2$. The relative frequency could even be 1 if the coin landed heads every time, or 0 if it landed tails every time. But while possible, it hardly ever happens that the relative frequency of heads in a long series of tosses differs greatly from $1/2$. Intuitively, such a large fluctuation is extremely unlikely for a fair coin. And this is precisely what is predicted by the theory of repeated trials, taken up in Chapter 2.

In the *frequency interpretation*, the probability of an event A is the expected or estimated relative frequency of A in a large number of trials. In symbols, the proportion of times A occurs in n trials, call it $P_n(A)$, is expected to be roughly equal to the theoretical probability $P(A)$ if n is large:

$$P_n(A) \approx P(A) \text{ for large } n$$

Under ideal circumstances, the larger the number of trials n, the more likely it is that this approximation will achieve any desired degree of accuracy. This idea is

FIGURE 2. Relative frequencies of heads in two long series of coin tosses. For a small number of trials, the relative frequencies fluctuate quite noticeably as the number of trials varies. But these fluctuations tend to decrease as the number of trials increases. Initially, the two sequences of relative frequencies look quite different. But after a while, both relative frequencies settle down around $1/2$. (The two series were obtained using a computer random number generator to simulate coin tosses.)

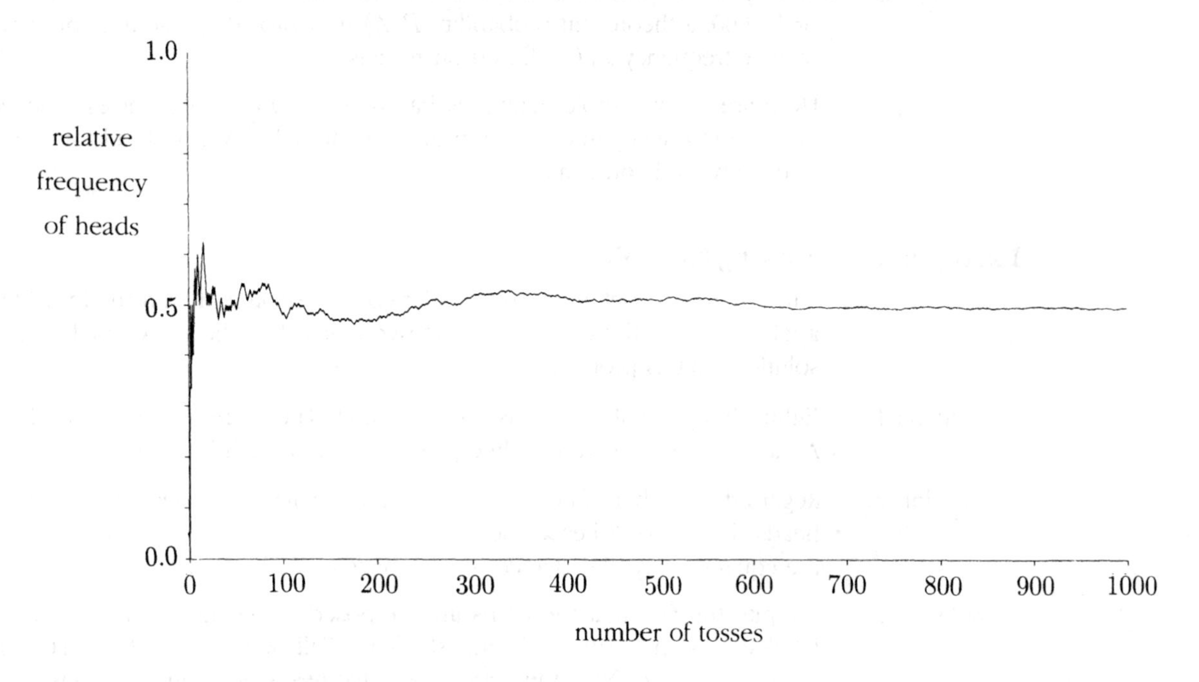

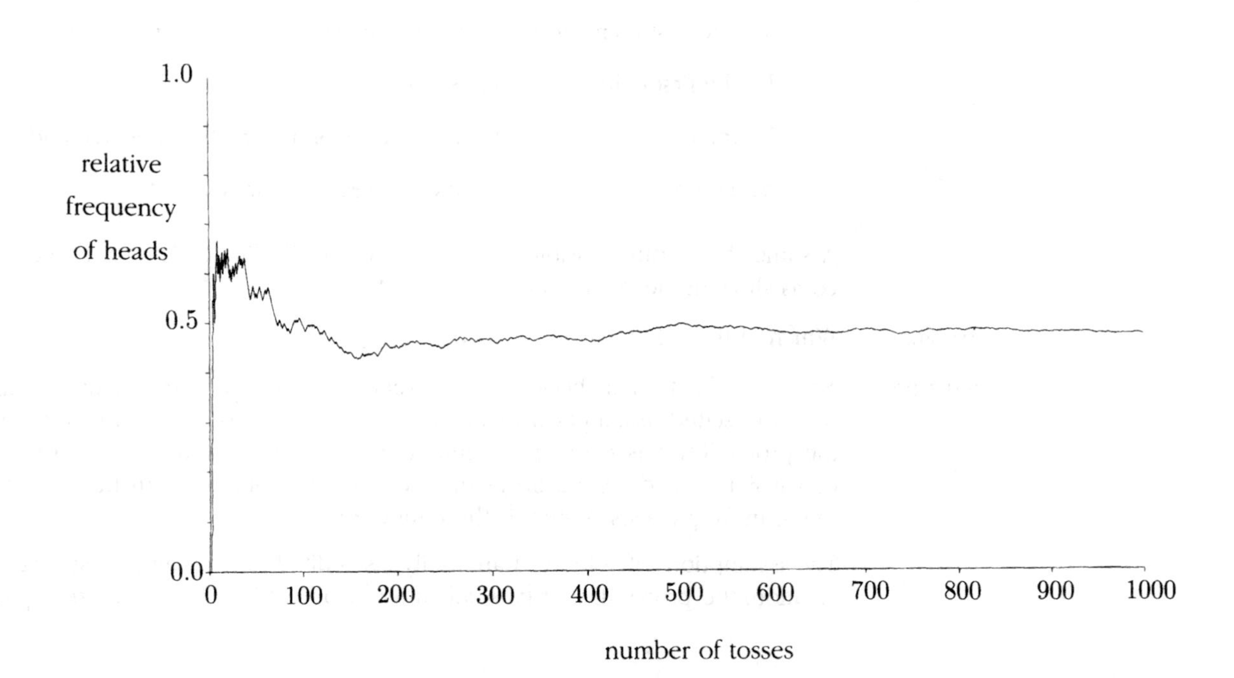

made precise in Chapter 2 by a mathematical result called the *law of large numbers*. The theoretical probability $P(A)$ may even be conceived theoretically as a limit of relative frequencies $P_n(A)$ as $n \to \infty$. While intuitively appealing, this idea can only be made precise in a theoretical framework allowing infinitely many trials, so it is not really practical. The practical point is that for large but finite values of n, say $n = 1000$ or $10,000$, a theoretical probability $P(A)$ may provide a useful approximation to a relative frequency $P_n(A)$ based on n trials.

Here are a few simple examples based on long-run frequencies. The first shows how the frequency interpretation dictates the right level of detail for an assumption of equally likely outcomes.

Example 2. Tossing two coins.

Suppose a cup containing two similar coins is shaken, then turned upside down on a table. What is the chance that the two coins show heads? Consider the following solutions to this problem.

Solution 1. Either they both show heads, or they don't. These are the two possible outcomes. Assuming these are equally likely, the chance of both heads is $1/2$.

Solution 2. Regard the number of heads showing on the coins as the outcome. There could be 0 heads, 1 head, or 2 heads. Now there are three possible outcomes. Assuming these are equally likely, the chance of both heads is $1/3$.

Solution 3. Despite the fact that the coins are supposed to be similar, imagine that they are labeled in some way to distinguish them. Call one of them the first coin and the other the second. Now there are four outcomes which might be considered:

hh: the first coin shows heads and the second coin shows heads;

ht: the first coin shows heads and the second coin shows tails;

th: the first coin shows tails and the second coin shows heads; and

tt: the first coin shows tails and the second coin shows tails.

Assume these four possible outcomes are equally likely. Then the event of both coins showing heads has a chance of $1/4$.

Problem. Which of the solutions above is correct?

Discussion. So far as the formal theory is concerned, they all are! Each solution starts from a clearly stated assumption of equally likely outcomes, then correctly determines the probability based on this assumption. The assumptions are different, and the conclusions are different. So at most one of the solutions can be consistent with long-run frequencies. Which is the right one?

The assumptions of Solution 1 are easily discredited. By the same reasoning as in that solution, the probability of two tails must also be $1/2$. That leaves zero probability for the event of a head and a tail, which is clearly ridiculous so far as long-run frequencies are concerned. Solution 2 looks quite plausible, and is not easy to fault by armchair reasoning. Solution 3 looks artificial in comparison. Why should it be necessary to distinguish between two similar coins?

On balance, these arguments seem to point to the $1/3$ of Solution 2 as the answer. But the reality check is the long-run frequency. As a matter of practical experiment, which you can try yourself, the long-run frequency turns out to be around $1/4$, no matter whether you can distinguish between the coins or not. So Solution 3 is the one which matches up with long-run frequencies.

Remark. There is a physical principle involved here, which is a useful guide for getting probabilities to match long-run frequencies. All macroscopic physical objects like coins, grains of sand, and so on, behave statistically as if they are distinguishable. So, if you want to calculate chances for rolling several dice or tossing several coins, you should always assume they are distinguishable when setting up the outcome space. Interestingly, however, physicists have found that atomic particles such as protons and electrons behave statistically as if they are genuinely indistinguishable.

The moral of the above example is that even if an assumption of equally likely outcomes is appropriate at some level of description, this level is not something which can be judged on mathematical grounds alone. It must be judged using some further interpretation of probability, such as the long-run frequency idea. Furthermore, there are examples like tossing a biased coin, or recording the sex of a newborn child, where long-run frequencies seem to stabilize around some more or less arbitrary decimal fraction between 0 and 1.

Example 3. Sex of children.

Table 1 shows that the relative frequency of boys among newborn children in the U.S.A. appears to be stable at around 0.513.

Observation of the sex of a child is comparable to a scheme with equally likely outcomes obtained by drawing at random with replacement from a box of 1000 tickets, containing 487 tickets marked *girl* and 513 tickets marked *boy*. This allows probabilities for births to be calculated as if they were probabilities for random sampling from a box of tickets. But the analogy is not complete. The individual tickets have no physical interpretation like the sides of a die or the pockets of a roulette wheel. And there seems to be no way to decide what the composition of the box should be without counting births. Still, the above data suggest a reasonable model for the outcome of a single birth: the outcome space $\{girl, boy\}$, with probability $p = 0.513$ for *boy* and $1 - p = 0.487$ for *girl*.

TABLE 1. Proportion of boys among live births to residents of the U.S.A.

Year	Number of births*	Proportion of boys
1974	3,159,958	0.5133340
1975	3,144,198	0.5130513
1976	3,167,788	0.5127982
1977	3,326,632	0.5128058
1978	3,333,279	0.5128266
1979	3,494,398	0.5126110
1980	3,612,258	0.5128692
1981	3,629,238	0.5125792

*Births to residents of the U.S.A., based on 100% of births in selected states, and a 50% sample in all others. Source: *Information Please Almanac, Atlas and Yearbook*, 1985.

Opinions

The notion of probabilities as an approximation to long-run frequencies makes good sense in a context of repeated trials. But it does not always make sense to think in terms of repeated trials. Consider, for example:

- the probability of a particular patient surviving an operation;
- the probability that a particular motorist is involved in an accident next year;
- the probability that a particular interest rate will be below 5% in a year's time;
- the probability of a major earthquake in Berkeley before the year 2000.

If you are the patient considering an operation, you want the doctor to tell you what he thinks *your* chances are. The notion of your undergoing repeated operations is absurd. Even if it is known that in similar operations in the past there was, say, a 10% fatality rate, this figure is irrelevant if the doctor knows that your state of health is much better, or you are much younger, or are different in some other respect from the population of patients on which the 10% figure is based. Rarely would it be possible for the doctor to know survival percentages for patients just like you. The more factors that are taken into account, the more difficult it is to obtain relevant data, the smaller the number of cases on which figures could be based. If enough factors were taken into account, your case would be unique. What then are you to make of it if the doctor says you have a 95% chance of surviving? Essentially, this is a matter of *opinion*. In the doctor's opinion, your chance of survival is 95%. Another doctor might have another opinion, say 98%. You might ask several opinions, then somehow form your own opinion as to your chances.

Similar considerations apply to the other examples above. In none of these examples does the relative frequency idea make much sense. Ultimately, probability statements of this kind come down to some kind of intuitive judgment of the uncertainties involved. Such judgments lead to the notion of ***subjective probabilities***, which may also be called ***probabilistic opinions***, or ***degrees of belief.*** This conception of probability corresponds well to everyday language, such as the following:

> It is unlikely that there will be an earthquake in Berkeley next year.
>
> If I toss a coin once, the probability that it will land heads is $1/2$.
>
> The chance of rain tomorrow is 30%.

Such statements have a superficial objective quality, since they make no reference to the person who is making them. But viewed as objective statements they are at best very hard to interpret, and at worst either meaningless or unverifiable. To give such statements meaning, it is simplest just to interpret them as expressions of probabilistic opinion. Intuitive comparison of probabilities can be helpful in formulating a probabilistic opinion. Comparisons can be made within a particular context, for example, by deciding that two or more events are equally likely, or that an event is twice as likely as another. Or comparisons can be made between different contexts. Comparison with a standard experiment like drawing tickets from a box can be a useful device. Which do you think is more likely? Event A, or getting a marked ticket on a draw at random from a box containing 20% marked tickets? If you think A is more likely, then you should assign a probability $P(A) \geq 20\%$. If you have trouble deciding which is more likely, ask yourself which option you would prefer: To win a prize of some kind if A occurs, or to win the same prize if a marked ticket is drawn?

Like the long-run frequency idea, the idea of subjective probability has its limitations. Subjective probabilities are necessarily rather imprecise. It may be difficult or impossible to pool the subjective probability opinions of different individuals about the same events. Assessment of subjective probabilities of events, regarded as having very small or very large probabilities, is very difficult, particularly if these events have important consequences for the person attempting to judge their probabilities.

Despite such difficulties, the idea of interpreting probabilities as subjective opinions about uncertainties is something many people find reasonable. As well as broadening the range of application of probabilistic ideas, the subjective interpretation gives insight into the mathematics of probability theory. For example, the notion of conditional probability, introduced in the next chapter, captures the idea of how your probabilistic opinion may change over time as you acquire new information or data.

Exercises 1.2

1. If you get a speeding ticket in the state of Washington, it states on the ticket: "If you believe you did not commit the infraction, you may request a hearing. At the hearing, the state must prove by a preponderance of the evidence (more likely than not) that you committed the infraction." What do you think the phrase "more likely than not" means? Does it refer to relative frequencies? to an opinion? if so, whose opinion?

2. If a bookmaker quotes payoff odds of 99 to 1 against a particular horse winning a race, does that suggest the chance that the horse will win is $1/100$, less than $1/100$, or more than $1/100$? Explain.

3. Suppose there are 10 horses in a race and a bookmaker quotes odds of r_i to 1 against horse i winning. Let $p_i = \frac{1}{r_i+1}$, $i = 1$ to 10, so each p_i is between 0 and 1. Let $\Sigma = p_1 + \cdots + p_{10}$.

    a) Do you expect that Σ is greater than, smaller than, or equal to 1? Why?

    b) Suppose Σ were less than 1. Could you take advantage of this? How? [*Hint*: By betting on all 10 horses in the race, a bettor can win a constant amount of money, regardless which horse wins.]

4. A gambler who makes 100 bets of \$1, each at payoff odds of 8 to 1, wins 10 of these bets and loses 90.

    a) How many dollars has the gambler gained overall?

    b) What is the gambler's average financial gain per bet?

    Suppose now that the gambler makes a sequence of \$1 bets at payoff odds of r_{pay} to 1. Define an *empirical odds ratio* $r_\#$ to be the gambler's number of losses divided by the number of wins. So, in the numerical example above, r_{pay} was 8, and $r_\#$ was $90/10 = 9$. Show that the gambler's average financial gain per bet is $\$(r_{\text{pay}} - r_\#)/(r_\# + 1)$. Explain carefully the connection between this formula and the house percentage formula in Exercise 1.1.11.

1.3 Distributions

From a purely mathematical point of view, probability is defined as a function of events. The events are represented as sets, and it is assumed that the probability function satisfies the basic rules of proportion. These are the rules for fractions or percentages in a population, and for relative areas of regions in a plane. To state the rules, we must first consider the representation of events as subsets of an outcome space.

Suppose an outcome space Ω is given, and that all events of interest are represented as subsets of Ω. Think of Ω as representing all ways that some situation might turn out. It is no longer assumed that Ω is necessarily a finite set, or that all possible outcomes are equally likely. But if A is an event, the subset of Ω corresponding to A is still the set of all ways that A might happen. This subset of Ω will also be denoted A. Thus events are identified with subsets of Ω.

TABLE 1. Translations between events and sets. To interpret the Venn diagrams in terms of events, imagine that a point is picked at random from the square. Each point in the square then represents an outcome, and each region of the diagram represents the event that the point is picked from that region.

Event language	Set language	Set notation	Venn diagram
outcome space	universal set	Ω	
event	subset of Ω	A, B, C, etc.	
impossible event	empty set	$\emptyset$	
not A, opposite of A	complement of A	A^c	A
either A or B or both	union of A and B	$A \cup B$	A B
both A and B	intersection of A and B	AB, $A \cap B$	A B
A and B are mutually exclusive	A and B are disjoint	$AB = \emptyset$	A B
if A then B	A is a subset of B	$A \subseteq B$	A B

The rules of probability involve logical relations between events. These are translated into corresponding relations between sets. For example, if C is the event which occurs if either A or B occurs (allowing the possibility that both A and B might occur), then the set of ways C can happen is tne union of the set of ways A can happen and the set of ways B can happen. In set notation, $C = A \cup B$. Table 1 gives a summary of such translations.

Partitions

Say that an event B is *partitioned* into n events $B_1, \ldots, B_n$ if $B = B_1 \cup B_2 \cup \cdots \cup B_n$, and the events $B_1, \ldots, B_n$ are mutually exclusive. That is to say, every outcome in B belongs to one and only one of the subsets B_i. Think of B as split up into separate cases $B_1, \ldots, B_n$. Figure 1 shows a subset B of the square is partitioned in three different ways. However B is partitioned into subsets, or broken up into pieces, the area in B is the sum of the areas of the pieces. This is the *addition rule* for area.

FIGURE 1. Partitions of a set B.

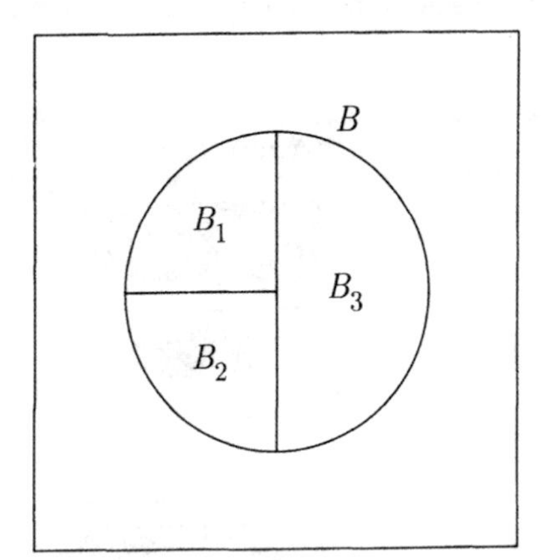

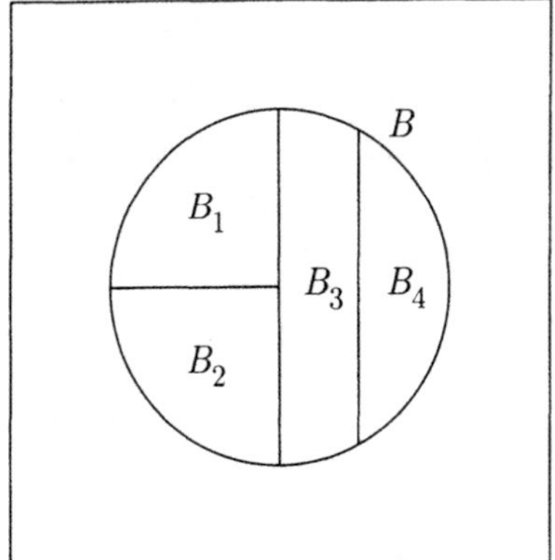

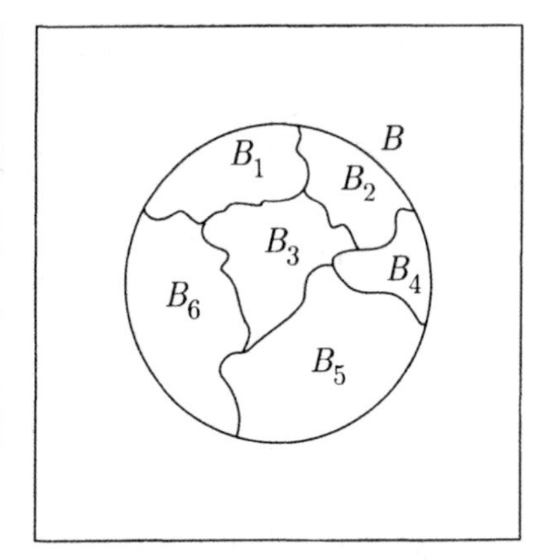

The addition rule is satisfied by other measures of sets instead of area, for example, length, volume, and the number or proportion of elements for finite sets.

The addition rule now appears as one of the three basic rules of proportion. No matter how probabilities are interpreted, it is generally agreed they must satisfy the same three rules:

Rules of Proportion and Probability

- **Non-negative:** $P(B) \geq 0$
- **Addition:** If $B_1, B_2, \ldots, B_n$ is a partition of B, then

$$P(B) = P(B_1) + P(B_2) + \cdots + P(B_n)$$

- **Total one:** $P(\Omega) = 1$

A *distribution* over Ω is a function of subsets of Ω satisfying these rules.

The term "distribution" is natural if you think of mass distributed over an area or volume Ω, and $P(A)$ representing the proportion of the total mass in the subset A of Ω. Now think of probability as some kind of stuff, like mass, distributed over a space of outcomes. The rules for probability are very intuitive if you think informally of an event B as something that might or might not happen, and of $P(B)$ as a measure of how likely it is that B will happen. It is agreed to measure probability on a scale of 0 to 1. The addition rule says that if something can happen in different ways, the probability that it happens is the sum of the probabilities of all the different ways it can happen.

Technical remark. When the outcome space Ω is infinite, it is usually assumed that there is a similar addition rule for partitions of an event into an infinite sequence of events. See Section 3.4. In a rigorous treatment of measures like probability, length or area, defined as functions of subsets of an infinite set Ω, it is necessary to describe precisely those subsets of Ω, called *measurable sets*, whose measure can be unambiguously defined by starting from natural assumptions about the measure of simple sets like intervals or rectangles, using the addition rule, and taking limits. See Billingsley's book *Probability and Measure* for details.

Here are some useful general rules of probability, derived from the basic rules and illustrated by Venn diagrams. In the diagrams, think of probability as defined by relative areas.

Complement Rule: The probability of the complement of A is

$$P(\text{not } A) = P(A^c) = 1 - P(A)$$

Proof. Because Ω is partitioned into A and A^c, and $P(\Omega) = 1$,

$$1 = P(A) + P(A^c)$$

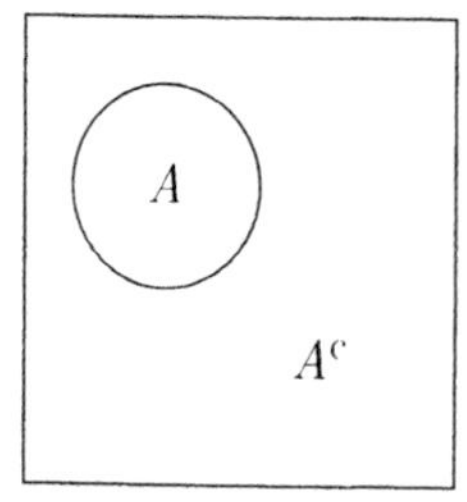

Remarks. Note that if $A = \Omega$, then $A^c = \emptyset$, the empty set, and $P(A) = 1$. So the rule of complements implies $P(\emptyset) = 0$. The empty set contains nothing. Also, for a set A, $P(A) = 1 - P(A^c)$ and $P(A^c) \geq 0$, so $P(A) \leq 1$. Thus probabilities are always between 0 and 1.

The next rule is a generalization of the rule of complements:

Difference Rule: If occurrence of A implies occurrence of B, then $P(A) \leq P(B)$, and the difference between these probabilities is the probability that B occurs and A does not:

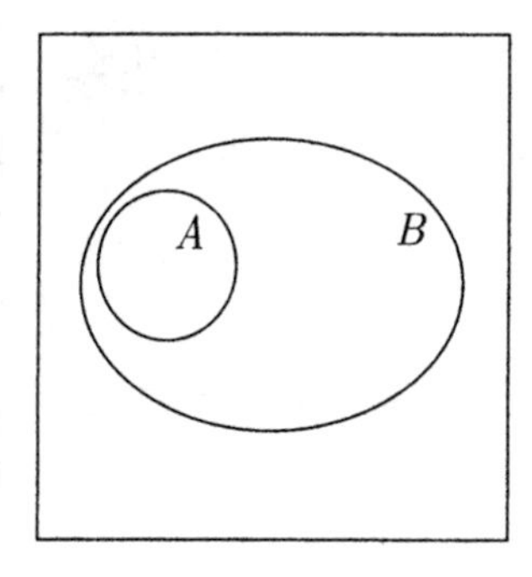

$$P(B \text{ and not } A) = P(BA^c) = P(B) - P(A)$$

Proof. In other words, the assumption is that every outcome in A is an outcome in B, so A is a subset of B. Since B can be partitioned into A and (B but not A),

$$P(B) = P(A) + P(BA^c)$$

by the addition rule. Now subtract $P(A)$ from both sides.

Inclusion–Exclusion: $P(A \cup B) = P(A) + P(B) - P(AB)$

Remarks. Here $A \cup B$ means A or B or both (union) while AB means both A and B (intersection, $A \cap B$). This is the modification of the addition rule for events A and B that overlap, as in the following diagram. The addition rule for mutually exclusive A and B is the special case when $AB = \emptyset$, so $P(AB) = 0$. The extension to three or more sets is given in the exercises.

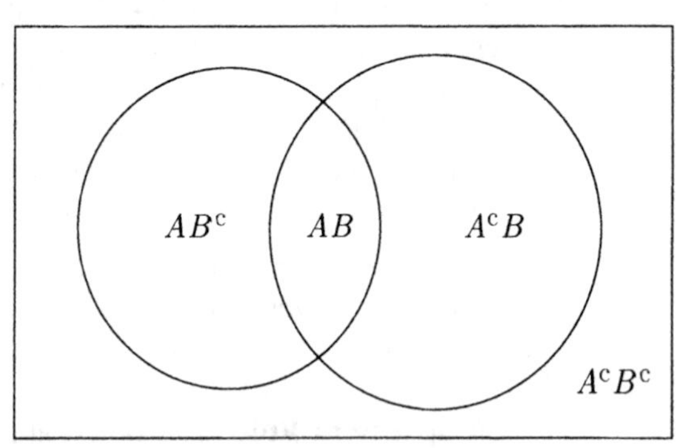

Proof. As the diagram shows, the sets AB^c, AB, and A^cB form a partition of $A \cup B$, so

$$P(A \cup B) = P(AB^c) + P(AB) + P(A^cB)$$

Similarly

$$P(A) = P(AB^c) + P(AB)$$

$$P(B) = P(A^cB) + P(AB)$$

so

$$P(A) + P(B) = P(AB^c) + 2P(AB) + P(A^cB)$$

This is the same expression as for $P(A \cup B)$, but $P(AB)$ is included twice. Subtracting $P(AB)$ excludes one of these terms, to give the inclusion–exclusion formula.

Example 1. Rich and famous.

In a certain population, 10% of the people are rich, 5% are famous, and 3% are rich and famous. For a person picked at random from this population:

Problem 1. What is the chance that the person is not rich?

Solution. Here probabilities are defined by proportions in the population. By the rule of complements

$$P(\text{not rich}) = 100\% - P(\text{rich}) = 100\% - 10\% = 90\%$$

Problem 2. What is the chance that the person is rich but not famous?

Solution. By the difference rule

$$\begin{aligned} P(\text{rich but not famous}) &= P(\text{rich}) - P(\text{rich and famous}) \\ &= 10\% - 3\% = 7\% \end{aligned}$$

Problem 3. What is the chance that the person is either rich or famous?

Solution. By the inclusion–exclusion formula,

$$\begin{aligned} P(\text{rich or famous}) &= P(\text{rich}) + P(\text{famous}) - P(\text{rich and famous}) \\ &= 10\% + 5\% - 3\% = 12\% \end{aligned}$$

Example 2. Numbered tickets.

Proportion $P(i)$ of the tickets in a box are numbered i, with this distribution:

number i	1	2	3	4	5	6
proportion $P(i)$	1/4	1/8	1/8	1/8	1/8	1/4

Problem. If a ticket is drawn at random from the box, what is the chance that the number on the ticket is 3 or greater?

Solution. Assuming all tickets in the box are equally likely to be drawn, by the addition rule:

$$P(3 \text{ or } 4 \text{ or } 5 \text{ or } 6) = P(3) + P(4) + P(5) + P(6) = \frac{1}{8} + \frac{1}{8} + \frac{1}{8} + \frac{1}{4} = \frac{5}{8}$$

In the above example, outcomes with unequal probabilities (corresponding to various numbers) were obtained by partitioning a set of equally likely outcomes (the individual tickets) into subsets of different sizes. It was then possible to work with the probability distribution over the smaller number of outcomes defined by the partition, using the addition rule. This is the key to problems such as the following where there is no natural analysis in terms of equally likely outcomes.

Example 3. Shapes.

A *shape* is a 6-sided die with faces cut as shown in the following diagram:

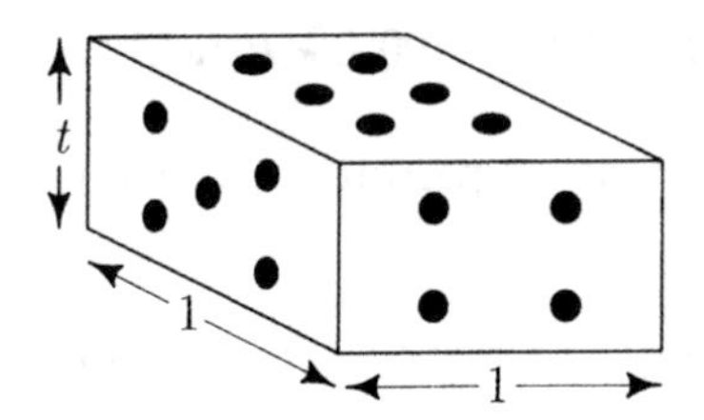

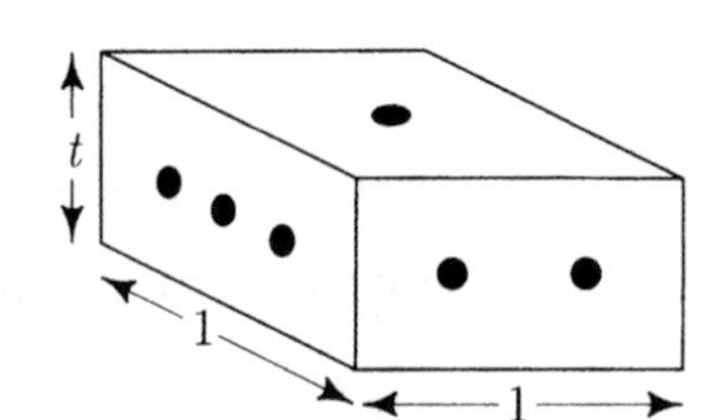

The faces showing 1 and 6 are square faces with side of length one unit, but the distance between these faces, or the *thickness* of the shape, is a length $t \leq 1$. So each of the faces 2, 3, 4, and 5 is a rectangle instead of a square. Such a die may land either flat (1 or 6), or on its side (2, 3, 4, or 5). As the thickness of the shape decreases from 1 to 0, it is intuitively clear that the chance that the shape lands flat increases continuously from 1/3 to 1. Suppose that the thickness t is such that the chance of the shape landing flat is 1/2. You could understand this to mean that over a long sequence of rolls, the shape landed flat about as often as it landed on its side.

Problem. What is the probability that such a shape shows number 3 or greater?

Solution. For $i = 1$ to 6, let $P(i)$ be the probability that the shape lands showing i. Using the addition rule,

$$1/2 = P(\text{flat}) = P(1) + P(6)$$
$$1/2 = P(\text{side}) = P(2) + P(3) + P(4) + P(5)$$

The symmetry of the shape suggests the assumptions:

$$P(1) = P(6) \quad \text{and} \quad P(2) = P(3) = P(4) = P(5)$$

These equations imply that the probabilities $P(i)$ are as displayed in the following table and in Figure 2.

number i	1	2	3	4	5	6
probability $P(i)$	1/4	1/8	1/8	1/8	1/8	1/4

The probability that the shape shows a number greater than or equal to 3 is then given by the addition rule:

$$P(3 \text{ or } 4 \text{ or } 5 \text{ or } 6) = P(3) + P(4) + P(5) + P(6) = \frac{1}{8} + \frac{1}{8} + \frac{1}{8} + \frac{1}{4} = \frac{5}{8}$$

FIGURE 2. **Histogram of the distribution in Example 3.** This is a bar graph showing the probabilities for the shape showing face i. The area of the bar over i is proportional to $P(i)$. By the addition rule for probabilities and areas, the probability that the shape shows a number greater than or equal to 3 is the shaded area relative to the total area, that is, $5/8$.

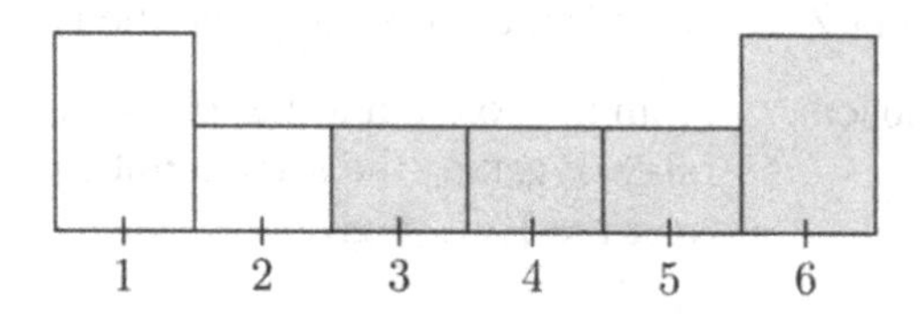

Discussion. Notice that the distribution of the number on the shape in Example 3 is identical to the distribution of a number drawn at random from the box of Example 2. The probability of getting a number greater than or equal to 3 is therefore the same in both examples. Similarly, for any subset B of $\{1, \ldots, 6\}$, the probability of getting an outcome in B is the same in both examples. The two procedures for obtaining a numerical outcome between 1 and 6, rolling the shape, and drawing a ticket from the box, are called *probabilistically equivalent.* In other words, the two outcomes have the *same distribution.* This means the set of possible outcomes and the distribution of probability over these outcomes is the same in both cases. It would not make sense, however, to say that the two procedures generated the *same outcome.* On the contrary, the two procedures would most likely produce two different numbers:

Example 4. Picking a number from a box and rolling a shape.

Suppose one number is obtained by drawing at random from the box of tickets in Example 2, and another number is obtained by rolling a shape as in Example 3.

Problem 1. What is the chance of the event that the number from the box is i and the number on the ticket is j?

Solution. Consider the following two procedures for obtaining a pair of numbers (i, j):

- Draw from the box of tickets to obtain i. Roll the shape to obtain j.
- Draw from the box of tickets to obtain i. Replace this ticket in the box, mix up the tickets in the box and draw again to obtain j.

The second procedure is called random sampling with replacement (Exercise 1.1.3). It is intuitively clear that these two procedures must be probabilistically equivalent. That is to say the probability of any event determined by the first pair must be the same as the probability of the corresponding event for the second pair. In particular, the probability that the box produces i and the shape rolls j must be the same as the probability of getting i on the first draw and j on the second draw in two draws at random with replacement from the box. To solve the problem, let us *assume* this probabilistic equivalence. The point is that for two draws at random with replacement the probability of getting particular numbers i on the first draw and j on the second draw can be found by the method of Section 1.1. Suppose there are N tickets in the box, and that all $N \times N = N^2$ possible pairs of tickets are equally likely in two draws at random with replacement. Since the number of tickets labeled i is $P(i)N$ for $P(i)$ displayed in Example 2, the number of ways to get (i, j) is $P(i)N \times P(j)N = P(i)P(j)N^2$. So the required probability is $P(i)P(j)N^2/N^2 = P(i)P(j)$.

Problem 2. What is the probability that the two numbers are different?

Solution. From the solution to the previous problem, for any *particular* number i, the probability of getting the same number i from the ticket and the shape is $P(i)^2$. Summing over $i = 1, \ldots, 6$ gives

$$P(\text{ticket and shape show the same unspecified number}) = \sum_{i=1}^{6} P(i)^2$$

By the complement rule

$$P(\text{ticket and shape show different numbers}) = 1 - \sum_{i=1}^{6} P(i)^2$$

$$= 1 - 2\left(\frac{1}{4}\right)^2 - 4\left(\frac{1}{8}\right)^2 = \frac{13}{16}$$

Discussion. The above example illustrates an important technique for solving probability problems. Look for a probabilistic equivalent of the original problem that is easier to understand. Then solve the equivalent problem. The solution of Problem 1 shows that the basic assumption made on intuitive grounds, that

> *the ticket–shape scheme is probabilistically equivalent to a ticket–ticket scheme for draws with replacement*

implies a *product rule* for calculating the probability of an intersection of two events, one determined by the ticket and the other by the die:

$$P(\text{ticket shows number } i \text{ and shape shows number } j)$$
$$= P(\text{ticket shows number } i)\, P(\text{shape shows number } j)$$

for all i and j. Events A and B such as these, with $P(AB) = P(A)P(B)$, are called *independent* events. The concept of independence is studied in Section 1.4. In language defined more formally in Section 3.1, the assumption of equivalence of the ticket–shape and ticket–ticket schemes can be restated as follows:

> *the number on the ticket and the number rolled by the shape are independent random variables with the same distribution.*

Named Distributions

The distribution on the set $\{1, \ldots, 6\}$ defined by the probabilities $P(1), \ldots, P(6)$ in the previous three examples is of no particular importance. It just illustrated numerically some general properties of a probability distribution over a finite set. There are some special distributions, however, that appear in a wide variety of contexts and are given names. Some of these *named distributions* are mentioned in the following paragraphs. Other named distributions appear throughout the book. There is a summary of the properties of the most important of these distributions on pages 476 to 488. Most named distributions have one or more *parameters* in their definition. These are constants appearing in the formula for the distribution which affect its shape and properties. Typically, the parameters are subject to some *constraints* such as non-negativity, so that the numbers defined by the formula satisfy the rules of probability.

Bernoulli (p) distribution. For p between 0 and 1, this is the distribution on $\{0, 1\}$ defined by the following distribution table:

possible outcome	0	1
probability	$1-p$	p

FIGURE 3. Histograms of some Bernoulli (p) distributions.

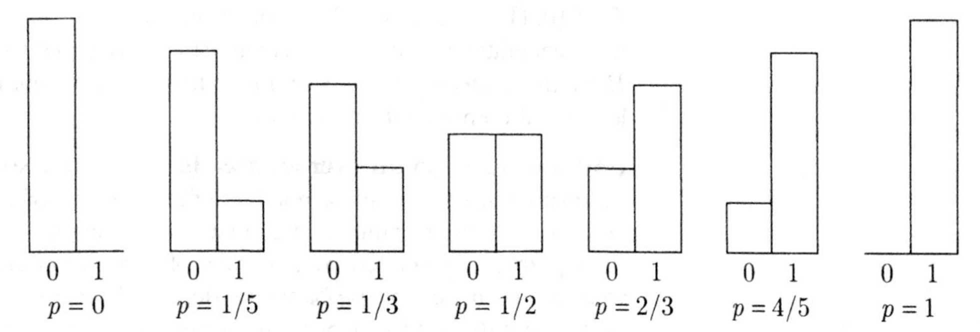

Think of any event A, for which you think it makes sense to consider the probability $P(A)$. For example, A might be the event of heads on a coin toss, perhaps for a biased coin. Now define an outcome to be 1 if A occurs, and 0 otherwise. If you like, imagine that you win a dollar if A occurs, nothing otherwise. Then the outcome is the number of dollars you win. This outcome, associated with the event A, is called the *indicator* of A. The distribution of the indicator of A is the Bernoulli (p) distribution for $p = P(A)$.

The number p is the parameter of the Bernoulli (p) distribution. The effect of increasing p from 0 to 1 is to shift the probability from being all concentrated at 0 to being all concentrated at 1, as shown by the histograms in Figure 3.

Uniform distribution on a finite set. This distribution, defined by an assumption of equally likely outcomes, appeared in many examples in Section 1.1. To be clear about exactly what uniform distribution is meant, it is essential to define clearly the *range* of the uniform distribution, that is, the precise set of outcomes assumed equally likely. If the range is a set of n possible outcomes, for instance $\{1, 2 \ldots, n\}$ or $\{0, 1, \ldots, n-1\}$, the probability of each possible outcome is $1/n$. The probability $P(B)$ of an outcome in the set B is then $P(B) = \#(B)/n$. Note that the uniform distribution on $\{0, 1\}$ is identical to the Bernoulli $(1/2)$ distribution. This is the distribution of the indicator of heads on a fair coin toss.

Uniform (a, b) distribution. This refers to the distribution of a point picked uniformly at random from the interval (a, b) where a and b are two numbers with $a < b$. The basic assumption is that probability is proportional to length. So for $a < x < y < b$ the probability that the point falls in the interval (x, y) is assumed to be $(y - x)/(b - a)$. By rescaling the interval (a, b) to the *unit interval* $(0, 1)$, problems involving the uniform (a, b) distribution are reduced to problems involving the uniform $(0, 1)$ or *standard uniform* distribution. See Section 4.1 for details. Most calculators and computer languages have a command, often called "RND", that produces a pseudo-random number with approximately uniform $(0, 1)$ distribution. These numbers are called *pseudo-random* because the results of successive calls of RND are in fact generated by application of a simple deterministic formula starting from some initial number in $(0, 1)$, called the *seed*. The formula has the property that for $0 < x < y < 1$ the long-run relative frequency of numbers in (x, y) is almost exactly equal to $y - x$. By the addition rule for long-run frequencies, for any subset B of $(0, 1)$ which is a finite union of intervals, the long-run frequency with which RND generates numbers in B is almost exactly equal to the probability assigned to B by the uniform $(0, 1)$ distribution (that is the *length* of B, which is the sum of lengths of component intervals of B).

Uniform distribution over an area in the plane. Now probabilities are defined by relative areas instead of relative lengths. Think of a point picked uniformly at random from the rectangular area in a Venn diagram. Long-run frequencies for pairs $(\text{RND}_1, \text{RND}_2)$ generated by two calls of a pseudo-random number generator are well approximated by probabilities derived from the uniform distribution on the *unit square* $(0, 1) \times (0, 1)$. Section 5.1 gives examples, and extensions of the idea to higher dimensions.

Empirical Distributions

Let $(x_1, x_2, \ldots, x_n)$ be a list of n numbers. Think of x_i as the ith measurement of some physical quantity like the length or weight of something, in a series of repeated measurements. The *empirical distribution* of the list of n numbers is the distribution on the line $(-\infty, \infty)$ defined by

$$P_n(a,b) = \#\{i : 1 \leq i \leq n,\ a < x_i < b\}/n$$

That is, $P_n(a,b)$ is the proportion of the n numbers in the list that lie in the interval (a,b). To give this distribution a probabilistic interpretation, imagine n tickets in a box with number x_i written on the ith ticket. Then for a ticket picked uniformly at random from the box, $P_n(a,b)$ is the probability that the number on the ticket drawn is in (a,b). So the empirical distribution of a list is the distribution of a number picked at random from the list.

The empirical distribution of a data list is displayed by a *histogram*, that is, a bar graph in which proportions in the list are represented by the areas of various bars.

FIGURE 4. **A data histogram.** Actual values of the data points are shown by marks on the horizontal axis. The area of the bar over each bin shows the proportion of data points in the bin.

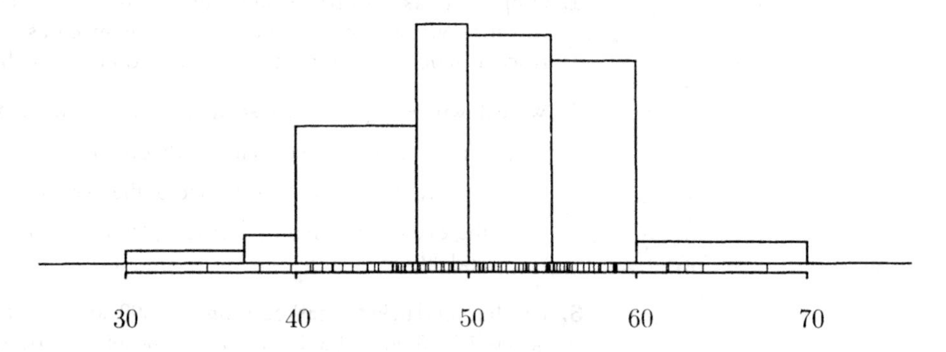

A histogram can be drawn as follows. First the interval of values is cut at some sequence of *cut points* $b_1 < b_2 < \cdots < b_m$, such that all the x_i are contained in (b_1, b_m), and none of the cut points equals any of the x_i. The cut points define $m-1$ subintervals (b_j, b_{j+1}), $1 \leq j \leq m-1$, called *bins*. The histogram is drawn by placing a rectangle over the jth bin with base the bin width $b_{j+1} - b_j$ and height

$$\frac{P_n(b_j, b_{j+1})}{(b_{j+1} - b_j)} = \frac{\#\{i : 1 \leq i \leq n, b_j < x_i < b_{j+1}\}}{n(b_{j+1} - b_j)}$$

This height is the *proportion of observations per unit length* in the jth bin. The area of the bar over the jth bin is the base times height, which is the proportion of observations in the jth bin:

$$(b_{j+1} - b_j) \times \frac{P_n(b_j, b_{j+1})}{(b_{j+1} - b_j)} = P_n(b_j, b_{j+1})$$

The total area of the histogram is the sum of the areas of these bars, which is

$$\sum_{j=1}^{m-1} P_n(b_j, b_{j+1}) = P_n(b_1, b_m) = 1$$

by the addition rule for proportions, and the choice of b_1 and b_m so that all the observations lie in (b_1, b_m).

A histogram smoothes out the data to display the general shape of an empirical distribution. Such a histogram often follows quite a smooth curve. This leads to the idea, developed in Section 4.1, of approximating empirical proportions by areas under a curve. The same idea is used in Section 2.2 to approximate probability histograms.

Exercises 1.3

1. Suppose a cake is divided into three portions, one for you, one for your friend, and one for your neighbor. If you get twice as much as your friend, and your friend gets twice as much as your neighbor, what proportion of the cake do you get?

2. Write down the expression in set notation corresponding to each of the following events.

a) the event which occurs if exactly one of the events A and B occurs;

b) the event which occurs if none of the events A, B, or C occurs;

c) the events obtained by replacing "none" in b) by "exactly one," "exactly two," and "three."

3. Five hundred tickets, marked 1 through 500, are sold at a high-school cake raffle. I have tickets 17, 93, and 202. My friend has tickets 4, 101, 102, and 398. One of the tickets will be chosen at random, and the owner of the winning ticket gets a cake. Make an outcome space for this situation, and indicate how each of the following events can be represented as a subset of your outcome space.

a) one of my tickets is the winner; b) neither my friend nor I win the raffle;

c) the number on the winning ticket is just 1 away from the number on one of my tickets.

4. Let $\Omega = \{0, 1, 2\}$ be the outcome space in a model for tossing a coin twice and observing the total number of heads. Say if the following events can be represented as subsets of Ω. If you say "yes," provide the subset; if you say "no," explain why:

a) the coin does not land heads both times;

b) on one of the tosses the coin lands heads, and on the other toss it lands tails;

c) on the first toss the coin lands heads, and on the second toss it lands tails;

d) the coin lands heads at least once.

5. Think of the set $\Omega = \{HHH, HHT, HTH, HTT, THH, THT, TTH, TTT\}$ as the outcome space for three tosses of a coin. For example, the subset $\{HHH, TTT\}$ corresponds to the event that all three tosses land the same way. Give similar verbal descriptions for the events described by each of the following subsets of Ω.

a) $\{HHH, HHT, HTH, HTT\}$ b) $\{HTH, HTT, TTT, TTH\}$
c) $\{HTT, HTH, HHT, HHH\}$ d) $\{HHH, HHT, HTH, THH\}$
e) $\{THT, HTT, TTH\}$ f) $\{HHT, HHH, TTH, TTT\}$

6. Suppose a word is picked at random from this sentence.

a) What is the distribution of the length of the word picked?

b) What is the distribution of the number of vowels in the word?

7. Shapes. Following Example 3, suppose the probability that the shape lands flat (1 or 6) is p for some $0 \leq p \leq 1$.

a) For each $k = 1, 2, \ldots, 6$ find a formula for $P(k)$ in terms of p.

b) Find a formula in terms of p for the probability that the number shown by the shape is 3 or more.

8. Let A and B be events such that $P(A) = 0.6$, $P(B) = 0.4$, and $P(AB) = 0.2$. Find the probabilities of: a) $A \cup B$ b) A^c c) B^c d) $A^c B$ e) $A \cup B^c$ f) $A^c B^c$

9. Events F, G, and H are such that

$$P(F) = 0.7, \quad P(G) = 0.6, \quad P(H) = 0.5,$$

$$P(FG) = 0.4, \quad P(FH) = 0.3, \quad P(GH) = 0.2, \quad P(FGH) = 0.1.$$

Find: (a) $P(F \cup G)$; (b) $P(F \cup G \cup H)$; (c) $P(F^c G^c H)$.

10. Events A, B, and C are defined in an outcome space. Find expressions for the following probabilities in terms of $P(A)$, $P(B)$, $P(C)$, $P(AB)$, $P(AC)$, $P(BC)$, and $P(ABC)$.

a) The probability that exactly two of A, B, C occur.

b) The probability that exactly one of these events occurs.

c) The probability that none of these events occur.

11. Inclusion–exclusion formula for 3 events. Write $A \cup B \cup C = (A \cup B) \cup C$ and use the inclusion–exclusion formula three times to derive the inclusion–exclusion formula for 3 events:

$$P(A \cup B \cup C) = P(A) + P(B) + P(C) - P(AB) - P(AC) - P(BC) + P(ABC)$$

12. Inclusion–exclusion formula for n events. Derive the inclusion–exclusion formula for n events

$$P\Big(\bigcup_{i=1}^{n} A_i\Big) = \sum_i P(A_i) - \sum_{i<j} P(A_i A_j) + \sum_{i<j<k} P(A_i A_j A_k) - \cdots + (-1)^{n+1} P(A_1 \ldots A_n)$$

by mathematical induction after showing that

$$P\left(\bigcup_{i=1}^{n+1} A_i\right) = P\left(\bigcup_{i=1}^{n} A_i\right) + P(A_{n+1}) - P\left(\bigcup_{i=1}^{n} A_i A_{n+1}\right)$$

13. Boole's inequality. The inclusion–exclusion formula gives the probability of a union of events in terms of probabilities of intersections of the various subcollections of these events. Because this expression is rather complicated, and probabilities of intersections may be unknown or hard to compute, it is useful to know that there are simple bounds. Use induction on n to derive *Boole's inequality*: $P(\bigcup_{i=1}^n A_i) \leq \sum_{i=1}^n P(A_i)$.

14. Show that $P(A \cap B) \geq P(A) + P(B) - 1$.

15. Use Boole's inequality and the fact that $(\bigcup_{i=1}^n A_i)^c = \bigcap_{i=1}^n A_i^c$ to show that

$$P(B_1 B_2 \cdots B_n) \geq \sum_{i=1}^{n} P(B_i) - (n-1)$$

16. Bonferroni's inequalities. According to Boole's inequality, the first sum in the inclusion–exclusion formula gives an upper bound on the probability of a union. This is the first of the series of *Bonferroni inequalities*. The next shows that the first sum minus the second is a lower bound. Show by using induction on n, and Boole's inequality, that:

a) $P(\cup_{i=1}^n A_i) \geq \sum_{i=1}^n P(A_i) - \sum_{i<j} P(A_i A_j)$.

b) Continuing like this, show that adding the third sum $\sum_{i<j<k} P(A_i A_j A_k)$ gives an upper bound, subtracting the fourth sum gives a lower bound, and so on. [*Hint.* In each case, use induction on n, and the previous inequality. For example, for the inequality that involves adding the third sum, use induction on n and the result of a).]

Note: The successive bounds do not always get better as more sums are introduced, despite the fact that the final formula, involving all n sums, is exact.

1.4 Conditional Probability and Independence

The first few examples of this section illustrate the idea of conditional probability in a setting of equally likely outcomes.

Example 1. Three coin tosses.

If you bet that 2 or more heads will appear in 3 tosses of a fair coin, you are more likely to win the bet given the first toss lands heads than given the first toss lands tails. To be precise, assume the 8 possible patterns of heads and tails in the three tosses, $\{hhh, hht, hth, htt, thh, tht, tth, ttt\}$, are equally likely. Then the *overall* or *unconditional* probability of the event

$$A = (2 \text{ or more heads in 3 tosses}) = \{hhh, hht, hth, thh\}$$

is $P(A) = 4/8 = 1/2$. But given that the first toss lands heads (say H), event A occurs if there is at least one head in the next two tosses, with a chance of $3/4$. So it is said that the conditional probability of A given H is $3/4$. The mathematical notation for the conditional probability of A given H is $P(A|H)$, read "P of A given H". In the present example

$$P(A|H) = 3/4$$

because $H = \{hhh, hht, hth, htt\}$ can occur in 4 ways, and just 3 of these outcomes make A occur. These 3 outcomes define the event $\{hhh, hht, hth\}$ which is the *intersection* of the events A and H, denoted A *and* H, $A \cap H$, or simply AH. Similarly, if the event H^c = "first toss lands tails" occurs, event A happens only if the next two tosses land heads, with probability $1/4$. So

$$P(A|H^c) = 1/4$$

Conditional probabilities can be defined as follows in any setting with equally likely outcomes.

Counting Formula for $P(A|B)$

For a finite set Ω of equally likely outcomes, and events A and B represented by subsets Ω, the *conditional probability of A given B* is

$$P(A|B) = \frac{\#(AB)}{\#(B)}$$

the proportion of outcomes in B that are also in A. Here $AB = A \cap B = A$ *and* B is the *intersection* of A and B.

Example 2. Tickets.

Problem. A box contains 10 capsules, similar except that four are black and six are white. Inside each capsule is a ticket marked either *win* or *lose*. The capsules are opaque, so the result on the ticket inside cannot be read without breaking open the capsule. Suppose a capsule is drawn at random from the box, then broken open to read the result. If it says *win*, you win a prize. Otherwise, you win nothing. The numbers of winning and losing tickets of each color are given in the diagram, which shows the tickets inside the capsules. Suppose that the capsule has just been drawn, but not yet broken to read the result. The capsule is black. Now what is the probability that you win a prize?

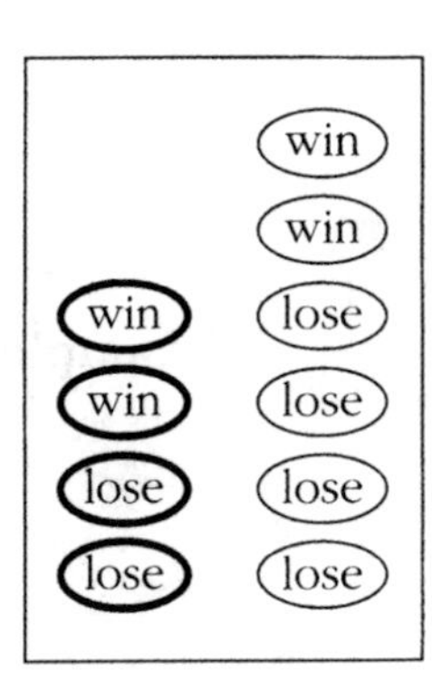

Solution. This conditional probability is the proportion of winners among black capsules:

$$P(\text{win}|\text{black}) = \frac{\#(\text{win and black})}{\#(\text{black})} = \frac{2}{4} = 0.5$$

Compare with the unconditional probability $P(\text{win}) = 4/10 = 0.4$

Example 3. Two-sided cards.

Problem. A hat contains three cards.

One card is black on both sides.

One card is white on both sides.

One card is black on one side and white on the other.

b/b

b/w

w/w

The cards are mixed up in the hat. Then a single card is drawn and placed on a table. If the visible side of the card is black, what is the chance that the other side is white?

Solution. Label the faces of the cards:

b_1 and b_2 for the black–black card;

w_1 and w_2 for the white–white card;

b_3 and w_3 for the black–white card.

Assume that each of these six faces is equally likely to be the face showing uppermost. Experience shows that this assumption does correspond to long-run frequencies, provided the cards are similar in size and shape, and well mixed up in the hat. The outcome space is then the set of six possible faces which might show uppermost:

$$\{b_1, b_2, b_3, w_1, w_2, w_3\}$$

The event {black on top} is identified as

$$\{\text{black on top}\} = \{b_1, b_2, b_3\}$$

Similarly,

$$\{\text{white on bottom}\} = \{b_3, w_1, w_2\}$$

Given that the event {black on top} has occurred, the face showing is equally likely to be b_1, b_2, or b_3. Only in the last case is the card white on the bottom. So the chance of white on bottom given black on top is

$$P(\text{white on bottom}|\text{black on top}) = \frac{\#(\text{white on bottom and black on top})}{\#(\text{black on top})} = \frac{1}{3}$$

Discussion. You might reason as follows: The card must be either the black–black card or the black–white card. These are equally likely possibilities, so the chance that the other side is white is 1/2. Many people find this argument convincing, but it is basically wrong. The assumption of equally likely outcomes, given the top side is black, is not consistent with long-run frequencies. If you repeat the experiment of drawing from the hat over and over, replacing the cards and mixing them up each time, you will find that over the long run, among draws when the top side is black, the bottom side will be white only about 1/3 of the time, rather than 1/2 of the time.

Frequency interpretation of conditional probability. This is illustrated by the previous example. If $P(A)$ approximates to the relative frequency of A in a long series of trials, then $P(A|B)$ approximates the relative frequency of trials producing A among those trials which happen to result in B. A general formula for $P(A|B)$, consistent with this interpretation, is found as follows. Start with the counting formula for $P(A|B)$ in a setting of equally likely outcomes, then divide both numerator and denominator by $\#(\Omega)$ to express $P(A|B)$ in terms of the unconditional probabilities $P(AB) = \#(AB)/\#(\Omega)$ and $P(B) = \#(B)/\#(\Omega)$:

$$P(A|B) = \frac{\#(AB)}{\#(B)} = \frac{\#(AB)/\#(\Omega)}{\#(B)/\#(\Omega)} = \frac{P(AB)}{P(B)}$$

General Formula for $P(A|B)$

$$P(A|B) = \frac{P(AB)}{P(B)}$$

If probabilities $P(A)$ are specified for subsets A of an outcome space Ω, then conditional probabilities given B can be calculated using this formula. This restricts the outcome space to B and renormalizes the distribution on B. In case the original distribution is defined by relative numbers, or relative areas, the same will be true of the conditional distribution given B, but with the restriction from Ω to B. To make a clear distinction, $P(A)$ or $P(AB)$ is called an *overall* or *unconditional* probability, and $P(A|B)$ a *conditional* probability.

Example 4. Relative areas.

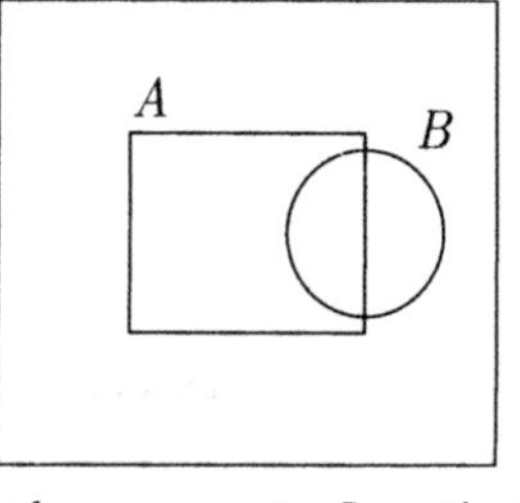

Suppose a point is picked uniformly at random from the big rectangle in the diagram. Imagine that information about the position of this point is revealed to you in two stages, by the answers to the following questions:

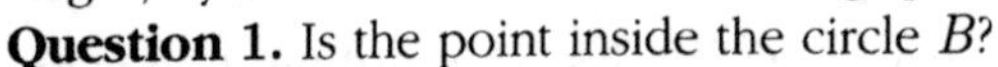

Question 1. Is the point inside the circle B?
Question 2. Is the point inside the rectangle A?

Problem. If the answer to Question 1 is yes, what is the probability that the answer to Question 2 will be yes?

Solution. The problem is to find the probability that the point is in the rectangle A given that it is in the circle B. By inspection of the diagram, approximately half the area inside B is inside A. So the required probability is

$$P(A|B) = \frac{P(AB)}{P(B)} = \frac{\text{Area}(AB)}{\text{Area}(B)} \approx 1/2$$

Remark. The formula for conditional probability in this case corresponds to the idea that given the point is in B, equal areas within B still have equal probabilities.

Tree Diagrams and the Multiplication Rule

In the above example a conditional probability was calculated from overall probabilities. But in applications there are usually many events A and B such that the conditional probability $P(A|B)$ and the overall probability $P(B)$ are more obvious than the overall probability $P(AB)$. Then $P(AB)$ is calculated using the following rearrangement of the general formula for conditional probability:

Multiplication Rule

$$P(AB) = P(A|B)P(B)$$

This rule is very intuitive in terms of the frequency interpretation. If, for example, B happens over the long run about 1/2 the time ($P(B) = 1/2$), and about 1/3 of the times that B happens A happens too ($P(A|B) = 1/3$), then A and B happens about 1/3 of $1/2 = 1/3 \times 1/2 = 1/6$ of the time ($P(AB) = P(A|B)P(B) = 1/6$).

The multiplication rule is often used to set up a probability model with intuitively prescribed conditional probabilities. Typically, A will be an event determined by some overall outcome which can be thought of as occurring by stages, and B will be some event depending just on the first stage. If you think of B happening before A it is more natural to rewrite the multiplication rule, with BA instead of AB and the two factors switched:

$$P(BA) = P(B)P(A|B)$$

In words, the chance of B followed by A is the chance of B times the chance of A given B.

Example 5. Picking a box, then a ball.

Problem. Suppose that there are two boxes, labeled odd and even. The odd box contains three balls numbered 1, 3, 5. The even box contains two balls labeled 2, 4. One of the boxes is picked at random by tossing a fair coin. Then a ball is picked at random from this box. What is the probability that the ball drawn is ball 3?

Solution. A scheme like this can be represented in a *tree diagram*. Each branch represents a possible way things might turn out. Probabilities and conditional probabilities are indicated along the branch.

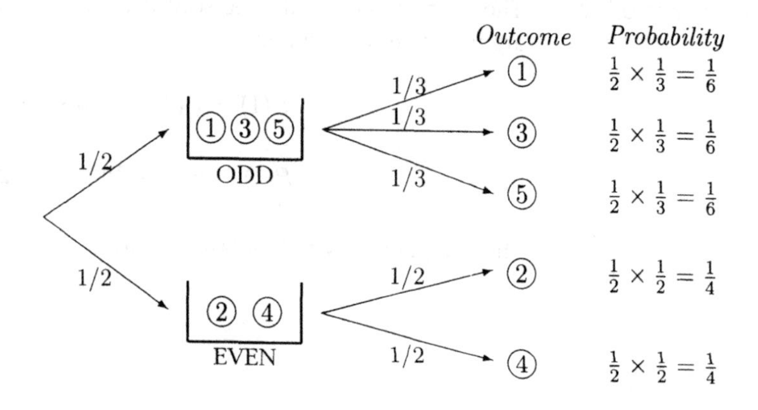

Because the box is chosen by a fair coin toss,

$$P(\text{odd}) = P(\text{even}) = 1/2$$

The only way to get 3 is to first pick the odd box, then pick 3. By assumption

$$P(3|\text{odd}) = 1/3$$

Now by the multiplication rule,

$$P(3) = P(\text{odd and } 3) = P(\text{odd})P(3\,|\,\text{odd}) = \frac{1}{2} \times \frac{1}{3} = \frac{1}{6}$$

This is the product of the probabilities along the path representing the outcome 3. The corresponding products along the other possible branches give the distribution displayed in the tree diagram.

This is a different representation of the same problem, using a Venn diagram.

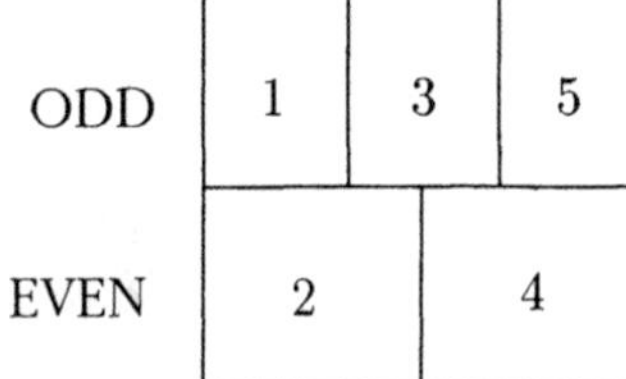

Remark 1. A naive approach to the above problem would be to assume that all outcomes were equally likely. But this would imply

$$P(\text{first box}) = P(\text{odd}) = 3/5$$

$$P(\text{second box}) = P(\text{even}) = 2/5$$

which is inconsistent with the box being chosen by a fair coin toss.

Remark 2. The problem could also be solved without conditional probabilities by a symmetry argument, assuming that

$$P(1) = P(3) = P(5) \quad \text{and} \quad P(2) = P(4)$$

$$P(1) + P(3) + P(5) = P(2) + P(4) = 1/2$$

These equations yield the same answer as above.

To summarize the method of the previous example:

Multiplication Rule in a Tree Diagram

After setting up a tree diagram whose paths represent joint outcomes, the multiplication rule is used to define a distribution of probability over paths. The probability of each joint outcome represented by a path is obtained by multiplying the probability and conditional probability along the path.

Example 6. Electrical components.

Suppose there are two electrical components. The chance that the first component fails is 10%. If the first component fails, the chance that the second component fails is 20%. But if the first component works, the chance that the second component fails is 5%.

Problem. Calculate the probabilities of the following events:

1. at least one of the components works;
2. exactly one of the components works;
3. the second component works.

Solution. Here is the tree diagram showing all possible performances of the first and second components. Probabilities are filled in using the above data and the rule of complements.

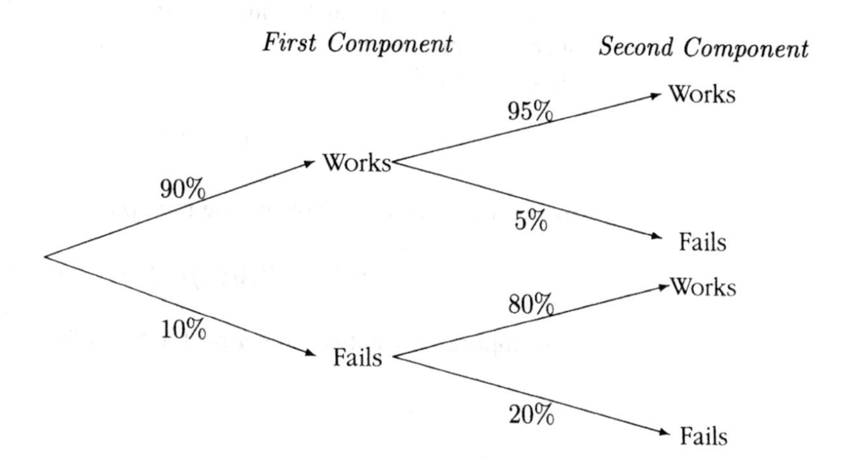

By inspection of the diagram,

$$\begin{aligned}
P(\text{at least one works}) &= 1 - P(\text{both fail}) \\
&= 1 - 0.1 \times 0.2 = 0.98 \\
P(\text{exactly one works}) &= P(\text{first works and second fails}) \\
&\quad + P(\text{first fails and second works}) \\
&= 0.9 \times 0.05 + 0.1 \times 0.8 = 0.125 \\
P(\text{second works}) &= P(\text{first works and second works}) \\
&\quad + P(\text{first fails and second works}) \\
&= 0.9 \times 0.95 + 0.1 \times 0.8 = 0.935
\end{aligned}$$

Averaging Conditional Probabilities

The last two parts of the previous example illustrate a rule of average conditional probabilities: for any events A and B, the overall probability $P(A)$ is the average of the two conditional probabilities $P(A|B)$ and $P(A|B^c)$ with weights $P(B)$ and $P(B^c)$:

$$P(A) = P(A|B)P(B) + P(A|B^c)P(B^c)$$

In the example, B and B^c were (first works) and (first fails), while A was (exactly one works) in one instance, and (second works) in the other. The formula gives the probability of A as the sum of products of probabilities along paths leading to A in the tree diagram. The event B defines a partition of the whole outcome space Ω into two events B and B^c, corresponding to two initial branches in the tree. There is a similar formula for any partition $B_1, ..., B_n$ of the whole outcome space Ω, corresponding to n initial branches of a tree. For any event A the events $AB_1, ..., AB_n$ form a partition of A, so

$$P(A) = P(AB_1) + \cdots + P(AB_n)$$

by the addition rule. Applying the multiplication rule to each term gives

$$P(A) = P(A|B_1)P(B_1) + \cdots + P(A|B_n)P(B_n)$$

This important result is summarized in the following box.

Rule of Average Conditional Probabilities

For a partition $B_1, \ldots, B_n$ of Ω,

$$P(A) = P(A|B_1)P(B_1) + \cdots + P(A|B_n)P(B_n)$$

In words: the overall probability $P(A)$ is the weighted average of the conditional probabilities $P(A|B_i)$ with weights $P(B_i)$.

Example 7. Sampling without replacement.

Problem. Suppose two cards are dealt from a well-shuffled deck of 52 cards. What is the probability that the second card is black?

Solution. A common response to this question is that you can't say. It depends on whether the first card is black or not. If the first card is black, the chance that the second is black is $25/51$, since no matter which black card the first one is, the second is equally likely to be any of the 51 remaining cards, and there are 25 black cards remaining. If the first card is red, the chance that the second is black is $26/51$, by similar reasoning. These are the *conditional* probabilities of black on the second card given black and red, respectively, on the first card. But the question does not refer to the first card at all. The *overall* probability of black on the second card is the *average* of these conditional probabilities:

$$\begin{aligned} P(\text{second black}) &= P(\text{second black}|\text{first black})P(\text{first black}) \\ &\quad + P(\text{second black}|\text{first red})P(\text{first red}) \\ &= \frac{25}{51} \cdot \frac{1}{2} + \frac{26}{51} \cdot \frac{1}{2} = \left(\frac{25+26}{51}\right) \times \frac{1}{2} = \frac{1}{2} \end{aligned}$$

Discussion. This can also be argued by symmetry. Since there are equal numbers of black and red cards in the deck, the assumptions made at the start are symmetric with respect to black and red. This makes

$$P(\text{second black}) = P(\text{second red})$$

Since

$$P(\text{second black}) + P(\text{second red}) = 1$$

this gives the answer of $1/2$. This argument shows just as well that if n cards are dealt, then $P(n\text{th card black}) = 1/2$, $P(n\text{th card an ace}) = 1/13$, and so on.

Independence

We have just seen that for any events A and B, $P(A)$ is the average of the conditional probabilities $P(A|B)$ and $P(A|B^c)$, weighted by $P(B)$ and $P(B^c)$. Suppose now that the chance of A does not depend on whether or not B occurs, and in either case equals p, say. In symbols:

$$P(A|B) = P(A|B^c) = p \tag{1}$$

Then also the unconditional probability of A is p:

$$P(A) = P(A|B)P(B) + P(A|B^c)P(B^c) = pP(B) + pP(B^c) = p$$

For example, A might be the event that a card dealt from a well-shuffled deck was an ace, B the event that a die showed a six. Such events A and B are called *independent*. Intuitively, independent events have no influence on each other. It would be reasonable to suppose that any event determined by a card dealt from a shuffled deck would be independent of any event determined by rolling a die. To be brief, the deal and the die roll would be called independent.

One more example: two draws at random from a population would be independent if done with replacement between draws, but *dependent* (i.e., not independent) if done without replacement.

Independence of events A and B can be presented mathematically in a variety of equivalent ways. For example, it was just shown that the definition (1) above (which assumes both $P(B) > 0$ and $P(B^c) > 0$), implies

$$P(A|B) = P(A) \tag{2}$$

A similar calculation shows that (2) implies (1). The formula $P(A|B) = P(AB)/P(B)$ shows (2) is equivalent to the following:

Multiplication Rule for Independent Events

$$P(AB) = P(A)P(B)$$

The multiplication rule is usually taken as the formal mathematical definition of independence, to include the case of events with probability 0 or 1. (Such an event is then, by definition, independent of every other event.)

The multiplication rule brings out the symmetry of independence. Assuming $P(A) > 0$, and using the fact that $AB = BA$ and $P(A)P(B) = P(B)P(A)$, the multiplication rule allows (2) to be turned around to

$$P(B|A) = P(B) \tag{3}$$

and (1) can be turned around similarly.

Assuming A and B are independent, all of these formulae hold also with either A^c substituted for A, B^c for B, or with both substitutions. This is obvious for (1), hence also true for the others. To spell out an example, since A splits into AB^c and AB,

$$\begin{aligned} P(AB^c) &= P(A) - P(AB) \\ &= P(A) - P(A)P(B) \quad \text{assuming the multiplication rule for } A \text{ and } B \\ &= P(A)(1 - P(B)) \\ &= P(A)P(B^c) \quad \text{by the rule of complements.} \end{aligned}$$

So the multiplication rule works just as well with B^c instead of B. The same goes for A^c instead of A.

Here the various probabilities determined by independent events A and B are illustrated graphically as proportions in a Venn diagram. Event A is represented by a rectangle lying horizontally, event B by a rectangle standing vertically.

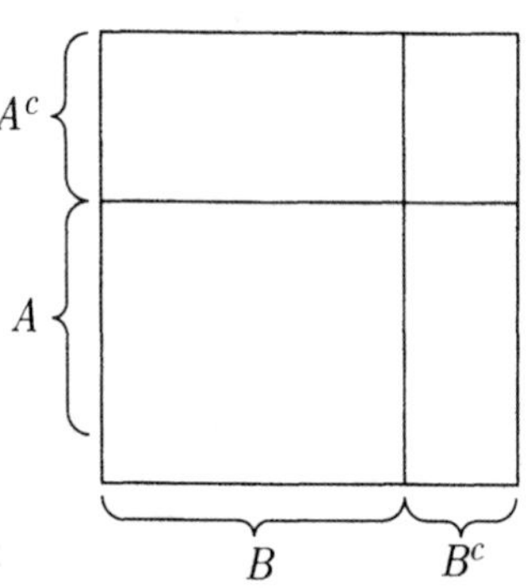

Example 8. **Reliability of two components in series.**

A system consists of two components C_1 and C_2, each of which must remain operative for the overall system to function. The components C_1 and C_2 are then said to be connected in series, and represented diagrammatically as follows:

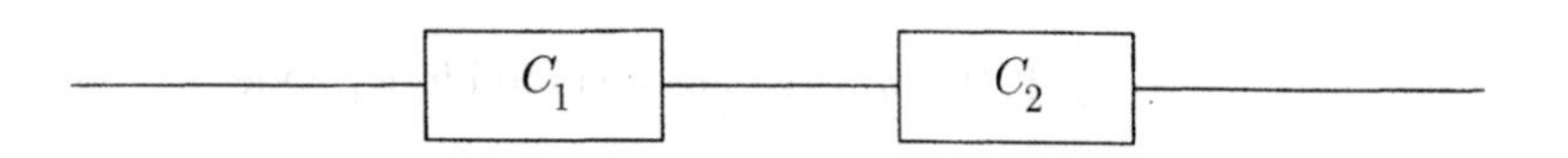

Let W_i be the event that component C_i works without failure for a given period of time, say one day. The event that the whole system operates without failure for one day is the event that both C_1 and C_2 operate without failure, that is, the event W_1W_2. The probabilities $P(W_1)$ and $P(W_2)$ are called the reliabilities of components C_1 and C_2. The probability $P(W_1W_2)$ is the reliability of the whole system. Suppose that the component reliabilities $P(W_1)$ and $P(W_2)$ are known from empirical data of past performances of similar components, say $P(W_1) = 0.9$ and $P(W_2) = 0.8$. If the particular components C_1 and C_2 have never been used together before, $P(W_1W_2)$ cannot be known empirically. But it may still be reasonable to assume that the events W_1 and W_2 are independent. Then the reliability of the whole system would be given by the formula

$$P(\text{system works}) = P(W_1W_2) = P(W_1)P(W_2) = 0.9 \times 0.8 = 0.72$$

Hopefully this number, 0.72, would give an indication of the long-run relative frequency of satisfactory performance of the system. But bear in mind that such a number is based on a theoretical assumption of independence which may or may not prove well founded in practice. The sort of thing which might prevent independence is the possibility of failures of both components due to a common cause, for example, voltage fluctuations in a power supply, the whole system being flooded, the system catching fire, etc. For the series system considered here such factors would tend to make the reliability $P(W_1W_2)$ greater than if W_1 and W_2 were independent, suggesting that the number, 0.72, would be too low an estimate of the reliability.

Example 9. Reliability of two components in parallel.

A method of increasing the reliability of a system is to put components in parallel, so the system will work if either of the components works. Two components C_1 and C_2 in parallel may be represented diagrammatically as follows:

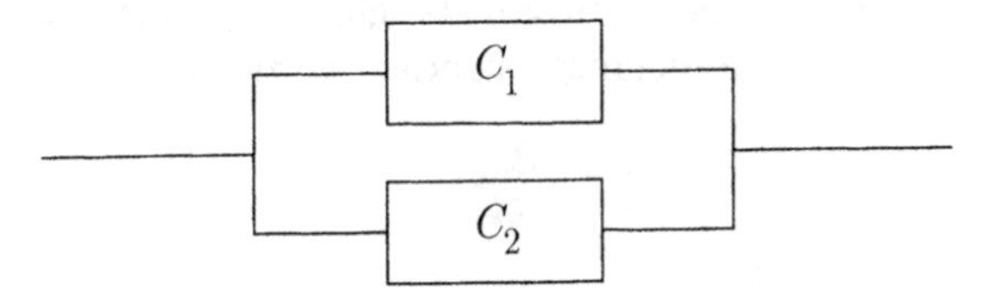

Suppose, as in the last example, that the individual components C_1 and C_2 have reliabilities $P(W_1)$ and $P(W_2)$, where W_1 is the event that C_1 works. The event that the whole system functions is now the event $W_1 \cup W_2$ that either C_1 or C_2 works. The complementary event of system failure is the event F_1F_2 that both C_1 and C_2 fail, where F_i is the complement of W_i. Thus the reliability of the whole system is

$$P(\text{system works}) = P(W_1 \cup W_2) = 1 - P(F_1F_2)$$

If W_1 and W_2 are assumed independent, so are F_1 and F_2. In that case

$$P(\text{system works}) = 1 - P(F_1)P(F_2)$$

For example, if the component reliabilities are $P(W_1) = 0.9$ and $P(W_2) = 0.8$ as before, then $P(F_1) = 0.1$ and $P(F_2) = 0.2$, and the system reliability is

$$P(\text{system works}) = 1 - (0.1)(0.2) = 0.98$$

This is a considerable improvement over the reliability of the individual components. The assumption of independent failures must be viewed with particular suspicion in parallel systems, as it tends to lead to exaggerated estimates of system reliabilities. Suppose, for example, that all failures of component C_1 and half the failures of component C_2 occur due to severe voltage fluctuation in a power supply common to C_1 and C_2. Then F_1 is the event of a voltage fluctuation, and it should be assumed that $P(F_1|F_2) = 0.5$ instead of the independence assumption $P(F_1|F_2) = 0.1$. With the new assumptions,

$$P(F_1F_2) = P(F_2)P(F_1|F_2) = (0.2)(0.5) = 0.1$$

$$P(\text{system works}) = 1 - P(F_1F_2) = 0.9$$

As a general rule, failures of both components due to a common cause will tend to decrease the reliability of a parallel system below the value predicted by an independence assumption.

Exercises 1.4

1. In a particular population of men and women, 92% of women are right handed, and 88% of men are right handed. Indicate whether each of the following statements is (i) true, (ii) false, or (iii) can't be decided on the basis of the information given.

a) The overall proportion of right handers in the population is exactly 90%.

b) The overall proportion of right handers in the population is between 88% and 92%.

c) If the sex ratio in the population is 1-to-1 then a) is true.

d) If a) is true then the sex ratio in the population is 1-to-1.

e) If there are at least three times as many women as men in the population, then the overall population of right handers is at least 91%.

2. A light bulb company has factories in two cities. The factory in city A produces two-thirds of the company's light bulbs. The remainder are produced in city B, and of these, 1% are defective. Among all bulbs manufactured by the company, what proportion are not defective and made in city B?

3. Suppose:
$P(\text{rain today})=40\%$; $P(\text{rain tomorrow})=50\%$; $P(\text{rain today and tomorrow})=30\%$.
Given that it rains today, what is the chance that it will rain tomorrow?

4. Two independent events have probabilities 0.1 and 0.3. What is the probability that

a) neither of the events occurs?

b) at least one of the events occurs?

c) exactly one of the events occurs?

5. There are two urns. The first urn contains 2 black balls and 3 white balls. The second urn contains 4 black balls and 3 white balls. An urn is chosen at random, and a ball is chosen at random from that urn.

a) Draw a suitable tree diagram.

b) Assign probabilities and conditional probabilities to the branches of the tree.

c) Calculate the probability that the ball drawn is black.

6. Suppose two cards are dealt from a deck of 52. What is the probability that the second card is a spade given that the first card is black?

7. Suppose A and B are two events with $P(A) = 0.5$, $\quad P(A \cup B) = 0.8$.

a) For what value of $P(B)$ would A and B be mutually exclusive?

b) For what value of $P(B)$ would A and B be independent?

8. A hat contains a number of cards, with

30% white on both sides;

50% black on one side and white on the other;

20% black on both sides.

The cards are mixed up, then a single card is drawn at random and placed on the table. If the top side is black, what is the chance that the other side is white?

9. Three high schools have senior classes of size 100, 400, and 500, respectively. Here are two schemes for selecting a student from among the three senior classes:

A: Make a list of all 1000 seniors, and choose a student at random from this list.

B: Pick one school at random, then pick a student at random from the senior class in that school.

Show that these two schemes are not probabilistically equivalent. Here is a third scheme:

C: Pick school i with probability p_i ($p_1 + p_2 + p_3 = 1$), then pick a student at random from the senior class in that school.

Find the probabilities p_1, p_2, and p_3 which make scheme C equivalent to scheme A.

10. Suppose electric power is supplied from two independent sources which work with probabilities 0.4, 0.5, respectively. If both sources are providing power enough power will be available with probability 1. If exactly one of them works there will be enough power with probability 0.6. Of course, if none of them works the probability that there will be sufficient supply is 0.

a) What are the probabilities that exactly k sources work for $k = 0, 1, 2$?

b) Compute the probability that enough power will be available.

11. Assume identical twins are always of the same sex, equally likely boys or girls. Assume that for fraternal twins the firstborn is equally likely to be a boy or a girl, and so is the secondborn, independently of the first. Assume that proportion p of twins are identical, proportion $q = 1-p$ fraternal. Find formulae in terms of p for the following probabilities for twins:

a) P(both boys)

b) P(firstborn boy and secondborn girl)

c) P(secondborn girl | firstborn boy)

d) P(secondborn girl | firstborn girl).

12. Give a formula for $P(F|G^c)$ in terms of $P(F)$, $P(G)$, and $P(FG)$ only.

1.5 Bayes' Rule

The rules of conditional probability, described in the last section, combine to give a general formula for updating probabilities called *Bayes' rule.* Before stating the rule in general, here is an example to illustrate the basic setup.

Example 1. Which box?

Suppose there are three similar boxes. Box i contains i white balls and one black ball, $i = 1, 2, 3$, as shown in the following diagram.

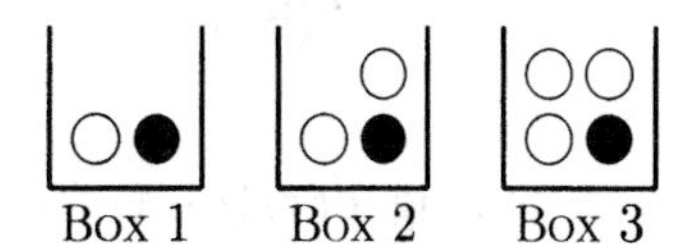

Suppose I mix up the boxes and then pick one at random. Then I pick a ball at random from the box and show you the ball. I offer you a prize if you can guess correctly what box it came from.

Problem. Which box would you guess if the ball drawn is white and what is your chance of guessing right?

Solution. An intuitively reasonable guess is Box 3, because the most likely explanation of how a white ball was drawn is that it came from a box with a large proportion of whites. To confirm this, here is a calculation of

$$P(\text{Box } i|\text{white}) = \frac{P(\text{Box } i \text{ and white})}{P(\text{white})} \qquad (i = 1, 2, 3) \qquad (*)$$

These are the chances that you would be right if you guessed Box i, given that the ball drawn is white. The following diagram shows the probabilistic assumptions:

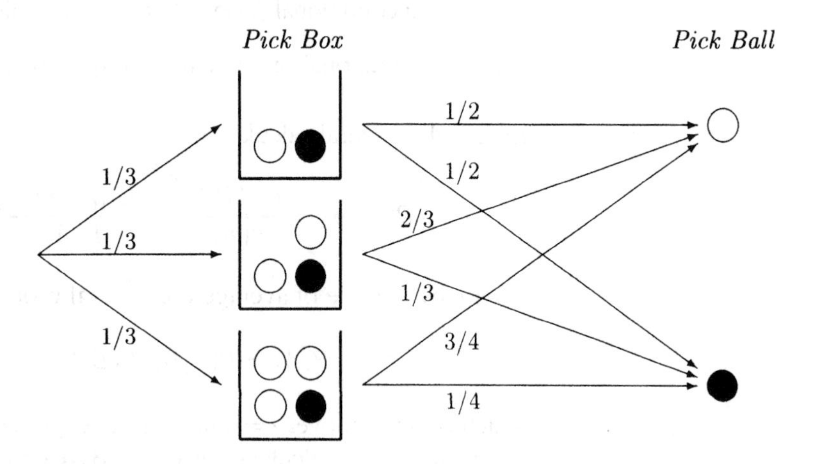

From the diagram, the numerator in $(*)$ is

$$P(\text{Box } i \text{ and white}) = P(\text{Box } i)P(\text{white}|\text{Box } i) = \frac{1}{3} \times \frac{i}{i+1} \quad (i = 1, 2, 3)$$

By the addition rule, the denominator in $(*)$ is the sum of these terms over $i = 1, 2, 3$:

$$P(\text{white}) = \frac{1}{3} \times \frac{1}{2} + \frac{1}{3} \times \frac{2}{3} + \frac{1}{3} \times \frac{3}{4} = \frac{23}{36} \quad \text{and}$$

$$P(\text{Box } i|\text{white}) = \frac{\frac{1}{3} \times \frac{i}{i+1}}{\frac{23}{36}} = \frac{12}{23} \times \frac{i}{i+1} \qquad (i = 1, 2, 3)$$

Substituting for $i/(i+1)$ for $i = 1, 2, 3$ gives the following numerical results:

i	1	2	3
$P(\text{Box } i\|\text{white})$	6/23	8/23	9/23

This confirms the intuitive idea that Box 3 is the most likely explanation of a white ball. Given a white ball, the chance that you would be right if you guessed this box would be $9/23 \approx 39.13\%$.

Suppose, more generally, that events $B_1, \ldots, B_n$ represent n mutually exclusive possible results of the first stage of some procedure. Which one of these results has occurred is assumed unknown. Rather, the result A of some second stage has been observed, whose chances depend on which of the B_i's has occurred. In the previous example A was the event that a white ball was drawn and B_i the event that it came from a box with i white balls. The general problem is to calculate the probabilities of the events B_i given occurrence of A (called ***posterior probabilities***), in terms of

(i) the unconditional probabilities $P(B_i)$ (called ***prior probabilities***);

(ii) the conditional probabilities $P(A|B_i)$ (called ***likelihoods***).

Here is the general calculation:

$$P(B_i|A) = \frac{P(AB_i)}{P(A)} = \frac{P(A|B_i)P(B_i)}{P(A)} \qquad \text{(multiplication rule)}$$

where, by the rule of average conditional probabilities, the denominator is

$$P(A) = P(A|B_1)P(B_1) + \cdots + P(A|B_n)P(B_n)$$

which is the sum over $i = 1$ to n of the expression $P(A|B_i)P(B_i)$ in the numerator. The result of this calculation is called Bayes' rule.

Bayes' Rule

For a partition $B_1, \ldots, B_n$ of all possible outcomes,

$$P(B_i|A) = \frac{P(A|B_i)P(B_i)}{P(A|B_1)P(B_1) + \cdots + P(A|B_n)P(B_n)} \quad (i = 1, \ldots, n)$$

It is better not to try to memorize this formula, as it is easily derived from the basic rules of conditional probability which are easier to remember. Rather, understand the sequence of steps by which it is derived. These are the same steps used to solve the balls and boxes problem.

Example 2. Which box? (continued).

Consider again the same three boxes as in the previous example. Suppose I pick a box. Then I pick a ball at random from the box and show you the ball. I offer you a prize if you can guess correctly what box it came from.

Problem. Which box would you guess if the drawn ball is white, and what is your chance of guessing right?

Discussion. The wording of this problem is identical to the wording of Example 1 above, except that the sentence "Suppose I mix up the boxes and then pick one at random" has been replaced by "Suppose I pick a box". A naive approach to the new problem is to suppose it is the same as the old one, with the answer:

$$\text{guess Box 3, with probability of being right} = 9/23$$

But this makes an *implicit assumption* that I am equally likely to pick any one of the three boxes. And the problem cannot be solved without assuming some values π_i for the probabilities that I pick box $i, i = 1, 2, 3$. These probabilities π_i are called *prior probabilities* because they refer to your opinion about which box I picked, prior to learning the color of the ball drawn. Once you have assigned these prior probabilities $\pi_i, i = 1, 2, 3$, the previous calculations can be repeated. From the prior probabilities π_i and the probabilities $i/(i + 1)$ of getting the observed result, given box i (the *likelihoods*), you can obtain the *posterior probabilities* by Bayes' rule:

$$P(\text{Box } i|\text{white}) = \frac{\pi_i(\frac{i}{i+1})}{\pi_1 \times \frac{1}{2} + \pi_2 \times \frac{2}{3} + \pi_3 \times \frac{3}{4}}$$

Thus, given that a white ball was drawn, to maximize your chance of guessing correctly you should guess box i for whichever i maximizes $\pi_i \left(\frac{i}{i+1}\right)$. Which i this is depends on the π_i. The probabilities in question are now clearly a matter of your *opinion* about how I picked the box. There remains the problem of how to assign the prior probabilities π_i. This is a tricky business, as it depends on psychological factors, such as whether or not you think I am deliberately trying to make it hard for you to guess, and if so what strategy you think I'm using. For further analysis, see Exercises 1.5.7 and 1.5.8.

In principle, every application of Bayes' rule is such as the above examples of guessing the box that produced a particular color of ball. There is always the problem of deciding what the prior probabilities should be. Most often the prior probabilities will only make sense in a subjective interpretation of probability. But in problems like the next example (false positives) the prior probabilities may be known as population proportions. This example is like a scheme with two boxes D and D^c:

Box D containing 95% balls labeled $+$ and 5% labeled $-$

Box D^c containing 2% balls labeled $+$ and 98% labeled $-$

If box D has prior probability 1%, and a draw from the box yields a $+$, what is the chance that the $+$ came from box D? As the solution shows, such extremely skewed priors and likelihoods may lead to surprising conclusions.

Example 3. False positives.

Problem. Suppose that a laboratory test on a blood sample yields one of two results, positive or negative. It is found that 95% of people with a particular disease produce a positive result. But 2% of people without the disease will also produce a positive result (a *false positive*). Suppose that 1% of the population actually has the disease. What is the probability that a person chosen at random from the population will have the disease, given that the person's blood yields a positive result?

Solution. Let $P(F)$ denote the proportion of people in the population with characteristic F. Then $P(F|G)$ is the proportion of those in the population with characteristic G who also have characteristic F. The desired probability is $P(D|+)$ where D indicates the disease, and + indicates a positive test result. The data in the problem indicate that

$$P(+|D) = 0.95, \;\; P(+|D^c) = 0.02, \;\; P(D) = 0.01, \;\; P(D^c) = 0.99.$$

Applying Bayes' rule with $A = +$, $B_1 = D$, $B_2 = D^c$, gives

$$\begin{aligned} P(D|+) &= \frac{P(+|D)P(D)}{P(+|D)P(D) + P(+|D^c)P(D^c)} \\ &= \frac{(.95)(.01)}{(.95)(.01) + (.02)(.99)} \\ &= \frac{95}{293} \approx 32\% \end{aligned}$$

Discussion. Thus only 32% of those persons who produce a positive test result actually have the disease. At first this result seems surprisingly low. The point is that because the disease is so rare, the number of true positives coming from the few people with the disease is comparable to the number of false positives coming from the many without the disease.

Interpretation of conditional probabilities. In applications of Bayes' rule it is important to keep in mind the interpretation of the various probabilities involved. Typically, the likelihoods $P(A|B_i)$ will admit a long-run frequency interpretation. If the prior probabilities $P(B_i)$ also have a long-run frequency interpretation, then so too will the conditional probability $P(B_i|A)$ given by Bayes' formula. In Example 3 there were two hypotheses $B_1 = D$ that a person was diseased and $B_2 = D^c$ that a person was not. The observed event was the event $A = +$ of a positive laboratory test. There the conditional probability $P(D|+)$ admitted an empirical interpretation, as that proportion of individuals in the population in question showing a positive test who actually had the disease. This conditional probability also admits a long-run frequency interpretation in terms of repeated sampling of that population, or some other population with the same characteristics assumed in the calculations. Among persons who produce a positive laboratory test, the long-run proportion with the disease will most likely be close to $P(D|+) \approx 32\%$.

There are many situations, however, where it is impossible to give a long-run frequency interpretation to the prior probabilities $P(B_i)$. The same must then be said of the posterior probabilities $P(B_i|A)$ which are calculated in terms of them, even if the likelihoods $P(A|B_i)$ have long-run frequency interpretations.

Calculations by Bayes' rule can often be simplified by noting that it is only the ratios $P(B_i)$ to $P(B_j)$ (the *prior odds ratios*) and the ratios $P(A|B_i)$ to $P(A|B_j)$ (the *likelihood ratios*) which matter. As you can check as an exercise, if the prior odds ratios are written as, say, R_i to R_j, and the likelihood ratios as, say, L_i to L_j, meaning that

$$P(B_i) = cR_i \qquad \text{for some constant } c$$

and

$$P(A|B_i) = dL_i \qquad \text{for some constant } d$$

then the *posterior odds ratios* $P(B_i|A)$ to $P(B_j|A)$ are simply R_iL_i to R_jL_j, and

$$P(B_i|A) = \frac{R_iL_i}{R_1L_1 + \cdots + R_nL_n}.$$

This is summarized by the following:

Bayes' Rule for Odds

posterior odds = prior odds × likelihoods.

Bayes' rule for odds shows clearly how the prior odds are just as important a factor as the likelihood ratio in computing the posterior odds. If the prior odds don't make sense in terms of long-run frequencies, neither will the posterior odds.

But even if the probabilities don't admit a long-run frequency interpretation, you might find it useful to regard the probabilities in Bayes' rule as subjective probabilities. Bayes' rule then dictates how opinions should be revised in the light of new information, to be consistent with the rules of probability. Here is a typical example.

Example 4. Diagnosis of a particular patient.

Problem. Suppose a doctor is examining a patient from the population in Example 3. This patient was not chosen at random. He walked into the doctor's office because he was feeling sick. After examining the patient, but not seeing the result of the blood test, the doctor's opinion is that there is a 30% chance that the patient has the disease. How should the doctor revise her opinion after seeing a positive blood test?

Solution. To be consistent with the rules of probability, the doctor should use Bayes' rule. Now the prior probabilities are

$$P(D) = 30\%, \quad P(D^c) = 70\%$$

while it might be reasonable to suppose that the likelihoods

$$P(+|D) = 95\%, \quad P(+|D^c) = 2\%$$

are the same as before. The posterior probability can be calculated as before, using Bayes' rule, but with the new prior probabilities. In terms of odds, the prior odds in favor of the disease are 3 to 7, the likelihood ratio in favor of the disease is 95 to 2, so the posterior odds in favor are 3×95 to 7×2, or 285 to 14. So given the positive blood test result, the doctor should revise her opinion and say that the patient has the disease with probability

$$\frac{285}{285 + 14} = \frac{285}{299} = 0.95317$$

Discussion. Notice how working with prior odds of 30 to 70 instead of 1 to 99 has a drastic effect on the conclusion. Provided the prior odds are not heavily against the disease, the evidence of the blood test carries a lot of weight. The likelihood ratio of 95 to 2 overwhelms the doctor's prior odds of 3 to 7, so there should be little doubt left in the doctor's mind after seeing the positive blood test. The puzzling question in this kind of application is how does the doctor come up with the odds of 3 to 7 after the medical examination? To come up with such odds, the doctor must make an intuitive judgment based on the whole complex of evidence gained from an examination of the patient. It seems impossible to adequately formalize this process mathematically. The theory does not help the doctor come up with a prior opinion, or explain how the doctor should revise an opinion in the light of complex information such as is gained from a medical examination. All the theory can do in this context is to suggest how an opinion should be revised in the light of a single additional piece of information, such as the result of a blood test.

Notice how the terms prior and posterior are relative terms, like today and tomorrow. The posterior distribution after today's test will be the prior distribution for tomorrow's test. So an opinion can be revised repeatedly using Bayes' rule. At each stage in this process, all probabilities should be computed conditionally on everything that has gone before.

Exercises 1.5

1. There are two boxes, the odd box containing 1 black marble and 3 white marbles, and the even box containing 2 black marbles and 4 white marbles. A box is selected at random, and a marble is drawn at random from the selected box.

a) What is the probability that the marble is black?

b) Given the marble is white, what is the probability that it came from the even box?

2. **Polya's urn scheme**. An urn contains 4 white balls and 6 black balls. A ball is chosen at random, and its color noted. The ball is then replaced, along with 3 more balls of the same color (so that there are now 13 balls in the urn). Then another ball is drawn at random from the urn.

a) Find the chance that the second ball drawn is white. (Draw an appropriate tree diagram.)

b) Given that the second ball drawn is white, what is the probability that the first ball drawn is black?

c) Suppose the original contents of the urn are w white and b black balls, and that after a ball is drawn from the urn, it is replaced along with d more balls of the same color. In part a), w was 4, b was 6, and d was 3. Show that the chance that the second ball drawn is white is $\frac{w}{w+b}$. [Note that the probability above does not depend on the value of d.]

3. A manufacturing process produces integrated circuit chips. Over the long run the fraction of bad chips produced by the process is around 20%. Thoroughly testing a chip to determine whether it is good or bad is rather expensive, so a cheap test is tried. All good chips will pass the cheap test, but so will 10% of the bad chips.

a) Given a chip passes the cheap test, what is the probability that it is a good chip?

b) If a company using this manufacturing process sells all chips which pass the cheap test, over the long run what percentage of chips sold will be bad?

4. A digital communications system consists of a transmitter and a receiver. During each short transmission interval the transmitter sends a signal which is to be interpreted as a zero, or it sends a different signal which is to be interpreted as a one. At the end of each interval, the receiver makes its best guess at what was transmitted. Consider the events:

$$T_0 = \{\text{Transmitter sends } 0\}, \quad R_0 = \{\text{Receiver concludes that a } 0 \text{ was sent}\},$$
$$T_1 = \{\text{Transmitter sends } 1\}, \quad R_1 = \{\text{Receiver concludes that a } 1 \text{ was sent}\}.$$

Assume that $P(R_0|T_0) = 0.99$, $P(R_1|T_1) = 0.98$, and $P(T_1) = 0.5$. Find:

a) the probability of a transmission error given R_1;

b) the overall probability of a transmission error.

c) Repeat a) and b) assuming $P(T_1) = 0.8$ instead of 0.5.

5. **False diagnosis.** The fraction of persons in a population who have a certain disease is 0.01. A diagnostic test is available to test for the disease. But for a healthy person the chance of being falsely diagnosed as having the disease is 0.05, while for someone with the disease the chance of being falsely diagnosed as healthy is 0.2. Suppose the test is performed on a person selected at random from the population.

   a) What is the probability that the test shows a positive result (meaning the person is diagnosed as diseased, perhaps correctly, perhaps not)?

   b) What is the probability that the person selected at random is one who has the disease but is diagnosed healthy?

   c) What is the probability that the person is correctly diagnosed and is healthy?

   d) Suppose the test shows a positive result. What is the probability that the person tested actually has the disease?

   e) Do the above probabilities admit a long-run frequency interpretation? Explain.

6. An experimenter observes the occurrence of an event A as the result of a particular experiment. There are three different hypotheses, H_1, H_2, and H_3, which the experimenter regards as the only possible explanations of the occurrence of A. Under hypothesis H_1, the experiment should produce the result A about 10% of the time over the long run, under H_2 about 1% of the time, and under H_3 about 39% of the time. Having observed A, the experimenter decides that H_3 is the most likely explanation, and that the probability that H_3 is true is

$$\frac{39\%}{10\% + 1\% + 39\%} = 78\%.$$

   a) What assumption is the experimenter implicitly making?

   b) Does the probability 78% admit a long-run frequency interpretation?

   c) Suppose the experiment is a laboratory test on a blood sample from an individual chosen at random from a particular population. The hypothesis H_i is that the individual's blood is of some particular type i. Over the whole population it is known that the proportion of individuals with blood of type 1 is 50%, the proportion with type 2 blood is 45%, and the remaining proportion is type 3. Revise the experimenter's calculation of the probability of H_3 given A, so that it admits a long-run frequency interpretation. Is H_3 still the most likely hypothesis given A?

7. **Guessing what box.** Consider a game as in Examples 1 and 2, where I pick one of the three boxes, then you guess which box I picked after seeing the color of a ball drawn at random from the box. Then you learn whether your guess was right or wrong. Suppose we play the game over and over, replacing the ball drawn and mixing up the balls between plays. Your objective is to guess the box correctly as often as possible.

a) Suppose you know that I pick a box each time at random (probability $1/3$ for each box). And suppose you adopt the strategy of guessing the box with highest posterior probability given the observed color, as described in Example 1, in case the observed color is white. About what proportion of the time do you expect to be right over the long run?

b) Could you do any better by another guessing strategy? Explain.

c) Suppose you use guessing strategy found in a), but I was in fact randomizing the choice of the box each time, with probabilities $(1/2, 1/4, 1/4)$ instead of $(1/3, 1/3, 1/3)$. Now how would your strategy perform over the long run?

d) Suppose you knew I was either randomizing with probabilities $(1/3, 1/3, 1/3)$, or with probabilities $(1/2, 1/4, 1/4)$. How could you learn which I was doing? How should you respond, and how would your response perform over the long run?

8. Optimal strategies for guessing what box. (Continuation of Exercise 7, due to David Blackwell.) The question now arises: What randomizing strategy should I use to make it as hard as possible for you to guess correctly? Consider what happens if I use the $(\frac{6}{23}, \frac{9}{23}, \frac{8}{23})$ strategy, and answer the following questions:

a) What box should you guess if you see a black ball?

b) What box should you guess if you see a white ball?

c) What is your overall chance of winning?

You should conclude that with this strategy, your chance of winning is at most $\frac{9}{23}$, no matter what you do. Moreover, you have a strategy which guarantees you this chance of winning, no matter what randomization I use. It is the following:
If black, guess 1 with probability $\frac{18}{23}$, 2 with probability $\frac{5}{23}$, and 3 with probability 0.
If white, guess 1 with probability 0, 2 with probability $\frac{11}{23}$, and 3 with probability $\frac{12}{23}$.

d) Check that using this strategy, you win with probability $\frac{9}{23}$, no matter what box I pick.

According to the above analysis, I can limit your chance of winning to $\frac{9}{23}$ by a good choice of strategy, and you can guarantee that chance of winning by a good choice of strategy. The fraction $\frac{9}{23}$ is called the *value* of the above game, where it is understood that the payoff to you is 1 for guessing correctly, 0 otherwise. Optimal strategies of the type discussed above and a resulting value can be defined for a large class of games between two players called zero-sum games. For further discussion consult books on game theory.

9. A box contains three "shapes", as described in Example 1.3.3. One of the shapes is a fair die, and lands flat with probability $1/3$. The other two shapes land flat with probabilities $1/2$ and $2/3$, respectively.

a) One of the three shapes will be chosen at random, and rolled. What is the chance that the number rolled is 6?

b) Given that the number rolled is 6, what is the chance that the fair die was chosen?

1.6 Sequences of Events

This section is concerned with how to calculate probabilities of events determined by a sequence of outcomes. All that is involved is repeated application of the basic addition and multiplication rules of probability.

The first step is a calculation of the probability of an intersection of three events A, B, and C. This event, which occurs if all three of the events occur may be written as $ABC = (AB)C$. The chance of this event can be computed by using the multiplication rule twice:

$$P(ABC) = P(AB)P(C|AB) = P(A)P(B|A)P(C|AB).$$

FIGURE 1. Tree diagram for the multiplication rule for three events.

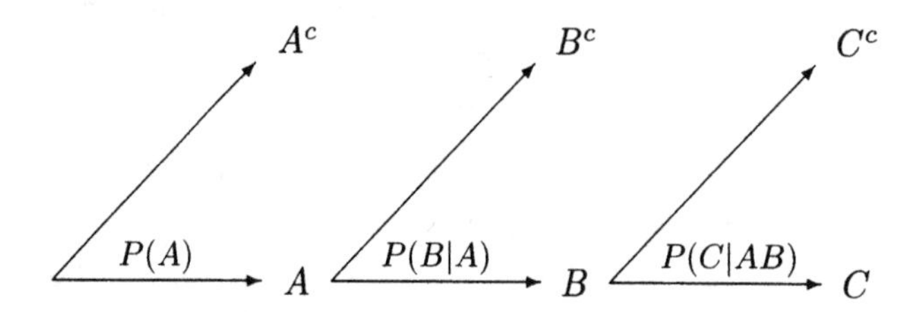

Repeating this argument shows that for n events, $A_1, \ldots, A_n$, the probability that every one of these events occurs is a product of n factors.

Multiplication Rule for *n* Events

$$P(A_1A_2A_3\cdots A_n) = P(A_1)P(A_2|A_1)P(A_3|A_1A_2)\cdots P(A_n|A_1A_2\cdots A_{n-1}).$$

In words, if $p_1 = P(A_1)$ is the probability of the first event, $p_2 = P(A_2|A_1)$ is the probability of the second event given that the first event has occurred, $p_3 = P(A_3|A_1A_2)$ is the probability of the third event given that the first two events have occurred, and so on, then the probability that n events $A_1, \ldots, A_n$ all occur is the product $p_1 \times p_2 \times \cdots \times p_n$.

This multiplication rule is used to specify the probabilities of paths in a tree diagram. Probabilities of various events of interest can then be found by adding the probabilities over appropriate sets of paths. This technique is illustrated by the following examples.

Example 1. Completion by stages.

Problem. A contractor is planning a construction project to be completed in three stages. The contractor figures that

(i) the chance that the first stage will be completed on time is 0.7.

(ii) given that the first stage is completed on time, the chance that the second stage will be completed on time is 0.8.

(iii) given that both the first and second stages are completed on time, the chance that the third stage will be completed on time is 0.9.

To be consistent, what should the contractor calculate is the chance that all three stages will be completed on time?

Solution. Let C_i be the event that the ith stage is completed on time, and let L_i be the event that stage i is late (the complement of C_i). The data of the problem are represented in the following tree diagram:

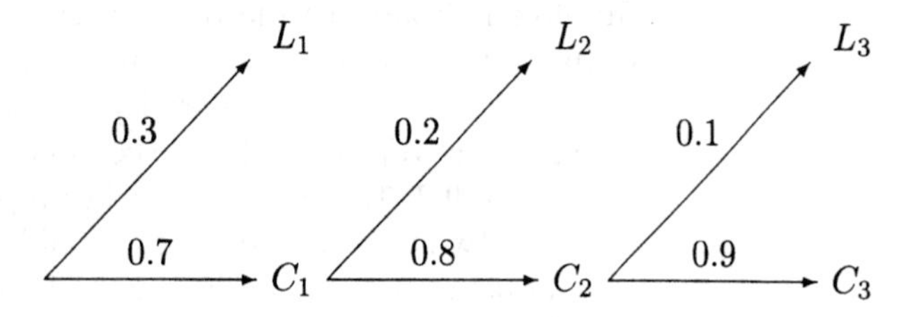

The event that all three stages are completed on time is the event $C_1C_2C_3$. By the multiplication rule,

$$P(C_1C_2C_3) = 0.7 \times 0.8 \times 0.9 = 0.504$$

Note. The data determine the probability of some other events, such as the event that the first and second stages are completed on time but the third is not, which is

$$P(C_1C_2L_3) = 0.7 \times 0.8 \times 0.1 = 0.056$$

But the data do not determine the probability of the event that the second stage is late, which is not represented in the diagram. To calculate this probability, it would be necessary to know $P(L_2|L_1)$, the chance that the second stage is late given that the first stage is late. Then $P(L_2)$ could be obtained by the rule of average conditional probabilities.

Example 2. **The geometric distribution.**

A symmetric die has proportion p of its faces painted white and proportion q of its faces painted black, where $q = 1 - p$. The die is rolled until the first time a white face shows up.

Problem 1. What is the chance that this takes three or less rolls?

Solution. Assume that, no matter how the die may have landed in previous rolls, the die shows white on each roll with probability p and black with probability q. The problem can then be represented as follows by a tree diagram.

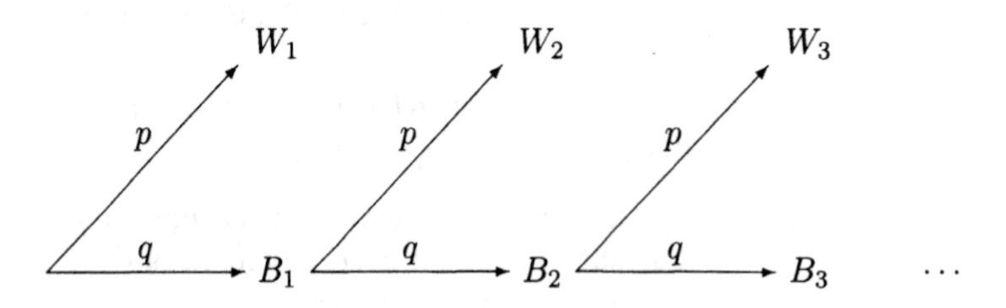

The dots indicate that the diagram could be continued in the same way for rolls 4, 5, 6, and so on, but the outcomes of these rolls are not relevant to the problem. The event {white in 3 or less rolls} is represented by three branches of the tree, the first ending at W_1 on roll 1, the second ending at W_2 on roll 2, and the third ending at W_3 on roll 3. These three branches represent three mutually exclusive ways that the event {white in 3 or less rolls} could happen. The probability of each branch is the product of probabilities along the branches. Thus

$$\begin{aligned} P(\text{white in 3 or less rolls}) &= P(W_1) + P(B_1W_2) + P(B_1B_2W_3) \\ &= p + qp + q^2p \\ &= (1 + q + q^2)p \end{aligned}$$

Problem 2. What is the chance that it takes four or more rolls to get a white face?

Solution. This looks as if you have to think about the part of the diagram labeled $\cdots$, representing what might happen if you rolled the die 4 times, 5 times, 6 times, and so on. But there is no need to face this infinite sequence of possible outcomes. The event that it takes 4 or more rolls to get a white face is the complement of the event that it takes three or less rolls to get a white face. Therefore

$$\begin{aligned} P(\text{4 or more rolls to get white}) &= 1 - P(\text{white in 3 or less rolls}) \\ &= 1 - (1 + q + q^2)p \end{aligned}$$

by the solution to the previous problem.

Discussion. If you substitute $p = 1 - q$ in this formula and simplify, it reduces to simply q^3. To understand why, notice that the event that it takes four or more rolls to get white is simply the event that the first three rolls are black. And the probability of this event is q^3, from the tree diagram. This gives the simplest solution to both problems above. As a numerical example, for an ordinary six-sided die, with face 6 white, and the rest black, so $p = 1/6$, $q = 5/6$,

$$P(4 \text{ or more rolls to get a six}) = q^3 = (5/6)^3 = 125/216 \approx 0.58$$

$$P(3 \text{ or less rolls to get a six}) = 1 - q^3 = 1 - (5/6)^3 = 91/216 \approx 0.42$$

The tree diagram shows that the distribution of the number of rolls required to get a white face is as follows:

number of rolls	1	2	3	$\cdots$	k	$\cdots$
probability	p	qp	q^2p	$\cdots$	$q^{k-1}p$	$\cdots$

This is the ***geometric distribution with parameter*** p, studied further in Section 3.4. Figure 2 shows the histogram of this distribution for $p = 1/6$, $q = 5/6$.

FIGURE 2. Geometric distribution of the number of fair die rolls to get a 6. Each bar of the histogram is 5/6 the height of the bar to its left.

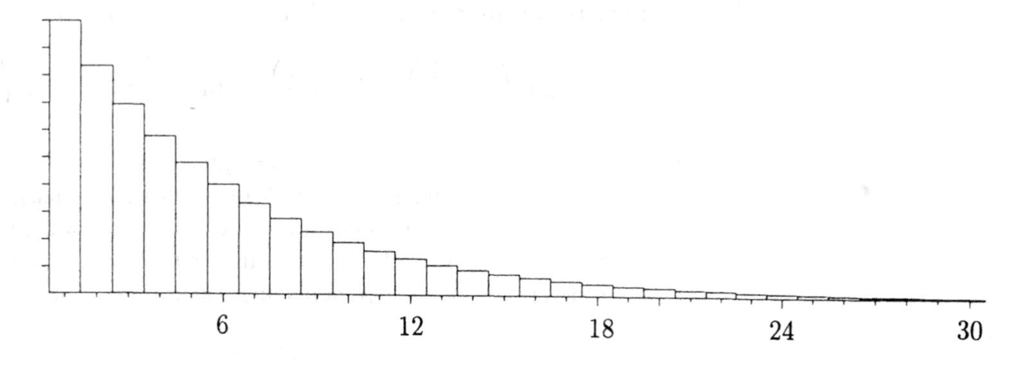

Example 3. The gambler's rule.

Problem. Suppose you play a game over and over again, each time with chance $1/N$ of winning the game, no matter what the results of previous games. How many times n must you play to have a better than 50% chance of at least one win in the n games?

Solution. It seems intuitive that n must be comparable in magnitude to N, but just what fraction of N is not clear without calculation. According to a very old gambler's rule, n is about $(2/3)N$. To check this, notice that

$$\begin{aligned} P(\text{at least one win in } n \text{ games}) &= 1 - P(\text{no win in } n \text{ games}) \\ &= 1 - \left(1 - \frac{1}{N}\right)^n \end{aligned}$$

We are looking for the least n such that

$$1 - \left(1 - \frac{1}{N}\right)^n > \frac{1}{2}, \quad \text{i.e.,} \quad \left(1 - \frac{1}{N}\right)^n < \frac{1}{2}$$

For small N you can find n by repeated multiplication by $(1-1/N)$ until the product is less than $1/2$, and check that the gambler's rule holds. For larger N this becomes tedious. It is more efficient to take logarithms, and to look for the least n such that

$$n \log\left(1 - \frac{1}{N}\right) < \log\left(\frac{1}{2}\right)$$

Keep in mind that both sides are now negative. To find this integer n, first find n^*, perhaps not an integer, such that

$$n^* \log\left(1 - \frac{1}{N}\right) = \log\left(\frac{1}{2}\right), \quad \text{that is,} \quad n^* = \log\left(\frac{1}{2}\right) / \log\left(1 - \frac{1}{N}\right)$$

So the desired n is the least integer greater than n^*. You can check that n^* is so close to $2N/3$ for small values of N that n is also the least integer greater than $2N/3$ for $N = 1, 2, \ldots, 27$. This rule breaks down for $N = 28$, but the fraction n/N stays quite close to $2/3$ as $N \to \infty$. To understand why, take logarithms to the base e. (See the appendix on exponents and logarithms.) Then there is the approximation

$$\log(1 + z) \sim z \quad \text{as} \quad z \to 0$$

Apply this to $z = -1/N$ as $N \to \infty$ to get

$$n \sim n^* \sim \log\left(\frac{1}{2}\right) / \left(-\frac{1}{N}\right) = N \log(2)$$

where the symbol $\sim$ indicates *asymptotic equivalence* as $N \to \infty$, meaning the ratio of the two sides tends to 1 as $N \to \infty$. So the asymptotic ratio of n to N is

$$\log(2) \approx 0.69 \approx 2/3$$

Example 4. Probability of a flush.

Problem. Suppose that a five-card hand is dealt from a well-shuffled deck of 52 cards. What is the probability that the hand is a flush (all cards of the same suit)?

Solution. A flush could be a flush of spades (S), a flush of hearts (H), a flush of diamonds (D), or a flush of clubs (C). These are four mutually exclusive and equally likely cases. The way to get a spade flush is suggested by the following diagram, with S_i representing the event that the ith card dealt is a spade:

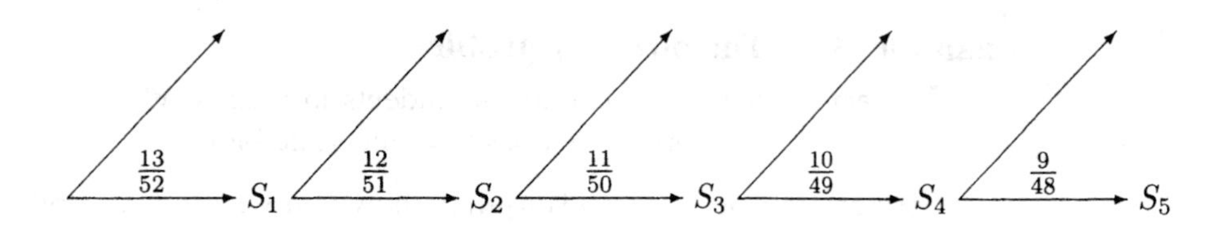

The conditional probabilities in the diagram were obtained from the usual assumptions of a well-shuffled deck:

(i) *the first card is equally likely to be any of the 52 cards in the deck;*

(ii) *given the first card, the second is equally likely to be any of the 51 left;*

(iii) *given the first two cards, the third is equally likely to be any of the 50 left;*

and so on. To illustrate, (iii) implies $P(S_3|S_1S_2) = 11/50$, because given that the first and second cards are spades, no matter what spades they are, there are 11 spades left among the 50 remaining cards in the deck. Using the multiplication rule,

$$P(\text{Spade flush}) = \frac{13}{52} \times \frac{12}{51} \times \frac{11}{50} \times \frac{10}{49} \times \frac{9}{48}$$

Therefore

$$P(\text{flush}) = 4P(\text{Spade flush}) = \frac{12}{51} \times \frac{11}{50} \times \frac{10}{49} \times \frac{9}{48} = 0.00198$$

Remark. The probability of any particular sequence of 5 cards can be calculated using the multiplication rule. You could think of this in terms of a huge tree diagram, with 52 branches for the first card, each of these branching into 51 possibilities for the second card, each of these branching into 50 possibilities for the third card, and so on. Each path in the tree would then represent a possible sequence of 5 cards. The probability of any particular sequence being dealt, for example $(J\heartsuit, K\spadesuit, 2\heartsuit, 3\diamondsuit, 5\diamondsuit)$, meaning the first card is the Jack of Hearts, the next is the King of Spades, and so on, would be

$$\frac{1}{52} \times \frac{1}{51} \times \frac{1}{50} \times \frac{1}{49} \times \frac{1}{48}$$

the same for all possible sequences (called *permutations*) of 5 of the 52 cards. This serves as the basic assumption for calculating probabilities of other types of card hands, by a counting method explained in Chapter 2.

Example 5. The birthday problem.

Problem. Suppose there are n students in a class. What is the probability that at least two students in the class have the same birthday?

Solution. The first step is to think how you would determine whether or not this event has occurred for a particular class of students. Here is a natural method. First order the students in some arbitrary way, say alphabetically, then go through the list of students' birthdays in that order, and check whether or not each birthday is one that has appeared previously. If you find a repeat birthday in this process, stop. There are at least two students in the class with the same birthday. But if you get right through the list of n students, with no repeats, then no two students in the class have the same birthday.

Let R_j be the event that the checking process stops with a repeat birthday at the jth student on the list, and let D_j be the event that the first j birthdays are different. The event B_n that there are at least two students in the class with the same birthday is the event $R_2 \cup R_3 \cup \cdots \cup R_n$ that the checking process stops with a repeat at some stage $j \leq n$ as you go through the list. The events $R_2, \ldots, R_n$ are represented in the following diagram. They are mutually exclusive, so

$$P(B_n) = P(R_2) + P(R_3) + \cdots + P(R_n)$$

But it is simpler to calculate the probability of B_n from its complement, which is D_n, the event that all n birthdays are different:

$$P(B_n) = 1 - P(D_n)$$

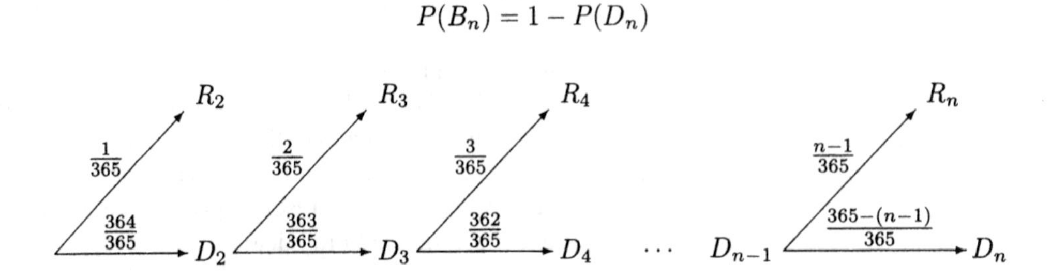

The conditional probabilities in the diagram are based on the following assumption:

No matter what the birthdays of the first $j-1$ students, the birthday of the jth student is equally likely to be any one of the 365 days of the year.

This ignores leap years, and seasonal variation in birth rates. But it can be shown that neither of these considerations affects the answer very much. Granted the assumption, we have

$$P(D_2) = \frac{364}{365} = \left(1 - \frac{1}{365}\right)$$

because no matter what the birthday of the first student, there are 364 out of 365 possible birthdays for the second student which would make the first and second students have different birthdays. If the first j birthdays are different, then so are the first i for every $i < j$, so $D_j \subset D_i$. Thus $D_2 D_3 \cdots D_j = D_j$, and

$$P(D_{j+1} | D_2 D_3 \cdots D_j) = P(D_{j+1} | D_j) = \frac{365 - j}{365} = \left(1 - \frac{j}{365}\right)$$

because given D_j, the first j students have j different birthdays, and no matter what these birthdays are, the next student must have one of the remaining $365-j$ birthdays for D_{j+1} to occur. Multiplying these conditional probabilities along the branch of the diagram through $D_2, D_3, \ldots, D_n$ gives

$$P(D_n) = P(D_2 D_3 \cdots D_n) = \left(1 - \frac{1}{365}\right)\left(1 - \frac{2}{365}\right) \cdots \left(1 - \frac{n-1}{365}\right)$$

where the last factor comes from taking $j = n - 1$ in the formula for $P(D_{j+1} | D_j)$.

Discussion. Figure 3 displays the graph of $P(B_n) = 1 - P(D_n)$ against n, obtained by this formula. The most amazing thing is how rapidly $P(B_n)$ increases as n increases. The least n such that $P(B_n) > 1/2$ is $n = 23$:

$$P(B_{23}) = 50.6\%$$

and $P(B_n)$ is up to about 94% by $n = 45$, and 99.8% by $n = 65$. Above $n = 70$, $P(B_n)$ is so close to 1 that there is no point in plotting the graph. The value of $P(B_n)$ is shown in the graph by the height of the dot above n on the horizontal scale. These dots are closely approximated by the smooth curve drawn just below the dots. This curve is obtained by calculating the product using logarithms and the tangent approximation $\log(1 + z) \sim z$ for small z, as in Example 3. Thus

$$
\begin{aligned}
\log P(D_n) &= \log(1 - 1/365) + \log(1 - 2/365) + \cdots + \log(1 - (n-1)/365) \\
&\approx -\frac{1}{365} - \frac{2}{365} - \cdots - \frac{n-1}{365} \\
&= -\frac{1}{365}(1 + 2 + \cdots + (n-1)) \\
&= -\frac{1}{365} \times \frac{1}{2} n(n-1) \qquad \text{so} \\
P(D_n) &\approx e^{-\frac{n(n-1)}{2 \times 365}} \qquad \text{and} \qquad P(B_n) \approx 1 - e^{-\frac{n(n-1)}{2 \times 365}}
\end{aligned}
$$

As the graph shows, this approximation is excellent over all values of n. The reason is that by the time n is large enough that $(n-1)/365$ is much greater than zero, so the tangent approximation is poor, both $P(D_n)$ and its exponential approximation are so close to zero that their difference is negligible anyway.

The histogram in the lower half of Figure 3 shows the distribution of how long it takes for the checking process to stop in a class of 70 or so students. The chance that this process stops after j steps is $P(R_j)$, the height of the bar over j. This is calculated as $P(R_j) = P(B_j) - P(B_{j-1})$ using the difference rule, because R_j occurs if and only if B_j occurs but B_{j-1} does not. Because these probabilities are differences in the birthday probabilities $P(B_j)$, and the step size in j is 1, the curve followed by this histogram is close to the derivative of the curve followed by the graphs of $P(B_n)$.

Generalization. The above example generalizes easily as follows. Consider a sequence of trials of some kind where each trial is equally likely to result in any one of N possible outcomes, no matter what the results of previous trials. For example, picking tickets at random with replacement from a box of N tickets, or repeatedly spinning a roulette wheel with N pockets. Let B_n be the event that the first repeat outcome appears by trial number n. Then by the same argument as in the birthday problem, where $N = 365$,

$$
\begin{aligned}
P(B_n) &= 1 - \left(1 - \frac{1}{N}\right)\left(1 - \frac{2}{N}\right) \cdots \left(1 - \frac{(n-1)}{N}\right) \\
&\approx 1 - e^{-\frac{n(n-1)}{2N}} \\
&\approx 1 - e^{-n^2/2N}
\end{aligned}
$$

if N is large. How large does n have to be to have at least a 50% chance of a repeat by n? Set

$$1 - e^{-n^2/2N} \approx 1/2 \quad \text{that is,}$$

$$n \approx \sqrt{2 \log 2}\sqrt{N} \approx 1.177\sqrt{N}$$

FIGURE 3. Probabilities in the birthday problem. See the discussion after Example 5.

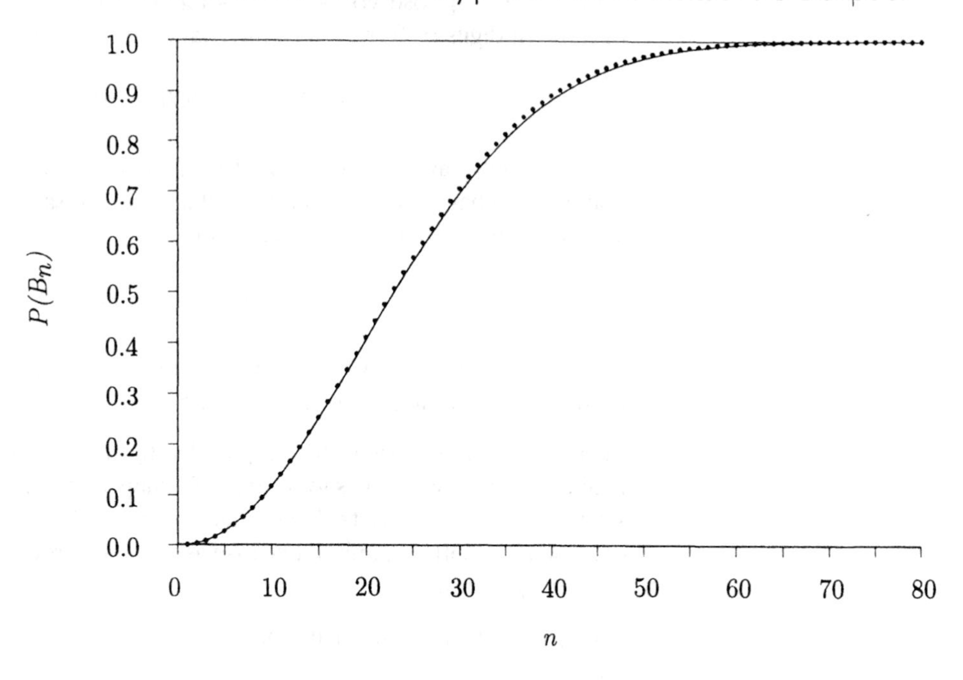

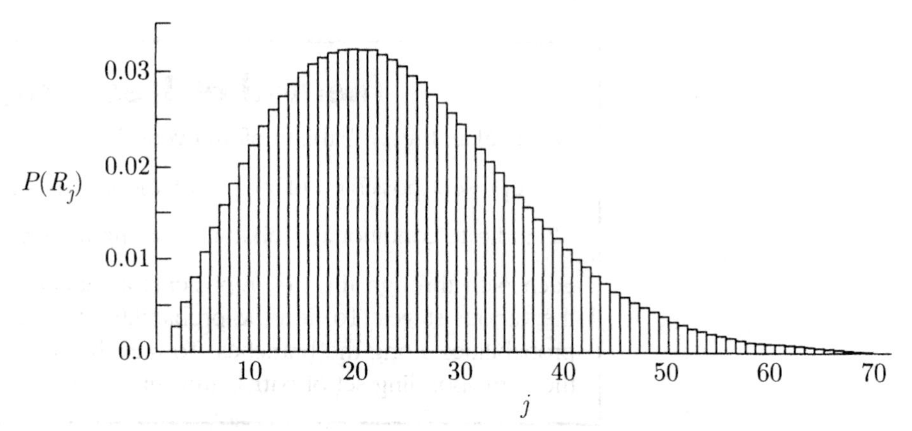

To illustrate, suppose you keep looking at six-figure numbers, drawn from a table of random digits or from a computer random number generator, such as

$$349221, \quad 512039, \quad 489583, \quad \dots$$

Then you only have to look at about 1177 such numbers before there is a 50% chance that you will have seen two numbers that were the same. And after you have looked at 5000 such numbers the chance has risen to

$$1 - e^{-5^2/2} \approx 0.9999962$$

Thus coincidences between random numbers should be expected to occur much more often than might naively be supposed.

These calculations should be compared with the gambler's rule, according to which it takes about $(2/3)N$ trials to have a 50% chance of repeating any particular number, say the first one. It takes far fewer trials to have a 50% chance of having a duplication of some unspecified number, because as the trials proceed, there are more and more ways to get a repeat.

The general method used in each of the above examples can be summarized as follows.

Method of Tree Diagrams

To calculate the probability of an event defined by a sequence of stages:

1. represent possible outcomes at various stages in a tree diagram;
2. indicate conditional probabilities along the branches of the tree.

Each path through the tree represents a sequence of possible outcomes for the various stages. To find the probability of a path, multiply the conditional probabilities along the branches. To find the probability of an event, identify the corresponding set of paths, and sum the probabilities of these paths.

The main art in using this method effectively is to make your tree represent the right amount of detail in the problem. There must be enough detail so that the event you want to calculate is represented by a set of paths in the tree, but not so much detail that you are overwhelmed by the size of the tree or the difficulty of figuring out exactly which paths correspond to your event. Typically, the best tree to work with is a reduction of some much larger tree of possibilities. The conditional probabilities along the branches of the reduced tree will then be obtained by reference to the larger tree, as in calculating the probability of a flush and the birthday problem. But once these conditional probabilities have been figured out, all calculations can be done with the reduced tree.

Independence

The idea of independence for several events is a natural extension of the idea for two events. For example, events A, B, and C are called ***independent*** if, first, the chance of B does not depend on whether or not A occurs:

$$P(B|A) = P(B|A^c) = P(B) \tag{1}$$

and, second, the chance of C does not depend on which of the events A and B occur and which do not:

$$P(C|AB) = P(C|A^cB) = P(C|AB^c) = P(C|A^cB^c) = P(C) \tag{2}$$

As in the case of independence of two events, described by (1) alone, these conditions combine to give a simple multiplication rule for probabilities of intersections:

Multiplication Rule for Three Independent Events

$$P(ABC) = P(A)P(B)P(C)$$

and the same for any number of the events replaced by their complements.

This multiplication rule, which is really a list of $2^3 = 8$ rules, one for each path in the tree describing the results of all 3 events, gives the simplest formal definition of independence of 3 events. Independence of n events is defined similarly by a list of 2^n multiplication rules. It is a special feature of the case $n = 2$ that just one product formula $P(AB) = P(A)P(B)$ implies the 3 others. For a larger number of events, independence is a very strong condition. This is because the probabilities of 2^n possible intersections are exactly determined by the probabilities of just n events. So while intuitive in theory, independence may be hard to check in practice.

Example 6. Chance of two or more.

Problem. Suppose that a gambler places a bet on the result of each of four different horse races. He judges that the outcomes of the races are independent, and that he has probability p_i of winning on the ith race. What is the probability that he wins two or more of his bets?

Solution. It is easier to calculate the probability of the complement

$$\begin{aligned} P(\text{wins at least two}) &= 1 - P(\text{wins 0 or 1}) \\ &= 1 - P(\text{wins 0}) - P(\text{wins 1}) \end{aligned}$$

where, in terms of the events $B_i = \{\text{wins } i\text{th bet}\}$,

$$P(\text{wins } 0) = P(B_1^c B_2^c B_3^c B_4^c) = q_1 q_2 q_3 q_4$$

where $q_i = 1 - p_i$, and

$$\begin{aligned} P(\text{win } 1) &= P(B_1 B_2^c B_3^c B_4^c \text{ or } B_1^c B_2 B_3^c B_4^c \text{ or } B_1^c B_2^c B_3 B_4^c \text{ or } B_1^c B_2^c B_3^c B_4) \\ &= p_1 q_2 q_3 q_4 + q_1 p_2 q_3 q_4 + q_1 q_2 p_3 q_4 + q_1 q_2 q_3 p_4. \end{aligned}$$

Example 7. Flow in a circuit.

Problem. Suppose that each of the switches S_i in the following circuits is closed with probability p_i, and open with probability $q_i = 1 - p_i$, $i = 1, \ldots, 5$.

Calculate the probability that a current will flow through the circuit, assuming that the switches act independently.

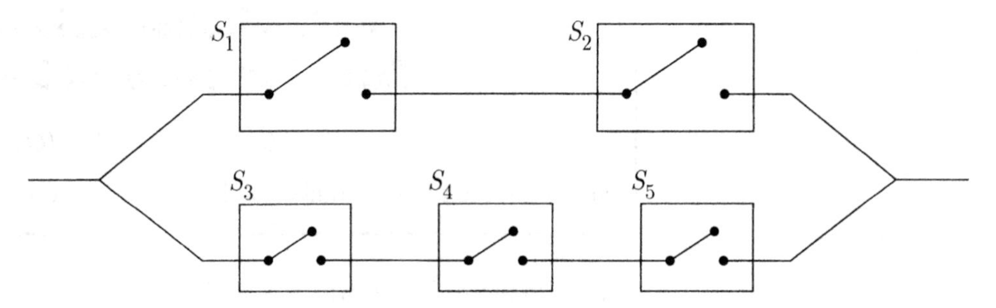

Solution.

$$\begin{aligned} P(\text{current flows}) &= P(\text{flows along top or flows along bottom}) \\ &= P(\text{flows along top}) + P(\text{flows along bottom}) \\ &\qquad - P(\text{flows along top and bottom}), \quad \text{where} \end{aligned}$$

$$P(\text{flows along top}) = P(S_1 \text{ closed and } S_2 \text{ closed}) = p_1 p_2$$

by the independence of S_1 and S_2. And

$$\begin{aligned} P(\text{flows along bottom}) &= P(S_3 \text{ closed and } S_4 \text{ closed and } S_5 \text{ closed}) \\ &= p_3 p_4 p_5 \end{aligned}$$

by the independence of S_3, S_4 and S_5. Also

$$P(\text{flows along top and bottom}) = P(\text{all switches closed}) = p_1 p_2 p_3 p_4 p_5$$

by the independence of all five switches. So,

$$P(\text{current flows}) = p_1 p_2 + p_3 p_4 p_5 - p_1 p_2 p_3 p_4 p_5$$

Discussion. This example shows how it is a good idea to break up a complicated looking problem into smaller and easier ones, leading to a quick solution. A more painful approach to the same problem would be to list all the ways the switches could be, which would allow current to flow, together with their probabilities. There are altogether $8 + 4 - 1 = 11$ possible ways. You can see this by counting the ways for {flows along top}, adding these to the ways for {flows along bottom}, and subtracting the ways for {flows along top and bottom}, which have been counted twice. This is the inclusion–exclusion rule for counting. If you now write down a product of five factors chosen from p_i and q_i, one product for each of the 11 ways, add the 11 products, substitute $q_i = 1 - p_i$, simplify the algebra, you should get the same result!

Here is a useful rule to use when breaking down problems into smaller pieces, as in the previous example.

> If $A_1, \ldots, A_n$ are mutually independent then every event determined by a subcollection of these events is independent of every event determined by a subcollection of the remaining events.

To illustrate, if $B_1, \ldots, B_{10}$ are independent, then the event $B_1 \cup B_2 \cup B_9^c$ is independent of the event $B_3^c B_5^c B_8$.

Pairwise independence. You might think that if $B_1, \ldots, B_n$ were events such that

$$B_i \text{ is independent of } B_j \text{ for every } i \neq j \qquad (*)$$

then $B_1, \ldots, B_n$ would be independent. But this turns out not to be the case. Condition $(*)$, called ***pairwise independence***, is weaker than the condition of independence for $n \geq 3$. The reason is that pairwise independence of three events B_1, B_2, and B_3 amounts to the three equations

$$P(B_1 B_2) = P(B_1)P(B_2)$$

$$P(B_1 B_3) = P(B_1)P(B_3)$$

$$P(B_2 B_3) = P(B_2)P(B_3)$$

But independence of B_1, B_2, and B_3 requires also the equation

$$P(B_1 B_2 B_3) = P(B_1)P(B_2)P(B_3)$$

This turns out not to be implied by pairwise independence, as the next example shows. But this list of four equations does suffice for independence, as you can show as an exercise.

Example 8. Pairwise independent but not independent.
Consider the possible results of tossing a fair coin twice, with probabilities proportional to areas in the following diagram. Let S be the event, determined by the two tosses, that both coins land the same way.

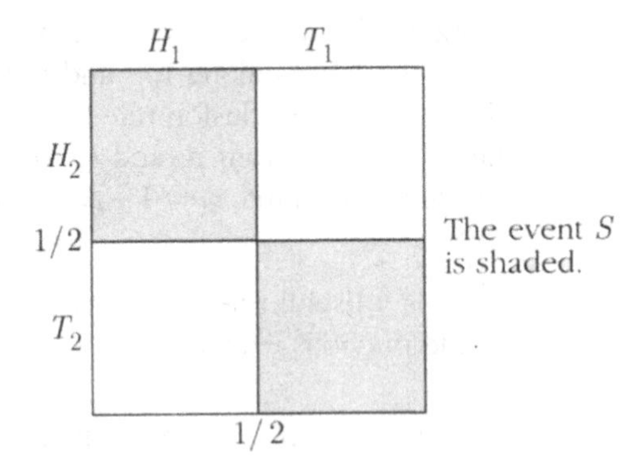

Then the events H_1, H_2, and S are not independent, because the event S is completely determined by H_1 and H_2:

$$P(S \mid H_1 H_2) = 1 \neq 1/2 = P(S)$$

But the events H_1, H_2, and S are pairwise independent. The events H_1 and H_2 are independent by assumption. Also,

$$P(H_1 S) = P(H_1 H_2) = 1/4 = P(H_1)P(S)$$

so H_1 and S are independent. And H_2 and S are independent for the same reason.

Exercises 1.6

1. There are twelve signs of the zodiac. How many people must be present for there to be at least a 50% chance that two or more of them were born under the same sign?

2. Suppose a batter's average (number of hits per at bat) is .300 over the season to date. What is the probability that the batter gets at least one hit in the next:

a) two at bats; b) three at bats; c) n at bats.

What assumptions are you making?

3. A biased coin lands heads with probability $2/3$. The coin is tossed three times.

a) Given that there was at least one head in the three tosses, what is the probability that there were at least two heads?

b) Use your answer in a) to find the probability that there was exactly one head, given that there was at least one head in the three tosses.

4. A typical slot machine in a Nevada casino has three wheels, each marked with twenty symbols at equal spacings around the wheel. The machine is engineered so that on each play the three wheels spin independently, and each wheel is equally likely to show any one of its twenty symbols when it stops spinning. On the central wheel, nine out of the twenty symbols are bells, while there is only one bell on the left wheel and one bell on the right wheel. The machine pays out the jackpot only if the wheels come to rest with each wheel showing a bell.

a) Calculate the probability of hitting the jackpot.

b) Calculate the probability of getting two bells but not the jackpot.

c) Suppose that instead there were three bells on the left, one in the middle, and three on the right. How would this affect the probabilities in a) and b)? Explain why the casino might find the $1-9-1$ machine more profitable than a $3-1-3$ machine.

5. Suppose you are one of n students in a class.

a) What is the chance that at least one other student has the same birthday as yours?

b) How large does the class have to be to make this probability at least $1/2$?

c) Explain the difference between this problem and the birthday problem.

6. Suppose you roll a fair six-sided die repeatedly until the first time you roll a number that you have rolled before.

a) For each $r = 1, 2, \ldots$ calculate the probability p_r that you roll exactly r times.

b) Without calculation, write down the value of $p_1 + p_2 + \cdots + p_{10}$. Explain.

c) Check that your calculated values of p_r have this value for their sum.

7. The ith switch in each of the following circuits is closed with probability p_i and open with probability q_i for each i. Assuming the switches function independently, find a formula in each case for the probability that a current can flow from left to right through the circuit.

a)

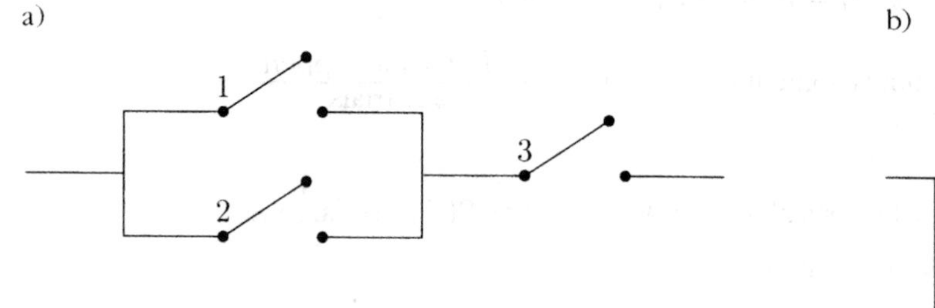

b)

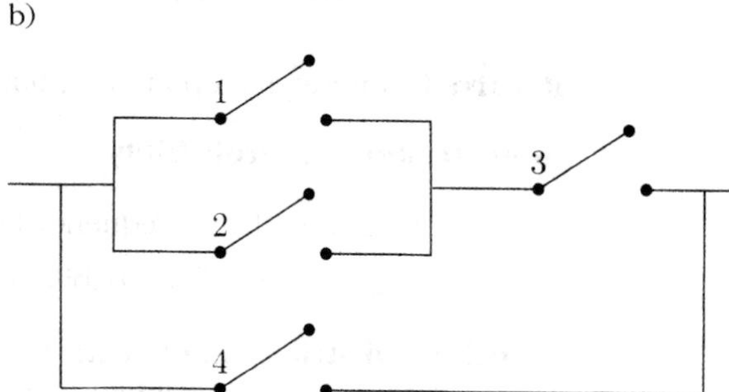

8. Suppose that the birthday of each of three people is equally likely to be any one of the 365 days of the year, independently of others. Let B_{ij} denote the event that person i has the same birthday as person j, where the labels i and j may be 1, 2, or 3.

a) Are the events B_{12} and B_{23} independent?

b) Are the events B_{12}, B_{23}, and B_{13} independent?

c) Are the events B_{12}, B_{23}, and B_{13} pairwise independent?

Introduction: Summary

Outcome space: A set of all possible outcomes of a situation or experiment, such that one and only one outcome must occur.

Events: Represented as subsets of an outcome space.

A and B, AB, $A \cap B$, *intersection*: event that both A and B occur.

A or B, $A \cup B$, *union*: event that either A or B (or both) occur.

$AB = \emptyset$, *disjoint*; *mutually exclusive*: no overlap, no intersection.

not A, A^c, *complement*: opposite of A: event that occurs if A does not.

$A \subset B$, *inclusion*: A is a part of B, A implies B, if A occurs then so does B.

Ω, *whole set, outcome space*: certain event, all possibilities, sure to happen.

$\emptyset$, *empty set, impossible event*: no way to happen.

partition of A: disjoint sets $A_1, \ldots, A_n$ with union A.

Rules of Probability and Proportion

- Non-negative: $P(A) \geq 0$
- Addition: $P(A) = \sum_{i=1}^{n} P(A_i)$ if $A_1, \ldots, A_n$ is a partition of A
- Total of 1: $P(\Omega) = 1$.
- Between 0 and 1: $0 \leq P(A) \leq 1$
- Empty set: $P(\emptyset) = 0$
- Complements: $P(A^c) = 1 - P(A)$
- Difference: $P(BA^c) = P(B) - P(A)$ if $A \subset B$
- Inclusion–Exclusion: $P(A \text{ or } B) = P(A) + P(B) - P(AB)$.

Relative frequency: Proportion of times something happens: $\dfrac{\#\text{of times it happens}}{\#\text{of trials}}$

Interpretations of Probability

- long-run relative frequency (statistical average): $P_n(A) \approx P(A)$ for large n.
- degree of belief (probabilistic opinion)

Probability distribution over Ω: Assignment of probabilities to events represented as subsets of Ω, satisfying rules of probability. A distribution over a finite set Ω can be specified with a *distribution table*:

outcome ω	a	b	c	$\cdots$
probability $P(\omega)$	$P(a)$	$P(b)$	$P(c)$	$\cdots$

The probabilities must sum to 1 over all outcomes.

Odds

Chance odds: ratio of probabilities, e.g., the following are equivalent: $P(A) = 3/10$; the odds of A are 3 in 10; the odds *in favor of* A are 3 to 7; the odds *against* A are 7 to 3.

Payoff odds: ratio of stakes: $\dfrac{\text{what you get}}{\text{what you bet}}$ (what you get does not include what you bet).

Fair odds rule: in a fair bet, payoff odds equal chance odds.

Conditional Probability

$P(A|B) =$ probability of A given B: probability of A with outcome space reduced to B. Compare with $P(A) =$ overall or unconditional probability of A.

Interpretations of conditional probability:

- *Intuitive/subjective:* chance of A if B is known to have occurred:
- *Long-run frequency:* long-run relative frequency of A's among trials that produce B.

As a function of A, for fixed B, conditional probabilities satisfy the rules of probability, e.g., $P(A^c|B) = 1 - P(A|B)$

Rules of Conditional Probability

Division: $P(A|B) = \dfrac{P(AB)}{P(B)}$ (note: $AB = BA$)

For probabilities defined by counting, $P(A|B) = \#(AB)/\#(B)$. Similarly for length, area, or volume instead of #.

Product: $P(AB) = P(A)P(B|A) = P(B)P(A|B)$

The following rules refer to a *partition* $B_1, \ldots, B_n$ of Ω, so $P(B_1) + \cdots + P(B_n) = 1$; for example, $B_1 = B$, $B_2 = B^c$ for any B.

Average rule: $P(A) = P(A|B_1)P(B_1) + \cdots + P(A|B_n)P(B_n)$

Bayes' rule: $P(B_i|A) = \dfrac{P(A|B_i)P(B_i)}{P(A)}$ where $P(A)$ is given by the weighted average formula.

Independence

Two trials are *independent* if learning the result of one does not affect chances for the other, e.g., two draws at random with replacement from a box of known composition.

The trials are *dependent* if learning the result of one does affect chances for the other, e.g., two draws at random without replacement from a box of known composition, or two draws at random with replacement from a box of random composition.

Independent events: A and B are such that

$$P(AB) = P(A)P(B) \iff P(A|B) = P(A) \quad \text{(learning } B \text{ occurs does not affect chances of } A)$$
$$\iff P(B|A) = P(B) \quad \text{(learning } A \text{ occurs does not affect chances of } B)$$

Independence of n events $A_1, \ldots, A_n$:

$$P(A_1A_2 \cdots A_n) = P(A_1) \cdots P(A_n),$$

and the same with any number of complements A_i^c substituted for A_i (2^n identities).

Review Exercises

1. A factory produces items in boxes of 2. Over the long run:

92% of boxes contain 0 defective items;

5% of boxes contain 1 defective item; and

3% of boxes contain 2 defective items.

A box is picked at random from production, then an item is picked at random from the box. Given that the item is defective, what is the chance that the second item in the box is defective?

2. A box contains 1 black ball and 1 white ball. A ball is drawn at random, then replaced in the box with an additional ball of the same color. Then a second ball is drawn at random from the three balls in the box. What is the probability that the first ball drawn was white, given that at least one of the two balls drawn was white?

3. Suppose I toss three coins. Two of them at least must land the same way. No matter whether they land heads or tails, the third coin is equally likely to land either the same way or oppositely. So the chance that all three coins land the same way is $1/2$. True or False? Explain!

4. There are two boxes.
Box 1 contains 2 red balls and 3 black balls.
Box 2 contains 8 red balls and 12 black balls.
One of the two boxes is picked at random, and then a ball is picked at random from the box.

a) Is the color of the ball independent of which box is chosen?

b) What if there were 10 black balls rather than 12 in Box 2, but the other numbers were the same?

5. To pass a test you have to perform successfully two consecutive tasks, one easy and one hard. The easy task you think you can perform with probability z, and the hard task you think you can perform with probability h, where $h < z$. You are allowed three attempts, either in the order (easy, hard, easy) or in the order (hard, easy, hard). Whichever order, you must be successful twice in a row to pass. Assuming that your attempts are independent, in what order should you choose to take the tasks in order to maximize your probability of passing the test?

6. Show that if A and B are independent, then so are A^c and B, A and B^c, and A^c and B^c.

7. A population of 50 registered voters contains 30 in favor of Proposition 134 and 20 opposed. An opinion survey selects a random sample of 4 voters from this population, as follows. One person is picked at random from the 50 voters, then another at random from the remaining 49, and so on, till 4 people have been picked.

a) What is the probability that there will be no one in favor of 134 in the sample?

b) What is the probability that there will be at least one person in favor?

c) What is the probability that exactly one pro 134 person will appear in the sample?

d) What is the probability that the majority of the sample will be pro 134? (Majority means strictly more than half.)

8. Cards are dealt from a well-shuffled standard deck until the first heart appears.

a) What is the probability that exactly 5 deals are required?

b) What is the probability that 5 or fewer deals are required?

c) What is the probability that exactly 3 deals were required, given that 5 or fewer were required?

9. Suppose events A, B, and C are independent with probabilities $1/5$, $1/4$, and $1/3$, respectively. Write down numerical expressions for the following probabilities:

a) $P(A \text{ and } B \text{ and } C)$

b) $P(A \text{ or } B \text{ or } C)$

c) $P(\text{exactly one of the three events occurs})$

10. The four major blood types are present in approximately the following proportions in the population of the U.S.A.

Type	A	B	AB	0
proportion	42%	10%	4%	44%

Note that each person's blood is exactly one of these four types. Type AB is a separate type, not the intersection of type A and type B.

a) If two people are picked at random from this population, what is the chance that their blood is of the same type? Of different types?

b) If four people are picked at random, let $P(k)$ be the chance that there are exactly k different blood types among them. Find $P(k)$ for $k = 1, 2, 3, 4$.

11. A hat contains n coins, f of which are fair, and b of which are biased to land heads with probability $2/3$. A coin is drawn from the hat and tossed twice. The first time it lands heads, and the second time it lands tails. Given this information, what is the probability that it is a fair coin?

12. Suppose n ordinary dice are rolled.

a) What is the chance that the dice show n different faces?

b) What is the chance that at least one number appears more than once?

13. Formula for $P(A \mid B)$ by conditioning on cases of B. Show if $B_1, \ldots, B_n$ is a partition of B, then

$$P(A|B) = P(A|B_1)P(B_1|B) + \cdots + P(A|B_n)P(B_n|B)$$

14. There are 100 boxes, and for each $i = 1, 2, \ldots, 100$, box i contains proportion $i/100$ of gold coins (the rest are silver). One box is chosen at random, then a coin is drawn at random from this box.

a) If the coin drawn is gold, which box would you guess was chosen? Why?

b) Suppose the boxes were not picked at random, but according to the following scheme. All the even-numbered boxes are equally likely, all the odd-numbered boxes are equally likely, but the chance of drawing an odd-numbered box is twice the chance of drawing an even-numbered box. If the coin drawn is gold, which box would you guess was chosen? [*Hint*: Write down the prior odds.]

15. There are three boxes, each with two drawers. Box 1 has a gold coin in each drawer, and box 2 has a silver coin in each drawer. Box 3 has a silver coin in one drawer and a gold coin in the other. One box is chosen at random, then a drawer is chosen at random from the box. Find the probability that box 1 is chosen, given that the chosen drawer yields a gold coin.

16. A dormitory has n students, all of whom like to gossip. One of the students hears a rumor, and tells it to one of the other $n - 1$ students picked at random. Subsequently, each student who hears the rumor tells it to a student picked at random from the dormitory (excluding, of course, himself/herself and the person from whom he/she heard the rumor). Let p_r be the probability that the rumor is told r times without coming back to a student who has already heard it from a dormitory-mate. So $p_1 = p_2 = 1$, and $p_n = 0$.

a) Find a formula for p_r for r between 3 and $n - 1$.

b) Estimate this probability for $n = 300$ and $r = 30$.

17. Some time ago I received the following letter:

"You may have previously received a letter notifying you that you had been a selectee in a recent sweepstake that we were conducting. According to our records, you have not claimed your gift.
We are always pleased when our bigger gifts are awarded because it's good publicity for our company. However, last year there were thousands of dollars worth of unclaimed gifts simply because the selectees failed to respond.
This letter is to inform you that one of the following people has won a New Datsun Sentra:

Collin Andrus	Oklahoma City, OK
James W. Pitman	Berkeley, CA
Larry Abbott	Burbank, CA

In compliance with the rules of the sweepstake, you are hereby notified that you are a selectee in Category I, which means you will receive one of the following:

1. R.C.A. Color TV;
2. 5 FT. Grandfather Clock;
3. Datsun Nissan Sentra.

To claim your gift, all you have to do is call toll free 1-800-643-3249 for an available time and date for you and your spouse to visit Heavenly Valley Townhouses and attend a sales representation tour on the many advantages that interval ownership has to offer."

According to small print on the back of the letter:

"The retail values and odds of receiving each gift are No. 1—$1/10,000$ ($329.95), No. 2—$9998/10,000$ ($249.95), No. 3—$1/10,000$($5,995.00)."

Let us assume that this is an honestly conducted sweepstake, and that each of the three individuals named above had originally a 1 in $10,000$ chance of winning the new Datsun. Now I know that the winner is one of these individuals, the rules of conditional probability imply that I have a one in three chance of winning the Datsun. True or false? Explain!

18. Suppose there are m equally likely possibilities for one stage, and n equally likely possibilities for another. Show that the two stages are independent if and only if all mn possible joint outcomes are equally likely.

19. A box contains 5 tickets numbered $1, 2, 3, 4$, and 5. Two tickets are drawn at random from the box. Find the chance that the numbers on the two tickets differ by two or more if the draws are made:

a) with replacement;

b) without replacement.

Repeat the problem with n tickets numbered $1, 2, \ldots, n$.

2 Repeated Trials and Sampling

This chapter studies a mathematical model for repeated trials, each of which may result in some event either happening or not happening. Occurrence of the event is called *success*, and non-occurrence called *failure*. For instance:

Nature of trial	Meaning of success	Meaning of failure	Probabilities p and q
Tossing a fair coin	head	tail	$1/2$ and $1/2$
Rolling a die	six	not six	$1/6$ and $5/6$
Rolling a pair of dice	double six	not double six	$1/36$ and $35/36$
Birth of a child	girl	boy	0.487 and 0.513

Suppose that on each trial there is success with probability p, failure with probability $q = 1-p$, and assume the trials are independent. Such trials are called *Bernoulli trials* or *Bernoulli* (p) *trials* to indicate the success probability p. The number of successes in n trials then cannot be predicted exactly. But if n is large we expect the number of successes to be about np, so the relative frequency of successes will, most likely, be close to p. The important questions treated in this chapter are: how likely? and how close? The answers to these questions, first discovered by the mathematicians James Bernoulli and Abraham De Moivre, around 1700, are the mathematical basis of the long-run frequency interpretation of probabilities.

The first step in Section 2.1 is to find a formula for the probability of getting k successes in n trials. This formula defines the *binomial probability distribution* over the possible numbers of successes from 0 to n. For large values of n, the histogram of the distribution turns out to follow a smooth curve quite closely.

2.1 The Binomial Distribution

The problem is to find a formula for the probability of getting k successes in n independent trials. This is solved by analysis of a tree diagram representing all possible results of the n trials, shown in Figure 1 for $n = 4$.

FIGURE 1. Tree diagram for derivation of the binomial distribution.

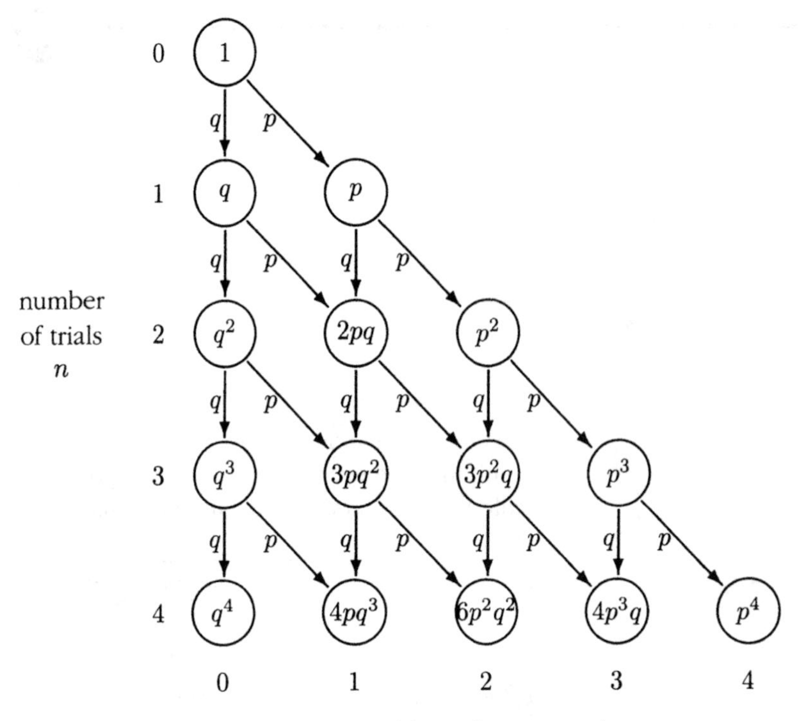

Each path down n steps through the tree diagram represents a possible outcome of the first n trials. The kth *node* in the nth row represents the event of k successes in n trials. The expression inside each node is its probability in terms of p and $1 - p = q$ (the probabilities of success and failure on each trial). This expression is the sum of the probabilities of all paths leading to this node. For example, in row 3 the probabilities of $k = 0, 1, 2, 3$ successes in $n = 3$ trials are the terms in the expansion

$$(p + q)^3 = q^3 + 3pq^2 + 3p^2q + p^3$$

For $k = 0$ or 3 there is only one path leading to k successes, hence the probability of q^3 or p^3 by the multiplication rule. For $k = 1$ the factor of 3 arises because there are three ways to get just one success in three trials, FFS, FSF, SFF, represented by the three paths through the diagram leading to the first node in row 3. The probabilities of these events are the terms qqp, qpq, and pqq in the expansion of $(q+p)^3$. These terms add to give the probability $3pq^2$ of $k=1$ success in 3 trials. Similarly, the probability of $k=2$ successes in 3 trials is $3p^2q$.

The tree diagram can be imagined drawn down to any number of trials n. To achieve k successes in n trials, the path must move down to the right k times, corresponding to the k successes, and straight down $n-k$ times, corresponding to the $n-k$ failures. The probability of every such path is the product of k factors of p, and $n-k$ factors of q, which is $p^k q^{n-k}$, regardless of the order of the factors. Therefore, the probability of k successes in the n trials is the sum of as many equal contributions of $p^k q^{n-k}$ as there are paths down through the diagram leading to the kth node of row n, or this number of paths times $p^k q^{n-k}$. This number of paths is denoted $\binom{n}{k}$ and called *n choose k*. So the probability of k successes in n trials is $\binom{n}{k}p^k q^{n-k}$. This conclusion and a formula for $\binom{n}{k}$ are summarized in the next box.

Binomial Distribution

For n independent trials, with probability p of success and probability $q=1-p$ of failure on each trial, the probability of k successes is given by the *binomial probability formula*:

$$P(k \text{ successes in } n \text{ trials}) = \binom{n}{k} p^k q^{n-k}$$

where $\binom{n}{k}$, called *n choose k*, is the number of different possible patterns of k successes and $n-k$ failures in n trials, given by the formula

$$\binom{n}{k} = \frac{n(n-1)\cdots(n-k+1)}{k(k-1)\cdots 1} = \frac{n!}{k!(n-k)!}$$

Here the $k!$ is *k factorial*, the product of the first k integers for $k \geq 1$, and $0! = 1$. For fixed n and p, as k varies, these binomial probabilities define a probability distribution over the set of $n+1$ integers $\{0, 1, \ldots, n\}$, called the *binomial (n,p) distribution*. This is the distribution of the number of successes in n independent trials, with probability p of success in each trial. The binomial (n,p) probabilities are the terms in the *binomial expansion*:

$$(p+q)^n = \sum_{k=0}^{n} \binom{n}{k} p^k q^{n-k}$$

Appendix 1 gives the background on counting and a derivation of the formula for $\binom{n}{k}$ in the box. The first expression for $\binom{n}{k}$ in the box is the simplest to use for numerical evaluations if $k < \frac{1}{2}n$. For example,

$$\binom{8}{3} = \frac{8 \times 7 \times 6}{3 \times 2 \times 1} = 8 \times 7 = 56$$

In this expression for $\binom{n}{k}$ there are always k factors in both the numerator and denominator. If $k > \frac{1}{2}n$, needless cancellation is avoided by first using symmetry:

$$\binom{n}{k} = \binom{n}{n-k}$$

as you can easily check. For instance, $\binom{9}{7} = \binom{9}{2} = \frac{9 \times 8}{2 \times 1} = 9 \times 4 = 36.$

To illustrate the binomial probability formula, the chance of getting 2 sixes and 7 non-sixes in 9 rolls of a die is therefore

$$\binom{9}{2}\left(\frac{1}{6}\right)^2\left(\frac{5}{6}\right)^7 = \frac{36 \times 5^7}{6^9} = 0.279$$

The convention $0! = 1$ makes the factorial formula for $\binom{n}{k}$ work even if k or n is 0. This formula is sometimes useful for algebraic manipulations. Because $n!$ increases so rapidly as a function of n, the factorial formula is awkward for numerical calculations of $\binom{n}{k}$. But for large values of n and k there are simple approximations to be described in the following sections.

The binomial expansion. Often called the *binomial theorem*, this is the expansion of $(p+q)^n$ as a sum of coefficients times powers of p and q. The coefficient $\binom{n}{k}$ of $p^k q^{n-k}$ is often called a *binomial coefficient.* For $p+q=1$ the binomial expansion of $(p+q)^n$ amounts to the fact that the probabilities in the binomial (n, p) distribution sum up to 1 over $k = 0$ to n:

$$\sum_{k=0}^{n} P(k \text{ successes in } n \text{ trials}) = \sum_{k=0}^{n} \binom{n}{k} p^k q^{n-k} = 1$$

This illustrates the addition rule for probabilities: as k varies from 0 to n, the $n+1$ events of getting, respectively,

$$0 \text{ successes, } 1 \text{ success, } 2 \text{ successes, } \ldots, n \text{ successes,}$$

in n trials, form a partition of all possible outcomes. For example, you can't get both 2 successes and 3 successes in 10 trials. And in n trials, you must get some number of successes between 0 and n.

The case of fair coin tossing. Then $p = q = 1/2$, so

$$p^k q^{n-k} = (1/2)^k (1/2)^{n-k} = (1/2)^n \qquad \text{and}$$

$$P(k \text{ heads in } n \text{ fair coin tosses}) = \binom{n}{k} / 2^n \qquad (0 \le k \le n)$$

All possible patterns of heads and tails of length n are equally likely in this case. So the above probability of k heads in n tosses is just the number of such patterns with k heads, namely $\binom{n}{k}$, relative to the total number of such patterns, namely 2^n. A consequence is that

$$\binom{n}{0} + \binom{n}{1} + \cdots + \binom{n}{n} = \sum_{k=0}^{n} \binom{n}{k} = 2^n$$

This is the binomial expansion of $(x+y)^n$ for $x = y = 1$.

Example 1. **Coin tossing and sex of children.**

Problem 1. Find the probability of getting four or more heads in six tosses of a fair coin.

Solution. $P(4 \text{ or more heads in } 6 \text{ tosses}) = P(4) + P(5) + P(6)$, where

$$P(k) = P(k \text{ heads in } 6 \text{ tosses}) = \binom{6}{k} / 2^6 \qquad \text{so}$$

$$P(4 \text{ or more heads in } 6 \text{ tosses}) = (15 + 6 + 1)/2^6 = 11/32$$

Problem 2. What is the probability that among five families, each with six children, at least three of the families have four or more girls?

Solution. Assume that each child in each family is equally likely to be a boy or a girl, independently of all other children. Then the chance that any particular family has four or more girls is $p = 11/32$, by the solution of the previous problem. Call this event a success in the present problem. Then the probability that at least 3 of the families have 4 or more girls is the probability of at least 3 successes in $n = 5$ trials, with probability $p = 11/32$ of success on each trial. So the required probability is

$$P(3 \text{ successes}) + P(4 \text{ successes}) + P(5 \text{ successes})$$

$$= \binom{5}{3}\left(\frac{11}{32}\right)^3\left(\frac{21}{32}\right)^2 + \binom{5}{4}\left(\frac{11}{32}\right)^4\left(\frac{21}{32}\right) + \binom{5}{5}\left(\frac{11}{32}\right)^5 = 0.226$$

Consecutive Odds Ratios

The binomial (n, p) distribution is most easily analyzed in terms of the chance of k successes relative to $k-1$ successes. These odds ratios are much simpler than the probabilities $P(k) = P(k \text{ successes})$. But the ratios determine the probabilities, so the whole distribution can be understood in terms of the consecutive odds ratios.

Consider first the case when $p = 1/2$. The nth row of Pascal's triangle displays the binomial $(n, 1/2)$ distribution as multiples of 2^{-n}. The numbers in this nth row first increase rapidly, then less rapidly. Then they level off, and start decreasing just as they have increased. This gives rise to the characteristic bell shape of the histogram of a symmetric binomial distribution.

FIGURE 2. **The binomial $(8, 1/2)$ distribution.** This is the distribution of the number of heads in eight fair coin tosses.

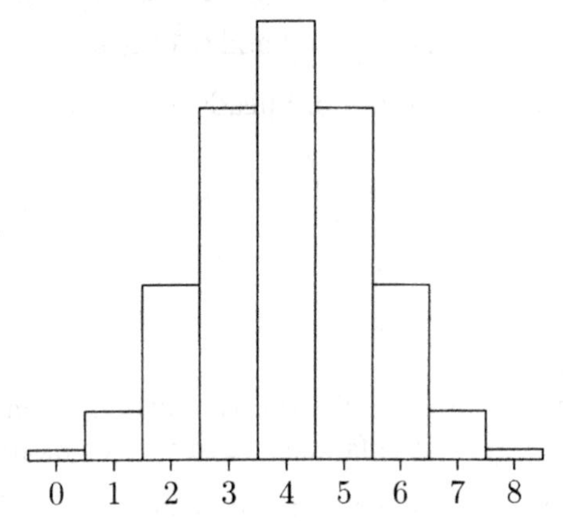

The aim now is to understand the shape of such a binomial distribution in terms of the ratio of the heights of consecutive bars. The numbers from the eighth row of Pascal's triangle are:

$$1 \quad 8 \quad 28 \quad 56 \quad 70 \quad 56 \quad 28 \quad 8 \quad 1$$

So the consecutive odds ratios are

$$\frac{8}{1} \quad \frac{28}{8} \quad \frac{56}{28} \quad \frac{70}{56} \quad \frac{56}{70} \quad \frac{28}{56} \quad \frac{8}{28} \quad \frac{1}{8}$$

which simplify to

$$\frac{8}{1} \quad \frac{7}{2} \quad \frac{6}{3} \quad \frac{5}{4} \quad \frac{4}{5} \quad \frac{3}{6} \quad \frac{2}{7} \quad \frac{1}{8}$$

So the ratios start big, and steadily decrease, crossing 1 in the middle. In the nth row of Pascal's triangle,

$$\binom{n}{0} \quad \binom{n}{1} \quad \binom{n}{2} \quad \binom{n}{3} \quad \cdots \quad \binom{n}{n-3} \quad \binom{n}{n-2} \quad \binom{n}{n-1} \quad \binom{n}{n}$$

the consecutive ratios decrease steadily as follows:

$$\frac{n}{1} \quad \frac{n-1}{2} \quad \frac{n-2}{3} \quad \cdots \quad \cdots \quad \frac{3}{n-2} \quad \frac{2}{n-1} \quad \frac{1}{n}$$

This simple pattern displays the special case $p = q = 1/2$ of the result stated in the following box:

Consecutive Odds for the Binomial Distribution

For independent trials with success probability p, the odds of k successes relative to $k-1$ successes are $R(k)$ to 1, where

$$R(k) = \frac{P(k \text{ successes in } n \text{ trials})}{P(k-1 \text{ successes in } n \text{ trials})} = \left[\frac{n-k+1}{k}\right]\frac{p}{q}$$

This follows from the binomial probability formula and the formula for $\binom{n}{k}$ by cancelling common factors. This simple formula for ratios makes it easy to calculate all the probabilities in a binomial distribution recursively.

Example 2. Computing all probabilities in a binomial distribution.

Problem 1. A pair of fair coins is tossed 8 times. Find the probability of getting both heads on k of these double tosses, for $k = 0$ to 8.

Solution. The chance of getting both heads on each double toss is $\frac{1}{2} \times \frac{1}{2} = \frac{1}{4}$. So the required probabilities form the binomial $(8, 1/4)$ distribution. The following table shows how simply these probabilities can be found, starting with $P(0)$ and then using the consecutive odds formula with $p/q = (\frac{1}{4})/(\frac{3}{4}) = \frac{1}{3}$.

Value of k	0	1	2	3	4	5	6	7	8
How $P(k)$ found	$(\frac{3}{4})^8$	$\frac{8}{1}\frac{1}{3}P(0)$	$\frac{7}{2}\frac{1}{3}P(1)$	$\frac{6}{3}\frac{1}{3}P(2)$	$\frac{5}{4}\frac{1}{3}P(3)$	$\frac{4}{5}\frac{1}{3}P(4)$	$\frac{3}{6}\frac{1}{3}P(5)$	$\frac{2}{7}\frac{1}{3}P(6)$	$\frac{1}{8}\frac{1}{3}P(7)$
Value of $P(k)$	.100	.267	.311	.208	.087	.023	.004	.0004	.00001

Notice how the ratios from Pascal's triangle first dominate the odds against a success ratio of 3 in the denominator, as the probabilities $P(k)$ increase for $k \leq 2$. Then for $k \geq 3$ the ratios from Pascal's triangle are smaller than the odds against success, and the probabilities $P(k)$ steadily decrease. Something similar happens, no matter what the values of n and p. See Figure 3 where this binomial $(8, 1/4)$ distribution is displayed along with other binomial (n, p) distributions for $n = 1$ to 8 and selected values of p.

What is the most likely number of successes in n independent trials with probability of success p on each trial? Intuitively, we expect about proportion p of the trials to be successes . In n trials, we therefore expect around np successes. So it is reasonable to guess that the most likely number of successes m, called the *mode* of the distribution, is an integer close to np. According to the following formula, the mode differs by at most 1 from np:

Most Likely Number of Successes (Mode of Binomial Distribution)

For $0 < p < 1$, the most likely number of successes in n independent trials with probability p of success on each trial is m, the greatest integer less than or equal to $np + p$:

$m = \text{int}\,(np + p)$ where int denotes the integer part function.

If $np + p$ is an integer, as in the case $p = 1/2$, n odd, then there are two most likely numbers, m and $m - 1$. Otherwise, there is a unique most likely number. In either case, the probabilities in the binomial (n, p) distribution are strictly increasing before they reach the maximum, and strictly decreasing after the maximum.

These features of the binomial distribution can be seen in Figure 3. Note the double maxima for $n = 3$, p a multiple of $1/4$, and $n = 7$, p a multiple of $1/8$. Check the formula in a few of these cases to see how it works.

Proof of the formula for the mode. Fix n and p, and consider the following statements about an integer k between 1 and n. Each statement may be true for some k and false for others. By manipulating inequalities and using the formula for consecutive odds, these statements (1) to (5) are logically equivalent:

$$P(k-1) \leq P(k) \tag{1}$$

$$1 \leq P(k)/P(k-1) \tag{2}$$

$$1 \leq \frac{(n-k+1)}{k} \frac{p}{1-p} \tag{3}$$

$$k(1-p) \leq (n-k+1)p \tag{4}$$

$$k \leq np + p \tag{5}$$

FIGURE 3. Histograms of some binomial distributions. The histogram in row n, column p shows the binomial (n, p) distribution for the number of successes in n independent trials, each with success probability p. In row n, the range of values shown is 0 to n. The horizontal scale changes from one row to the next, but equal probabilities are represented by equal areas, even in different histograms.

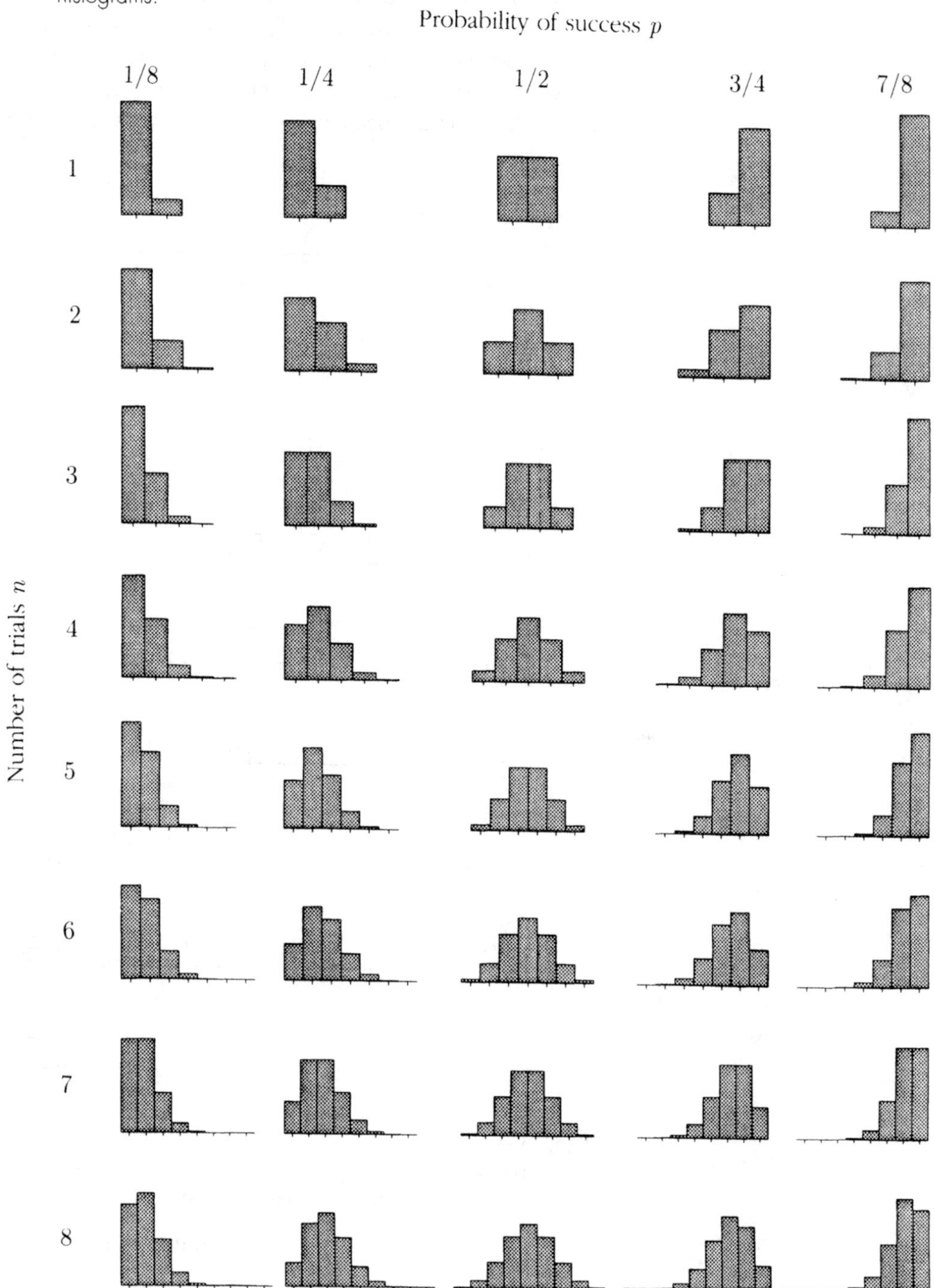

FIGURE 4. Distribution of the number of heads in n coin tosses. Histograms of the binomial (n, 1/2) distribution are shown for $n = 10$ to 100 by steps of 10. Each histogram is a bar graph of the probability of k successes $P(k)$ as a function of k, plotted with the same horizontal and vertical scale. Notice the following features: as n increases the distribution shifts steadily to the right, so as always to be centered on the expected number $n/2$; each distribution is symmetric about $n/2$; as n increases the distribution gradually spreads out, covering a wider range of values; still, the range of values on which the probability is concentrated becomes a smaller and smaller fraction of the whole range of possible values from 0 to n; and apart from these variations in height and width, the histograms all appear to follow the same bell-shaped curve.

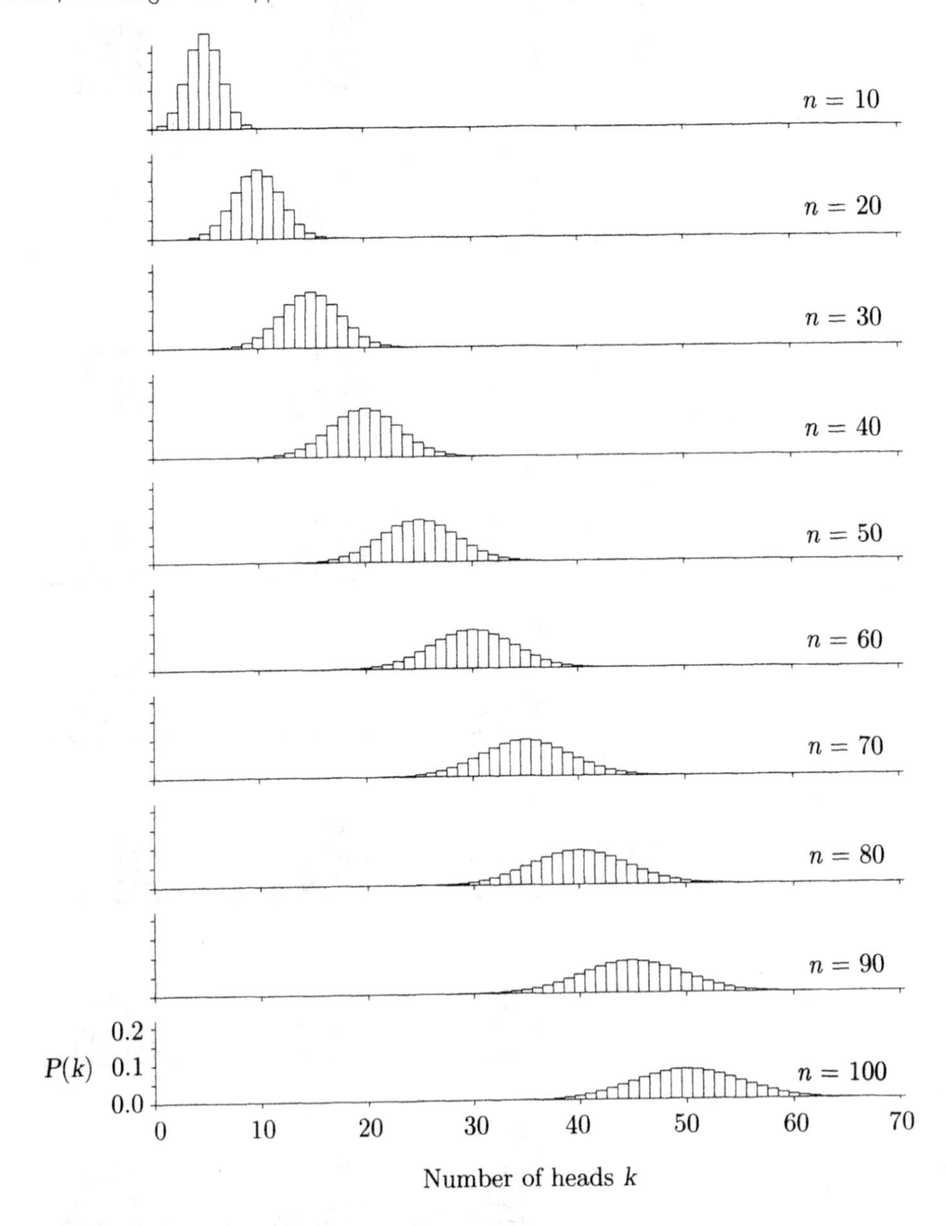

FIGURE 5. **Distribution of the number of successes in 100 trials.** Histograms of the binomial $(100, p)$ distribution are shown for $p = 10\%$ to 90% by steps of 10%. Each histogram is a bar graph of the probability of k successes $P(k)$ as a function of k, plotted with the same horizontal and vertical scale. Notice the following features: as p increases the distribution shifts steadily to the right, so as always to be centered around the expected number $100p$; the distribution is most spread out for $p = 50$; for all values of p the distribution concentrates on a range of numbers that is small in comparison to $n = 100$; and apart from these variations in height and width, and slight skewness toward the edges, the histograms all follow a symmetric bell-shaped curve quite closely.

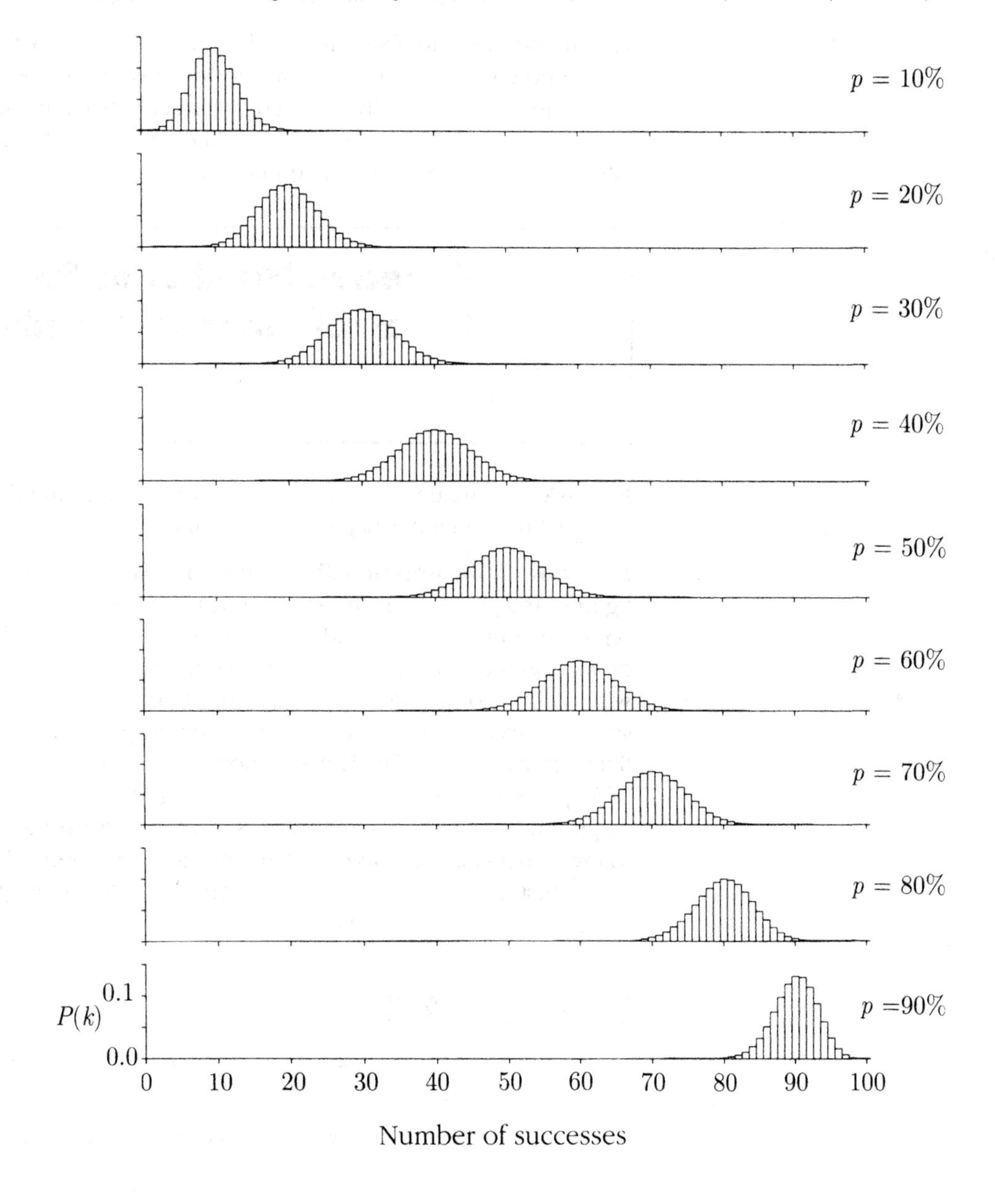

Let m be the largest k attaining the maximum value of $P(k)$ over all $0 \leq k \leq n$. By definition of m, $P(m-1) \leq P(m) > P(m+1)$. That is,

$$m \leq np + p < m + 1$$

by the equivalence of (1) and (5) for $k = m$ and $k = m + 1$. Thus m is the greatest integer less than or equal to $np + p$. (Strictly speaking, the cases $m = 0$ and $m = n$ should be considered separately, but the conclusion is the same.) □

The mean. The number np, which is always close to the mode of the binomial distribution, is called the *expected number of successes*, or the *mean* of the binomial (n, p) distribution, usually denoted μ (Greek letter mu). In case the mean μ is an integer, it turns out that μ is the most likely number of successes. But if μ is not an integer, μ is not even a possible number of successes.

Expected Number of Successes (Mean of Binomial Distribution)

$$\mu = np$$

Remark. For the time being this formula is taken as the definition of the mean of a binomial distribution. Chapter 3 gives a more general, consistent definition.

Behavior of the binomial distribution for large n. This is displayed in the last two figures. As a general rule, for large values of n, the binomial distribution concentrates on a range of values around the expected value np which, while becoming larger on an absolute numerical scale, becomes smaller on a relative scale in comparison with n. Put another way, as n increases, it becomes harder to predict the number of successes exactly, but easier to predict the proportion of successes, which will most likely be close to p. This is made more precise by the *square root law* and the *law of large numbers*, discussed in the following sections. Apart from slight variations in height and width, and some slight skewness toward the edges, all the histograms follow a bell-shaped curve of roughly the same form. This is the famous *normal curve*, first discovered by De Moivre, around 1730, as an approximation to binomial distribution for large values of n.

Exercises 2.1

1. a) How many sequences of zeros and ones of length 7 contain exactly 4 ones and 3 zeros?

   b) If you roll 7 dice, what is the chance of getting exactly 4 sixes?

2. Suppose that in 4-child families, each child is equally likely to be a boy or a girl, independently of the others. Which would then be more common, 4-child families with 2 boys and 2 girls, or 4-child families with different numbers of boys and girls? What would be the relative frequencies?

3. Suppose 5 dice are rolled. Assume they are fair and the rolls are independent. Calculate the probability of the following events:
$A =$ (exactly two sixes); $B =$ (at least two sixes); $C =$ (at most two sixes);
$D =$ (exactly three dice show 4 or greater); $E =$ (at least 3 dice show 4 or greater).

4. A die is rolled 8 times. Given that there were 3 sixes in the 8 rolls, what is the probability that there were 2 sixes in the first five rolls?

5. Given that there were 12 heads in 20 independent coin tosses, calculate

a) the chance that the first toss landed heads;

b) the chance that the first two tosses landed heads;

c) the chance that at least two of the first five tosses landed heads.

6. A man fires 8 shots at a target. Assume that the shots are independent, and each shot hits the bull's eye with probability 0.7.

a) What is the chance that he hits the bull's eye exactly 4 times?

b) Given that he hit the bull's eye at least twice, what is the chance that he hit the bull's eye exactly 4 times?

c) Given that the first two shots hit the bull's eye, what is the chance that he hits the bull's eye exactly 4 times in the 8 shots?

7. You roll a die, and I roll a die. You win if the number showing on your die is strictly greater than the one on mine. If we play this game five times, what is the chance that you win at least four times?

8. For each positive integer n, what is the largest value of p such that zero is the most likely number of successes in n independent trials with success probability p?

9. The chance of winning a bet on 00 at roulette is $1/38 = 0.026315$. In 325 bets on 00 at roulette, the chance of six wins is 0.104840. Use this fact, and consideration of odds ratios, to answer the following questions without long calculations.

a) What is the most likely number of wins in 325 bets on 00, and what is its probability?

b) Find the chance of ten wins in 325 bets on 00.

c) Find the chance of ten wins in 326 bets on 00.

10. Suppose a fair coin is tossed n times. Find simple formulae in terms of n and k for

a) $P(k-1 \text{ heads} \mid k-1 \text{ or } k \text{ heads})$;

b) $P(k \text{ heads} \mid k-1 \text{ or } k \text{ heads})$.

11. 70% of the people in a certain population are adults. A random sample of size 15 will be drawn, with replacement, from this population.

a) What is the most likely number of adults in the sample?

b) What is the chance of getting exactly this many adults?

12. A gambler decides to keep betting on red at roulette, and stop as soon as she has won a total of five bets.

a) What is the probability that she has to make exactly 8 bets before stopping?

b) What is the probability that she has to make at least 9 bets?

13. Genetics. Hereditary characteristics are determined by pairs of *genes.* A gene pair for a particular characteristic is transmitted from parents to offspring by choosing one gene at random from the mother's pair, and, independently, one at random from the father's. Each gene may have several forms, or *alleles.* For example, human beings have an allele (B) for brown eyes, and an allele (b) for blue eyes. A person with allele pair BB has brown eyes, and a person with allele pair bb has blue eyes. A person with allele pair Bb or bB will have brown eyes—the allele B is called *dominant* and b *recessive.* So to have blue eyes, one must have the allele pair bb. The alleles don't "mix" or "blend".

a) A brown-eyed (BB) woman and a blue-eyed man plan to have a child. Can the child have blue eyes?

b) A brown-eyed (Bb) woman and a blue-eyed man plan to have a child. Find the chance that the child has brown eyes.

c) A brown-eyed (Bb) woman and a brown-eyed (Bb) man plan to have a child. Find the chance that the child has brown eyes.

d) A brown-eyed woman has brown-eyed parents, both Bb. She and a blue-eyed man have a child. Given that the child has brown eyes, what is the chance that the woman carries the allele b?

14. Genetics. In certain pea plants, the allele for tallness (T) dominates over the allele for shortness (s), and the allele for purple flowers (P) dominates over the allele for white flowers (w) (see Exercise 13). According to the *principle of independent assortment,* alleles for the two characteristics (flower color and height) are chosen independently of each other.

a) A (TT, PP) plant is crossed with a (ss, ww) plant. What will the offspring look like?

b) The offspring in part a) is self-fertilized, that is, crossed with itself. Write down the possible genetic combination (of flower color and height) that the offspring of this fertilization can have, and find the chance with which each such combination occurs.

c) Ten (Ts, Pw) plants are self-fertilized, each producing a new plant. Find the chance that at least 2 of the new plants are tall with purple flowers.

15. Consider the mode m of the binomial (n, p) distribution. Use the formula $m = int\,(np+p)$ to show the following:

a) If np happens to be an integer, then $m = np$.

b) If np is not an integer, then the most likely number of successes m is one of the two integers to either side of np.

c) Show by examples that m is not necessarily the closest integer to np. Neither is m always the integer above np, nor the integer below it.

2.2 Normal Approximation: Method

The figures of the previous section illustrate the general fact that no matter what the value of p, provided n is large enough, binomial (n, p) histograms have roughly the same bell shape. As n and p vary, the binomial (n, p) distributions differ in where they are centered, and in how spread out they are. But when the histograms are suitably scaled they all follow the same curve provided n is large enough. This section concerns the practical technique of using areas under the curve to approximate binomial probabilities. This can be understood without following the derivation of the curve in the next section.

The *normal curve* has equation

$$y = \frac{1}{\sqrt{2\pi}\sigma} e^{-\frac{1}{2}(x-\mu)^2/\sigma^2} \qquad (-\infty < x < \infty)$$

The equation involves the two fundamental constants $\pi = 3.14159265358\ldots$, and $e = 2.7182818285\ldots$, the base of natural logarithms. The curve has two *parameters*, the *mean* μ, and the *standard deviation* σ. Here μ can be any real number positive or negative, while σ can be any strictly positive number. The mean μ indicates where the curve is located, while the standard deviation σ marks a horizontal scale. You can check by calculus that the curve is symmetric about the point marked μ, concave on either side of μ, out to the points of inflection $\mu - \sigma$ and $\mu + \sigma$, where it switches to become convex (Exercise 15).

FIGURE 1. The normal curve.

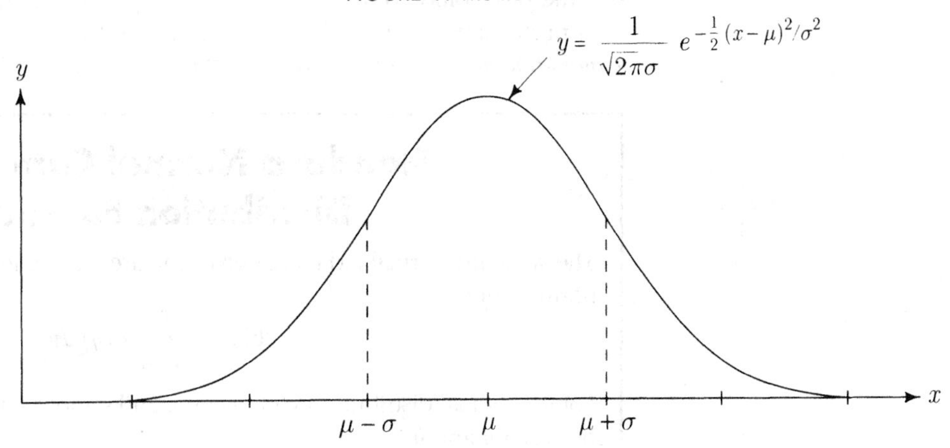

Think of the normal curve as a continuous histogram, defining a probability distribution over the line by relative areas under the curve. Then μ indicates the general location of the distribution, while σ measures how spread out the distribution is. The constant $1/\sqrt{2\pi}\sigma$ is put in the definition of the curve by convention, so that the total area under the curve is 1. This is shown by calculus in Section 5.3. See also Chapter 4 for a general treatment of continuous probability distributions like the normal.

The Normal Distribution

The normal distribution with mean μ and standard deviation σ is the distribution over the x-axis defined by areas under the normal curve with these parameters.

The equation of the normal curve with parameters μ and σ, can be written as

$$y = \frac{1}{\sqrt{2\pi}\sigma} e^{-\frac{1}{2}z^2}$$

where $z = (x - \mu)/\sigma$ measures the number of standard deviations from the mean μ to the number x, as shown in Figure 2. We say that z is x ***in standard units***. The ***standard normal distribution*** is the normal distribution with mean 0 and standard deviation 1. This is the distribution defined by areas under the ***standard normal curve*** $y = \phi(z)$ where

$$\phi(z) = \frac{1}{\sqrt{2\pi}} e^{-\frac{1}{2}z^2}$$

is called the ***standard normal density function***. The standard normal distribution is the distribution on the standard unit or z-scale derived from a normal distribution with arbitrary parameters μ and σ on the x-scale. As shown in Figure 2, the probability to the left of x in the normal distribution with mean μ and standard deviation σ is the probability to the left of $z = (x - \mu)/\sigma$ in the standard normal distribution. This probability is denoted $\Phi(z)$. This function of z is called the ***standard normal cumulative distribution function***, or standard normal c.d.f. for short.

Standard Normal Cumulative Distribution Function

The standard normal c.d.f $\Phi(z)$ gives the area to the left of z under the standard normal curve:

$$\Phi(z) = \int_{-\infty}^{z} \phi(y)\, dy$$

For the normal distribution with mean μ and standard deviation σ, the probability between a and b is

$$\Phi\left(\frac{b-\mu}{\sigma}\right) - \Phi\left(\frac{a-\mu}{\sigma}\right)$$

Because the function $e^{-\frac{1}{2}z^2}$ does not have a simple indefinite integral, there is no simple exact formula for $\Phi(z)$. But $\Phi(z)$ has been calculated numerically. Values of $\Phi(z)$ are tabulated in Appendix 5 for $z \geq 0$.

FIGURE 2. **A normal distribution and the standard normal c.d.f.** The top graph shows the curve that defines the normal distribution with mean μ and standard deviation σ. The lower graph shows the standard normal c.d.f. $\Phi(z)$, the probability in the normal distribution to the left of z on the standard unit scale. The area shaded under the normal curve is $\Phi(z)$ for a particular value z between -1 and 0. This area appears as a height in the graph of the normal c.d.f. $\Phi(z)$.

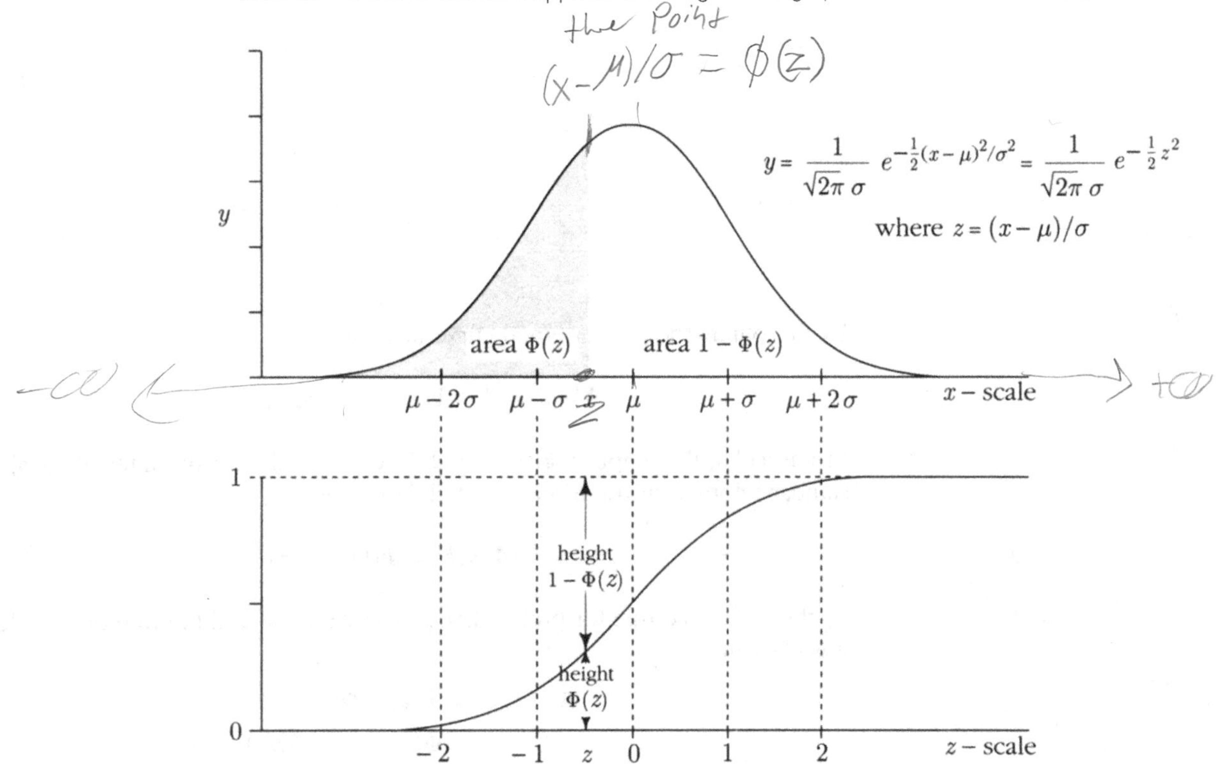

Remark. Instead of using the normal table, you may prefer to program an approximate formula for $\Phi(z)$ on a calculator. A formula, good enough for most purposes, is

$$\Phi(z) \approx 1 - \tfrac{1}{2}\left(1 + c_1 z + c_2 z^2 + c_3 z^3 + c_4 z^4\right)^{-4} \quad (z \geq 0)$$

$$\text{where} \quad c_1 = 0.196854 \quad c_2 = 0.115194$$
$$c_3 = 0.000344 \quad c_4 = 0.019527$$

For every value of $z \geq 0$, the absolute error of this approximation is less than 2.5×10^{-4} [Abramowitz and Stegun, *Handbook of Mathematical Functions*].

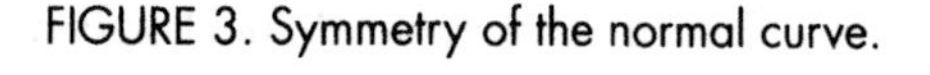

FIGURE 3. Symmetry of the normal curve.

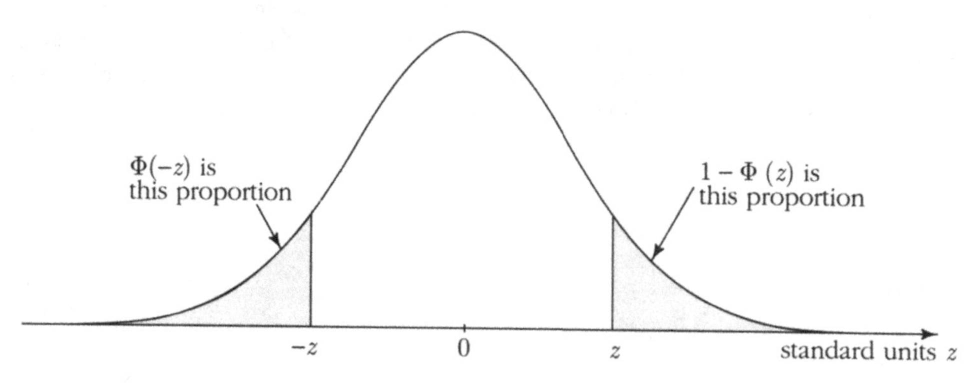

By the symmetry of the normal curve (see Figure 3),

$$\Phi(-z) = 1 - \Phi(z) \qquad (-\infty < z < \infty)$$

In particular, this implies $\Phi(0) = 1/2$. The probability of the interval (a, b) for the standard normal distribution, denoted $\Phi(a, b)$, is

$$\Phi(a, b) = \Phi(b) - \Phi(a)$$

by the difference rule for probabilities. From Figure 3 and the rule of complements, it is clear that

$$\begin{aligned} \Phi(-z, z) &= \Phi(z) - \Phi(-z) \\ &= \Phi(z) - (1 - \Phi(z)) \\ &= 2\Phi(z) - 1 \end{aligned}$$

These formulae are used constantly when working with the normal distribution. But, to avoid mistakes, it is best not to memorize them. Rather sketch the standard normal curve each time. Remember the symmetry of the curve, and the definition of $\Phi(z)$, as the proportion of area under the curve to the left of z. Then the formulae are obvious from the diagram. There are three standard normal probabilities which are worth remembering:

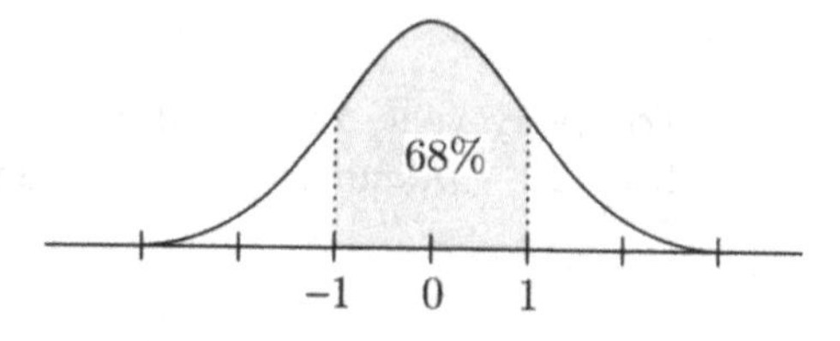

$\Phi(-1, 1) \approx 68\%$, the probability within one standard deviation of the mean,

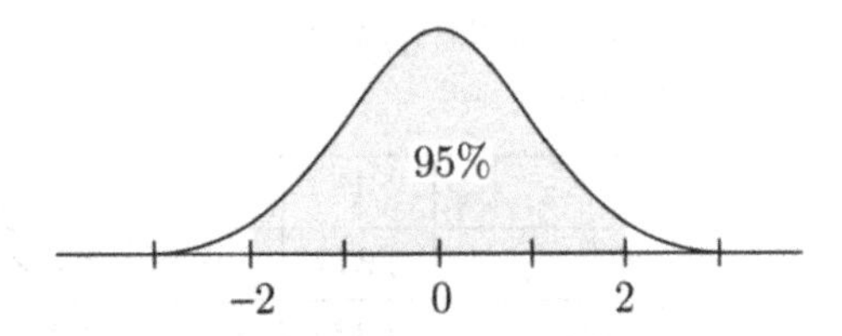

$\Phi(-2, 2) \approx 95\%$, the probability within two standard deviations of the mean,

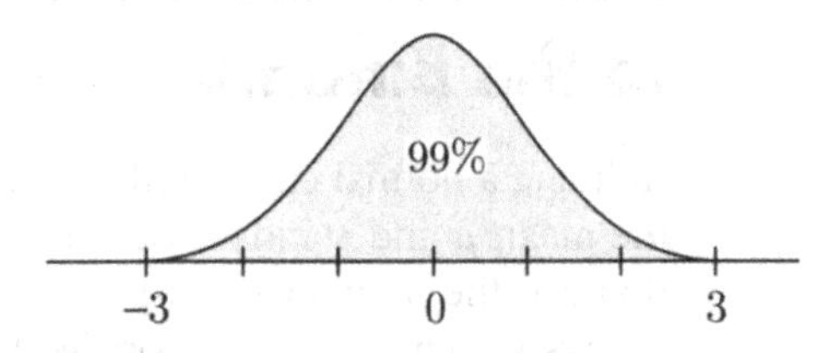

$\Phi(-3, 3) \approx 99.7\%$, the probability within three standard deviations of the mean.

From these probabilities you can easily find $\Phi(a, b)$ for several other intervals. For example,

$$\Phi(0, 1) = \tfrac{1}{2}\Phi(-1, 1) \approx \tfrac{1}{2}68\% = 34\%$$

$$\Phi(2, \infty) = \tfrac{1}{2}(1 - \Phi(-2, 2)) \approx \tfrac{1}{2}(100\% - 95\%) = 2.5\%$$

The probability $\Phi(-z, z)^c$ beyond z standard deviations from the mean in a normal distribution is

$$\Phi(-z, z)^c = 1 - \Phi(-z, z) = 2(1 - \Phi(z)) < 2\phi(z)/z$$

as shown in Table 1 for $z = 1$ to 6. The factor $\exp\left(-\frac{1}{2}z^2\right)$ in the definition of $\phi(z)$ makes $\phi(z)$ extremely small for large z. The above inequality, left as an exercise, shows that $\Phi(-z, z)^c$ is even smaller for $z \geq 2$.

Not too much significance should be placed on the extremely small probabilities $\Phi(-z, z)^c$ for z larger than about 3. The point is that the normal distribution is mostly applied as an approximation to some other distribution. Typically the errors involved in such an approximation, though small, are orders of magnitude larger than $\Phi(-z, z)^c$ for $z > 3$.

TABLE 1. **Standard normal probability outside $(-z, z)$.** The probability $\Phi(-z,z)^c$ is tabulated along with $2\phi(z)/z$, which is larger than $\Phi(-z,z)^c$ for all z, and a very good approximation to it for large z.

z	1	2	3	4	5	6
$\Phi(-z,z)^c$	0.317	0.046	2.7×10^{-3}	6.3×10^{-5}	5.7×10^{-7}	1.97×10^{-9}
$2\phi(z)/z$	0.484	0.054	2.9×10^{-3}	6.7×10^{-5}	5.9×10^{-7}	2.03×10^{-9}

The Normal Approximation to the Binomial Distribution

In fitting a normal curve to the binomial (n,p) distribution the main question is how the mean μ and standard deviation σ are determined by n and p. As noted in Section 2.1, the number $\mu = np$, called the mean of the binomial (n,p) distribution, is always within ± 1 of the most likely value, $m = \text{int}(np + p)$. So $\mu = np$ is a convenient place to locate the center. How to find the right value of σ is less obvious. As explained in the next section, provided $\sqrt{npq}$ is sufficiently large, good approximations to binomial probabilities are obtained by areas under the normal curve with mean $\mu = np$ and $\sigma = \sqrt{npq}$. Later, in Section 3.3, it will be explained how this formula for σ is consistent with the right general definition of the standard deviation of a probability distribution.

FIGURE 4. **A binomial histogram, with the normal curve superimposed.** Both the x scale (number of successes) and the z scale (standard units) are shown.

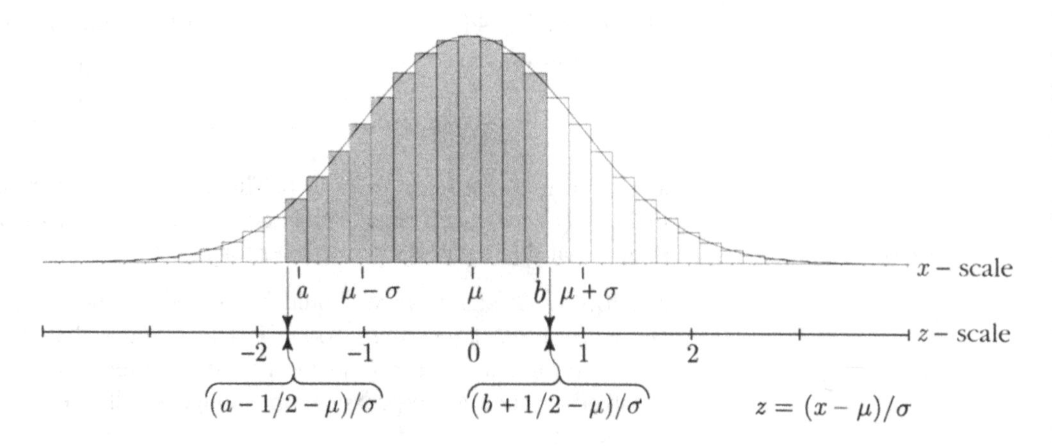

Let $P(a \text{ to } b)$ be the probability of getting between a and b successes (inclusive) in n independent trials with success probability p. Then, from Figure 4, we see that:

$$\begin{aligned}
P(a \text{ to } b) &= \text{proportion of area under the binomial } (n, p) \text{ histogram} \\
&\quad\ \text{between } a - \tfrac{1}{2} \text{ and } b + \tfrac{1}{2} \\
&\approx \text{proportion of area under the normal curve } \frac{1}{\sqrt{2\pi}\,\sigma} e^{-\frac{1}{2}(x-\mu)^2/\sigma^2} \\
&\quad\ \text{between } x = a - \tfrac{1}{2} \text{ and } b + \tfrac{1}{2} \\
&= \text{proportion of area under the normal curve } \frac{1}{\sqrt{2\pi}} e^{-\frac{1}{2}z^2} \\
&\quad\ \text{between } z = (a - \tfrac{1}{2} - \mu)/\sigma \text{ and } z = (b + \tfrac{1}{2} - \mu)/\sigma.
\end{aligned}$$

In terms of the standard normal c.d.f. Φ, this gives the following:

Normal Approximation to the Binomial Distribution

For n independent trials with success probability p

$$P(a \text{ to } b \text{ successes}) \approx \Phi\left(\frac{b + \frac{1}{2} - \mu}{\sigma}\right) - \Phi\left(\frac{a - \frac{1}{2} - \mu}{\sigma}\right)$$

where $\mu = np$ is the *mean*, and $\sigma = \sqrt{npq}$ is the *standard deviation.*

Use of $a - \frac{1}{2}$ and $b + \frac{1}{2}$ in the normal approximation rather than a and b is called the *continuity correction.* This correction is essential to obtain good approximations for small values of $\sqrt{npq}$. For large $\sqrt{npq}$ it makes little difference unless a and b are very close.

Example 1. 100 fair coin tosses.

Problem. Find, approximately, the chance of getting 50 heads in 100 tosses of a fair coin.

Solution. Here $n = 100$, $p = 1/2$, so $\mu = 50$, $\sigma = 5$. The normal approximation above with $a = b = 50$ gives

$$\begin{aligned}
P(50) &\approx \Phi\left((50 + \tfrac{1}{2} - 50)/5\right) - \Phi\left((50 - \tfrac{1}{2} - 50)/5\right) \\
&= \Phi(0.1) - \Phi(-0.1) \\
&= 2\Phi(0.1) - 1 = 2 \times 0.5398 - 1 = 0.0796 \quad \text{(exact value 0.0795892)}
\end{aligned}$$

Continuation. Other probabilities can be computed in the same way—for example

$$
\begin{aligned}
P(45 \text{ to } 55) &\approx \Phi\left((55\tfrac{1}{2} - 50)/5\right) - \Phi\left((44\tfrac{1}{2} - 50)/5\right) \\
&= \Phi(1.1) - \Phi(-1.1) \\
&= 2\Phi(1.1) - 1 = 2 \times 0.8643 - 1 \\
&= 0.7286 \qquad \text{(exact value 0.728747)} \\
P(40 \text{ to } 60) &\approx 2\Phi(2.1) - 1 = 2 \times 0.9821 - 1 \\
&= 0.9642 \qquad \text{(exact value 0.9648)} \\
P(35 \text{ to } 65) &\approx 2\Phi(3.1) - 1 = 2 \times 0.9990 - 1 \\
&= 0.9980 \qquad \text{(exact value 0.99821)}
\end{aligned}
$$

Fluctuations in the number of successes. For any fixed p, the normal approximation to the binomial (n, p) distribution gets better and better as n gets larger. So, in a large number of independent trials with success probability p, the typical size of the random fluctuations in the number of successes is of the order of $\sigma = \sqrt{npq}$. For example,

$$
\begin{aligned}
P(\mu - \sigma \text{ to } \mu + \sigma \text{ successes in } n \text{ trials}) &\approx 68\% \\
P(\mu - 2\sigma \text{ to } \mu + 2\sigma \text{ successes in } n \text{ trials}) &\approx 95\% \\
P(\mu - 3\sigma \text{ to } \mu + 3\sigma \text{ successes in } n \text{ trials}) &\approx 99.7\%
\end{aligned}
$$

It can be shown that for fixed p, as $n \to \infty$, each probability on the left approaches the exact value of the corresponding proportion of area under the normal curve.

Fluctuations in the proportion of successes. While the typical size of random fluctuations of the *number* of successes in n trials away from the expected number np is a moderate multiple of $\sqrt{npq}$, the typical size of random fluctuations in the *relative frequency* of successes about the expected proportion p is correspondingly of order $\sqrt{npq}/n = \sqrt{pq/n}$. Since $\sqrt{pq} \le \frac{1}{2}$ for all $0 < p < 1$, and $1/\sqrt{n} \to 0$ as $n \to \infty$, this makes precise the rate at which we can expect relative frequencies to stabilize under ideal conditions.

Square Root Law

For large n, in n independent trials with probability p of success on each trial:

- the *number* of successes will, with high probability, lie in a relatively small interval of numbers, centered on np, with width a moderate multiple of $\sqrt{n}$ on the numerical scale;
- the *proportion* of successes will, with high probability, lie in a small interval centered on p, with width a moderate multiple of $1/\sqrt{n}$.

Numerical computations show that the square root law also holds for small values of n, but its most important implications are for large n. In particular, it implies the following mathematical confirmation of our intuitive idea of probability as a limit of long-run frequencies:

Law of Large Numbers

If n is large, the proportion of successes in n independent trials will, with overwhelming probability, be very close to p, the probability of success on each trial. More formally:

- for independent trials, with probability p of success on each trial, for each $\epsilon > 0$, no matter how small, as $n \to \infty$,

$$P(\text{proportion of successes in } n \text{ trials differs from } p \text{ by less than } \epsilon) \to 1$$

Confidence Intervals

The normal approximation is the basis of the statistical method of *confidence intervals*. Suppose you think that you are observing the results of a sequence of independent trials with success probability p, but you don't know the value of p. For example, you might be observing whether or not a biased die rolled a six (success) or not six (failure). Suppose in n trials you observe that the relative frequency of successes is $\hat{p}$. If n is large, it is natural to expect that the unknown probability p is most likely fairly close to $\hat{p}$. For example, since

$$\Phi(-4, 4) \approx 99.99\%$$

the above results state that if n is large enough, no matter what p is, it is 99.99% certain that the observed number of successes, $n\hat{p}$, differs from np by less than $4\sqrt{npq}$, so the relative frequency $\hat{p}$ will differ from p by less than $4\sqrt{pq/n}$, which is at most $2/\sqrt{n}$. Having observed the value of $\hat{p}$, it is natural to suppose that this overwhelmingly likely event has occurred, which implies that p is within $2/\sqrt{n}$ of $\hat{p}$. The interval $\hat{p} \pm 2/\sqrt{n}$, within which p can reasonably be expected to lie, is called a 99.99% *confidence interval* for p.

Example 2. Estimating the bias on a die.

Problem. In a million rolls of a biased die, the number 6 shows $180,000$ times. Find a 99.99% confidence interval for the probability that the die rolls six.

Solution. The observed relative frequency of sixes is $\hat{p} = 0.18$. So a 99.99% confidence interval for the probability that the die rolls six is

$$0.18 \pm 2/\sqrt{1,000,000} \quad \text{or} \quad (0.178, 0.182)$$

Remark. This procedure of going $\pm 2/\sqrt{n}$ from the observed $\hat{p}$ to make the confidence interval is somewhat conservative, meaning the coverage probability will be even higher than 99.99% for large n. This is due to neglecting the factor $\sqrt{pq} \leq 0.5$ and so overestimating the standard deviation $\sigma = \sqrt{npq}$ in case p is not 0.5, as the above $\hat{p}$ would strongly suggest. The usual statistical procedure is to estimate $\sqrt{pq}$ by $\sqrt{\hat{p}(1-\hat{p})}$, which is $\sqrt{0.18 \times 0.82} = 0.384$ in the above example. This reduces the length of the interval by a factor of $0.384/0.5 = 77\%$ in this case.

The most important thing to note in this kind of calculation is how the length of the confidence interval depends on n through the square root law. Suppose the confidence interval is $\hat{p} \pm c/\sqrt{n}$, for some constant c. No matter what c is, to reduce the length of the confidence interval by a factor of f requires an increase of n by a factor of f^2. So to halve the length of a confidence interval, you must quadruple the number of trials.

Example 3. Random sampling.

Problem. Two survey organizations make 99% confidence intervals for the proportion of women in a certain population. Both organizations take random samples with replacement from the population; the first uses a sample of size 350 while the second uses a sample of size 1000. Which confidence interval will be shorter, and by how much?

Solution. The interval based on the larger sample size will be shorter. The size of the second sample is $1000/350 = 2.86$ times the size of the first, so the length of the second interval is $1/\sqrt{2.86}$ times the length of the first, that is, 0.59 times the length of the first.

Example 4. How many trials?

Suppose you estimate the probability p that a biased coin lands heads by tossing it n times and estimating p by the proportion $\hat{p}$ of the times the coin lands heads in the n tosses.

Problem. How many times n must you toss the coin to be at least 99% sure that $\hat{p}$ will be: a) within 0.1 of p? b) within .01 of p?

Solution. First find z such that $\Phi(-z, z) = 99\%$,

$$\text{i.e.,} \quad 2\Phi(z) - 1 = 0.99 \quad \text{i.e.,} \quad \Phi(z) = 0.995$$

Inspection of the table gives $z \approx 2.575$. For large n, $\hat{p}$ will with probability at least 99% lie in the interval $p \pm 2.575\sqrt{pq}/\sqrt{n}$. Since $\sqrt{pq} \leq 0.5$, the difference between

$\hat{p}$ and p will then be less than

$$2.575 \times 0.5/\sqrt{n}$$

For a), set this equal to 0.1 and solve for n:

$$2.575 \times 0.5/\sqrt{n} = 0.1$$

$$n = \left(\frac{2.575 \times 0.5}{0.1}\right)^2 = 165.77$$

So 166 trials suffice for at least 99% probability of accuracy to within 0.1.

b) By the square root law, to increase precision by a factor of 10, requires an increase in the number of trials by $10^2 = 100$. So about 16,577 trials would be required for 99% probability of accuracy to within .01.

How good is the normal approximation? As a general rule, the larger the standard deviation $\sigma = \sqrt{npq}$, and the closer p is to $1/2$, the better the normal approximation to the binomial (n, p) distribution. The approximation works best for $p = 1/2$ due to the symmetry of the binomial distribution in this case. For $p \neq 1/2$ the approximation is not quite as good, but as the graphs at the end of Section 2.1 show, as n increases the binomial distribution becomes more and more symmetric about its mean. It is shown in the next section that the shape of the binomial distribution approaches the shape of the normal curve as $n \to \infty$ for every fixed p with $0 < p < 1$.

How good the normal approximation is for particular n and p can be measured as follows. Let $N(a \text{ to } b)$ denote the normal approximation with continuity correction to a binomial probability $P(a \text{ to } b)$. Define $W(n,p)$, the ***worst error*** in the normal approximation to the binomial(n,p) distribution, to be the biggest absolute difference between $P(a \text{ to } b)$ and $N(a \text{ to } b)$, over all integers a and b with $0 \leq a \leq b \leq n$:

$$W(n,p) = \max_{0 \leq a \leq b \leq n} |P(a \text{ to } b) - N(a \text{ to } b)|$$

Numerical calculations show that $W(n, 1/2)$ is less than 0.01 for all $n \geq 10$, and less than 0.005 for all $n \geq 20$. Such a small error of approximation is negligible for most practical purposes. For $p \neq 1/2$ there is a systematic error in the normal approximation because an asymmetric distribution is approximated by a symmetric one. A refinement of the normal approximation described in the next paragraph shows that

$$W(n,p) \approx \frac{1}{10} \frac{|1-2p|}{\sqrt{npq}} \tag{1}$$

where the error of the approximation is negligible for all practical purposes provided $\sigma = \sqrt{npq}$ is at least about 3. This formula shows clearly how the larger σ, and the closer p is to $1/2$, the smaller $W(n,p)$ tends to be. Because $|1-2p| \leq 1$ for all $0 \leq p \leq 1$, even if p is close to 0 or 1, the worst error is small provided σ is large enough. For $\sigma \geq 3$ the worst error is about $1/10\sigma$ for p close to 0 or 1 and large n. Numerical calculations confirm the following consequences of (1): the worst error $W(n,p)$ is

- less than 0.01 for $n \geq 20$ and p between 0.4 and 0.6
- less than 0.02 for $n \geq 20$ and p between 0.3 and 0.7
- less than 0.03 for $n \geq 25$ and p between 0.2 and 0.8
- less than 0.05 for $n \geq 30$ and p between 0.1 and 0.9

The systematic error in the normal approximation of magnitude about $1/10\sigma$ can be reduced to an error that is negligible in comparison by the *skewness correction* explained in the next paragraph. This method gives satisfactory approximations to binomial probabilities for arbitrary n and p with $\sigma \geq 3$. For p close to 0 or 1, and $\sigma \leq 3$, a better approximation to the binomial distribution is provided by using the Poisson distribution described in the next section.

The skew-normal approximation. Figures 5 and 7 show how the histogram of the binomial $(100, 1/10)$ distribution is slightly skewed relative to its approximating normal curve. The histogram is better approximated by adding to the standard normal curve $\phi(z)$ a small multiple of the curve $y = \phi'''(z)$, where

$$\phi'''(z) = (3z - z^3)\phi(z)$$

is the third derivative of $\phi(z)$ (Exercise 16), as graphed in Figure 6. By careful analysis of the histogram of a binomial (n,p) distribution plotted on a standard units scale, it can be shown that for $p \neq 1/2$ adding the right small multiple of the anti-symmetric function $\phi'''(z)$ to the symmetric function $\phi(z)$ gives a curve which respects the slight asymmetry of the binomial histogram, and so follows it much more closely than the plain normal curve $\phi(z)$. The resulting *skew-normal curve* has equation

$$y = \phi(z) - \frac{1}{6}\text{Skewness}\,(n,p)\,\phi'''(z) \quad \text{where} \tag{2}$$

$$\text{Skewness}\,(n,p) = (1-2p)/\sqrt{npq} = (1-2p)/\sigma$$

is a number called the *skewness* of the binomial(n,p) distribution, which measures its degree of asymmetry. The skewness is 0 if $p = 1/2$, when the distribution is perfectly symmetric about $n/2$. The skewness positive for $p < 1/2$ when the distribution is called *skewed to the right*, and negative for $p > 1/2$ when the distribution is *skewed to the left*. The meaning of these terms is made precise by the way the binomial histogram follows the skew-normal curve (2) more closely than it does the

FIGURE 5. **Normal curve approximating the binomial $(100, 1/10)$ histogram.** Notice how the bars are slightly above the normal curve just to the left of the mean, and slightly below the curve just to the right of the mean. Further away from the mean, the bars lie below the curve in the left tail, and above the curve in the right tail.

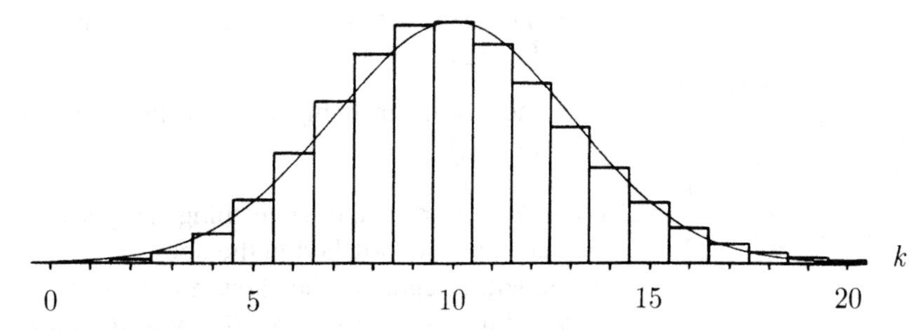

FIGURE 6. **Graph of $\phi'''(z) = (3z - z^3)\phi(z)$.** Note how the function is positive in the intervals $(-\infty, -\sqrt{3})$ and $(0, \sqrt{3})$, and negative in the intervals $(-\sqrt{3}, 0)$ and $(\sqrt{3}, \infty)$. The zeros are at 0 and $\pm\sqrt{3}$. The z-scale is the standard unit scale derived from the histogram in Figures 5 and 7.

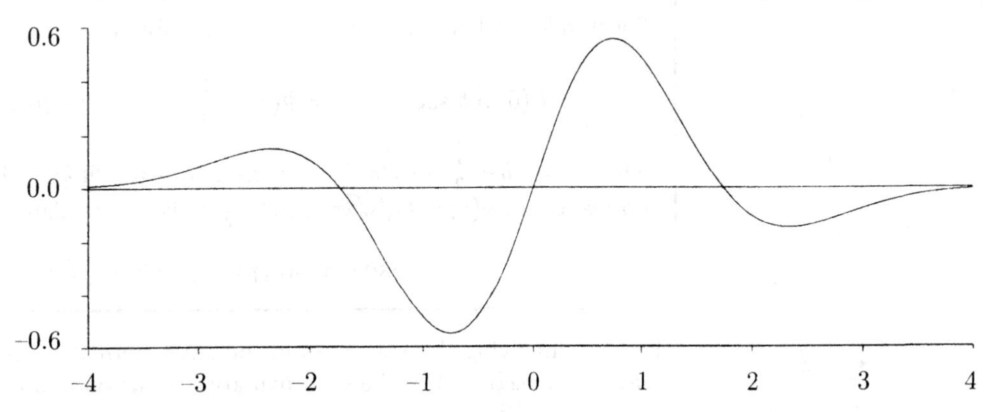

FIGURE 7. **Skew–normal curve approximating the binomial $(100, 1/10)$ histogram.** Refer to Example 5. Both the normal curve $y = \phi(z)$ and the skew–normal curve $y = \phi(z) - (2/45)\phi'''(z)$ are shown. The skew–normal curve follows the binomial histogram much more closely. The difference between the normal and skew–normal curves is $2/45$ times the curve $\phi'''(z)$ graphed in Figure 6.

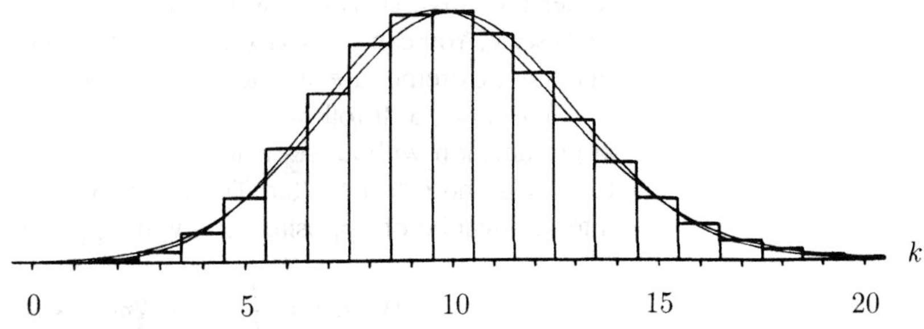

plain normal curve. Figure 7 illustrates how in the case $p < 1/2$ when the binomial histogram is skewed to the right, there are numbers $z_- < z_0 < z_+$ on the standard units scale, with $z_0 \approx 0$ and $z_\pm \approx \pm\sqrt{3}$, (the three zeros of $\phi'''(z)$) such that

- the histogram is lower than the normal curve on the intervals $(-\infty, z_-)$ and (z_0, z_+)
- the histogram is higher than the normal curve on the intervals (z_-, z_0) and (z_+, ∞)

For $1/2 < p < 1$, the same thing happens, except that the words "higher" and "lower" must be switched in the above description. The distribution is then skewed to the left. Integrating the skew-normal curve (2) from $-\infty$ to the point z on the standard unit scale (Exercise 16) gives the following:

Skew-Normal Approximation to the Binomial Distribution

For n independent trials with success probability p,

$$P(0 \text{ to } b \text{ successes}) \approx \Phi(z) - \frac{1}{6}\text{Skewness}\,(n,p)(z^2-1)\phi(z)$$

where $z = (b + \frac{1}{2} - \mu)/\sigma$ for $\mu = np$ and $\sigma = \sqrt{npq}$, $\Phi(z)$ is the standard normal c.d.f., $\phi(z) = (1/\sqrt{2\pi})\exp(-\frac{1}{2}z^2)$ is the standard normal curve, and

$$\text{Skewness}(n,p) = (1-2p)/\sqrt{npq}$$

The term involving the skewness in the skew-normal approximation is called the *skewness correction*. The skew-normal approximation to an interval probability

$$P(a \text{ to } b) = P(0 \text{ to } b) - P(0 \text{ to } a-1)$$

is found by using the above approximation twice and taking the difference. The resulting normal approximation with skewness correction to $P(a \text{ to } b)$ differs from the plain normal approximation $N(a \text{ to } b)$ by 1/6 of the skewness times the area under the curve $\phi'''(z)$ between points corresponding to a and b on the standard units scale. You can show (Exercise 16) that this area is always between ± 0.577, and that these extremes are attained over the intervals from $z = -\sqrt{3}$ to $z = 0$, and from $z = 0$ to $z = \sqrt{3}$. It follows that for $p \neq 1/2$, the worst error $W(n,p)$ in the normal approximation without skewness correction occurs for $a \approx \mu - \sqrt{3}\sigma$ and $b \approx \mu$, or for $a \approx \mu$ and $b \approx \mu + \sqrt{3}\sigma$. The errors of the normal approximation for these two intervals will be of opposite signs with approximately equal magnitudes of

$$W(n,p) \approx \frac{1}{6} \times |1-2p|/\sigma \times 0.577 \approx |1-2p|/10\sigma$$

Thus the skew-normal approximation implies this simple estimate for the worst error in the plain normal approximation, and shows the intervals on which such an error is to be expected. This formula shows the plain normal approximation is rather rough for σ in the range from 3 to 10 and p close to 0 or 1. Numerical calculations show that provided $\sigma \geq 3$ (no matter what p) the skew-normal approximation gives interval probabilities correct to two decimal places (error at most 0.005) which is adequate for most practical purposes. For fixed p, as $n \to \infty$, the skewness of the binomial distribution converges to 0, so in the limit of large n the skewness correction can be ignored, just like the continuity correction, which is of the same order of magnitude $1/\sigma$.

Example 5. Distribution of the number of 0's in 100 random digits.

Consider the distribution of the random number of times a particular digit, say 0, appears among 100 random digits picked independently and uniformly at random from the set of 10 digits $\{0, 1, \ldots, 9\}$. This is the binomial$(100, 1/10)$ distribution which is displayed in Figure 7, along with the approximating normal and skew-normal curves. The mean is $\mu = 100 \times 1/10 = 10$, the standard deviation is $\sigma = \sqrt{npq} = \sqrt{100 \times (1/10) \times (9/10)} = 3$, and the skewness is $(1-2p)/\sqrt{npq} = (1 - (2/10))/3 = 4/15$. From (2), the skew-normal curve approximating the shape of the binomial histogram has equation $y = \phi(z) - \frac{2}{45}(3z - z^3)\phi(z)$, as graphed in Figure 7. The probability of 4 or fewer 0's is

$$P(0 \text{ to } 4) = \sum_{k=0}^{4} \binom{100}{k} \left(\frac{1}{10}\right)^k \left(\frac{9}{10}\right)^{100-k} = 0.024$$

by exact calculation, correct to three decimal places. The normal approximation to this probability is $\Phi(z)$ for $z = (4\frac{1}{2} - 10)/3 = -11/6$, i.e., $\Phi(-11/6) = 0.033$, which is not a very good approximation. The skew-normal approximation, which is not much harder to compute, is

$$\begin{aligned} &\Phi(z) - \frac{1}{6}\text{Skewness}(100, 1/10)(z^2 - 1)\phi(z) \\ &= 0.033 - \frac{1}{6}\frac{4}{15}\left(\left(\frac{-11}{6}\right)^2 - 1\right)\frac{1}{\sqrt{2\pi}}\exp\left(-\frac{1}{2}\left(\frac{-11}{6}\right)^2\right) \\ &= 0.026 \end{aligned}$$

which differs from the exact value by only 0.002. Similar calculations yield the numbers displayed in Table 2. The numbers are correct to three decimal places. The ranges selected, 0 to 4, 5 to 9, 10 to 15, and 16 to 100, are the ranges over which the normal approximation is first too high, then too low, too high, and too low again. The normal approximation is very rough in this example, but the skew-normal approximation is excellent.

TABLE 2. **Approximations to the binomial $(100, 1/10)$ distribution.** The probability $P(a \text{ to } b)$ of from a and b successes (inclusive) in 100 independent trials, with probability $1/10$ of success on each trial, is shown along with approximations using the normal and skew–normal curves.

value range	exact probability	skew–normal approximation	normal approximation
0 – 4	0.024	0.026	0.033
5 – 9	0.428	0.425	0.400
10 – 15	0.509	0.508	0.533
16 – 100	0.040	0.041	0.033

Exercises 2.2

1. Let H be the number of heads in 400 tosses of a fair coin. Find normal approximations to: a) $P(190 \leq H \leq 210)$; b) $P(210 \leq H \leq 220)$; c) $P(H = 200)$; d) $P(H = 210)$.

2. Recalculate the approximations above for a biased coin with $P(\text{heads}) = 0.51$.

3. A fair coin is tossed repeatedly. Consider the following two possible outcomes:

55 or more heads in the first 100 tosses
220 or more heads in the first 400 tosses

a) Without calculation, say which of these outcomes is more likely. Why?

b) Confirm your answer to a) by a calculation.

4. Suppose that each of 300 patients has a probability of $1/3$ of being helped by a treatment independent of its effect on the other patients. Find approximately the probability that more than 120 patients are helped by the treatment.

5. Suppose you bet a dollar on red, 25 times in a row, at roulette. Each time you win a dollar with probability $18/38$, lose with probability $20/38$. Find, approximately, the chance that after 25 bets you have at least as much money as you started with.

6. To estimate the percent of district voters who oppose a certain ballot measure, a survey organization takes a random sample of 200 voters from a district. If 45% of the voters in the district oppose the measure, estimate the chance that:

a) exactly 90 voters in the sample oppose the measure;

b) more than half the voters in the sample oppose the measure.

[Assume that all voters in the district are equally likely to be in the sample, independent of each other.]

7. City A has a population of 4 million, and city B has 6 million. Both cities have the same proportion of women. A random sample (with replacement) will be taken from each city, to estimate this proportion. In each of the following cases, say whether the two samples give equally good estimates; and if you think one estimate is better than the other, say how much better it is.

a) A 0.01% sample from each city.

b) A sample of size 400 from each city.

c) A 0.1% sample from city A, and a 0.075% sample from city B.

8. Find, approximately, the chance of getting 100 sixes in 600 rolls of a die.

9. **Airline overbooking.** An airline knows that over the long run, 90% of passengers who reserve seats show up for their flight. On a particular flight with 300 seats, the airline accepts 324 reservations.

a) Assuming that passengers show up independently of each other, what is the chance that the flight will be overbooked?

b) Suppose that people tend to travel in groups. Would that increase or decrease the probability of overbooking? Explain your answer.

c) Redo the calculation a) assuming that passengers always travel in pairs. Check that your answers to a), b), and c) are consistent.

10. A probability class has 30 students. As part of an assignment, each student tosses a coin 200 times and records the number of heads. Approximately what is the chance that no student gets exactly 100 heads?

11. **Batting averages.** Suppose that a baseball player's long-run batting average (number of hits per time at bat) is .300. Assuming that each time at bat yields a hit with a consistent probability, independently of other times, what is the chance that the player's average over the next 100 times at bat will be

a) .310 or better? b) .330 or better? c) .270 or worse?

d) Suppose the player tends to have periods of good form and periods of bad form. Would different times at bat then be independent? Would that tend to increase or decrease the above chances?

e) Suppose the player actually hits .330 over the 100 times at bat. Would you be convinced that his form had improved significantly? or could the improvement just as well be due to chance?

12. A fair coin is tossed $10,000$ times. Find a number m such that the chance of the number of heads being between $5000 - m$ and $5000 + m$ is approximately $2/3$.

13. A pollster wishes to know the percentage p of people in a population who intend to vote for a particular candidate. How large must a random sample with replacement be in order to be at least 95% sure that the sample percentage is within one percentage point of p?

14. Wonderful Widgets Inc. has developed electronic devices which work properly with probability 0.95, independently of each other. The new devices are shipped out in boxes containing 400 each.

a) What percentage of boxes contains 390 or more working devices?

b) The company wants to guarantee, say, that k or more devices per box work. What is the largest k such that at least 95% of the boxes meet the warranty?

15. **First two derivatives of the normal curve.** Let $\phi'(z)$, $\phi''(z)$ be the first and second derivatives of the standard normal curve $\phi(z) = (1/\sqrt{2\pi})\exp(-\frac{1}{2}z^2)$. Show that:

a) $\phi'(z) = -z\phi(z)$

b) $\phi''(z) = (z^2 - 1)\phi(z)$

c) Sketch the graphs of $\phi(z)$, $\phi'(z)$, $\phi''(z)$ on the same scale for z between -4 and 4. What are the graphs like outside of this range?

d) Use b) and the chain rule of calculus to find the second derivative at x of the normal curve with parameters μ and σ^2.

e) Use the result of d) to verify the assertions in the sentence above Figure 1 on page 93.

16. **Third derivative of the normal curve.**

a) Show that $\phi(z)$ has third derivative $\phi'''(z) = (-z^3 + 3z)\phi(z)$

b) Show that $\int_{-\infty}^{x} \phi'''(z)dz = \phi''(x)$, and hence

$$\int_{-\infty}^{-\sqrt{3}} \phi'''(z)dz = -\int_{\sqrt{3}}^{\infty} \phi'''(z)dz = 2\phi(\sqrt{3}) \approx 0.178$$

and

$$-\int_{-\sqrt{3}}^{0} \phi'''(z)dz = \int_{0}^{\sqrt{3}} \phi'''(z)dz = \phi(0) + 2\phi(\sqrt{3}) \approx 0.577$$

c) Show that $\int_a^b \phi'''(z)dz$ lies between $\pm[\phi(0) + 2\phi(\sqrt{3})]$ for every $a < b$. [*Hint:* No more calculation required. Consider the graph of $\phi'''(z)$ and the interpretation of the integral in terms of areas.]

17. **Standard normal tail bound.** Show that $1 - \Phi(z) < \phi(z)/z$ for positive z by the following steps.

a) Show that

$$1 - \Phi(z) = \int_{z}^{\infty} \phi(x)dx.$$

(This integral cannot be evaluated by calculus.)

b) Show that multiplying the integrand by x/z gives a new integral whose value is strictly larger.

c) Evaluate the new integral.

2.3 Normal Approximation: Derivation (Optional)

This section is more mathematical than the previous and following ones and can be skipped at first reading. Its main aim is to derive the formula for the normal curve by study of binomial probabilities for large n. The basic idea is that for any p with $0 < p < 1$, as n increases the binomial (n, p) distribution becomes better and better approximated by a normal distribution with parameters $\mu = np$ and $\sigma = \sqrt{npq}$. Why this happens is the subject of this section.

Recall first the calculus definition of e, the base of natural logarithms, as the unique number such that the function $y = \log_e x$ has derivative

$$\frac{d}{dx} \log_e x = \frac{1}{x}$$

Here $y = \log_e x$ means $x = e^y$. In the following, all logarithms are to the base e: log means $\log_e$. See Appendix 4 for further background on exponentials and logarithms. Since $\log(1) = 0$ and the derivative of $\log x$ at $x = 1$ is $1/1 = 1$,

$$\log(1 + \delta) \approx \delta \quad \text{for small } \delta$$

with an error of approximation which becomes negligible in comparison to δ as $\delta \to 0$. This simple approximation makes e the preferred or *natural* base of logarithms, and makes e turn up in almost any limit of a product of an increasing number of factors. The emergence of the normal curve from the binomial probability formula is a case in point.

Let $H(k) = P(k)/P(m)$ be the height at k of a binomial histogram scaled to have maximum height 1 at $k = m$, where $m = \text{int}(np+p)$ is the mode. Note that $H(m) = 1$. The normal approximation will now be derived by a sequence of steps, starting with an approximation for $H(k)$. Consider for illustration the distribution of the number of heads in 100 fair coin tosses:

FIGURE 1. Binomial $(100, 0.5)$ histogram. Bar graph of $H(k) = P(k)/P(m)$.

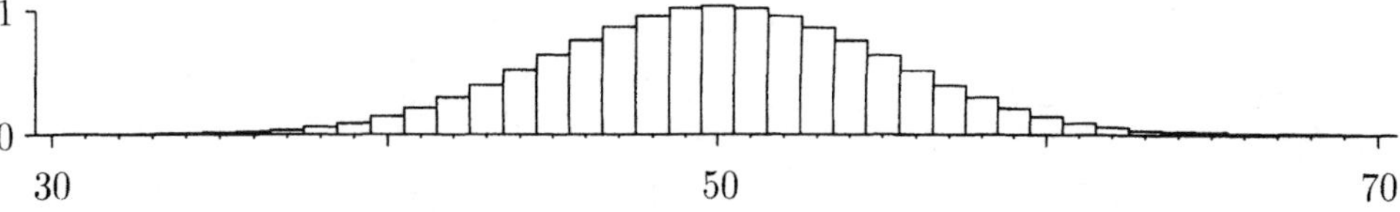

The histogram heights $H(k)$ can be found by multiplying the consecutive odds ratios

$$R(k) = H(k)/H(k-1) = P(k)/P(k-1) = \frac{n-k+1}{k}\frac{p}{q}$$

FIGURE 2. Binomial $(100, 0.5)$ consecutive odds, histogram, and their logarithms. These graphs are drawn to scale. You can see how $\log R(k)$ is nearly linear with a gentle *slope* of about $-1/25$. Because $\log H(k)$ is a sum of increments of this nearly linear function (see equal shaded areas for $k = 59$), its graph is nearly parabolic. By approximation of the area in the top graph with a right-angled triangle with sides $(k - 50)$ and slope $\times$ $(k - 50)$, the area is $\log H(k) \approx \frac{1}{2}$ slope $\times$ $(k - 50)^2 \approx -\frac{1}{2}(k - \mu)^2/\sigma^2$ for $\mu = 50, \sigma = 5 = \sqrt{25}$. This is formula (1).

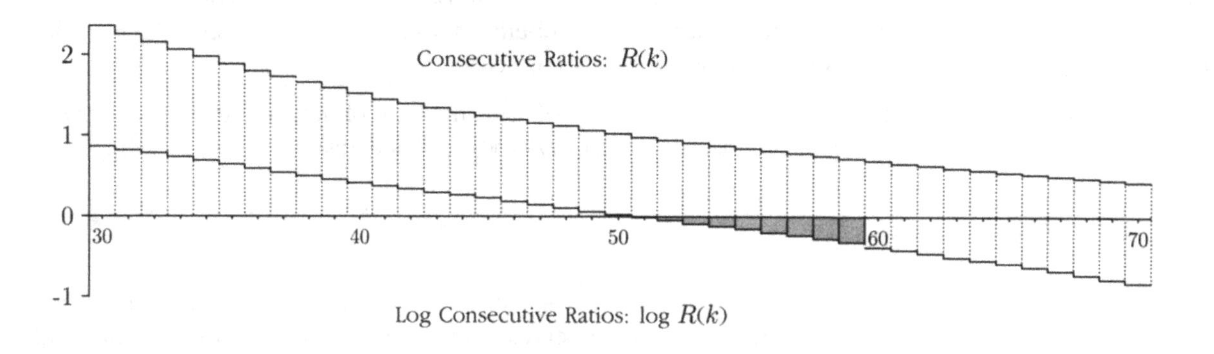

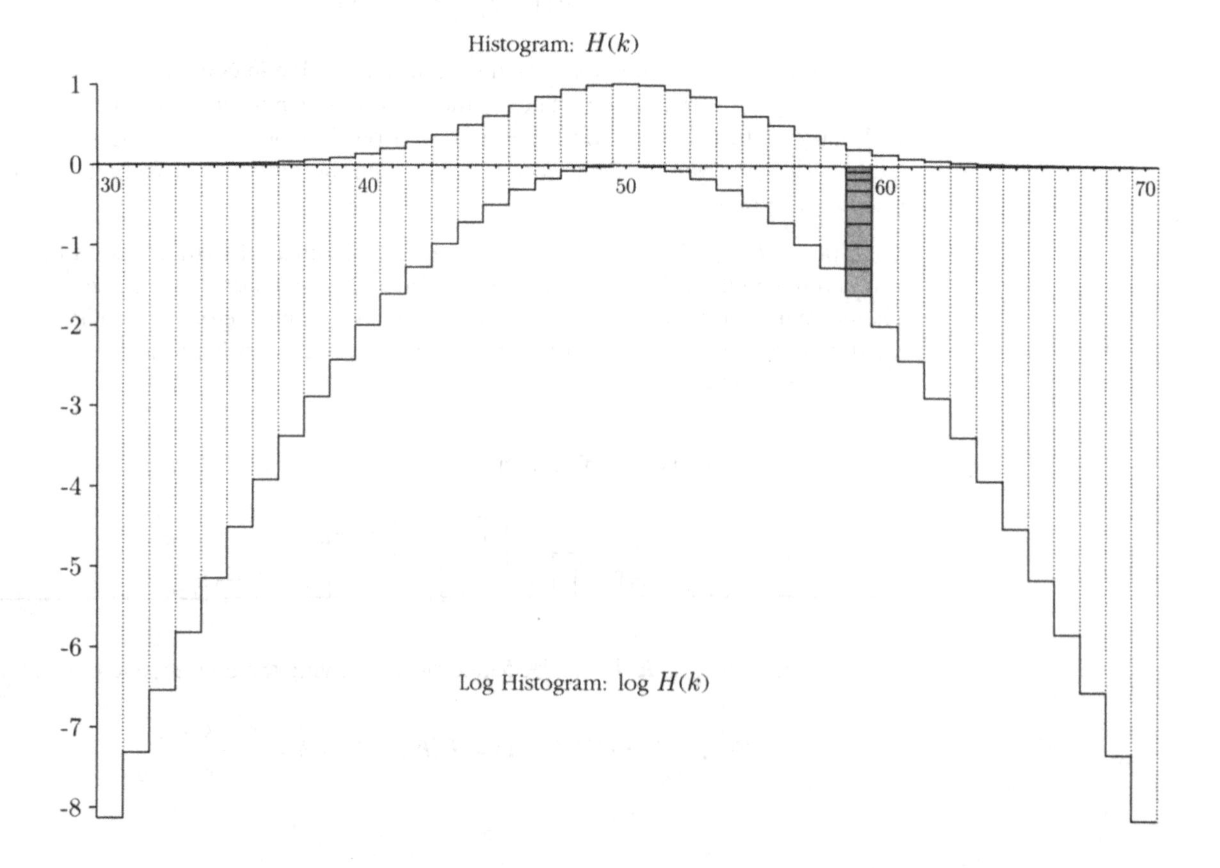

For $k > m$, $H(k)$ is the product of $(m - k)$ consecutive ratios

$$H(k) = H(m)\frac{P(m+1)}{P(m)}\frac{P(m+2)}{P(m+1)}\cdots\frac{P(k)}{P(k-1)} = R(m+1)R(m+2)\cdots R(k)$$

and there is a similar expression for $k < m$. The key to the normal approximation is that as the ratios $R(k)$ decrease for values of k near m, crossing near m from more than 1 to less than 1, they do so *very slowly*, and due to the formula for $R(k)$, *almost linearly*.

This is shown in a particular case in Figure 2, and is true no matter what the value of p, provided n is large enough. As n gets larger, the consecutive odds ratios $R(k)$ decrease more and more slowly near $k = m$. Consequently, as n increases, $R(k)$ stays close to 1 over a wider and wider range of numbers k. This means that for large n, for a wide range of k near $m \approx np$, $H(k)$ is the product of factors that are all very close to 1. The way to handle this product is to take logs to the base e:

$$\log H(k) = \log R(m+1) + \cdots + \log R(k) \qquad \text{as graphed in Figure 2.}$$

Now write $k = m + x \approx np + x$, $k + 1 \approx k$, assume x is small in comparison to npq, and use $\log(1+\delta) \approx \delta$ for small δ to justify the following approximation:

$$\begin{aligned}\log R(k) = \log\left(\frac{n-k+1}{k}\cdot\frac{p}{q}\right) &\approx \log\left(\frac{(n-np-x)p}{(np+x)q}\right)\\ &= \log\left(1-\frac{px}{npq}\right) - \log\left(1+\frac{qx}{npq}\right)\\ &\approx -\frac{px}{npq} - \frac{qx}{npq} = \frac{-x}{npq} = -\frac{(k-m)}{npq}\end{aligned}$$

This shows that if $x = k - m$ is kept small in comparison to n, then $\log R(k)$ is an approximately linear function of k, as in Figure 2, with slope approximately $-1/npq$. Adding up these approximations, using $1 + 2 + \cdots + x = \frac{1}{2}x(x+1) \approx \frac{1}{2}x^2$, gives

$$\log H(k) \approx -\frac{1}{npq} - \frac{2}{npq} - \cdots - \frac{(k-m)}{npq} \approx -\frac{1}{2}\frac{(k-m)^2}{npq} \approx -\frac{1}{2}\frac{(k-np)^2}{npq}$$

This is illustrated by the roughly triangular area shaded in Figure 2. A similar argument works for $k < m$. So for the heights $H(k) = P(k)/P(m)$ of the binomial (n, p) histogram there is a preliminary form of the normal approximation:

$$H(k) \approx e^{-\frac{1}{2}(k-\mu)^2/\sigma^2} \tag{1}$$

where $\mu = np$ is the *mean* and $\sigma = \sqrt{npq}$ is the *standard deviation*.

The argument shows this approximation will be good provided $|k - m|$ is small in comparison with npq. A more careful argument shows that this range of k is really all that matters. Now approximate $P(k)$ instead of $H(k)$:

$$P(k) = H(k)P(m) = H(k)/H(0 \text{ to } n) \quad \text{where } H(0 \text{ to } n) = H(0) + \cdots + H(n) \qquad (2)$$

Here $H(0 \text{ to } n)$, the total area under the binomial (n, p) histogram with maximum height 1, can be approximated by the total area under the approximating normal curve (1), which is an integral:

$$\begin{aligned} H(0 \text{ to } n) &\sim \int_{-\infty}^{\infty} e^{-\frac{1}{2}(x-\mu)^2/\sigma^2} dx \\ &= \sigma \left[\int_{-\infty}^{\infty} e^{-\frac{1}{2}z^2} dz \right] \quad \text{by the calculus change of variable } (x-\mu)/\sigma = z, \quad dx = \sigma dz \\ &= \sigma\sqrt{2\pi} \quad \text{as shown by calculus in Section 5.3} \end{aligned}$$

It can be shown that the relative error of approximation can be made arbitrarily small, no matter what the values of n and p, provided that $\sigma = \sqrt{npq}$ is sufficiently large. Now combine this with (1) and (2):

$$P(k) \approx \frac{1}{\sqrt{2\pi}\sigma} e^{-\frac{1}{2}(k-\mu)^2/\sigma^2} \quad \text{where} \quad \mu = np, \quad \sigma = \sqrt{npq} \qquad (3)$$

The precise meaning of the $\approx$ involved here is somewhat technical. As $\sigma \to \infty$, both sides tend to zero. But the *relative* error of approximation tends to 0 provided $(k - \mu)/\sigma$ remains bounded. See Feller's book *An Introduction to Probability Theory and its Applications*, Vol. I, for more details.

The equation of the normal curve appears in formula (3) as a function of k. The probability of an interval of numbers is now approximated by replacing relative areas under the histogram by relative areas under the approximating curve.

What makes the normal curve a better and better approximation as $n \to \infty$, is that for large n, as k moves away from m, the histogram heights $H(k)$ approach zero before the consecutive ratios $R(k)$ differ significantly from 1. In the expression

$$\log H(k) = \log R(m + 1) + \cdots + \log R(k)$$

a large number of terms on the right, each nearly zero, add up to a total $\log H(k)$ which is significantly different from 0.

Probability of the Most Likely Number of Successes

A consequence of the normal approximation (3) for $k = m$, closely related to the square root law discussed in the previous section, is that the most likely value $m = \text{int}(np + p)$ in the binomial (n, p) distribution has probability

$$P(m) \sim \frac{1}{\sqrt{2\pi}\sigma} = \frac{1}{\sqrt{2\pi npq}} \quad \text{as} \quad n \to \infty \tag{5}$$

For fixed p, as $n \to \infty$, the relative error in this approximation tends to 0. In particular, no matter what the success probability p, the probability of the most likely number of successes in n independent trials tends to zero as $n \to \infty$, like a constant divided by $\sqrt{n}$. For fixed n, the approximation is always best for p near $\frac{1}{2}$, and worst for p close to 0 or 1 when the binomial distribution is skewed and the normal approximation not so accurate. In particular, if $p = \frac{1}{2}$, so $m = \frac{n}{2}$ if n is even, $\frac{n}{2} \pm \frac{1}{2}$ if n is odd,

$$P(m \text{ heads in } n \text{ fair coin tosses}) = \binom{n}{m} 2^{-n} \sim \sqrt{\frac{2}{n\pi}} \quad \text{as} \quad n \to \infty \tag{6}$$

As you can check on a pocket calculator, the asymptotic formula gives excellent results even for quite small values of n, and the relative error of the approximation decreases as n increases. According to the asymptotic formula, this relative error tends to 0 as $n \to \infty$. As $n \to \infty, 1/\sqrt{n} \to 0$, so the chance of getting exactly as many heads as tails tends to zero as the number of tosses tends to ∞.

To understand why this is so, recall the basis of the normal approximation. For large n the binomial (n, p) probabilities are distributed almost uniformly if you look close to the center of the distribution. The consecutive odds ratios are very close to one over an interval containing nearly all the probability. Still, these ratios conspire over larger distances to produce the gradual decreasing trend of the histogram away from its maximum, following the normal curve. By a distance of $4\sigma = 2\sqrt{n}$ or so from the center the histogram has almost vanished. And nearly all the probability must lie in this interval. Because a total probability of nearly 1 is distributed smoothly over an interval of length about $4\sqrt{n}$, the probabilities of even the most likely numbers in the middle cannot be much greater than $1/\sqrt{n}$. Thus even the most likely value m has a probability $P(m)$ which tends to zero as $n \to \infty$ like a constant over $\sqrt{n}$. See the exercises for another derivation of this, and a different evaluation of the constant, which leads to a remarkable infinite product formula for π.

Exercises 2.3

1. Suppose you knew the consecutive odds ratios $R(k) = P(k)/P(k-1)$ of a distribution $P(0), \ldots, P(n)$. Find a formula for $P(k)$ in terms of $R(1), \ldots, R(n)$. Thus the consecutive odds ratios determine a distribution.

2. A fair coin is tossed 10,000 times. The probability of getting exactly 5000 heads is closest to:

0.001, 0.01, 0.1, 0.2, 0.5, 0.7, 0.9, 0.99, 0.999.

Pick the correct number and justify your choice.

3. Equalizations in coin tossing. Let $P(k \text{ in } n)$ be the probability of exactly k heads in n independent fair coin tosses. Let $n = 2m$ be even, and consider $P(m \text{ in } 2m)$, the chance of getting m heads and m tails in $2m$ tosses. Derive the following formulae:

a) $P(m-1 \text{ in } 2m) = P(m+1 \text{ in } 2m) = P(m \text{ in } 2m)\left(1 - \dfrac{1}{m+1}\right)$

b) $P(m+1 \text{ in } 2m+2) = \frac{1}{4}P(m-1 \text{ in } 2m) + \frac{1}{2}P(m \text{ in } 2m) + \frac{1}{4}P(m+1 \text{ in } 2m)$

c) By a) and b)

$$\frac{P(m+1 \text{ in } 2m+2)}{P(m \text{ in } 2m)} = 1 - \frac{1}{2(m+1)}$$

Check this also by cancelling factorials in the binomial formula.

d) By repeated application of c),

$$P(m \text{ in } 2m) = \left(1 - \frac{1}{2\times 1}\right)\left(1 - \frac{1}{2\times 2}\right)\cdots\left(1 - \frac{1}{2\times m}\right)$$

e) $0 < P(m \text{ in } 2m) < e^{-\frac{1}{2}\left(\frac{1}{1}+\frac{1}{2}+\cdots+\frac{1}{m}\right)} < \dfrac{1}{\sqrt{m}}$

f) $P(m \text{ in } 2m) \to 0$ as $m \to \infty$. The bound of $1/\sqrt{m}$ is of the right order of magnitude, as shown by both the following calculations and the normal approximation. Let $\alpha_m = P(m \text{ in } 2m)$. Then verify the following:

$$\frac{(m+1/2)\alpha_m^2}{(m-1+1/2)\alpha_{m-1}^2} = 1 - \frac{1}{4m^2}$$

g)

$$2(m+1/2)\alpha_m^2 = \left(1 - \frac{1}{2^2}\right)\left(1 - \frac{1}{4^2}\right)\cdots\left(1 - \frac{1}{(2m)^2}\right)$$

$$= \frac{1}{2}\cdot\frac{3}{2}\cdot\frac{3}{4}\cdot\frac{5}{4}\cdot\frac{5}{6}\cdot\frac{7}{6}\cdots\frac{(2m-1)}{2m}\cdot\frac{(2m+1)}{2m}$$

h) $\alpha_m \sim K/\sqrt{m}$ as $m \to \infty$, where

$$2K^2 = 2\lim_{m\to\infty}\left(m + \frac{1}{2}\right)\alpha_m^2 = \frac{1}{2}\cdot\frac{3}{2}\cdot\frac{3}{4}\cdot\frac{5}{4}\cdot\frac{5}{6}\cdot\frac{7}{6}\cdots$$

Deduce by comparison with the normal approximation that the value of the infinite product is $2/\pi$.

2.4 Poisson Approximation

Even if n is very large, if p is close enough to 0 or 1 the standard deviation $\sigma = \sqrt{npq}$ is small. The binomial (n, p) distribution then does not follow the normal curve at all closely. By switching consideration from successes to failures, if necessary, we need only consider the case when p is nearly 0 and q is nearly 1. Then the standard deviation $\sigma = \sqrt{npq}$ is

$$\sigma = \sqrt{\mu q} \approx \sqrt{\mu} \qquad \text{where } \mu = np \text{ is the mean.}$$

If, for example, $\mu = 1$, so we are considering n trials with probability $p = 1/n$ of success on each trial, then $\sigma \approx 1$. The normal approximation will be very bad no matter how large n is. This is because the normal curve is symmetric, while the binomial distribution is not even approximately symmetric, due to the impossibility of negative values.

Example 1. The binomial (10, 1/10) distribution.

This is the distribution of the number of black balls obtained in 10 random draws with replacement from a box containing 1 black ball and 9 white ones.

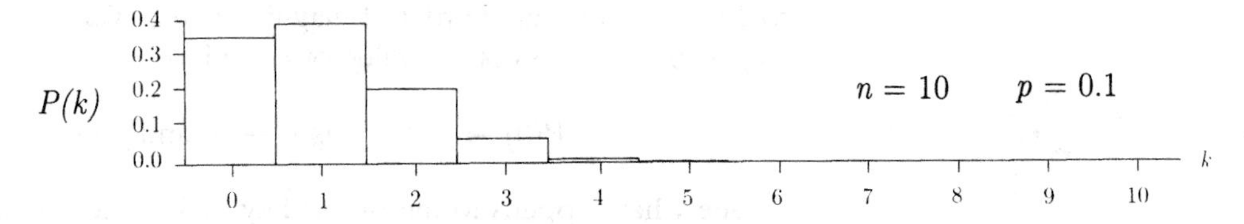

Example 2. The binomial (100, 1/100) distribution.

This is the distribution of the number of black balls obtained in 100 random draws with replacement from a box containing 1 black ball and 99 white ones.

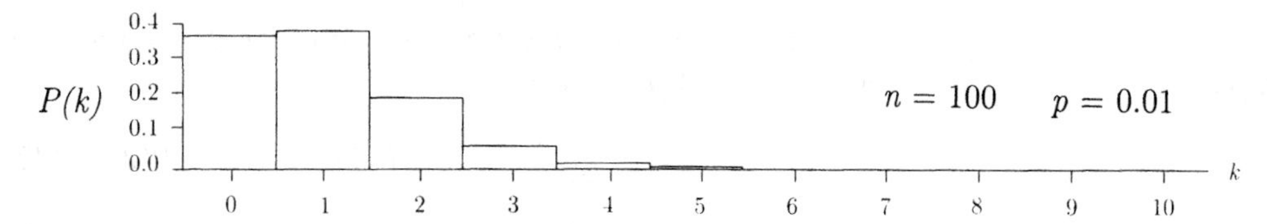

Example 3. The binomial (1000, 1/1000) distribution.

Now take 1000 random draws with replacement from a box with 1 black ball and 999 white ones. This is the distribution of the number of black balls drawn:

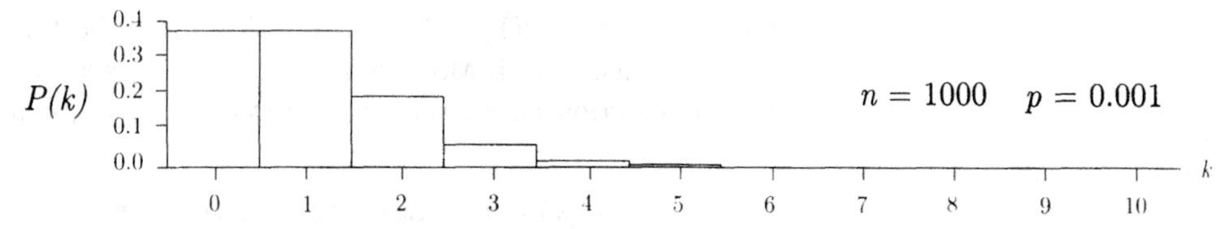

As these examples show, binomial distributions with parameters n and $1/n$ are always concentrated on a small number of values near the mean value $\mu = 1$, with a shape which approaches a limit as $n \to \infty$ and $p = 1/n \to 0$. This limit corresponds to sampling more and more times with replacement from a box containing a smaller and smaller proportion of black balls. If μ, the expected number of black balls in the sample, is kept constant, the binomial (n, p) distribution with mean $\mu = np$ approaches a limit as $n \to \infty$ and $p \to 0$. This limit distribution, called the *Poisson distribution* with *parameter* μ, provides useful approximations to binomial probabilities in case n is large and p is so small that the normal approximation is bad.

The limit involved here is essentially the same as for the gambler's rule of Section 1.6. As in that example, the chance of getting zero successes in n trials with probability p of success on each trial is

$$P(0) = (1-p)^n \approx (e^{-p})^n = e^{-np} = e^{-\mu}$$

by the exponential approximation

$$1 - p \approx e^{-p} \qquad \text{if} \quad p \approx 0$$

It can be shown that no matter what the value of n, the error in this approximation to $P(0)$ is of the same order of magnitude as p. Consequently, this error tends to 0 as $p \to 0$, regardless of the value of n, and

$$P(0) \to e^{-\mu} \qquad \text{as } n \to \infty \text{ and } p \to 0 \text{ with } np \to \mu$$

To see what happens to the probability of k successes under the same conditions, look at the consecutive odds ratio:

$$R(k) = \frac{P(k)}{P(k-1)} = \frac{n-k+1}{k}\,\frac{p}{1-p} = \frac{np}{k}\,\frac{(1-(k-1)/n)}{1-p} \approx \frac{\mu}{k}$$

if n is large and p is small. In particular, if $\mu = 1$ as in the examples above, the first two odds ratios are

$$R(1) \approx 1/1 \qquad R(2) \approx 1/2$$

as apparent in the histograms. In the limit as $n \to \infty$ the binomial $(n, 1/n)$ distribution approaches a distribution with

$$P(0) = e^{-1}$$

and odds ratios $R(1) = 1$, $R(2) = 1/2$. This is the Poisson (μ) distribution defined below in case $\mu = 1$. More generally, for any fixed value of $\mu = np$, as $n \to \infty$ and $p \to 0$, the consecutive odds ratio $R(k)$ tends to μ/k, and

$$P(k) = P(0)R(1)R(2)\cdots R(k) \to e^{-\mu}\frac{\mu}{1}\cdot\frac{\mu}{2}\cdots\frac{\mu}{k} = e^{-\mu}\frac{\mu^k}{k!}$$

To summarize, we have the following:

Poisson Approximation to the Binomial Distribution

If n is large and p is small, the distribution of the number of successes in n independent trials is largely determined by the value of the mean $\mu = np$, according to the *Poisson approximation*

$$P(k \text{ successes}) \approx e^{-\mu}\frac{\mu^k}{k!}$$

Remark. It can be shown that the accuracy of the approximation depends largely on the value of p, and hardly at all on the value of n. Roughly speaking, absolute errors in using this approximation will be of the same order of magnitude as p.

Example 4. Defectives in a sample.

Problem 1. Suppose that over the long run a manufacturing process produces 1% defective items. What is the chance of getting two or more defective items in a sample of 200 items produced by the process?

Solution. Assume each item is defective with probability p, independently of other items. The long-run percentage of defectives would then be $100p\%$, so we can estimate $p = 1/100$. The number of defectives in a sample size of 200 then has binomial $(200, 1/100)$ distribution, with mean $\mu = 200 \times 1/100 = 2$. Using the Poisson approximation

$$\begin{aligned} P(2 \text{ or more defectives}) &= 1 - P(0) - P(1) \\ &\approx 1 - e^{-2}\frac{2^0}{0!} - e^{-2}\frac{2^1}{1!} \\ &= 1 - 3e^{-2} = 0.594 \end{aligned}$$

A check on the Poisson approximation. As a check on the approximation

$$P(k \text{ successes in } n \text{ trials}) \approx e^{-\mu}\mu^k/k! \qquad \text{where } \mu = np,$$

sum both sides from $k = 0$ to n to obtain

$$1 \approx e^{-\mu}\sum_{k=0}^{n}\frac{\mu^k}{k!}$$

FIGURE 1. Poisson distributions. Notice how when μ is small the distribution is piled up on values near zero. As μ increases, the distribution shifts to the right and spreads out, gradually approaching the normal distribution in shape as $\mu \to \infty$. This can be shown by a variation of the argument in Section 2.3.

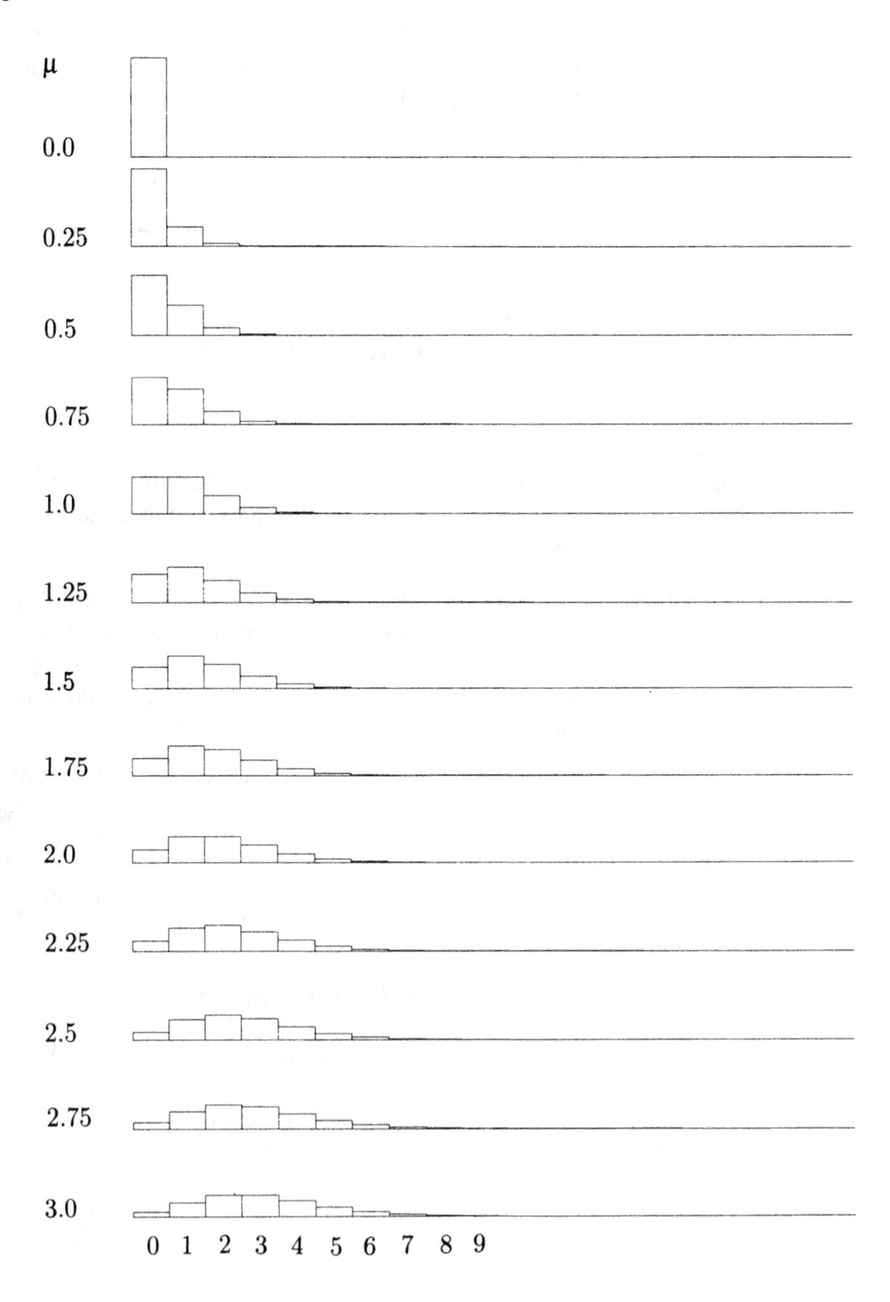

For fixed μ, as $n \to \infty$ and $p = \mu/n \to 0$, this approximation becomes better and better. The limit of the sum is a well-known formula for e^{μ}:

$$\lim_{n\to\infty} \sum_{k=0}^{n} \frac{\mu^k}{k!} = \sum_{k=0}^{\infty} \frac{\mu^k}{k!} = e^{\mu}$$

and

$$e^{-\mu}e^{\mu} = e^{-\mu+\mu} = e^0 = 1$$

See Appendix 4 for further details.

This calculation does show that the limiting probabilities $P_\mu(k) = e^{-\mu}\mu^k/k!$ form a probability distribution on $\{0, 1, 2, \ldots\}$, meaning that

$$P_\mu(k) \geq 0 \text{ and } \sum_{k=0}^{\infty} P_\mu(k) = 1.$$

This kind of distribution over an infinite set of possible values is discussed more generally in Section 3.4. More about the Poisson distribution can be found in Section 3.5.

The Poisson (μ) Distribution

The *Poisson distribution with parameter* μ or *Poisson* (μ) *distribution* is the distribution of probabilities $P_\mu(k)$ over $\{0, 1, 2, \ldots\}$ defined by

$$P_\mu(k) = e^{-\mu}\mu^k/k! \qquad (k = 0, 1, 2, \ldots)$$

Exercises 2.4

1. Sketch the histograms of binomial distributions with the following parameters (n, p):

a) $(10^6, 10^{-6})$; b) $(10^6, 2 \times 10^{-6})$; c) $(3284, 10^{-4})$; d) $(1000, 0.998)$.

2. Find Poisson approximations to the probabilities of the following events in 500 independent trials with probability 0.02 of success on each trial:

a) 1 success; b) 2 or fewer successes; c) more than 3 successes.

3. The chance of getting 25 or more sixes in 100 rolls of a die is 0.022. If you rolled 100 dice once every day for a year, find the chance that you would see 25 or more sixes:

a) at least once; b) at least twice.

4. Repeat the previous problem for the event of getting 30 or more sixes in 100 die rolls, which has probability 0.00068.

5. Suppose that each week you buy a ticket in a lottery which gives you a chance of $1/100$ of a win. You do this each week for a year. What is the chance that you get k wins during the year, approximately? Calculate as a decimal for $k = 0, 1, 2$.

6. A box contains 1000 balls, of which 2 are black and the rest are white.

a) Which of the following is most likely to occur in 1000 draws with replacement from the box?

fewer than 2 black balls, exactly 2 black balls, more than 2 black balls

b) If two series of 1000 draws are made at random from this box, what, approximately, is the chance that they produce the same number of black balls?

7. Let X be the number of successes in 25 independent trials with probability $1/10$ of success on each trial. Let m be the most likely value of S.

a) Find m.

b) Find $P(S = m)$ correct to 3 decimal places.

c) What is the value of the normal approximation to $P(S = m)$?

d) What is the value of the Poisson approximation to $P(S = m)$?

e) Repeat a) for $n = 2500$ trials instead of 25. Which would now give the better approximation to $P(S = m)$, the normal or the Poisson approximation? Find $P(S = m)$ approximately using the best approximation.

f) Repeat e) for 2500 trials and $p = 1/1000$ instead of $p = 1/10$.

8. Mode of the Poisson distribution. Use consecutive odds ratios to find the largest k that maximizes the Poisson (μ) probability $P_\mu(k)$. For what values of μ is there a double maximum? What are the two values of k in that case? Is there ever a triple maximum?

9. A cereal company advertises a prize in every box of its cereal. In fact, only about 95% of their boxes have prizes in them. If a family buys one box of this cereal every week for a year, estimate the chance that they will collect more than 45 prizes. What assumptions are you making?

10. Let N be a fixed large integer. Consider n independent trials, each of which is a success with probability $1/N$. Recall that the gambler's rule (see Example 1.6.3) says that if $n \approx \frac{2}{3}N$, the chance of at least one success in n trials is about $1/2$. Show that if $n \approx \frac{5}{3}N$, then the chance of at least two successes is about $1/2$.

2.5 Random Sampling

Random sampling is a statistical technique for gaining information about the composition of a large population from the composition of a random sample from the population. Suppose that each element of the population can be classified into one of two categories, say "good" and "bad". Of course, the designation of which elements are good is quite arbitrary, and will depend on the problem at hand. In practical problems the fraction of good elements in the population will be unknown. The problem is to estimate this fraction based on the composition of the sample, and to know how accurate this estimate is likely to be. The natural estimate of the fraction of good elements in the population is the fraction of good elements in the sample. That is to say, population percentages are estimated by sample percentages. The accuracy of this estimate depends on exactly what procedure was used to obtain the sample. The ideal is to obtain a sample that is as representative as possible of the whole population. This ideal is approached by picking the sample at random. Provided the sample size is large enough, the proportion in the sample will most likely be close to the proportion in the population.

Sampling with Replacement

Suppose n individuals are drawn one by one at random from a population of size N, with replacement between draws. On each draw it is assumed that each of the N individuals has the same chance of being chosen, and the successive draws are assumed independent. So all N^n possible sequences of choices are equally likely. This might be done, for example, by drawing tickets from a box, with replacement of the tickets and mixing between draws. There is no restriction on the sample size n. In principle, the procedure can be repeated indefinitely.

Consider now the distribution of the number of good elements in a sample of size n with replacement from a population of G good and B bad elements, with $G + B = N$. This is the distribution of the number of successes in n independent trials, with probability $p = G/N$ of success in each trial, that is to say the binomial (n, p) distribution for $p = G/N$. Provided the sample size n is large enough, this binomial distribution with parameters n and $p = G/N$ is well approximated by the normal curve with parameters $\mu = np$ and $\sigma = \sqrt{npq}$. According to the law of large numbers, if n is sufficiently large, the proportion of good elements in the sample is likely to be close to the proportion $p = G/N$ of good elements in the population. By the normal approximation, if n is sufficiently large, the number of good elements in the sample will lie in the range $np \pm 2\sqrt{npq}$ with probability about 95%. So if n is sufficiently large, the proportion of good elements in the sample will lie in the range $p \pm 2\sqrt{pq/n}$ with probability about 95%. Since $\sqrt{pq} \leq 1/2$, this means that

$$P(p - 1/\sqrt{n} \leq \text{sample proportion} \leq p + 1/\sqrt{n}) \geq 95\%$$

If the proportion of good elements in a population is not known, the result above can be used to estimate the unknown proportion by the method of confidence intervals. If the sample size is large, then with probability greater than 95% the sample proportion of good elements will lie within $1/\sqrt{n}$ of the population proportion. So if the observed proportion of good elements in a large sample is $\hat{p}$, guess that the population proportion lies in the range $\hat{p} \pm 1/\sqrt{n}$. The interval $\hat{p} \pm 1/\sqrt{n}$ is an *approximate* 95% *confidence interval* for the unknown population proportion.

Sampling Without Replacement

In this procedure, elements in a population of size N are drawn one by one at random as before, but without replacement between draws. The sample size n is now restricted to $n \leq N$. At each stage it is assumed that no matter what elements have been drawn so far, all remaining elements are equally likely on the next draw. Equivalently, all possible orderings of n of the N elements are assumed equally likely.

The number of different possible orderings of n out of N elements is denoted by $(N)_n$, a symbol which can be read "*N order n*". As explained in Appendix 1, the product rule for counting gives the formula

$$(N)_n = N(N-1)\cdots(N-n+1)$$

where there are n factors in the product. Compare with N to the power n:

$$N^n = N \cdot N \cdots N \qquad (n \text{ factors})$$

which is the larger number of possible samples with replacement, and N choose n:

$$\binom{N}{n} = (N)_n/n!$$

which is the smaller number of different unordered samples or subsets of size n. This is just the formula for $\binom{n}{k}$ of Section 2.1 with N instead of n and n instead of k. When rewritten in the form

$$(N)_n = \binom{N}{n} n!$$

this formula can be understood as follows: Each of the $\binom{N}{n}$ possible unordered samples of size n can be ordered in $n!$ different ways to obtain $n!$ different ordered samples of size n. Thus $(N)_n$, the number of ordered samples of size n, is $\binom{N}{n}$ times $n!$ by the product rule of counting.

Consider now the distribution of the number of good elements in a sample of size n without replacement from a population of G good and B bad elements with $G+B=N$. The problem is to find the chance of getting g good and b bad elements in the sample, for $0 \leq g \leq n$ and $g+b=n$. Thinking in terms of an ordered random sample, one way to get g good and b bad in the sample is if the first g elements in the sample are good and the last b are bad. Either by the product rule for conditional probabilities, or by the product rule for counting, the chance of this event is

$$\frac{G}{N} \cdot \frac{G-1}{N-1} \cdots \frac{G-g+1}{N-g+1} \cdot \frac{B}{N-g} \cdot \frac{B-1}{N-g-1} \cdots \frac{B-b+1}{N-g-b+1} = \frac{(G)_g (B)_b}{(N)_n}$$

This is the chance of just one of $\binom{n}{g}$ different possible patterns of g good and b bad elements in an ordered sample of size n. But the chance of any other pattern of g good and b bad, for example, the first b elements bad and the next g elements good, is just the same, because the same factors then appear in a different order. Thus, multiplying the above expression by $\binom{n}{g}$ gives the chance of g good and b bad elements appearing in an unspecified pattern, as in the second formula of the following box:

Sampling With and Without Replacement

Suppose a population of size N contains G good and B bad elements, with $N = G + B$. For a sample of size $n = g + b$, where $0 \le g \le n$, the probability of getting g good elements and b bad elements is

- **for sampling with replacement**

$$P(g \text{ good and } b \text{ bad}) = \binom{n}{g} \frac{G^g B^b}{N^n}$$

- **for sampling without replacement**

$$P(g \text{ good and } b \text{ bad}) = \binom{n}{g} \frac{(G)_g (B)_b}{(N)_n} = \frac{\binom{G}{g}\binom{B}{b}}{\binom{N}{n}}$$

The formula for sampling with replacement is just the usual binomial formula written in a way that parallels with the first formula for sampling without replacement. The second formula for sampling without replacement follows from the first by cancellation after using the formula $\binom{M}{m} = (M)_m/m!$ three times. This expression can also be derived another way, by working in the outcome space of all $\binom{N}{n}$ possible unordered samples. Since there are $n!$ ordered samples corresponding to each unordered sample, each possible unordered sample has the same chance

$$\frac{n!}{(N)_n} = 1 \Big/ \binom{N}{n}$$

And $\binom{G}{g}\binom{B}{b}$ is the number of possible unordered samples with g good and b bad elements, by yet another application of the product rule of counting. The good

FIGURE 1. **Some hypergeometric distributions.** The histograms display the distribution of the number of good elements in a sample of size n without replacement from a population of $N = 10$ elements, containing G good elements and $B = 10 - G$ bad ones, for $n = 2, 4, 6, 8$ (different columns) and $G = 2, 4, 6, 8$ (different rows). Each horizontal scale is marked by ticks at $0, 1, \ldots, 10$

$n = 4$ $n = 6$ $n = 8$

$G = 2$
$B = 8$

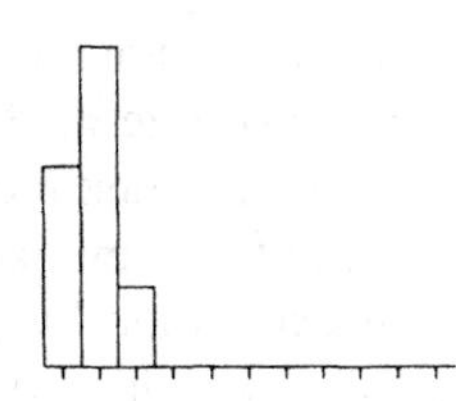

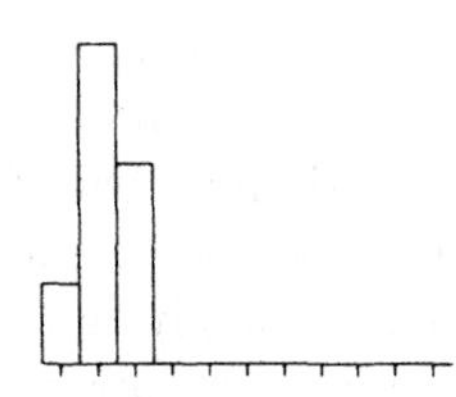

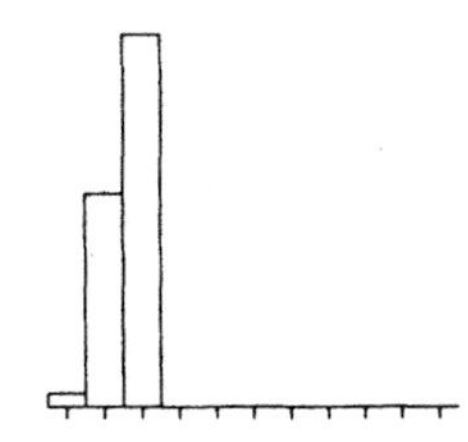

$G = 4$
$B = 6$

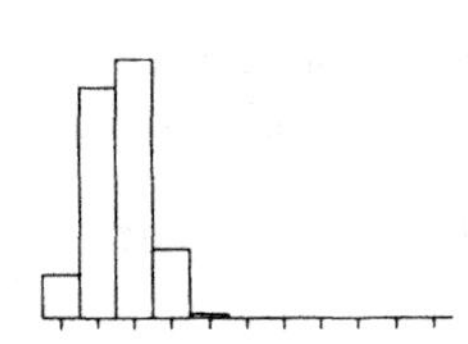

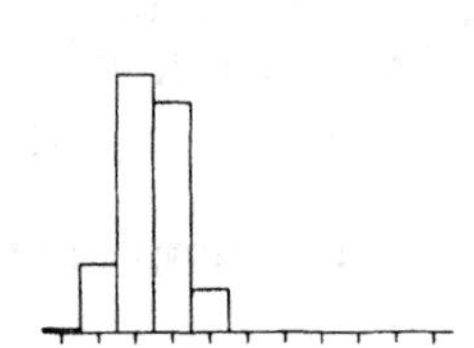

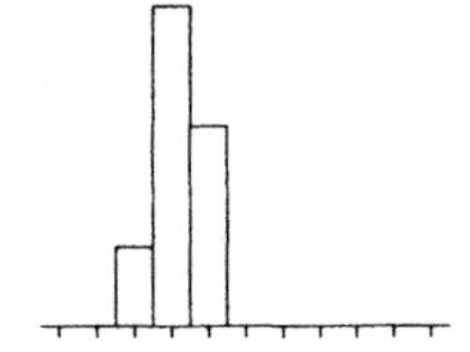

$G = 6$
$B = 4$

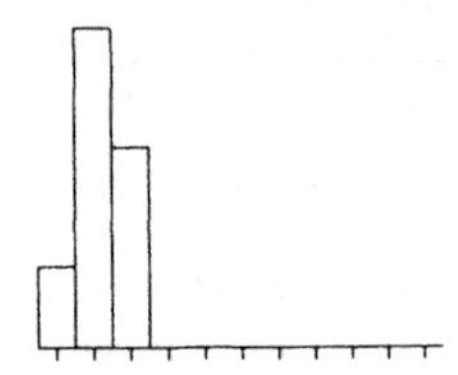

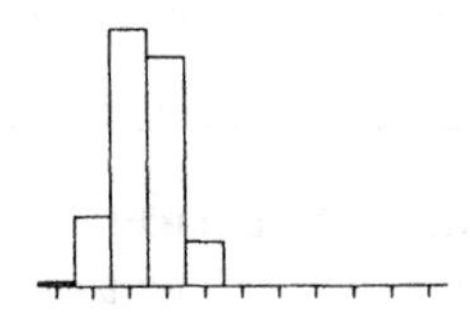

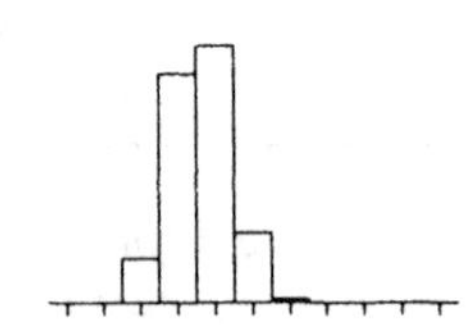

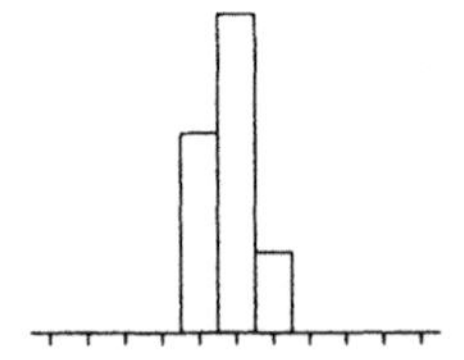

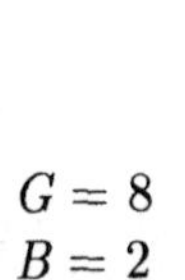

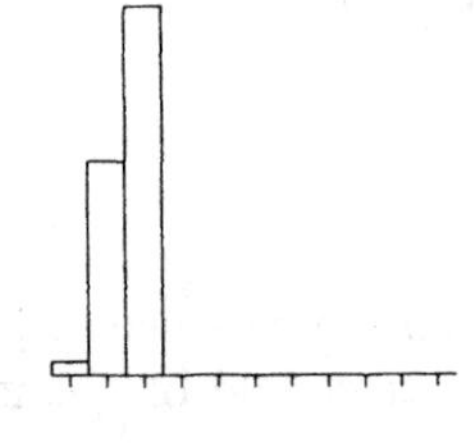

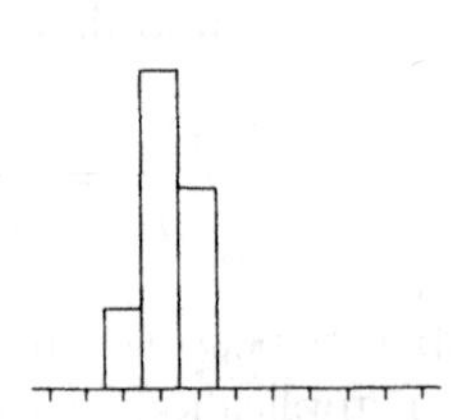

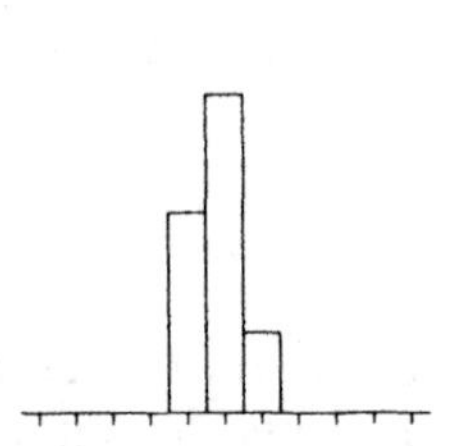

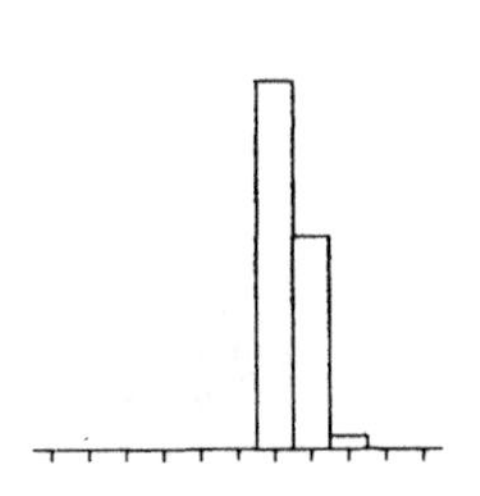

elements can be chosen in $\binom{G}{g}$ ways, and no matter how these are chosen, the bad ones may be chosen in $\binom{B}{b}$ ways. This method of counting unordered samples is what is used to calculate the probabilities of various poker hands. See Exercise 12.

The hypergeometric distribution. This is the name of the distribution of the number of good elements in a sample of size n without replacement from a population of G good and $N-G$ bad elements. The distribution has three parameters, n, N and G. The probability that this distribution assigns to $g \in \{0, 1, \ldots, n\}$ is the probability $P(g \text{ good and } b \text{ bad})$ for sampling without replacement, as in the box, for $b = n - g$ and $B = N - G$. Note that this probability may be zero for some g between 0 and n. (See Exercise 11). The fact that these probabilities add up to 1, and so define a distribution on $\{0, 1, \ldots, n\}$, is not obvious from the formula, but it follows at once from the rules of probability: as g varies from 0 to n the events of getting g good elements and b bad elements in sampling without replacement form a partition of the whole outcome space.

Binomial approximation to the hypergeometric distribution. If N, G, and B are large in comparison to n, g, and b, the formulae for sampling with and without replacement give nearly identical probabilities. More precisely, for fixed n, b, and g, and $N \to \infty$, $G \to \infty$, and $B \to \infty$, the ratio of the two probabilities tends to 1. This follows from the fact that for any fixed n,

$$(N)_n / N^n \to 1 \quad \text{as} \quad N \to \infty$$

In practice, this makes the binomial distribution a useful approximation to the more complicated hypergeometric distribution. The approximation is quite intuitive, because if the sample size is small in comparison to the population size there is very little chance of a duplicate in sampling with replacement. The chance of getting a duplicate in sampling with replacement is just $1 - (N)_n/N^n \approx 0$ if $n \ll \sqrt{N}$ (see the birthday problem of Section 1.6). And given that there are no duplicates, the sample with replacement is just like a sample without replacement, in the sense that all orderings are equally likely.

Normal approximation to the hypergeometric distribution. This is discussed in Section 3.6.

Exercises 2.5

1. Suppose you take a random sample of 10 tickets without replacement from a box containing 20 red tickets and 30 blue tickets.

a) What is the chance of getting exactly 4 red tickets?

b) Repeat a) for sampling with replacement.

2. Three cards are dealt from a standard deck of 52 cards, containing 26 red cards and 26 black cards. Write down the probability that:

a) the first card is red and the second two black;

b) exactly one of the cards dealt is red;

c) at least one of the cards dealt is red.

3. A deck of cards is shuffled and dealt to four players, with each receiving 13 cards. Find:

a) the probability that the first player holds all the aces;

b) the probability that the first player holds all the aces given that she holds the ace of hearts;

c) the probability that the first player holds all the aces given that she holds at least one;

d) the probability that the second player holds all the aces given that he holds all the hearts.

4. A population of $100,000$ people consists of 40% men and 60% women. A random sample of size 100 is drawn from this population without replacement. Write down an expression for the probability that there are at least 45 men in the sample. Approximately what is the value of this probability?

5. Suppose 55% of a large population of voters actually favor candidate A. How large a random sample must be taken for there to be a 99% chance that the majority of voters in the sample will favor candidate A?

6. In a hand of 13 cards drawn randomly from a pack of 52, find the chance of:

a) no court cards (J, Q, K, A);

b) at least one ace but no other court cards;

c) at most one kind of court card.

7. A box contains 50 black balls and 30 red balls. Four balls are drawn at random from the box, one after the other, without replacement. Find the chance that:

a) all four balls are black;

b) exactly three balls are black;

c) the first red ball appears on the last draw.

8. In a raffle with 100 tickets, 10 people buy 10 tickets each. If there are 3 winning tickets drawn at random find the probability that:

a) one person gets all 3 winning tickets;

b) there are 3 different winners;

c) some person gets two winners and someone else gets just one.

9. A lot of 50 items is inspected by the following two-stage plan.

(i) A first sample of 5 items is drawn without replacement. If all are good the lot is passed; if two or more are bad the lot is rejected.

(ii) If the first sample contains just one bad item, a second sample of 10 more items is drawn without replacement (from the remaining 45 items) and the lot is rejected if two or more of these are bad. Otherwise it is accepted.

Suppose there are 10 bad items in the lot.

a) What is the probability that the second sample is drawn and contains more than one bad item?

b) Write down an expression for the probability that the lot is accepted.

10. Suppose a population of N elements consists of G good, B bad, and I indifferent elements, with $B + G + I = N$. If a random sample of size n is drawn with replacement from this population, explain why the chance that the sample contains k_1 good elements, k_2 bad elements, and k_3 indifferent elements, where $k_1 + k_2 + k_3 = n$, is

$$\frac{n!}{k_1!k_2!k_3!}(G/N)^{k_1}(B/N)^{k_2}(I/N)^{k_3}$$

11. Range of the hypergeometric distribution. For $1 \leq n \leq N$ and $0 \leq G \leq N$, describe the set of g with $0 \leq g \leq n$ such that there is strictly positive probability of getting g good elements in a random sample of size n without replacement from a population of G good and $N - G$ bad elements. Explain why the formula for the probability in question gives the correct value, (possibly 0) for all $0 \leq g \leq n$.

12. Poker hands. Assume all $\binom{52}{5}$ hands equally likely. Find the probability of being dealt:

a) a straight flush (5 consecutive cards of the same suit);

b) four of a kind (ranks a, a, a, a, b);

c) a full house (ranks a, a, a, b, b);

d) a flush (5 of the same suit, not a straight flush);

e) a straight (5 consecutive ranks, not a flush);

f) three of a kind (ranks a, a, a, b, c);

g) two pairs (ranks a, a, b, b, c);

h) a pair (ranks a, a, b, c, d);

i) none of the above.

13. A factory which produces chips in lots of ten thousand uses the following scheme to check the quality of its product. From each lot of chips produced, a random sample of size 500 is taken. If the sample contains 10 or less defectives, the lot is passed. If the sample contains more than 10 defectives, another random sample of size 500 is chosen from the lot. If this sample contains 10 or less defectives, the lot is passed. Otherwise, the lot is rejected. If a lot actually contains 5% defectives, find the chance that it will pass. [Approximate by sampling with replacement, and use the normal curve.]

Repeated Trials and Sampling: Summary

Binomial Probability Formula

$$P(k \text{ successes in } n \text{ trials}) = \binom{n}{k} p^k q^{n-k} \qquad \text{for independent trials with}$$

$$p = \text{probability of success on each trial,}$$
$$q = 1 - p = \text{probability of failure on each trial.}$$

For fixed n, as k varies from 0 to n, these probabilities define the *binomial* (n, p) *distribution* on $\{0, 1, \ldots, n\}$. That the probabilities add to 1 amounts to the

Binomial Theorem: $(p+q)^n = \sum_{k=0}^{n} \binom{n}{k} p^k q^{n-k}$

$$\text{Here,} \quad \binom{n}{k} = \frac{n!}{k!(n-k)!} = \frac{n(n-1)\cdots(n-k+1)}{k(k-1)\cdots 1}$$

$= \text{binomial coefficient called } n$ *choose* k

$= \text{number of ways to pick } k \text{ places out of } n$

$= \text{number of subsets of } k \text{ of a set of } n$

$= \text{number in row } n\text{, column } k \text{ of Pascal's triangle}$

$$\text{Note: } \binom{n}{n} = \binom{n}{0} = 1$$

Recursion Formula for Pascal's Triangle

$$\binom{n}{k} = \binom{n-1}{k-1} + \binom{n-1}{k} \qquad (\text{for } 0 < k < n, \quad n = 1, 2, \ldots)$$

Symmetry of Pascal's Triangle

$$\binom{n}{k} = \binom{n}{n-k}$$

Consecutive Ratios in Pascal's Triangle

$$\binom{n}{k} \Big/ \binom{n}{k-1} = \frac{n-k+1}{k}$$

Consecutive Ratios in the Binomial (n, p) Distribution

$$R(k) = \frac{P(k)}{P(k-1)} = \frac{(n-k+1)}{k}\frac{p}{q}$$

Mode of Binomial (n, p) Distribution: m = most likely value = $\text{int}(np + p)$

Normal Approximation to the Binomial Distribution

$$P(k) \approx \frac{1}{\sigma}\phi\left(\frac{k-\mu}{\sigma}\right)$$

where $\mu = np$ is the *mean,*

$\sigma = \sqrt{npq}$ is the *standard deviation,*

$z = (k-\mu)/\sigma$ is k in *standard units,*

$\phi(z) = \frac{1}{\sqrt{2\pi}}e^{-\frac{1}{2}z^2}$ is the *standard normal density function.*

$$P(a \text{ to } b) \approx \Phi\left(\frac{b + \frac{1}{2} - \mu}{\sigma}\right) - \Phi\left(\frac{a - \frac{1}{2} - \mu}{\sigma}\right)$$

where $\Phi(z) = \int_{-\infty}^{z} \phi(x)dx$ is the *standard normal c.d.f.*

This approximation should be used only if $\sigma \geq 3$. The larger σ, the better.

$$\Phi(-z) = 1 - \Phi(z)$$
$$\Phi(a, b) = \Phi(b) - \Phi(a)$$
$$\Phi(-b, b) = 2\Phi(b) - 1$$
$$P(\mu - \sigma \text{ to } \mu + \sigma \text{ success in } n \text{ trials}) \approx \Phi(-1, 1) \approx 68\%$$
$$P(\mu - 2\sigma \text{ to } \mu + 2\sigma \text{ success in } n \text{ trials}) \approx \Phi(-2, 2) \approx 95\%$$
$$P(\mu - 3\sigma \text{ to } \mu + 3\sigma \text{ success in } n \text{ trials}) \approx \Phi(-3, 3) \approx 99.7\%$$

Square Root Law for Independent Trials: The deviation from the expected number of successes np will most likely be a small multiple of $\sigma = \sqrt{npq} \leq \frac{1}{2}\sqrt{n}$.

$$P\left(p - \tfrac{1}{\sqrt{n}} \leq \text{sample proportion} \leq p + \tfrac{1}{\sqrt{n}}\right) \geq 95\% \qquad \text{for large } n.$$

Poisson Approximation to the Binomial Distribution

If p is close to zero

$$P(k) \approx e^{-\mu}\mu^k/k! \qquad \text{where } \mu = np$$

Random Sampling: See box on page 125.

Review Exercises

1. Ten dice are rolled. Write down numerical expressions for

a) the probability that exactly 4 dice are sixes.

b) the probability that exactly 4 dice are sixes given that none of the dice is a five.

c) the probability of 4 sixes, 3 fives, 2 fours, and a three.

d) the probability that none of the first three dice is a six given 4 sixes among the ten dice.

2. A fair die is rolled 36 times. Approximate the probability that 12 or more sixes appear.

3. Suppose I roll a fair die, then toss as many coins as there are spots on the die.

a) What is the probability that exactly three heads appear among the coins?

b) Given three heads appear, what is the probability that the die showed 4?

4. A fair coin is tossed 10 times. Given that at least 9 of the tosses resulted in tails, what is the probability that exactly 9 of the tosses resulted in tails?

5. A thumb tack was tossed 100 times, and landed point up on 40 tosses and point down on 60 tosses. Given this information, what is the probability that the first three tosses landed point down?

6. Four numbers are drawn at random from a box of ten numbers $0, 1, \ldots, 9$. Find the probability that the largest number drawn is a six:

a) if the draws are made with replacement;

b) if the draws are made without replacement.

7. 10^6 fair coins are tossed. Find a number k such that the chance that the number of heads is between $500,000 - k$ and $500,000 + k$ is approximately 0.96.

8. Suppose you and I each roll ten dice. What is the probability that we each roll the same number of sixes?

9. In a certain town, 10% of the families have no children, 10% have one child, 40% have two children, 30% have three children, and 10% have four children. Assume that births are independent of each other, and equally likely to produce male or female.

a) One family is picked at random from all of the families in this town. What is the probability that there are at least two children in the family?

b) One family is picked at random from all of the families in this town. Guess the size of the family, given that it has at least two girls. Give reasons for your guess.

c) A family is picked at random from among the families with four children. Then a child is picked at random from the selected family. What is the chance that the child picked is a girl with at least one brother?

10. **Lie detectors.** According to a newspaper report, in 2 million lie detector tests, 300,000 were estimated to have produced erroneous results. Assuming these figures to be correct, answer the following:

a) If ten tests were picked at random from these 2 million tests, what would be the chance that at least one of them produced an erroneous result? Sketch the histogram of the distribution of the number of erroneous results among these ten tests.

b) Suppose these 2 million tests were done on a variety of machines. If a machine were picked at random, then ten tests picked at random from these tests performed on that machine, would it be reasonable to suppose that the chance that at least one of them produced an erroneous result would be the same as in a)? Explain.

11. Consider two machines, A and B, each producing the same items. Each machine produces a large number of these items every day. However, production per day from machine B, being newer, is twice that of A. Further the rate of defectives is 1% for B and 2% for A. The daily output of the machines is combined and then a random sample of size 12 taken. Find the probability that the sample contains 2 defective items. What assumptions are you making?

12. In poker, a hand containing face values of the form (x, x, y, z, w) is called one pair.

a) If I deal a poker hand, what is the probability that I get one pair?

b) I keep dealing independent poker hands. Write an expression for the probability that I get my 150th 'one pair' on or after the 400th deal.

c) Approximately what is the value of the probability in b)?

13. A seed manufacturer sells seeds in packets of 50. Assume that each seed germinates with a chance of 99%, independently of all others. The manufacturer promises to replace, at no cost to the buyer, any packet that has 3 or more seeds that do not germinate. What is the chance that the manufacturer has to replace more than 40 of the next 4000 packets sold?

14. a) If Ted and Jim are among 10 people arranged randomly in a line, what is the chance that they stand next to each other?

b) What if the ten people are arranged at random in a circle?

c) Generalize to find the chance of k particular people ending up all together if n people are arranged at random in a line or a circle.

15. Draws are made at random with replacement from a box of colored balls with the following composition:

color	red	blue	green	yellow
proportion	0.1	0.2	0.3	0.4

Write down and justify unsimplified expressions for the probabilities of the following events:

a) exactly 5 yellow balls appear in 20 draws;

b) exactly 2 red, 4 blue, 6 green and 8 yellow balls appear in 20 draws;

c) the number of draws required to produce 3 red balls is 25.

16. Eight cards are drawn from a well-shuffled deck of 52 cards. What is the probability that the 8 cards contain: a) 4 aces; b) 4 aces and 4 kings;

c) 4 of a kind (any kind, including the possibility of 4 of two kinds).

17. If four dice are rolled, what is the probability of:

a) four of a kind; b) three of a kind; c) two pairs?

18. Seven dice are rolled. Write down unsimplified expressions for the probabilities of each of the following events:

a) exactly three sixes;

b) three of one kind and four of another;

c) two fours, two fives, and three sixes; d) each number appears;

e) the sum of the dice is 9 or more.

19. In a World Series, teams A and B play until one team wins four games. Suppose all games are independent, and that on each game, the probability that team A beats team B is $2/3$.

a) What is the probability that team A wins the series in four games?

b) What is the probability that team A wins the series, given team B won games 1 and 2?

20. A computer communication channel transmits words of n bits using an error-correcting code which is capable of correcting errors in up to k bits. Here each bit is either a 0 or a 1. Assume each bit is transmitted correctly with probability p and incorrectly with probability q independently of all other bits.

a) Find a formula for the probability that a word is correctly transmitted.

b) Calculate the probability of correct transmission for $n = 8$, $k = 2$, and $q = 0.01$.

21. Suppose a single bit is transmitted by repeating it n times and the message is interpreted by majority decoding. For example, for $n = 5$, if the message received is 10010, it is concluded that a 0 was sent. Assuming n is odd and each bit in the message is transmitted correctly with probability p, independently of the other bits, find a formula for the probability that the message is correctly received.

22. Suppose that, on average, 3% of the purchasers of airline tickets do not appear for the departure of their flight. Determine how many tickets should be sold for a flight on an airplane which has 400 seats, such that with probability 0.95 everybody who appears for the departure of the flight will have a seat. What assumptions are you making?

23. Ten percent of the families in a town have no children, twenty percent have one child, forty percent have two children, twenty percent have three, and ten percent have four. Assume each child in a family is equally likely to be a boy or a girl, independently of all the others. A family is picked at random from this town. Given that there is at least one boy in the family, what is the chance that there is also at least one girl?

24. In a large population, the distribution of the number of children per family is as follows:

Number of children n	0	1	2	3	4	5
Proportion families with n children	0.15	0.2	0.3	0.2	0.1	0.05

Assume that each child in a family is a boy or a girl with probability $1/2$, independently.

a) If a family is picked at random, what is the chance that it contains exactly two girls?

b) If a child is picked at random from the children of this population, what is the chance that the child comes from a family with exactly two girls?

25. At Wimbledon, men's singles matches are played on a "best of five sets" basis, that is, players A and B play until one of them has won 3 sets. Suppose each set is won by A with probability p, independently of all previous sets.

a) For each $i = 3, 4, 5$, find a formula in terms of p and $q = 1 - p$ that player A wins in exactly i sets.

b) In terms of p and q, what is the probability that player A wins the match?

c) Given that player A won the match, what is the probability (in terms of p and q) that the match lasted only three sets?

d) Compute the probability in c) for the case $p = 2/3$.

e) Do you think the assumption of independence made above is reasonable?

26. Suppose 3 points are picked at random from 10 points equally spaced around the circumference of a circle.

a) What is the probability that two particular adjacent points, say A and B, are both among the 3 points picked at random?

b) What is the probability that among the 3 points picked at random there is least one pair of adjacent points?

27. A university schedules its final examinations in 18 "examination groups", so that courses held at different times are in different examination groups. The examination times are spread over 6 days, with 3 examinations each day. Suppose all students take 4 examinations. About what proportion of students will have their 4 examinations on different days? [You need to make some assumptions—state what the assumptions are.]

28 **The matching problem.** There are n letters addressed to n people at n different addresses. The n addresses are typed on n envelopes. A disgruntled secretary shuffles the letters and puts them in the envelopes in random order, one letter per envelope.

a) Find the probability that at least one letter is put in a correctly addressed envelope. [*Hint*: Use the inclusion-exclusion formula of Exercise 1.3.12]

b) What is this probability approximately, for large n?

29. **Cosmic wimpout.** In this game five dice are rolled. Four of the dice have the same set of symbols and numbers on their faces. The numbers are 5 and 10, and let us call the symbols A, B, C, and D. The fifth die is the same, except symbol D is replaced by a different symbol W, indicating a wild roll. In one version of the game, the following kinds of rolls count for a score:

- any roll that shows one or more numbers;
- any roll that shows a triple of symbols, where the wild symbol W can count as any symbol you like, e.g., WAABC scores a triple, the W counting as A;
- a roll that shows W together with one of each of the other symbols A, B, C, and D.

Any other combination fails to score, and is called a wimpout. Calculate the probability of a wimpout.

30. Stirling's formula. Use logarithms and calculus to derive an approximation of the form

$$n! \sim C\left(\frac{n}{e}\right)^n \sqrt{n}$$

for some constant C. Now compare with the normal approximation to the probability of m heads and m tails in $2m$ fair coin tosses to deduce that $C = \sqrt{2\pi}$.

31. The normal approximation works reasonably well whenever the area under the normal curve over the range of the binomial distribution is close to one. Show that if $\sqrt{npq} \geq 3$, then at least 99% of the area under the normal curve is between 0 and n by showing:

a) $np - 3\sqrt{npq} \geq 0$; b) $nq - 3\sqrt{npq} \geq 0$; c) $0 \leq np \pm 3\sqrt{npq} \leq n$.

32. Call a card hand of h cards a *straight* if the denominations can be arranged as $d, d+1, \ldots, d+h$, for some $1 \leq d \leq 13-h$, where $d=1$ represents ace, $d=11$ for jack, 12 for queen, and 13 for king (so aces only count low). Call the hand a *flush* if all h cards are of the same suit. Assume for simplicity that a straight flush counts both as a straight and as a flush. For which h is a straight more likely than a flush?

33. a) How could you simulate a biased coin landing heads with probability $p = 1/3$ if you only had available a fair coin?

b) How could you simulate fair coin tossing if you only had available a coin with unknown bias p strictly between 0 and 1?

34. a) Explain why if you and I each toss m fair coins, the chance that we both get the same (unspecified) number of heads equals the chance that we get exactly m heads and m tails between us.

b) If I toss m fair coins, and you toss $m+1$ fair coins, what is the chance that you get strictly more heads than I do?

35. At roulette, the chance of winning a bet on a single number is $1/38$.

a) Write down a numerical expression for the chance of winning between 20 and 35 bets (inclusive) out of 1000 bets on a single number. Do not evaluate this expression.

b) Should the normal curve be used to approximate the chance in a)? (Give a reason.) If yes, find the normal approximation. If no, use some better method of approximation.

36. An efficient way of computing probabilities in the binomial (n, p) distribution to any desired degree of accuracy is to use the following method. Let $m = \text{int}(np + p)$, and fix some small number $\epsilon > 0$. Starting from $H(m) = 1$, find the histogram heights

$H(k) = P(k)/P(m)$ for $k = m+1, m+2, \ldots$ by repeatedly multiplying consecutive odds ratios, until $k = b$ say, the least $k > m$ such that $H(k) < \epsilon$. Find $H(k)$ for $a \le k < m$ similarly, where a is the greatest $k < m$ such that $H(k) < \epsilon$.

a) Show that the binomial (n, p) probability $P(a \text{ to } b)$ is at least $1 - \epsilon$.

[*Hint*: Use the fact that the consecutive odds ratios are decreasing to show $P(b+j) \le \epsilon P(m+j)$ for $j = 1, 2, \ldots$, hence $P(b+1 \text{ to } n) \le \epsilon P(m+1 \text{ to } n)$. Bound the left tail similarly. This argument was discovered by N. Bernoulli around 1700.]

b) For $a \le k \le b$ let $P(k \,|\, a \text{ to } b) = H(k)/\Sigma$ where Σ is the sum of the $H(j)$ over $a < j < b$. Deduce from a) that for $a \le k \le b$

$$(1-\epsilon)P(k \,|\, a \text{ to } b) \le P(k) \le P(k \,|\, a \text{ to } b)$$

So $P(k \,|\, a \text{ to } b)$ computed as above is an approximation to $P(k)$ with a relative error of at most ϵ. The computer run time to compute $P(k \,|\, a \text{ to } b)$ for every $a \le k \le b$ is approximately $K(b-a)$ for some constant K depending on the speed of the computer.

c) Use the normal approximation to find an approximate formula for the run time in terms of n, p, K, and ϵ which will be asymptotically correct as $n \to \infty$.

d) If it takes my computer 2 seconds to compute this approximation to the distribution of the number of reds in 100 spins of a roulette wheel with $\epsilon = 0.001$, approximately how long should it take my computer to approximate the distribution for 1000 spins with the same ϵ?

37. Integrals related to equalizations in coin tossing. Let $I_n = \int_{-\pi/2}^{\pi/2} \cos^n(x)dx$.

a) Show that for $n = 2, 3, \ldots$

$$I_n = \frac{n-1}{n} I_{n-2} \quad \text{and} \quad \sqrt{\frac{2\pi}{n+1}} < I_n < \sqrt{\frac{2\pi}{n}}.$$

b) Referring to Exercise 2.3.3, show that these formulae yield much sharper bounds on $\alpha_m = P(m \text{ in } 2m)$, the probability of exactly m heads in $2m$ fair coin tosses, as well as the value of $K = \lim_{m\to\infty} \sqrt{m}\alpha_m$.

c) Use $\cos(x) \approx 1 - \frac{1}{2}x^2$ for $x \approx 0$ and an exponential approximation to deduce that $I_n \approx C/\sqrt{n}$ for large n where

$$C = \int_{-\infty}^{\infty} \phi(z)dz$$

Compare with the estimates of I_n in a) to conclude that $C = \sqrt{2\pi}$.

3 Random Variables

This chapter extends the ideas of mean, standard deviation, and normal approximation to distributions more general than the binomial. This involves sums and averages of randomly produced numbers. Random variables, introduced in Section 3.1, provide a good notation for this purpose. The concept of the expectation or mean of a random variable is the subject of Section 3.2. Then standard deviation and the normal approximation appear in Section 3.3. In these first three sections, attention is restricted to random variables with a finite number of possible values. The ideas are extended to random variables with an infinite sequence of possible values in Section 3.4, then to random variables with a continuous distribution in the following chapters.

3.1 Introduction

The number of heads in four tosses of a coin could be any one of the possible values $0, 1, 2, 3, 4$. The term *random variable* is now introduced for something like the number of heads, which might be one of several possible values, with a distribution of probabilities over this set of values. Typically, capital letters X, Y, Z, etc., are used to denote random variables. For example, X might stand for "the number obtained by rolling a die", Y for "the number of heads in four coin tosses", and Z for "the suit of a card dealt from a well-shuffled deck". This is not really a new idea, rather a compact notation for the familiar idea of something or other picked at random according to a probability distribution.

The *range* of a random variable X is the set of all *possible values* that X might produce. This section only considers random variables with a finite range. But infinite ranges will appear in later sections. Usually, and unless otherwise specified in the following development, the range of a random variable is assumed to be a set of numbers. In case not, the nature of the range can be indicated by a change in terminology. For example, *random pair, random sequence,* or *random permutation.* In the following table, Z might be called a *random suit.*

TABLE 1. Some random variables and their ranges.

Random variable	Description	Range
X	Number on a die	$\{1,2,3,4,5,6\}$
Y	Number of heads in 4 coin tosses	$\{0,1,2,3,4\}$
Z	Suit of a card	$\{\spadesuit, \heartsuit, \clubsuit, \diamondsuit\}$

Distribution of X

A statement about a random variable, such as "$X \leq 3$", defines an event. The event occurs if the statement is true, and does not occur if the statement is false.

TABLE 2. Some events determined by X, the number on a die.

Verbal description of event	Notation	Subset of range	Probability
1. Number on the die is less than or equal to 3	$X \leq 3$	$\{1,2,3\}$	$1/2$
2. Number on the die is 6	$X = 6$	$\{6\}$	$1/6$
3. Number on the die is less than or equal to x	$X \leq x$	$\{1,2,\ldots,x\}$	$x/6$
4. Number on the die is x	$X = x$	$\{x\}$	$1/6$
5. Number on the die is in the subset B	$X \in B$	B	$\#(B)/6$

In lines 3 and 4 of the table, x denotes an arbitrary element of the range of X. In line 5, B is a generic subset of the range of X. Events defined by statements about a random variable X are called *events determined by* X. Every such event can be written as "$X \in B$" where B is the set of possible values of X for which the statement is true. The probability of this event is written $P(X \in B)$, or simply $P(B)$. The notation $P(B)$ is familiar as the probability of getting a value in B. The notation $P(X \in B)$ shows this probability refers to the random variable X. As B varies over subsets of the range of X, these probabilities must form a distribution, called *the distribution of* X. Assuming that X has only a finite number of possible values, the distribution of X is determined by the probabilities of individual values,

$$P(X = x) \qquad x \in \text{range of } X$$

via the addition rule

$$P(X \in B) = \sum_{x \in B} P(X = x)$$

Here it is assumed that the random variable X has a uniquely specified value, no matter what happens. So the events $(X = x)$ as x varies over the range of X are mutually exclusive and exhaustive, and their probabilities must add up to 1. By similar reasoning, $P(X \in B)$ is obtained by summing just over those values x in B. The probabilities $P(X = x)$ can be displayed in a distribution table or histogram, or given by a formula.

Dummy variables. There is nothing sacred about the use of the symbol x as a generic possible value of X. You could just as well use k or i or any other lowercase letter. For example, if X is the number of heads in four coin tosses, it makes perfect sense to write both

$$P(X = k) = \binom{4}{k} 2^{-4} \qquad (k = 0, \ldots, 4)$$

$$P(X \leq 2) = \sum_{i=0}^{2} \binom{4}{i} 2^{-4}$$

Here k and i are called ***dummy variables***. It is a useful convention to reserve capital letters for random variables, small letters for dummy variables. Often a matching lowercase letter is used to denote a generic possible value for an uppercase random variable. But this is not always convenient. So be prepared for statements like

$$P(X = v) = P(Y = v)$$

which means that X and Y have the same chance of being equal to v.

Functions

Often a random variable of interest, X say, is expressed as a function of another random variable W:

$$X = g(W)$$

Here g is a function defined on the range of W with values in the range of X. Such a function is a deterministic rule. The rule is that if W has value w, then X has value $g(w)$, uniquely determined by w, for every possible value w of W. Put another way,

X gives a less detailed description of what is happening than W. The distribution of X can be derived from that of W, because any event defined by X can be written in terms of $g(W)$ and hence in terms of W. As the next example shows, this is just a new way to say something familiar.

Example 1. Number of heads.

Let X be the number of heads in two tosses of a fair coin. The distribution of X is the binomial distribution with parameters $n = 2$ and $p = 1/2$, as discussed in Section 2.1:

x	0	1	2
$P(X = x)$	1/4	1/2	1/4

The probabilities of 1/4, 1/2, and 1/4 were obtained from the natural outcome space for two coin tosses: $\{hh, ht, th, tt\}$, by assuming the two tosses were independent. Let W represent which of these outcomes appeared. Once the random outcome W of both tosses becomes known, the number of heads X is completely determined by $X = g(W)$ where g is the function defined by the following table:

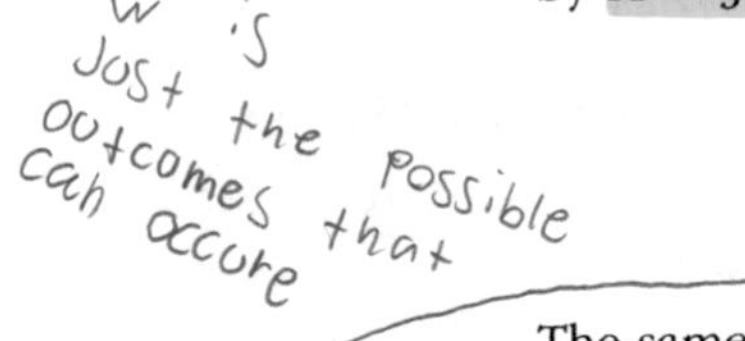

Outcome of tosses w	tt	th	ht	hh
Number of heads $g(w)$	0	1	1	2

The same relationship is displayed in the following diagram:

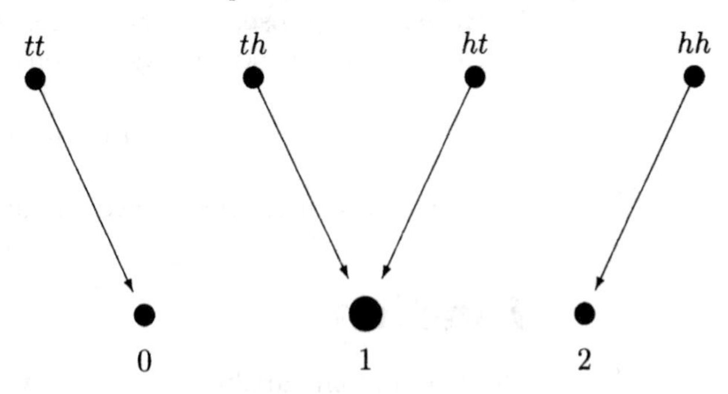

As the blobs and arrows suggest, the probability of each possible value x of X is the sum of the probabilities of those w for which $g(w) = x$. For $x = 2$ and 0 there is a unique w giving $g(w) = x$, so $P(X = x) = 1/4$ for these x. But there are two outcomes w giving $g(w) = 1$, so $P(X = 1) = 1/4 + 1/4 = 1/2$.

The distribution of $X = g(W)$. As in the last example with two trials, the number of successes in n trials can be regarded as a function of the detailed sequential outcome of all trials. To get the probability of a particular number k of successes, add the probabilities of all sequences giving rise to k successes. The same method gives a general formula for the distribution of $X = g(W)$ in terms of the distribution of W. Keep in mind that while a function g must assign to each w a unique value of x, many values of w may be assigned the same x. The event $(X = x)$ is the event that W has a value w such that $g(w) = x$. By the addition rule for probability, $P(X = x)$ is the sum of the probabilities $P(W = w)$ over all w such that $g(w) = x$:

$$P(X = x) = P(g(W) = x) = \sum_{w:g(w)=x} P(W = w)$$

Given a random variable X, new random variables are created by common numerical functions, for example,

$$2X \qquad 3X - 5 \qquad X^2 \qquad |X - 2|$$

To illustrate, if the value of X turns out to be -3, the values of these four variables are

$$-6 \qquad -14 \qquad 9 \qquad 5$$

Assuming the distribution of X is known, the probability of an event determined by a function of X is often found most simply by manipulating the statement of the event. The result of the manipulation is that the event in question occurs precisely when X falls in some set of values. To illustrate, suppose X has uniform distribution on the 19 integers $\{-9, -8, \ldots, 8, 9\}$. Then

$$\begin{aligned}
P(2X \leq 5) &= P(X \leq 5/2) = 12/19 \\
P(3X - 5 \leq 5) &= P(X \leq 10/3) = 13/19 \\
P(X^2 \leq 5) &= P(-\sqrt{5} \leq X \leq \sqrt{5}) = 5/19 \\
P(|X - 2| \leq 5) &= P(-5 \leq X - 2 \leq 5) \\
&= P(-3 \leq X \leq 7) = 11/19
\end{aligned}$$

Events like the last one turn up in prediction problems. If you try predicting the value of X by guessing that X is 2, then $|X - 2|$ is how far off your prediction is. And $P(|X - 2| \leq 5)$, found above, is the chance that your prediction is off by 5 or less.

Technical remark. In a more mathematical development of these ideas, it is necessary to say precisely what kind of mathematical object is a random variable. In the usual treatment, a random variable X is, by definition, a numerical function $X(w)$ defined on some basic space of possible outcomes w, where a probability distribution is given. For example, X representing a number of heads as in Example 1 would be the function $X(w)$, denoted $g(w)$ in that example, giving the number of heads as a function of a more complete description of the outcome. Then $P(X \in B) = P(\{w : X(w) \in B\})$ defines the distribution of X in terms of probability on the basic outcome space. With this formalism, a function h defined on the range of X defines another random variable $h(X)$, the composition of h and X, which is the function whose value for outcome w is $h(X(w))$.

Joint Distributions

Given two random variables X and Y defined in the same setting, we can consider their *combined* or *joint* outcome (X, Y) as a random pair of values. By definition, (X, Y) has value (x, y) if X has value x and Y has value y. Thus the event that $((X, Y) = (x, y))$ is the intersection of the events $(X = x)$ and $(Y = y)$, and is usually denoted $(X = x, Y = y)$. So commas mean intersections in statements about random variables.

The range of the joint outcome (X, Y) is the set of all ordered pairs (x, y) with x in the range of X, y in the range of Y, and $P(X = x, Y = y) > 0$. If the range of X is represented by points on a horizontal line, and the range of Y by points on a vertical line, then the range of (X, Y) is represented by a set of points in the plane. Alternatively, the range of (X, Y) may be represented by a set of paths through a tree diagram, as in Chapter 1.

The distribution of (X, Y) is called the ***joint distribution*** of X and Y. This distribution is determined by the probabilities

$$P(x, y) = P(X = x, Y = y)$$

which must satisfy

$$P(x, y) \geq 0 \quad \text{and} \quad \sum_{\text{all } (x,y)} P(x, y) = 1$$

Example 2. Two draws at random without replacement.

Let X and Y be the first and second draws made at random without replacement from a box containing three tickets numbered 1, 2, and 3. Assuming all six possible pairs of draws are equally likely, the joint distribution of X and Y is displayed as follows. The entry at position (x, y) is $P(x, y) = P(X = x, Y = y)$, the chance that the first draw is x and the second is y. Contrary to convention for matrices, here the first index x is for columns, increasing from left to right, and the second index y is for rows, increasing from bottom to top. This is to make the table consistent with conventional (x, y) co-ordinates in the plane, as in Figure 1 on page 148.

TABLE 3. Joint distribution table for (X, Y)

		possible values for X			distn. of Y
		1	2	3	(row sums)
possible values for Y	3	1/6	1/6	0	1/3
	2	1/6	0	1/6	1/3
	1	0	1/6	1/6	1/3
	distn. of X (column sums)	1/3	1/3	1/3	1 (total sum)

As in this example, the distribution of X can be obtained using the following:

Marginal Probabilities

$$P(X = x) = \sum_{\text{all } y} P(x, y)$$

where the sum is over all possible y in the range of Y.

This is just the basic addition rule for probabilities, since the events $(X = x, Y = y)$ form a partition of $(X = x)$ as y varies over the range of Y. The sum is over all entries in column x of the distribution table. These sums can be displayed as above to show the distribution table for X in a row along the bottom margin of the table. Similarly, the distribution of Y defined by

$$P(Y = y) = \sum_{\text{all } x} P(x, y)$$

can be displayed in a column on the right margin of the table. For this reason, when a joint distribution of X and Y is considered, the distribution of X and the distribution of Y are often called *marginal distributions.*

Same random variable or same distribution? In the last example, while the two random variables X and Y have *identical distributions*, it would be wrong to say they were equal. Indeed, for the two draws without replacement,

$$P(X = Y) = 0$$

so X is certainly not equal to Y. A second example: if X is the number of heads in ten tosses of a fair coin, and Y is the number of tails in those ten tosses, then X

and Y have identical distributions. Still, X and Y are not equal, since, for instance, $X = 6$ makes $Y = 4$. However X equals $10 - Y$, because no matter what the pattern of heads and tails, the number of heads is 10 minus the number of tails. That is to say X is certain to equal $10 - Y$, or $P(X = 10 - Y) = 1$. The next box summarizes this distinction.

Random Variables with the Same Distribution

Random variables X and Y have the *same* or *identical distribution* if X and Y have the same range, and for every value v in this range,

$$P(X = v) = P(Y = v).$$

Change of Variable Principle

If X has the same distribution as Y, then any statement about X has the same probability as the corresponding statement about Y, and $g(X)$ has the same distribution as $g(Y)$, for any function g. For example,

$$P(a \leq X \leq b) = P(a \leq Y \leq b) \text{ for all } a \text{ and } b,$$

and X^2 has the same distribution as Y^2.

Equality of Random Variables

Random variables X and Y are *equal*, written $X = Y$, if $P(X = Y) = 1$. In particular, if no matter what the outcome, the value of X equals the value of Y, then $X = Y$.

If two random variables are equal, then they have the same distribution. But random variables with the same distribution need not be equal.

The change of variable principle is an immediate consequence of the definition of equality in distribution. A later subsection on symmetry shows how the change of variable principle can be used to avoid unnecessary calculations.

Technical remark. The definition of equality of X and Y allows X and Y to differ on some exceptional set of outcomes that is assigned probability zero. This flexibility in the definition is of little significance for random variables with a finite range, but is convenient for random variables with infinite range, considered in later sections.

Computing probabilities from a joint distribution. Once the joint distribution of X and Y has been calculated, the probability of any event defined in terms of X and Y can be found. Simply sum the probabilities $P(x, y)$ over the relevant set of pairs (x, y):

Probabilities of Events Determined by X and Y

The probability that X and Y satisfy some condition is the sum of $P(x,y)$ over all pairs (x,y) satisfying that condition. For instance

$$P(X < Y) = \sum_{(x,y):x<y} P(x,y) = \sum_{\text{all } x} \sum_{y:y>x} P(x,y)$$

$$P(X = Y) = \sum_{(x,y):x=y} P(x,y) = \sum_{\text{all } x} P(x,x)$$

Distribution of a function of X and Y. The distribution of any function of (X,Y), for example

$$X+Y \qquad X-Y \qquad XY \qquad \min(X,Y) \qquad \max(X,Y)$$

can be obtained from the joint distribution of X and Y. For example,

$$P(X+Y=z) = \sum_{(x,y):x+y=z} P(x,y) = \sum_{\text{all } x} P(x, z-x).$$

There is a similar formula for any function $g(X,Y)$: the probability $P[g(X,Y)=z]$ is the sum of $P(x,y)$ over all pairs (x,y) with $g(x,y)=z$.

Example 3. Sum of the draws.

Problem. Calculate the distribution of $X+Y$ for two random draws X and Y from a box containing $\{1,2,3\}$: (a) without replacement, (b) with replacement.

Solution. (a) From the joint distribution table given earlier for draws without replacement, the possible values of the sum $S = X+Y$ are 3, 4, and 5. By inspection of the table, each possible value s for S corresponds to exactly two possible pairs (x,y), each with probability $\frac{1}{6}$. Hence the distribution of S is given by the following table:

s	3	4	5
$P(S=s)$	1/3	1/3	1/3

(b) If the draws are made with replacement, then the joint probabilities are

$$P(x,y) = 1/9 \qquad (1 \le x \le 3, \quad 1 \le y \le 3)$$

Now, there is one possible pair adding to 2, two possible pairs adding to 3, three adding to 4, two to 5, and one to 6. Thus for draws with replacement the distribution of S is given by the table:

FIGURE 1. Distributions for sampling with and without replacement from $\{1,2,3\}$. Refer to Example 3. In each case the joint distribution of (X,Y) is represented by a pattern of blobs, with the area of the blob over (x,y) proportional to $P(x,y)$. The distributions of X, Y, and $X+Y$ are displayed similarly around the edges of the joint pattern. Probabilities in these distributions are obtained by adding probabilities from the joint distribution as indicated by the arrows.

Sampling without replacement

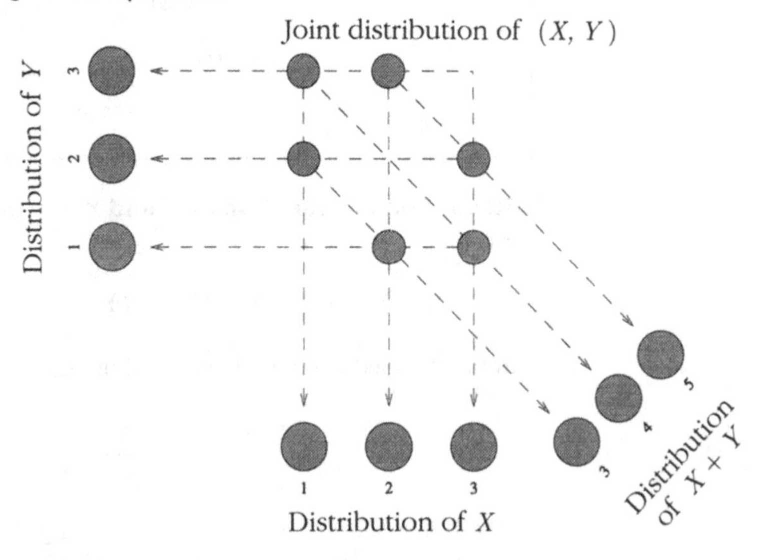

Sampling with replacement

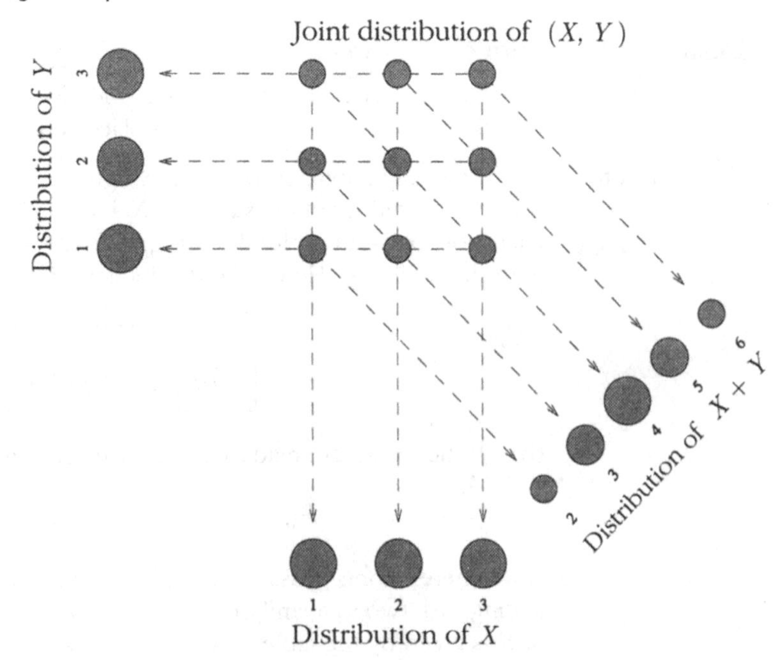

s	2	3	4	5	6
$P(S = s)$	1/9	2/9	3/9	2/9	1/9

These calculations are illustrated in Figure 1.

Discussion. The example shows that to find the distribution of a function of two random variables X and Y, such as their sum, you must think in terms of the joint distribution of X and Y. Knowing that both X and Y are uniform on $\{1, 2, 3\}$ is not enough to determine the distribution of $X + Y$.

Example 4. Minimum and maximum.

Let X be the minimum and Y the maximum of three digits picked at random without replacement from $\{0, 1, \ldots, 9\}$.

Problem 1. Find the joint distribution of X and Y.

Solution. Because the sampling is done without replacement the three digits drawn must be distinct, so the only possible pairs of values for the minimum and maximum are integers x and y with

$$0 \le x \le 7, \quad x + 2 \le y, \quad y \le 9,$$

as marked in the diagram on the left. To find the chance of such a pair, forget the order in which the digits come, and think about what subset of three digits is chosen from $\{0, 1, \ldots, 9\}$. Every subset has the same chance of being chosen, 1 in $\binom{10}{3}$, or $1/120$. To illustrate, for a minimum of 4 and a maximum of 7 there are just 2 possible subsets, $\{4, 5, 7\}$ and $\{4, 6, 7\}$. In this case, there are $7 - 4 - 1 = 2$ ways to pick the intermediate number. So

Values of Y = max

Values of X = min

$$P(X = 4, Y = 7) = (7 - 4 - 1)/120$$

In general, for a minimum of x and a maximum of y, there are $y - x - 1$ possible subsets, one for each possible value of the third number between x and y. Therefore, for possible pairs x and y as above

$$P(X = x, Y = y) = (y - x - 1)/120$$

Problem 2. Find the distribution of $Z = Y - X$, the maximum minus the minimum.

Solution. The possible values of Z are clearly $2, 3, \ldots, 9$. Any one of these possible values, z say, must come from a (min, max) pair (x, y) with a difference of $y - x = z$. Every such pair (x, y) has the same probability

$$(y - x - 1)/120 = (z - 1)/120$$

For $z = 9$ there is one pair $(0, 9)$, for $z = 8$ there are 2 pairs $(0, 8)$ and $(1, 9)$, and so on. In general, there are $10 - z$ possible pairs (x, y) with $y - x = z$. Therefore,

$$P(Z = z) = (10 - z)(z - 1)/120 \qquad (z = 2, 3, \ldots, 9)$$

To check, the sum of these probabilities from $z = 2$ to 9 is

$$((8\times1) + (7\times2) + (6\times3) + (5\times4) + (4\times5) + (3\times6) + (2\times7) + (1\times8))\,/120 = 1$$

Conditional Distributions

The basic rules of probability imply that for any given event A, and any random variable Y, the collection of conditional probabilities

$$P(Y \in B \,|\, A) = \frac{P[(Y \in B) \text{ and } A]}{P(A)}$$

defines a probability distribution as B varies over subsets of the range Y. This distribution is called the *conditional distribution of* Y *given* A. Intuitively, this is the appropriate revision of the distribution of Y given the information that event A has occurred. For Y with a finite range the conditional distribution of Y given A is specified by the conditional probabilities

$$P(Y = y \,|\, A) \quad \text{for } y \in \text{range of } Y.$$

The rules of a probability distribution imply $P(Y \in B \mid A) = \sum_{y \in B} P(Y = y \mid A)$. Most often the conditional distribution of Y given A is considered for each A of the form $(X = x)$ for some random variable X.

Conditional Distribution of Y Given X = x

For each possible value x of X, as y varies over the range of Y the probabilities $P(Y = y \mid X = x)$ define a probability distribution over the range of Y. This probability distribution, which may depend on the given value x of X, is called the *conditional distribution of* Y *given* $X = x$.

The given value x of X can be thought of as a *parameter* in the distribution of Y given $X = x$. If the joint distribution of X and Y is tabulated, then for given x the conditional probabilities $P(Y = y \mid X = x)$, are found from the joint distribution table by lifting out column x of the table and renormalizing the probabilities in this column by their sum, which is $P(X = x)$. Similarly, for given y, the probabilities $P(X = x \mid Y = y)$ for x in the range of X are found by lifting out row y from the table of joint probabilities and renormalizing this row of probabilities by their sum, which is $P(Y = y)$.

If the marginal (*unconditional*) distribution of X is known, together with the conditional distribution of Y given $X = x$ for all possible values x of X, the joint distribution of X and Y is found using the

Multiplication Rule

$$P(X = x, Y = y) = P(X = x)P(Y = y \mid X = x)$$

In this section conditional distributions serve only to motivate the following definition of independent random variables. See Section 6.1 (which can be read immediately) for a detailed discussion of conditional distributions for dependent random variables.

Independence

Intuitively, random variables X and Y are independent when the probabilities for various values of Y are unaffected by conditioning on the value of X. This is just a restatement in terms of random variables of the relation of independence between draws, trials, etc., as discussed in Chapter 1. For calculations with independent random variables, the simplest definition of independence is the following one using the product rule:

Independent Random Variables

Random variables X and Y are *independent* if

$$P(X = x, Y = y) = P(X = x)P(Y = y) \quad \text{for all } x \text{ and } y$$

If X and Y are independent random variables, then every event determined by X is independent of every event determined by Y:

$$P(X \in A, Y \in B) = P(X \in A)P(Y \in B)$$

Conceptually, independence means that conditioning on a given value of X does not affect the distribution of Y, and vice-versa. Thus the above definition of independence can be re-expressed as follows in terms of conditional distributions:

Conditional Distributions and Independence

The following three conditions are equivalent:

- X and Y are independent;
- the conditional distribution of Y given $X = x$ does not depend on x;
- the conditional distribution of X given $Y = y$ does not depend on y.

Example 5. Independent or not?

Problem. A box of 10 tickets contains some number r of red tickets. The rest are green. A sample of 100 tickets is drawn at random with replacement. Then a second sample of 100 tickets is drawn at random with replacement. Let X_1 be the number of red tickets in the first sample, and X_2 the number in the second sample. Are X_1 and X_2 independent?

Solution 1. If you regard r as known, then no matter how many red tickets you see in the first 100 draws, the second 100 draws is still a random sample with replacement from r red and $10 - r$ green tickets. Thus X_1 and X_2 are independent random variables, each with binomial distribution with parameters $n = 100$ and $p = r/10$.

Solution 2. On the other hand, if you don't know r, it seems intuitively obvious that X_1 and X_2 are dependent. For if you saw 53 reds in the first 100 draws, you would be inclined to guess there were around 5 red tickets in the box, and expect to see around 50% red on the next 100 draws. Whereas if you saw 17 reds in the first 100 draws, you would guess that 2 of the 10 tickets were red, and expect to see only 20% or so red on the next 100 draws. Thus, knowing the value of X_1 affects the chances of events determined by X_2, so X_1 and X_2 are dependent.

Discussion. Which solution is correct? It depends on whether r is regarded as a known constant, as in Solution 1, or the value of a random variable, R say, as in Solution 2. Solution 2 can be made more precise by assuming that *conditionally* on the event $(R = r)$ the random variables X_1 and X_2 are independent, with binomial $(100, r/10)$ distribution, just as if r were known as in Solution 1. But unconditionally these variables will be dependent, for the reasons given in Solution 2. Does it make sense to think of r as the value of a random variable R? With a frequency interpretation of probability, it makes sense only if the way the composition of the box was determined is regarded as somehow repeatable. The probabilities $P(R = r)$ for $0 \leq r \leq 10$ would then be long-run frequencies of different compositions. With a subjective interpretation of probability, $P(R = r)$ might be assigned according to your own opinion about the unknown number of reds in the box, even if there is no notion of repetitions.

Several Random Variables

The joint distribution of several random variables $X_1, X_2, \ldots, X_n$ is defined just as for two random variables by the *joint probabilities*

$$P(x_1, \ldots, x_n) = P(X_1 = x_1, \ldots, X_n = x_n)$$

for all possible values x_i of each X_i. Note that the commas signify an *intersection* of events. So $P(x_1, \ldots, x_n)$ is the probability that X_i has value x_i for every $1 \le i \le n$. This concept will now be illustrated by a number of examples.

Random permutations. A permutation of $\{1, 2, \ldots, n\}$ is a sequential ordering of the n numbers with no repeats. A *random permutation* of $\{1, 2, \ldots, n\}$ is a permutation picked uniformly at random from all $n!$ possible permutations of $\{1, 2, \ldots, n\}$. There are many ways to generate a random permutation. For example,

- Suppose tickets numbered $1, 2, \ldots, n$ are placed in a box and drawn one by one at random without replacement. Let X_i be the number of the ith ticket drawn, $1 \le i \le n$. Then $(X_1, X_2, \ldots, X_n)$ is a random permutation of $\{1, 2, \ldots, n\}$.
- Suppose cards numbered $1, 2, \ldots, n$ are thoroughly shuffled. Let Y_i be the number of the ith card from the top of the deck. Then $(Y_1, Y_2, \ldots, Y_n)$ is a random permutation of $\{1, 2, \ldots, n\}$.

Example 6. Joint distribution of a random permutation.

Problem 1. Describe the joint distribution of a random permutation of $\{1, 2, \ldots, n\}$, that is the common joint distribution of $(X_1, X_2, \ldots, X_n)$ and $(Y_1, Y_2, \ldots, Y_n)$.

Solution. Informally the answer is just "the uniform distribution over all $n!$ possible permutations of $\{1, \ldots, n\}$". To illustrate for $n = 3$, (X_1, X_2, X_3) is equally likely to be any one of the $3! = 6$ permutations

$$(1,2,3),\ (1,3,2),\ (2,1,3),\ (2,3,1),\ (3,1,2),\ (3,2,1)$$

and so is (Y_1, Y_2, Y_3). To state this in a formula for a general n, the joint probabilities

$$P(x_1, \ldots, x_n) = P(X_1 = x_1, \ldots, X_n = x_n) = P(Y_1 = x_1, \ldots, Y_n = x_n)$$

are given by

$$P(x_1, \ldots, x_n) = \begin{cases} 1/n! & \text{if } (x_1, \ldots, x_n) \text{ is a permutation of } \{1, 2, \ldots, n\} \\ 0 & \text{otherwise} \end{cases}$$

Discussion. Note that $P(x_1, \ldots, x_n)$ is a symmetric function of $(x_1, \ldots, x_n)$, as defined in Section 3.6, because for any rearrangement of the order of terms in a sequence, the original sequence is a permutation if and only if the rearranged sequence is a permutation. This symmetry property, studied further in Section 3.6, explains the simple solutions of both the next problem and the problem of Example 1.4.7.

Problem 2. For each $1 \le j \le n$, find the distribution of X_j for $(X_1, X_2, \ldots, X_n)$ a random permutation of $\{1, 2, \ldots, n\}$.

Solution. For each $1 \le x \le n$, the probability $P(X_j = x)$ is the number of permutations with x in the jth place, divided by $n!$. But if value x is fixed in the jth place, the values in the remaining $n-1$ places can be any permutation of the set $\{1, 2, \ldots, n\}$ with x deleted. Since there are $(n-1)!$ such permutations, whatever $x \in \{1, 2, \ldots, n\}$, $P(X_j = x) = (n-1)!/n! = 1/n$. Conclusion: for every $1 \le j \le n$, the distribution of X_j is uniform on $\{1, 2, \ldots, n\}$.

Independence of several variables. Random variables $X_1, \ldots, X_n$ are *independent* if their joint probabilities are products of their marginal probabilities:

$$P(x_1, x_2, \ldots, x_n) = P(X_1 = x_1)P(X_2 = x_2) \cdots P(X_n = x_n)$$

for all possible values x_i of each X_i. Summing these probabilities over all $(x_1, \ldots, x_n)$ such that $x_i \in A_i$ shows that events of the form $(X_i \in A_i)$ determined by independent random variables X_i are independent:

$$P(X_1 \in A_1, X_2 \in A_2, \ldots, X_n \in A_n) = P(X_1 \in A_1)P(X_2 \in A_2) \cdots P(X_n \in A_n)$$

Here for each i the set A_i can be any subset of the range of possible values of X_i. The results of the next three paragraphs are consequences of this formula.

Functions of independent random variables are independent. If $X_j, 1 \le j \le n$, are independent random variables, then so are the random variables Y_j defined by $Y_j = f_j(X_j)$ for arbitrary functions f_j defined on the range of X_j.

Disjoint blocks of independent random variables are independent. For example, if $X_1, X_2, \ldots, X_6$ are independent, then (X_1, X_2), (X_3, X_4), and (X_5, X_6) are three independent random pairs. These properties can be combined:

Functions of disjoint blocks of independent random variables are independent. For example, if $X_1, \ldots, X_5$ are independent positive random variables, then so are Y_1, Y_2, and Y_3 defined by $Y_1 = 5X_3 + \sqrt{X_5}$, $\quad Y_2 = X_4 X_2$, $\quad Y_3 = X_1$.

Repeated trials. Independent random variables with the same distribution, for example, repeated draws at random with replacement from some population, or repeated rolls of a die (perhaps biased) are called *repeated trials*. Independent trials that result in one of two possible outcomes, say success or failure, with constant probability p of success on each trial, as studied in Chapter 2, are called *Bernoulli*(p) trials. The number of successes S_n in n Bernoulli trials can be represented as

$$S_n = X_1 + X_2 + \cdots + X_n$$

where X_i is the *indicator* of success on trial i, that is to say the random variable that is 1 if trial i is a success and 0 if trial i is a failure. The sum simply counts the number of 1's, that is the number of successes in the n trials. The sequence $X_1, X_2, \dots, X_n$ is a sequence of n independent random variables, each with the Bernoulli(p) distribution on $\{0, 1\}$ defined at the end of Section 1.3. The Bernoulli(p) distribution of each X_i is the special case $n = 1$ of the binomial (n, p) distribution of the number of successes S_n in n trials, analyzed in Chapter 2. The next two sections show how the representation of S_n as the sum of n independent indicator variables leads to extensions of the law of large numbers and the normal approximation described in Chapter 2 to sums of independent random variables X_i with any common distribution over a finite set of possible values.

Here is the generalization of the binomial distribution that describes the joint distribution of counts in any finite number m of categories in independent trials.

Multinomial Distribution

Let N_i denote the number of results in category i in a sequence of independent trials with probability p_i for a result in the ith category on each trial, $1 \le i \le m$, where $p_1 + \cdots + p_m = 1$. Then for every m-tuple of non-negative integers $(n_1, n_2, \dots, n_m)$ with sum n

$$P(N_1 = n_1, N_2 = n_2, \dots, N_m = n_m) = \frac{n!}{n_1! n_2! \cdots n_m!} p_1^{n_1} p_2^{n_2} \cdots p_m^{n_m}$$

The product of powers of the p_i represents the probability of any *particular* sequence of results with n_i results in category i for each $1 \le i \le m$, while the ratio of factorials

$$\frac{n!}{n_1! n_2! \cdots n_m!} = \binom{n}{n_1}\binom{n - n_1}{n_2} \cdots \binom{n - n_1 - \cdots - n_{m-1}}{n_m}$$

called a *multinomial coefficient* is the number of different possible arrangements of symbols in a row of symbols made from n_1 symbols 1, n_2 symbols 2, ..., and n_m symbols m. A symbol i at place j in the row represents a result in category i on trial j. The derivation of this formula parallels the derivation of the binomial formula in Section 2.1, which is the special case $m = 2$. The multinomial distribution provides a natural example of a joint distribution of m variables $N_1, \dots, N_m$ that are not independent, due to the constraint that $N_1 + \cdots + N_m = n$.

Example 7. Fours, fives, and sixes.

Suppose a fair die is rolled 10 times, and the numbers of rolls of four, five, and six are recorded.

Solution. From the multinomial distribution for $n = 10$ trials, $m = 4$ categories ("four", "five", "six", and "other") with probabilities $1/6, 1/6, 1/6$ and $3/6$, the required probability is

$$P(N_{\text{four}} = 1, N_{\text{five}} = 2, N_{\text{six}} = 3, N_{\text{other}} = 4) = \frac{10!}{1!2!3!4!}\left(\frac{1}{6}\right)^1\left(\frac{1}{6}\right)^2\left(\frac{1}{6}\right)^3\left(\frac{3}{6}\right)^4$$

Symmetry

Symmetry arguments often simplify probability calculations. The basic idea is to recognize when probabilities of different events must be equal by symmetry.

Symmetry about 0. The distribution of X is *symmetric about* 0 if

$$P(X = -x) = P(X = x) \qquad \text{for all } x$$

A histogram displaying the distribution of X is then symmetric about 0 in the usual sense of reflection through the vertical axis. Equivalently, since $P(X = -x) = P(-X = x)$ for all x, $P(-X = x) = P(X = x)$ for all x. That is to say

$$-X \text{ has the same distribution as } X$$

Then for all a

$$P(X \geq a) = P(-X \leq -a) = P(X \leq -a)$$

Here the first equality holds because the two events $(X \geq a)$ and $(-X \leq -a)$ are identical (multiplication by -1: note the reversal of the inequality). Also the probability $P(-X \leq -a)$ equals $P(X \leq -a)$ because any statement about $-X$ has the same probability as the corresponding statement about X, by the equality in distribution of $-X$ and X (change of variable principle).

Example 8. Symmetry about 0 for sums of independent random variables.

Let $S_n = X_1 + \cdots + X_n$ where $X_1, \ldots, X_n$ are independent, and each X_i has a distribution that is symmetric about 0.

Problem. Show for every a

$$P(S_n \leq -a) = P(S_n \geq a)$$

Solution. In other words, the problem is to show that the distribution of S_n is symmetric about 0. Since, by assumption, $-X_i$ has the same distribution as X_i, and the X_i are independent, it follows that $(-X_1, \ldots, -X_n)$ has the same joint distribution as $(X_1, \ldots, X_n)$. This uses the fact that functions of independent random variables are independent (applied to $f(X_i) = -X_i$). Adding the coordinates of the two sequences $(-X_1, \ldots, -X_n)$ and $(X_1, \ldots, X_n)$ shows that $-S_n = (-X_1) + \cdots + (-X_n)$ has the same distribution as S_n. That is to say, the distribution of S_n is symmetric about 0.

Discussion. Note the use of the following form of the change of variable principle for sequences of random variables: if $(X_1, \ldots, X_n)$ and $(Y_1, \ldots, Y_n)$ have the same joint distribution, then $g(X_1, \ldots, X_n)$ and $g(Y_1, \ldots, Y_n)$ have the same distribution for any function g of n variables. For instance, $X_1 + \cdots + X_n$ and $Y_1 + \cdots + Y_n$ have the same distribution. This fact was used in the example for $Y_i = -X_i$. Note also how the reasoning did not involve any explicit summation of probabilities in the joint distribution of $(X_1, \ldots, X_n)$, which would be necessary to find a formula for the distribution of S_n. This is the point of a symmetry argument: to show two probabilities are equal without calculating either of them.

Symmetry about b. The distribution of a random variable Y with a finite number of numerical values is *symmetric about b* if

$$P(Y = b + x) = P(Y = b - x) \qquad \text{for all } x$$

Equivalently, the distribution of $Y - b$ is symmetric about 0. Then for every c

$$P(Y \leq b - c) = P(Y \geq b + c)$$

Symmetry for a sum of independent random variables. If Y_i has distribution symmetric about b_i, and the Y_i are independent, then $Y_1 + \cdots + Y_n$ has distribution symmetric about $b_1 + \cdots + b_n$. This follows from the result of the previous example applied to $X_i = Y_i - b_i$.

Example 9. Sum of 101 random digits.

Let S_{101} denote the sum of 101 independent random digits, each picked uniformly at random from $\{0, 1, \ldots, 9\}$.

Problem. Find $P(S_{101} \leq 454)$.

Solution. Here $S_{101} = Y_1 + \cdots + Y_{101}$ for Y_i that are independent, and the distribution of each Y_i is symmetric about $4\frac{1}{2}$. So the distribution of S_{101} is symmetric about $101 \times (4\frac{1}{2}) = 454.5$. Therefore

$$P(S_{101} \leq 454) = P(S_{101} \leq 454.5 - .5) = P(S_{101} \geq 454.5 + .5) = P(S_{101} \geq 455)$$

But since S_{101} has integer values, $P(S_{101} \leq 454) + P(S_{101} \geq 455) = 1$, which forces $P(S_{101} \leq 454) = \frac{1}{2}$.

Discussion. For S_n the sum of n digits the argument shows that the distribution of S_n is symmetric about $(4\frac{1}{2})n$ for every n. For odd n, say $n = 2m+1$, this symmetry can be used just as above to identify a probability in the distribution of S_{2m+1} that is exactly $1/2$:

$$P(S_{2m+1} \leq 9m + 4) = P(S_{2m+1} \geq 9m + 5) = \tfrac{1}{2}$$

For odd n the histogram of S_n has bars of equal height at the integers $(4\frac{1}{2})n \pm 1/2$, $(4\frac{1}{2})n \pm 3/2, \ldots$, so the distribution splits perfectly into two equal halves. For even n the histogram of S_n has a bar exactly on the point of symmetry $(4\frac{1}{2})n$, and equal bars at $(4\frac{1}{2})n \pm 1$, $(4\frac{1}{2})n \pm 2, \ldots$. Then the distribution of S_n does not split into equal halves to the right and left of $(4\frac{1}{2})n$, because there is a lump of probability right on the point of symmetry which cannot be split in two. It can be shown that for even n the central probability $P[S_n = (4\frac{1}{2})n]$ is actually the largest individual probabilty in the distribution of S_n. It will be seen in Section 3.3 that for large n the distribution of S_n follows a normal curve very closely. This is similar to what happens for large n to the binomial $(n, 1/2)$ distribution of $X_1 + \cdots + X_n$ for X_i picked at random from $\{0, 1\}$. It follows that as in the binomial case, for large even n the distribution of the sum of n digits has central term $P[S_n = (4\frac{1}{2})n]$ that converges to zero very slowly, like a constant over $\sqrt{n}$. For very large $n = 2m$ this term can be ignored, so

$$P(S_{2m} \leq 9m) = P(S_{2m} \geq 9m) \approx \tfrac{1}{2}$$

The approximate probability $\frac{1}{2}$ is less than the true probability by

$$P(S_{2m} = 9m)/2 \sim c/\sqrt{m}$$

where the constant c can be shown using the normal approximation to be equal to $1/\sqrt{33\pi}$, and "$\sim$" means that the ratio of the two sides tends to 1 as $m \to \infty$. (See Exercise 3.3.31).

Exercises 3.1

1. Let X be the number of heads in three tosses of a fair coin.

a) Display the distribution of X in a table. b) Find the distribution of $|X - 1|$.

2. Let X and Y be the numbers obtained in two draws at random from a box containing four tickets 1, 2, 3, and 4. Display the joint distribution table for X and Y:

a) for sampling with replacement; b) for sampling without replacement.

Calculate $P(X \leq Y)$ from the table in each case.

3. Suppose a fair die is rolled twice. Let S be the sum of the numbers on the two rolls.

a) What is the range of S? b) Find the distribution of S.

4. Let X_1 and X_2 be the numbers obtained on two rolls of a fair die. Let $Y_1 = \max(X_1, X_2)$, $Y_2 = \min(X_1, X_2)$. Display joint distribution tables for a) (X_1, X_2); b) (Y_1, Y_2).

5. Find the distribution of X_1X_2 for X_1 and X_2 as in Exercise 4.

6. A fair coin is tossed three times. Let X be the number of heads on the first two tosses, Y the number of heads on the last two tosses.

a) Make a table showing the joint distribution of X and Y.

b) Are X and Y independent? c) Find the distribution of $X + Y$.

7. Let A, B, and C be events that are independent, with probabilities a, b, and c. Let N be the random number of events that occur.

a) Express the event $(N = 2)$ in terms of A, B, and C. b) Find $P(N = 2)$.

8. A hand of five cards contains two aces and three kings. The five cards are shuffled and dealt one by one, until an ace appears.

a) Display in a table the distribution of the number of cards dealt.

b) Suppose that dealing is continued until the second ace appears. Again display the distribution of the number of cards dealt.

c) Explain why the probabilities in the second table are just those in the first in a different order. (*Hint*: Think about dealing off the bottom of the deck!)

9. A box contains 8 tickets. Two are marked 1, two marked 2, two marked 3, and two marked 4. Tickets are drawn at random from the box without replacement until a number appears that has appeared before. Let X be the number of draws that are made. Make a table to display the probability distribution of X.

10. Blocks of Bernoulli trials. In $n + m$ independent Bernoulli (p) trials, let S_n be the number of successes in the first n trials, T_m the number of successes in the last m trials.

a) What is the distribution of S_n? Why?

b) What is the distribution of T_m? Why?

c) What is the distribution of $S_n + T_m$? Why?

d) Are S_n and T_m independent? Why?

11. Binomial sums. Let U_n have binomial(n, p) distribution and let V_m have binomial(m, p) distribution. Suppose U_n and V_m are independent.

a) Find the distribution of $U_n + V_m$ without calculation by a simple argument that refers to the solution of Exercise 10.

b) Compare the result of part a) to a calculation of $P(U_n + V_m = k)$ for $0 \le k \le n+m$ from the joint distribution of U_n and V_m, and hence prove the identity

$$\sum_{j=0}^{n} \binom{n}{j}\binom{m}{k-j} = \binom{n+m}{k}$$

c) Derive the identity in part b) by a counting argument. [*Hint*: Classify the subsets of size k of $\{1, \ldots, n+m\}$ by how many elements of $\{1, \ldots, n\}$ they contain.]

d) Derive the identity in part b) in another way by finding the coefficient of $p^k q^{n+m-k}$ in $(p+q)^{n+m} = (p+q)^n(p+q)^m$ in two different ways.

e) Simplify the sum $\sum_{j=0}^{n} \binom{n}{j}^2$.

12. Grouping multinomial categories. Suppose that counts $(N_1, \ldots, N_m)$ are the numbers of results in m categories in n repeated trials. So $(N_1, \ldots, N_m)$ has multinomial distribution with parameters n and $p_1, \ldots, p_m$, as in the box above Example 7. Let $1 \leq i < j \leq m$. Answer the following questions with an explanation, but no calculation.

a) What is the distribution of N_i? b) What is the distribution of $N_i + N_j$?

c) What is the joint distribution of N_i, N_j, and $n - N_i - N_j$?

13. A box contains $2n$ balls of n different colors, with 2 of each color. Balls are picked at random from the box with replacement until two balls of the same color have appeared. Let X be the number of draws made.

a) Find a formula for $P(X > k)$, $\quad k = 2, 3, \ldots$

b) Assuming n is large, use an exponential approximation to find a formula for k in terms of n such that $P(X > k)$ is approximately $1/2$. Evaluate k for n equal to one million.

14. In a World Series, teams A and B play until one team has won four games. Assume that each game played is won by team A with probability p, independently of all previous games.

a) For $g = 4$ through 7, find a formula in terms of p and $q = 1-p$ for the probability that team A wins in g games.

b) What is the probability that team A wins the World Series, in terms of p and q?

c) Use your formula to evaluate this probability for $p = 2/3$.

d) Let X be a binomial $(7, p)$ random variable. Explain why $P(\text{A wins}) = P(X \geq 4)$ using an intuitive argument. Verify algebraically that this is true.

e) Let G represent the number of games played. What is the distribution of G? For what value of p is G independent of the winner of the series?

15. Let X and Y be independent, each uniformly distributed on $\{1, 2, \ldots, n\}$. Find:

a) $P(X = Y)$; b) $P(X < Y)$; c) $P(X > Y)$;

d) $P(\max(X, Y) = k)$ for $1 \leq k \leq n$;

e) $P(\min(X, Y) = k)$ for $1 \leq k \leq n$; f) $P(X + Y = k)$ for $2 \leq k \leq 2n$.

16. Discrete convolution formula. Let X and Y be independent random variables with non-negative integer values. Show that:

a) $P(X + Y = n) = \sum_{k=0}^{n} P(X = k)P(Y = n - k)$.

b) Find the probability that the sum of numbers on four dice is 8, by taking X to be the sum on two of the dice, Y the sum on the other two.

17. Let X be the number of heads in 20 fair coin tosses, Y a number picked uniformly at random from $\{0, 1, \ldots, 20\}$, independently of X. Let $Z = \max(X, Y)$.

a) Find a formula for $P(Z = k)$, $\quad k = 0, \ldots, 20$.

b) Without calculating out $P(Z = k)$ exactly, sketch the histogram of Z, and explain its unusual shape.

18. Three dice are rolled.

a) What is the probability that the total number of spots showing is 11 or more? [*Hint*: No long calculations!]

b) Find a number m such that if five dice are rolled, the probability that the total number of spots showing is m or more is the same as this probability of 11 or more spots from three dice.

19. Sum of biased dice. Let S be the sum of numbers obtained by rolling two biased dice with possibly different biases described by probabilities $p_1, \ldots, p_6$, and $r_1, \ldots, r_6$, all assumed to be nonzero.

a) Find formulae for $P(S = k)$ for $k = 2$, 7, and 12.

b) Show that $P(S = 7) > P(S = 2)\frac{r_6}{r_1} + P(S = 12)\frac{r_1}{r_6}$.

c) Deduce that no matter how the two dice are biased, the numbers 2, 7, and 12 cannot be equally likely values for the sum. In particular, the sum cannot be uniformly distributed on the numbers from 2 to 12.

d) Do there exist positive integers a and b and independent non-constant random variables X and Y such that $X+Y$ has uniform distribution on the set of integers $\{a, a+1, \ldots, a+b\}$?

20. Pairwise independence. Let $X_1, \ldots, X_n$ be a sequence of random variables. Suppose that X_i and X_j are independent for every pair (i, j) with $1 \leq i < j \leq n$. Does this imply $X_1, \ldots, X_n$ are independent? Sketch a proof or counterexample.

21. Sequential independence. Let $X_1, \ldots, X_n$ be a sequence of random variables. Suppose that for every $1 \leq m \leq n-1$ the random sequence $(X_1, \ldots, X_m)$ is independent of the next random variable X_{m+1}. Does this imply $X_1, \ldots, X_n$ are independent? Sketch a proof or give a counterexample.

22. Suppose that random variables X and Y, each with a finite number of possible values, have joint probabilities of the form

$$P(X = x, Y = y) = f(x)g(y)$$

for some functions f and g, for all (x, y).

a) Find formulae for $P(X = x)$ and $P(Y = y)$ in terms of f and g.

b) Use your formulae to show that X and Y are independent.

23. Suppose X and Y are two random variables such that $X \geq Y$.

a) For a fixed number T, which would be greater, $P(X \leq T)$ or $P(Y \leq T)$?

b) What if T is a random variable?

24. Suppose a box contains tickets, each labeled by an integer. Let X, Y, and Z be the results of draws at random with replacement from the box: Show that, no matter what the distribution of numbers in the box,

a) $P(X + Y \text{ is even}) \geq 1/2$; b) $P(X + Y + Z \text{ is a multiple of } 3) \geq 1/4$.

3.2 Expectation

The *mean* or *expected value* of a random variable X is a number derived from the distribution of X the same way that the *mean* or *average* $\bar{x}$ of a list of numbers $(x_1, \ldots, x_n)$ is derived from the empirical distribution of the list:

$$\bar{x} = (x_1 + \cdots + x_n)/n = \sum_{\text{all } x} x P_n(x) \tag{1}$$

where $P_n(x)$ is the proportion of the n values x_k that are equal to x. These proportions $P_n(x)$, which sum to 1 over all x, define the empirical distribution of the list (see the end of Section 1.3). To illustrate, the average of the list $(1, 0, 8, 6, 6, 1, 6)$ of $n = 7$ numbers is

$$(1+0+8+6+6+1+6)/7 = 0 \times \frac{1}{7} + 1 \times \frac{2}{7} + 6 \times \frac{3}{7} + 8 \times \frac{1}{7} = 4$$

The second formula for $\bar{x}$ in (1) is a *weighted average* of values x with weights $P_n(x)$. This formula is obtained in general, just as in the example, by grouping terms with a common x-value. The weighted average formula for $\bar{x}$ suggests the following definition:

Mean of a Distribution

The *mean* μ of a probability distribution $P(x)$ over a finite set of numerical values x is the average of the values x weighted by their probabilities:

$$\mu = \sum_{\text{all } x} x \, P(x)$$

The center of gravity. If you think of a distribution of mass instead of probability, the mean is the *center of gravity.* Think of a histogram of the distribution as a shape cut from a rigid material of constant thickness and density. The mean value is then a balance point for the histogram. The shape balances when supported at the mean, tips over to the right when supported at a point to the left of the mean, and tips to the left when supported to the right of the mean. This is due to the principle of moments in mechanics.

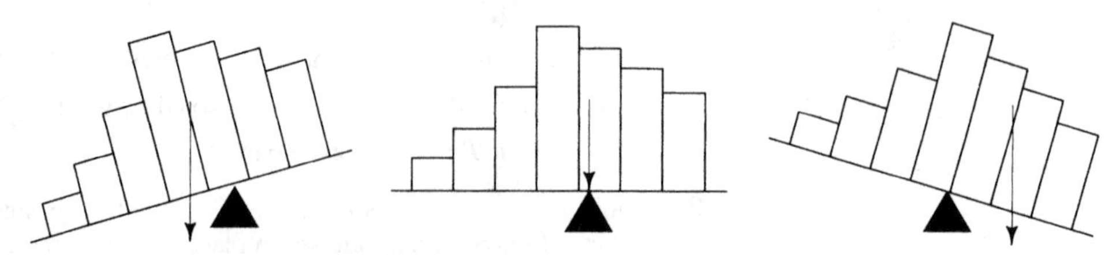

Mean of the binomial distribution. It is shown later in this section that the general definition of the mean μ of a distribution is consistent with the formula $\mu = np$

for the binomial (n, p) distribution, used in Chapter 2. In n independent trials with probability p of success on each trial, you expect to get around $\mu = np$ successes. So it is natural to say that the expected number of successes in n trials is np. This suggests the following definition of the expected value $E(X)$ of a random variable X. For X the number of successes in n trials, this definition makes $E(X) = np$. See Example 7.

Definition of Expectation

The *expectation* (also called *expected value*, or *mean*) of a random variable X, is the mean of the distribution of X, denoted $E(X)$. That is

$$E(X) = \sum_{\text{all } x} x\, P(X = x)$$

the *average of all possible values of X, weighted by their probabilities.*

Example 1. Random sampling.

Suppose n tickets numbered $x_1, \ldots, x_n$ are put in a box and a ticket is drawn at random. Let X be the x-value on the ticket drawn. Then $E(X) = \bar{x}$, the ordinary average of the list of numbers in the box. This follows from the above definition, and the weighted average formula (1) for $\bar{x}$, because the distribution of X is the empirical distribution of x-values in the list:

$$P(X = x) = P_n(x) = \#\{i : 1 \leq i \leq n \text{ and } x_i = x\}/n$$

Example 2. Two possible values.

If X takes two possible values, say a and b, with probabilities $P(a)$ and $P(b)$, then

$$E(X) = aP(a) + bP(b)$$

where $P(a) + P(b) = 1$. This weighted average of a and b is a number between a and b, proportion $P(b)$ of the way from a to b. The larger $P(a)$, the closer $E(X)$ is to a; and the larger $P(b)$, the closer $E(X)$ is to b.

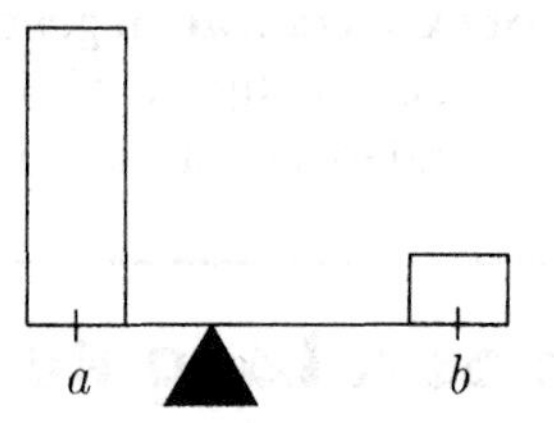

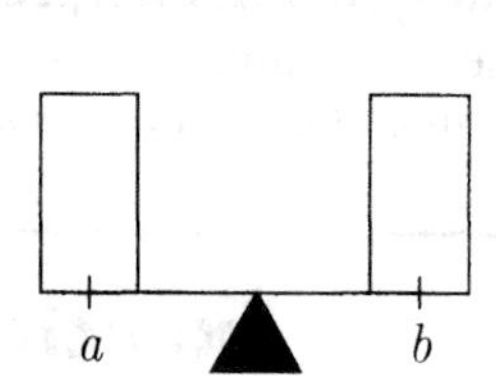

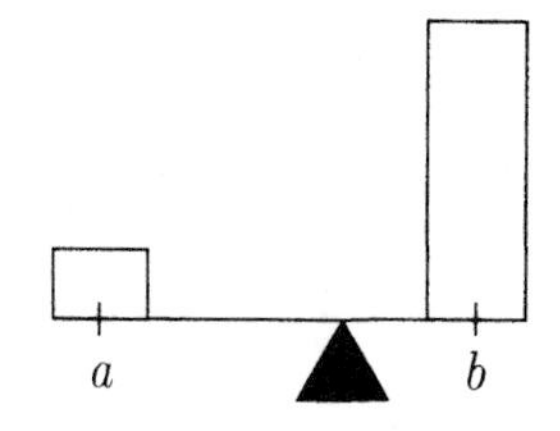

Example 3. Indicators.

This is the special case of the previous example for $a = 0$ and $b = 1$. Suppose $X = I_A$ is the *indicator* of event A. Since I_A has value 1 if A occurs, 0 otherwise, the events $(I_A = 1)$ and A are identical by definition. So

$$E(I_A) = 0P(I_A = 0) + 1P(I_A = 1) = P(A)$$

Indicators may seem trivial at first. But they combine to produce more interesting random variables by sums and products. Examples follow later in this section.

Example 4. Rolling a die.

Suppose X is the number produced by rolling a fair die. The definition of $E(X)$ makes

$$\begin{aligned} E(X) &= 1P(X = 1) \ + \ 2P(X = 2) \ + \ \cdots \ + \ 6P(X = 6) \\ &= 1 \times \frac{1}{6} \ + \ 2 \times \frac{1}{6} \ + \ 3 \times \frac{1}{6} \ + \ 4 \times \frac{1}{6} \ + \ 5 \times \frac{1}{6} \ + \ 6 \times \frac{1}{6} = 3.5 \end{aligned}$$

Of course, you should not expect a single die roll to be 3.5. But if you roll the die a large number of times you should expect the average of the rolls to be close to 3.5. To see why, calculate the sum of the rolls by grouping terms of the same value:

$$\text{sum of the rolls} = 1 \times (\text{number of 1's}) + \cdots + 6 \times (\text{number of 6's})$$

Dividing by the total number of rolls now gives

$$\text{average of the rolls} = 1 \times (\text{proportion of 1's}) + \cdots + 6 \times (\text{proportion of 6's})$$

Assuming a large number of independent rolls, each of these proportions is likely to be very close to $1/6$, by the law of large numbers. The average of the rolls will then be close to $E(X) = 3.5$. If the die were biased, with probability p_i of rolling number i, the same reasoning shows the long-run average is likely to be very close to

$$E(X) = 1p_1 + 2p_2 + 3p_3 + 4p_4 + 5p_5 + 6p_6$$

The long-run interpretation of expectation. In general, the long-run argument in the last example leads to the conclusion in the next box. A more precise formulation of this idea, a law of averages for independent trials, is given in Section 3.3.

Expectation as a Long-Run Average

If probabilities for values of X are approximate long-run frequencies, then $E(X)$ is approximately the long-run average value of X.

Because expectation approximates a long-run average (and because Example 1 equates an expectation and an average), the properties of expectation described in this section parallel properties of the ordinary average of a list of numbers. A summary of these properties of averages and expectations is displayed on pages 180 – 181.

Comparison of the mean with other measures of location. The mean is one way to locate a central point in the distribution of X. But there are other ways, for example, the *mode* and the *median*. A mode is the most likely possible value of X (there may be more than one). And a median is a number m such that both $P(X \leq m)$ and $P(X \geq m)$ are at least $1/2$. There may be more than one median. For example, if X is the number on a fair die, every integer between 1 and 6 is a mode of X, and every number between 3 and 4 is a median of X. The mean, the mode, and the median may be quite different. But if the distribution is symmetric about some point m, and has a single mode, the three quantities all equal m. Of all measures of location, the mean is most important in theory. This is due to the close connection between means and long-run averages, and the fact shown later in this section that the mean of the sum of two random variables is the sum of the means. There is no such simple rule for modes or medians.

FIGURE 1. Mean, mode, and median.

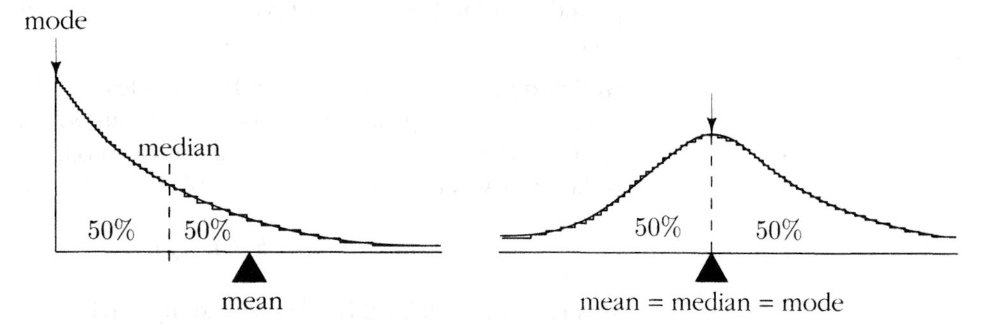

Gambling Interpretation of Expectation: The Fair Price

Suppose you bet on an outcome of some kind. You pay a fixed amount $\$b$ to place the bet, and the return from the bet is the random amount $\$X$. For example, you might pay \$4 to buy a return of $\$X$ where X is the number produced by a fair die roll. Suppose you made a long series of such bets, with independent repetitions of whatever random mechanism generates X, for example successive rolls of the die, or successive spins of a roulette wheel. After n repetitions, you have paid out $\$nb$ to place the bets. The return from your bets is the sum $S_n = X_1 + \cdots + X_n$, where X_i is the return from the ith bet. The basic assumption is that the X_i are independent with the same distribution as X. By the law of large numbers, the long-run proportion of trials that yield x is approximately $P(X = x)$. So over the n trials you should expect to see the return x about $nP(X = x)$ times. The total or *gross return* from a large number n of bets (not subtracting the price of the bets) should therefore be around

$$\$\sum_x x\, n\, P(X = x) = \$nE(X).$$

To summarize:

Over the long run, for a series of independent bets with returns like $\$X$, the average gross return per bet will probably be close to $\$E(X)$.

If you pay the same price $\$b$ to bet each time, your long-run *net return per bet* from a large number of bets will probably be about $\$(E(X) - b)$. To illustrate, if you pay $4 for the return of $\$X$ for X the number on a fair die roll, over the long run you should expect to lose about 50 cents a game. Such considerations lead to the following interpretation of $E(X)$ as a *fair price*:

$\$E(X)$ is the fair price to pay for a return of $\$X$. This price makes wins and losses tend to cancel out over the long run.

Precise information about the degree of cancellation of wins and losses to be expected over the long run is provided by the normal approximation in the next section.

Indicator variables and fair odds. The idea of a fair price is a generalization of the fair odds rule presented in Section 1.1. Suppose you pay the price $\$b$ to get a return of $1 if an event A occurs, and no return otherwise. The return from your bet is then $\$I_A$ where I_A is the indicator of A. The fair price for this return is $\$b$ where

$$b = E(I_A) = P(A).$$

This restates the fair odds rule (see Example 1.1.4).

The Addition Rule

Let $\$X$ and $\$Y$ be the returns from two bets on an outcome of some kind, for instance the returns from two stakes placed on different groups of numbers for a single spin of a roulette wheel. The combined return from the two bets is $\$(X + Y)$. It is quite intuitive that the fair price for this combination of two bets is

$$\$E(X + Y) = \$E(X) + \$E(Y),$$

the sum of the fair prices of the individual bets. This is the fundamental *addition rule* of expectation stated in the following box, and derived from the definition of expectation on page 177:

Addition Rule for Expectation

For any two random variables X and Y defined in the same setting,

$$E(X+Y) = E(X) + E(Y)$$

no matter whether X and Y are independent or not. Consequently, for a sequence of random variables $X_1, \ldots, X_n$, however dependent,

$$E(X_1 + \cdots + X_n) = E(X_1) + \cdots + E(X_n)$$

In calculations the definition of expectation

$$E(X) = \sum_{\text{all } x} x\, P(X = x)$$

is useful only if the formula for $P(X = x)$ allows an easy evaluation of the sum over all x of $xP(X = x)$. This happens only in the simplest examples. But even if the distribution of X is hard to compute, it is often possible to write X as a sum of simpler variables whose expectations are easily found. Then the expectation of X is found by the addition rule.

Example 5. Sum of dice.

Problem. Let T_n be the sum of numbers from n dice. Find $E(T_n)$.

Solution. Let $X_1, \ldots, X_n$ be the numbers obtained from the n die rolls. Then

$$\begin{aligned} T_n &= X_1 + \cdots + X_n, \qquad \text{so} \\ E(T_n) &= E(X_1) + \cdots + E(X_n) \qquad \text{by the addition rule} \\ &= 3.5 + \cdots + 3.5 \qquad (n \text{ terms}) \\ &= (3.5)n \end{aligned}$$

Discussion. Despite the fact that the distribution of T_n becomes more and more difficult to calculate exactly as n increases, the formula for $E(T_n)$ is simple. As a check, $E(T_2)$ can be found from its distribution:

$$\begin{aligned} E(T_2) = {} & 2 \times \frac{1}{36} + 3 \times \frac{2}{36} + 4 \times \frac{3}{36} + 5 \times \frac{4}{36} + 6 \times \frac{5}{36} + 7 \times \frac{6}{36} \\ & + 8 \times \frac{5}{36} + 9 \times \frac{4}{36} + 10 \times \frac{3}{36} + 11 \times \frac{2}{36} + 12 \times \frac{1}{36} \\ = {} & 7. \end{aligned}$$

The Method of Indicators

The idea of the method of indicators is that the random variable X that counts the number of events of some kind that occur can be represented as the sum of the indicators of these events. Then, by the addition rule for expectation, $E(X)$ is just the sum of the probabilities of the events. This is illustrated by the following two examples. First, it is worth restating the result of Example 3:

Expectation of an Indicator

The expectation of the indicator of an event is the probability of the event:

$$E(I_A) = P(A)$$

Example 6. **Working components.**

Suppose a system has n components, and that at a particular time the jth component is working with probability $p_j, j = 1, \ldots, n$. Let X be the number of components working at that time.

Problem. Find a formula for $E(X)$.

Solution. No matter which components work and which do not, the total number X that work can be found by adding 1 for each component that works and 0 for each component that does not. This is an expression for X in terms of indicators. Let I_j be the indicator random variable, which is 1 if the jth component is working, 0 otherwise. Then, as illustrated in Figure 2 for the case $n = 3$,

$$X = I_1 + I_2 + \cdots + I_n$$

FIGURE 2. Venn diagram for the number of working components. Here $n = 3$. The event that a particular component works is represented by the area inside a circle. These can overlap in any way.

Values of I_1, I_2, I_3

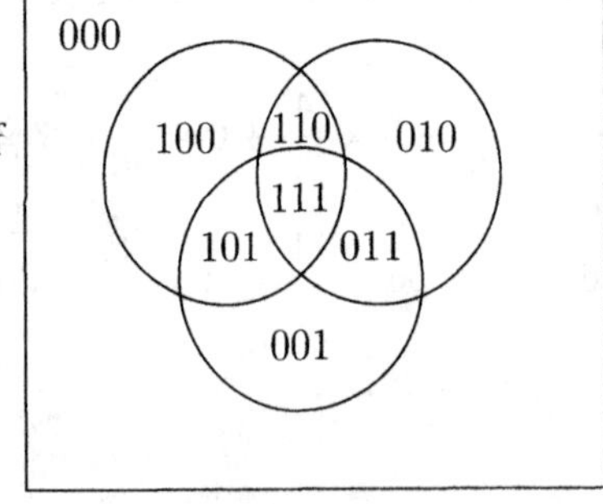

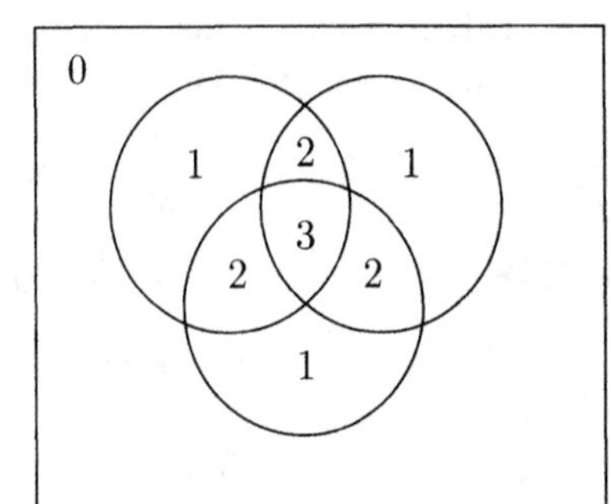

Values of $X = I_1 + I_2 + I_3$

Now take expectations of both sides. By the addition rule, and the fact that the expectation of I_j is p_j,

$$E(X) = p_1 + p_2 + \cdots + p_n$$

Discussion. You might think this problem could not be solved without further assumptions. True, the distribution of X cannot be found without assumptions about the dependence between the components. But due to the addition rule, $E(X)$ is the same, no matter what the dependence.

Example 7. Mean of the binomial distribution.

Suppose X is the number of successes in n independent trials with probability p of success in each trial, so X has binomial (n, p) distribution, as in Chapter 2.

Problem. Derive the formula $\mu = np$ for the mean of the binomial (n, p) distribution from the general definition of mean in this section.

Solution. As in the previous example, the total number of successes in the n trials can be written as a sum of indicators $X = I_1 + \cdots + I_n$ where I_j is the indicator of success on trial j, so $E(I_j) = p$ for each j, and the expected number of successes is

$$\begin{aligned} E(X) &= p + p + \cdots + p \qquad (n \text{ terms}) \\ &= np \end{aligned}$$

Discussion. This is not so obvious from the definition of $E(X)$:

$$E(X) = \sum_{\text{all } x} xP(X = x) = \sum_{x=0}^{n} x \binom{n}{x} p^x (1-p)^{n-x}$$

The calculation by the method of indicators implies that this expression must simplify to np. You can check this by algebra using the binomial theorem.

The general method. Examples 6 and 7 both illustrate the method of indicators. The general idea is that a random variable X with possible values $\{0, 1, \ldots, n\}$ can always be represented as counting the number of events that occur in some list of n events, say $A_1, \ldots, A_n$. Then X is called a *counting variable*. A suitable definition of the events A_j is usually clear from a verbal description of X. For instance,

- if X is the number of components that work among n components, let A_j be the event that the jth component works (Example 6).
- if X is the number of successes in n trials, let A_j be the event of success on trial j, for $1 \le j \le n$ (Example 7).
- if X is the number of aces in a 5-card poker hand, let A_j be the event that the jth card dealt is an ace, $1 \le j \le 5$ (Example 8).

The statement

$$X \text{ is the number of events } A_j \text{ that occur}$$

is expressed mathematically by the identity of random variables

$$X = I_1 + I_2 + \cdots + I_n \tag{2}$$

where I_j is the indicator of A_j. To illustrate, for X the number of aces in 5 cards, if the first and third cards are aces and the rest are not, this equation reads

$$2 = 1 + 0 + 1 + 0 + 0$$

while if the first three cards are aces and the last two are not, the equation reads

$$3 = 1 + 1 + 1 + 0 + 0$$

The point is that the number of aces can be found this way by adding zeros and ones, *no matter what the arrangement of the cards.* An equality like this, that holds by definition of the variables involved no matter what the outcome, is an *identity of random variables.* Take the expectation of both sides of (2), use the addition rule for expectation, and the fact that $E(I_j) = P(A_j)$ by definition of I_j as the indicator of A_j, to obtain the following generalization of the result of Examples 6 and 7:

Expected Number of Events that Occur

If X is the number of events that occur among some collection of events $A_1, \ldots, A_n$, then

$$E(X) = P(A_1) + P(A_2) + \cdots + P(A_n) \tag{3}$$

Usually it is easy to find $P(A_j)$, and add the results to find $E(X)$, as in Examples 6 and Example 7.

Example 8. The number of aces.

Let X be the number of aces in a 5-card poker hand. The probability that any particular card is an ace is $4/52$ (Examples 1.4.7 and 3.1.6), so the expected number of aces among 5 cards dealt from a well-shuffled deck is

$$E(X) = 5 \times 4/52 = 5/13$$

Compare with the method of computing $E(X)$ directly from the definition of $E(X)$ in terms of the distribution of X which was found in Section 2.5:

$$P(X = x) = \sum_{x=0}^{4} xP(X = x) = \sum_{x=0}^{4} x \binom{4}{x}\binom{48}{5-x} \Big/ \binom{52}{5}$$

You can check the second method gives the same answer. But the first method is quicker.

When to use the method of indicators. The examples show how the method of indicators can be used to find $E(X)$ for a counting variable X in either of the following circumstances:

- The probabilities $P(X = x)$ are known, but given by a formula that makes the expression $E(X) = \sum_x xP(X = x)$ hard to simplify.
- The nature of the dependence between the events A_j is either unknown, or known but so complicated that it is difficult to obtain a formula for $P(X = x)$.

The exact distribution of X depends in a fairly complicated way on the probabilities of various intersections of events being counted (Review Exercise 35). But, no matter what the dependence, the mean of the distribution is always given by the simple formula (3) for $E(X)$. There is usually more than one way to write a counting variable X as the sum of indicators of some collection of events. To find $E(X)$, all you need is one such collection of events whose probabilities you can calculate.

The tail sum formula for expectation of a counting variable. Every random variable with possible values $\{0, 1, \ldots, n\}$, however defined, is a counting variable representing number of events that occur in some list of n events $A_1, \ldots, A_n$. To see this, let A_j be the event $(X \geq j)$. If $X = x$ for $0 \leq x \leq n$, then A_j occurs for $1 \leq j \leq x$, and A_j does not occur for $x < j \leq n$. So if $X = x$ the number of events A_j that occur is precisely x. The resulting formula for $E(X)$ obtained by the method of indicators is displayed in the following box. Example 9 below gives an application.

Tail Sum Formula for Expectation

For X with possible values $\{0, 1, \ldots, n\}$,

$$E(X) = \sum_{j=1}^{n} P(X \geq j)$$

Alternative proof of the tail sum formula. Define $p_j = P(X = j)$. Then the expectation $E(X) = 1p_1 + 2p_2 + 3p_3 + \cdots + np_n$ is the following sum:

$$
\begin{array}{l}
p_1 \\
+p_2 + p_2 \\
+p_3 + p_3 + p_3 \\
\cdots \quad \cdots \quad \cdots \quad \cdots \\
+p_n + p_n + p_n + \cdots + p_n
\end{array}
$$

By the addition rule of probabilities, and the assumption that the only possible values of X are $\{0, 1, \ldots, n\}$, the sum of the first column of p's is $P(X \geq 1)$, the sum of the second column is $P(X \geq 2)$, and so on. The sum of the jth column is $P(X \geq j)$, $1 \leq j \leq n$. The whole sum is the sum of the column sums. □

Example 9. **Expectation of a minimum.**

Suppose that four dice are rolled.

Problem 1. Let M be the minimum of four numbers rolled. Find $E(M)$.

Solution. For any $1 \leq j \leq 6$, the event $(M \geq j)$ means that each X_i is at least j, where X_i is the number on the ith die. Thus

$$P(M \geq j) = P(X_1 \geq j, X_2 \geq j, X_3 \geq j, X_4 \geq j) = \left(\frac{6 - j + 1}{6}\right)^4$$

by independence of the X's, and fact that there are $6 - j + 1$ possible values for each X between j and 6. The tail sum formula gives

$$E(M) = P(M \geq 1) + P(M \geq 2) + \cdots + P(M \geq 6)$$
$$= \left(\frac{6}{6}\right)^4 + \left(\frac{5}{6}\right)^4 + \left(\frac{4}{6}\right)^4 + \left(\frac{3}{6}\right)^4 + \left(\frac{2}{6}\right)^4 + \left(\frac{1}{6}\right)^4 \approx 1.755$$

Discussion. The point of using the tail sum formula in this example is that the tail probabilities $P(M \geq j)$ are simpler than the individual probabilities

$$P(M = m) = P(M \geq m) - P(M \geq m + 1)$$

If you substitute this in the definition $E(M) = \sum_m mP(M = m)$, and simplify, you will find the coefficient of $P(M \geq j)$ is 1 for each j from 1 to n. That is the substance of the tail sum formula.

Problem 2. Let S be the sum of the largest three numbers among four dice. Find $E(S)$.

Solution. Notice that $S = T - M$, where T is the sum of all four numbers, and M is the minimum number. From Example 5, $E(T) = 4 \times (3.5) = 14$, and the value of $E(M)$ was just found. Since by the addition rule for expectation,

$$E(T) = E(T - M) + E(M) = E(S) + E(M)$$

$$E(S) = E(T) - E(M) = 14 - 1.755 = 12.245$$

Remark. It is much harder to find $E(S)$ via the distribution of S.

When is the sum of indicators an indicator? A sum of 0's and 1's is 0 or 1 if and only if there is at most a single 1 among all the terms. For events A_j with indicators I_j, this means that $\sum_j I_j$ is an indicator variable if and only if at most one of the events A_j can occur, that is, if and only if the events A_j are *mutually exclusive*. Then $\sum_j I_j$ is the indicator of the event $\bigcup_j A_j$ that at least one of the events A_j occurs. So in this case the result of the method of indicators is just the addition rule for probabilities:

$$P(\textstyle\bigcup_j A_j) = \sum_j P(A_j) \text{ if the } A_j \text{ are mutually exclusive.}$$

Boole's inequality. In general, for possibly overlapping events A_j, the above equality is replaced by *Boole's inequality* of Exercise 1.3.13:

$$P(\textstyle\bigcup_j A_j) \leq \sum_j P(A_j)$$

If X is the number of events A_j that occur, the left side is $P(X \geq 1)$, and the right side is $E(X)$. So Boole's inequality can be restated as follows: for any counting random variable X,

$$P(X \geq 1) \leq E(X)$$

This follows from the addition rule of probabilities and the definition of $E(X)$:

$$\begin{aligned} P(X \geq 1) &= p_1 + p_2 + p_3 + \cdots + p_n \\ &\leq p_1 + 2p_2 + 3p_3 + \cdots + np_n = E(X) \end{aligned}$$

To illustrate, Example 8 showed the expected number of aces among 5 cards is $5/13$. So the probability of at least one ace among 5 cards is at most $5/13 \approx 0.385$. The exact probability of at least one ace among 5 cards is $1 - \binom{48}{5}/\binom{52}{5} \approx 0.341$. In this case the upper bound of Boole's inequality is quite close to the exact probability of the union of events, because the probability of two or more aces in 5 cards is rather small (about 0.042). In other words, the events $A_1, \ldots, A_5$ do not overlap very much.

A generalization of Boole's inequality, called Markov's inequality, is illustrated by the following example:

Example 10. **Bounding a tail probability.**

Problem. For a non-negative random variable X with mean $E(X) = 3$, what is the largest that $P(X \geq 100)$ could possibly be?

Solution. The constraint that X is non-negative, i.e., $X \geq 0$, means that $P(X \geq 0) = 1$. In other words, all the probability in the distribution of X is in the interval $[0, \infty)$. Think of balancing a distribution of mass at 3, with all the mass in $[0, \infty)$. How can you get as much mass as possible in the interval $[100, \infty)$? Intuitively, the best you can do is to put some of the mass at 100 and the rest at 0 (as far to the left as allowed by the non-negativity constraint). This distribution balances at 3 if the proportion at 100 is $3/100$. This shows $P(X \geq 100)$ can be as large as $3/100$, and suggests it cannot be larger. Here is a proof. In the sum

$$\sum_{\text{all } x} xP(X = x) = 3$$

the terms with $x \geq 100$ contribute

$$\sum_{x \geq 100} xP(X = x) \geq \sum_{x \geq 100} 100P(X = x) = 100P(X \geq 100)$$

while all the terms are non-negative by the assumption that $X \geq 0$. This then gives $3 \geq 100P(X \geq 100)$, or $P(X \geq 100) \leq 3/100$.

Discussion. With arbitrary $E(X)$ and a instead of 3 and 100, this proves the following inequality. The point is that if $X \geq 0$, meaning all the possible values of X are non-negative, or $P(X \geq 0) = 1$, then knowing $E(X)$ puts a bound on how large the tail probability $P(X \geq a)$ can be.

Markov's Inequality

If $X \geq 0$, then $P(X \geq a) \leq \dfrac{E(X)}{a}$ for every $a > 0$.

Expectation of a Function of a Random Variable

Recall from Section 3.1 that if X is a random variable with a finite set of possible values, and $g(x)$ is a function defined on this set of possible values, then $g(X)$ is also a random variable. Examples of typical functions of a random variable X, whose expectations may be of interest, are X, X^2, X^k for some other power k, $\log(X)$ (assuming $X > 0$), e^X, or z^X for some other number z. The notation $g(X)$ is used for a generic function of X.

Expectation of a Function of X

Typically, $E[g(X)] \neq g[E(X)]$. Rather

$$E[g(X)] = \sum_{\text{all } x} g(x)P(X = x) \tag{4}$$

This formula is valid for any numerical function g defined on the set of possible values of X. In particular, for $g(x) = x^k$ with $k = 1, 2, \dots$ the number

$$E(X^k) = \sum_{\text{all } x} x^k P(X = x)$$

derived from the distribution of X is called the *kth moment of* X.

The point of formula (4) is that it expresses $E[g(X)]$ directly in terms of the distribution of X, without consideration of the set of possible values of $g(X)$ or the distribution of $g(X)$ over these values. This is an important shortcut in many calculations.

Proof of the formula for $E[g(x)]$. Look at the sum $\sum_{\text{all } x} g(x)P(X = x)$, which is claimed to equal $E[g(X)]$. Group the terms according to the value y of $g(x)$. The terms from x with $g(x) = y$ have sum

$$\sum_{x:g(x)=y} g(x)P(X = x) = \sum_{x:g(x)=y} yP(X = x) = yP(g(X) = y)$$

Now summing over all y gives $E[g(X)]$. □

Constant factors. If X is a random variable, then so is cX for any constant c. This is $g(X)$ for $g(x) = cx$. Apply the formula for $E[g(X)]$ and factor the c out of the sum to see that $E(cX) = cE(X)$. So constants can be pulled outside the expectation operator.

Constant random variables. It is sometimes useful to think of a constant c as a random variable with just one possible value c. Of course, the expected value of a constant random variable is its constant value.

Linear functions. The expectation of a linear function of X is determined by the mean or first moment of X:

$$E(aX + b) = E(aX) + E(b) = aE(X) + b$$

This is immediate from the addition rule and the last two paragraphs. Linear functions $g(x) = ax + b$ are exceptional in that $E[g(X)] = g(E(X))$, a rule that is false for a general function g.

Moments. The *first moment* of X is just the mean or expectation of X. The *second moment* of X is $E(X^2)$, sometimes called the *mean square* of X. The term *moment* is borrowed from mechanics where similar averages with respect to a distribution of mass rather than probability have physical interpretations (principle of moments, moment of inertia). The moments of X are features of the distribution of X. Two random variables with the same distribution have the same moments. The first two moments of a distribution are by far the most important. The first moment gives a central value in the distribution. It will be seen in the next section that a quantity called *variance* derived from the first two moments gives an indication of how spread out the distribution is. Third moments are used to describe the degree of asymmetry of a distribution. Higher moments of X are hard to interpret intuitively. But they play an important part in theoretical calculations beyond the scope of this book.

It will be seen in the next section that

$$E(X^2) \neq [E(X)]^2$$

except in the trivial case when X is a constant random variable.

Example 11. **Uniform distribution on three values.**

If X is uniformly distributed on $\{-1, 0, 1\}$, then X has mean

$$E(X) = -1 \times \frac{1}{3} + 0 \times \frac{1}{3} + 1 \times \frac{1}{3} = 0$$

so $[E(X)]^2 = 0$. But, by the formula for $E[g(X)]$ with $g(X) = X^2$, the second moment of X is

$$E(X^2) = (-1)^2 \times \frac{1}{3} + 0^2 \times \frac{1}{3} + 1^2 \times \frac{1}{3} = \frac{2}{3} \neq [E(X)]^2 = 0$$

Quadratic functions. The first two moments of X determine the expectation of any quadratic function of X. For instance, the quantity $E[(X - b)^2]$ for a constant b, which arises in a prediction problem considered below, is found by expanding $(X - b)^2 = X^2 - 2bX + b^2$ and using the rules of expectation to obtain

$$E[(X - b)^2] = E[X^2 - 2bX + b^2] = E(X^2) - 2bE(X) + b^2$$

Functions of two or more random variables. The proof of the formula for $E[g(X)]$ shows that this formula is valid for any numerical function g of a random variable X with a finite number of possible values, even if these values are not numerical. In particular, substituting a random pair (X, Y) instead of X gives a formula for the expectation of $g(X, Y)$ for a generic numerical function g of two variables:

$$E[g(X, Y)] = \sum_{\text{all } (x,y)} g(x, y) P(X = x, Y = y)$$

Proof of the addition rule. Think of X, Y, and $X+Y$ as three different functions of X and Y, two random variables with a joint distribution specified by probabilities $P(x,y) = P(X=x, Y=y)$. By three applications of the formula for $E[g(X,Y)]$,

$$E(X) = \sum_{\text{all }(x,y)} xP(x,y)$$

$$E(Y) = \sum_{\text{all }(x,y)} yP(x,y)$$

$$E(X+Y) = \sum_{\text{all }(x,y)} (x+y)P(x,y)$$

Add the expressions for $E(X)$ and $E(Y)$ and simplify to get the expression for $E(X+Y)$. Conclusion: The addition rule $E(X)+E(Y) = E(X+Y)$.

Expectation of a product. As in the proof of the addition rule, view XY, the product of X and Y, as a function of (X,Y) to obtain

$$E(XY) = \sum_x \sum_y xyP(X=x, Y=y)$$

where the double sum is a sum over all pairs (x,y) of possible values for (X,Y). This formula holds regardless of whether or not X and Y are independent. If X and Y are independent, the formula can be simplified as follows:

$$E(XY) = \sum_x \sum_y xyP(X=x)P(Y=y)$$

$$= \left[\sum_x xP(X=x)\right]\left[\sum_y yP(Y=y)\right] = [E(X)][E(Y)]$$

This yields the following:

Multiplication Rule for Expectation

If X and Y are **independent** then

$$E(XY) = [E(X)][E(Y)]$$

This multiplication rule will be used in the next section. Note well the assumption of independence. In contrast to the addition rule, the multiplication rule does not hold in general for dependent random variables. For example, if $X = Y$, the left side becomes $E(X^2)$ and the right side becomes $[E(X)]^2$. These two quantities are typically not equal (Example 11).

Expectation and Prediction

Suppose you want to predict the value of a random variable X. What is the best predictor of X? To define "best" you must decide on a criterion and a class of predictors. The simplest prediction problem is to predict the value of X by a constant, say b. Think in terms of losing some amount $L(x,b)$ if you predict b and the value of X is actually x. The function $L(x,b)$ is called a *loss function* in decision theory. It seems reasonable to try to pick b so as to minimize the *expected loss*, or *risk* $r(b) = E[L(X,b)]$

Example 12. **Right or wrong.**

Suppose that $L(x,b) = 0$ if $x = b$, and 1 otherwise. So you are penalized nothing if you get the value of X right, and penalized by one unit if you get the value of X wrong.

Problem. What is the best predictor?

Solution. $E[L(X,b)] = 0P(X = b) + 1P(X \neq b) = 1 - P(X = b)$.
So choosing b to minimize expected loss for this loss function is the same as choosing b to maximize $P(X = b)$. That is to say, b should be a mode of the distribution of X. Many probability distributions have a unique mode. But every possible value of a uniformly distributed random variable is a mode.

Example 13. **Absolute error.**

Suppose $L(x,b) = |x - b|$. So the penalty is the absolute value of the difference between the actual value and the predicted value. Now there is a bigger penalty for bigger mistakes. The expected loss is

$$r(b) = E(|X - b|) = \sum_x |x - b| P(X = x)$$

by the formula for $E[g(X)]$ applied to $g(x) = |x - b|$ for fixed b.

Problem. Find b that minimizes $r(b)$.

Solution. This time the solution is the median. To see why, look for a fixed x at the derivative

$$\frac{d}{db}|x - b| = \begin{cases} -1 & \text{if } b < x \\ 1 & \text{if } b > x \end{cases}$$

The sum defining $r(b)$ is over all possible values of X, say $x_1 < x_2 < \cdots < x_n$. So provided that $b \neq x_k$ for any k, the function $r(b)$ has the derivative

$$\begin{aligned}
\frac{dr(b)}{db} &= \sum_{x<b} 1P(X=x) + \sum_{x>b}(-1)P(X=x) \\
&= P(X<b) - P(X>b) \\
&= 2P(X \le x_k) - 1 \qquad \text{if} \quad x_k < b < x_{k+1}
\end{aligned}$$

So the function $r(b)$ is piecewise linear for b between x_k and x_{k+1}, decreasing if $P(X \le x_k) < 1/2$, increasing if $P(X \le x_k) > 1/2$, and flat if $P(X \le x_k) = 1/2$. So a b is minimizing if and only if $P(X \le b) \ge 1/2$ and $P(X \ge b) \ge 1/2$. Such a value b is a *median* of the distribution of X. A median always exists, but it may not be unique.

FIGURE 3. Risk functions for a die roll X with uniform distribution on $\{1, \ldots, 6\}$.
Left: Graph of the risk function $r(b) = E(|X - b|)$ for absolute error. (Refer to Example 13.) In this example, every number in the interval $[3, 4]$ is a median for X. Numbers in this interval are equally good as predictors of X according to the criterion of minimizing the expected absolute error, and better than any other number. **Right**: The risk function $r(b) = E[(X - b)^2]$ for quadratic loss function. (Refer to Example 14.) Now $E(X) = 3.5$ is the unique best predictor.

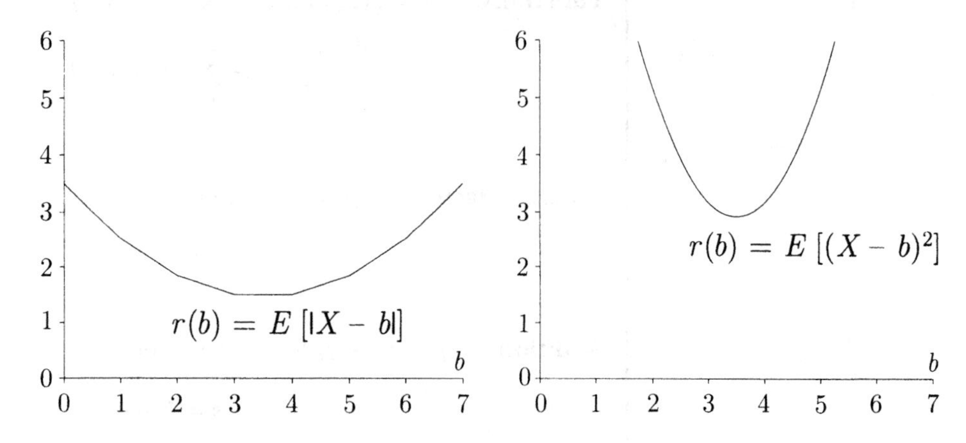

Example 14. **Squared error.**

Suppose now the penalty is *squared error*, using the *quadratic loss function* $L(x, b) = (x - b)^2$.

Problem. Find b that is the best constant predictor of X for this quadratic loss function.

Solution. This time the answer is just the mean. Now

$$r(b) = E[(X - b)^2] = E(X^2) - 2bE(X) + b^2$$

$$\frac{dr(b)}{db} = -2E(X) + 2b$$

so $b = E(X)$ gives the *unique* best predictor of X for the quadratic loss function.

Properties of Averages

Definition. The average of a list of numbers $x_1, \ldots, x_n$ is

$$\bar{x} = (x_1 + \cdots + x_n)/n = \sum_{\text{all } x} x P_n(x)$$

where $P_n(x)$ is the proportion of the n values x_k that are equal to x (empirical distribution of the list).

Constants. If $x_k = c$ for every k, then

$$\bar{x} = c$$

Indicators. If every number x_k in a list is either a zero or a one, then

$$\bar{x} = \text{proportion of ones in the list}$$

Functions. If $y_k = g(x_k)$ for each k, typically $\bar{y} \neq g(\bar{x})$. But

$$\bar{y} = \sum_{\text{all } x} g(x) P_n(x)$$

Constant factors. If $y_k = c x_k$ for every k, where c is constant, then

$$\bar{y} = c\bar{x}$$

Addition. If $s_k = x_k + y_k$ for each k, then

$$\bar{s} = \bar{x} + \bar{y}$$

Multiplication. If $z_k = x_k y_k$ for each k, typically $\bar{z} \neq \bar{x}\bar{y}$.

Properties of Expectation

Definition. The expectation of a random variable X is

$$E(X) = \sum_{\text{all } x} xP(X = x)$$

(average of values of X weighted by their probabilities).

Constants. The expectation of a constant random variable is its constant value

$$E(c) = c$$

Indicators. If I_A is the indicator of an event A, so $I_A = 1$ if A occurs, 0 otherwise, then

$$E(I_A) = P(A)$$

Functions. Typically, $E[g(X)] \neq g[E(X)]$, but

$$E[g(X)] = \sum_{\text{all } x} g(x)P(X = x)$$

Constant factors. For a constant c,

$$E(cX) = cE(X)$$

Addition. The expectation of a sum of random variables is the sum of the expectations:

$$E(X + Y) = E(X) + E(Y) \qquad \text{even if } X \text{ and } Y \text{ are dependent.}$$

Multiplication. Typically, $E(XY) \neq E(X)E(Y)$. But

$$E(XY) = E(X)E(Y) \qquad \text{if } X \text{ and } Y \text{ are independent.}$$

Exercises 3.2

1. Suppose that 10% of the numbers in a list are 15, 20% of the numbers are 25, and the remaining numbers are 50. What is the average of the numbers in the list?

2. One list of 100 numbers contains 20% ones and 80% twos. A second list of 100 numbers contains 50% threes and 50% fives. A third list is obtained by taking each number in the first list and adding the corresponding number in the second list.

a) What is the average of the third list? Or is this not determined by the information given?

Repeat a) with adding replaced by b) subtracting c) multiplying by d) dividing by.

3. What is the expected number of sixes appearing on three die rolls? What is the expected number of odd numbers?

4. Suppose all the numbers in a list of 100 numbers are non-negative, and the average of the list is 2. Prove that at most 25 of the numbers in the list are greater than 8.

5. In a game of Chuck-a-Luck, a player can bet \$1 on any one of the numbers $1, 2, 3, 4, 5,$ and 6. Three dice are rolled. If the player's number appears k times, where $k \geq 1$, the player gets \$$k$ back, plus the original stake of \$1. Otherwise, the player loses the \$1 stake. Some people find this game very appealing. They argue that they have a $1/6$ chance of getting their number on each die, so at least a $1/6+1/6+1/6 = 50\%$ chance of doubling their money. That's enough to break even, they figure, so the possible extra payoff in case their number comes up more than once puts the game in their favor.

a) What do you think of this reasoning?

b) Over the long run, how many cents per game should a player expect to win or lose playing Chuck-a-Luck?

6. Let X be the number of spades in 7 cards dealt from a well-shuffled deck of 52 cards containing 13 spades. Find $E(X)$.

7. In a circuit containing n switches, the ith switch is closed with probability p_i, $i = 1, \ldots, n$. Let X be the total number of switches that are closed. What is $E(X)$? Or is it impossible to say without further assumptions?

8. Suppose $E(X^2) = 3, \quad E(Y^2) = 4, \quad E(XY) = 2.$ Find $E\left[(X+Y)^2\right]$.

9. Let X and Y be two independent indicator random variables, with $P(X = 1) = p$ and $P(Y = 1) = r$. Find $E[(X - Y)^2]$ in terms of p and r.

10. Let A and B be independent events, with indicator random variables I_A and I_B.

a) Describe the distribution of $(I_A + I_B)^2$ in terms of $P(A)$ and $P(B)$.

b) What is $E(I_A + I_B)^2$?

11. There are 100 prize tickets among 1000 tickets in a lottery. What is the expected number of prize tickets you will get if you buy 3 tickets? What is a simple upper bound for the probability that you will win at least one prize? Compare with the actual probability. Why is the bound so close?

12. Show that if a and b are constants with $P(a \le X \le b) = 1$, then $a \le E(X) \le b$.

13. Suppose a fair die is rolled ten times. Find numerical values for the expectations of each of the following random variables:

a) the sum of the numbers in the ten rolls;

b) the sum of the largest two numbers in the first three rolls;

c) the maximum number in the first five rolls;

d) the number of multiples of three in the ten rolls;

e) the number of faces which fail to appear in the ten rolls;

f) the number of different faces that appear in the ten rolls;

14. A building has 10 floors above the basement. If 12 people get into an elevator at the basement, and each chooses a floor at random to get out, independently of the others, at how many floors do you expect the elevator to make a stop to let out one or more of these 12 people?

15. Predicting demand. Suppose that a store buys b items in anticipation of a random demand Y, where the possible values of Y are non-negative integers y representing the number of items in demand. Suppose that each item sold brings a profit of $\$\pi$, and each item stocked but unsold brings a loss of $\$\lambda$. The problem is to choose b to maximize expected profit.

a) Show that this problem is the same as the problem of finding the predictor b of Y which minimizes over all integers the expected loss, with loss function

$$L(y,b) = \begin{cases} -\pi y + \lambda(b-y) & \text{if} \quad y \le b \\ -\pi b & \text{if} \quad y > b \end{cases}$$

b) Let $r(b) = E[L(Y,b)]$. Use calculus to show that $r(b)$ is minimized over all the real numbers b, and hence over all the integers b, at the least integer y such that $P(Y \le y) \ge \pi/(\lambda + \pi)$. *Note.* If $\pi = \lambda$, this is the median. If $\pi/(\lambda+\pi) = k\%$, this y is called the kth percentile of the distribution of Y.

16. Aces. A standard deck of 52 cards is shuffled and dealt. Let X_1 be the number of cards appearing before the first ace, X_2 the number of cards between the first and second ace (not counting either ace), X_3 the number between the second and third ace, X_4 the number between the third and fourth ace, and X_5 the number after the last ace. It can be shown that each of these random variables X_i has the same distribution, $i = 1, 2, \ldots, 5$, and you can assume this to be true.

a) Write down a formula for $P(X_i = k)$, $0 \le k \le 48$.

b) Show that $E(X_i) = 9.6$. [*Hint*: Do not use your answer to a).]

c) Are $X_1, \ldots, X_5$ pairwise independent? Prove your answer.

17. A box contains 3 red balls, 4 blue balls, and 6 green balls. Balls are drawn one-by-one without replacement until all the red balls are drawn. Let D be the number of draws made. Calculate: a) $P(D \le 9)$; b) $P(D = 9)$; c) $E(D)$.

18. Suppose that X is a random variable with just two possible values a and b. For $x = a$ and b find a formula for $p(x) = P(X = x)$ in terms of a, b and $\mu = E(X)$.

19. A collection of tickets comes in four colors: red, blue, white, and green. There are twice as many reds as blues, equal numbers of blues and whites, and three times as many greens as whites. I choose 5 tickets at random with replacement. Let X be the number of different colors that appear.

a) Find a numerical expression for $P(X \geq 4)$.

b) Find a numerical expression for $E(X)$.

20. Show that the distribution of a random variable X with possible values 0, 1, and 2 is determined by $\mu_1 = E(X)$ and $\mu_2 = E(X^2)$, by finding a formula for $P(X = x)$ in terms of μ_1 and μ_2, $\quad x = 0, 1, 2$.

21. Indicators and the inclusion–exclusion formula. Let I_A be the indicator of A. Show the following:

a) the indicator of A^c, the complement of A, is $I_{A^c} = 1 - I_A$;

b) the indicator of the intersection AB of A and B is the product of I_A and I_B: $I_{AB} = I_A I_B$;

c) For any collection of events $A_1, \ldots, A_n$, the indicator of their union is

$$I_{A_1 \cup A_2 \cup \cdots \cup A_n} = 1 - (1 - I_{A_1})(1 - I_{A_2}) \cdots (1 - I_{A_n})$$

d) Expand the product in the last formula and use the rules of expectation to derive the inclusion–exclusion formula of Exercise 1.3.12.

22. Success runs in independent trials. Consider a sequence of $n \geq 4$ independent trials, each resulting in success (S) with probability p, and failure (F) with probability $1 - p$. Say a *run of three successes* occurs at the beginning of the sequence if the first four trials result in SSSF; a run of three successes occurs at the end of the sequence if the last four trials result in FSSS; and a run of three successes elsewhere in the sequence is the pattern FSSSF. Let $R_{3,n}$ denote the number of runs of three successes in the n trials.

a) Find $E(R_{3,n})$.

b) Define $R_{m,n}$, the number of success runs of length m in n trials, similarly for $1 \leq m \leq n$. Find $E(R_{m,n})$.

c) Let R_n be the total number of non-overlapping success runs in n trials, counting runs of any length between 1 and n. Find $E(R_n)$ by using the result of b).

d) Find $E(R_n)$ another way by considering for each $1 \leq j \leq n$ the number of runs that start on the jth trial. Check that the two methods give the same answer.

3.3 Standard Deviation and Normal Approximation

If you try to predict the value of a random variable X by its mean $E(X) = \mu$, you will be off by the random amount $X - \mu$. It is often important to have an idea of how large this deviation is likely to be. Because

$$E(X - \mu) = E(X) - \mu = 0$$

it is necessary to consider either the absolute value or the square of $X - \mu$ to get an idea of the size of the deviation without regard to sign. Because the algebra is easier with squares than with absolute values, it is natural to first consider $E[(X - \mu)^2]$, then take a square root to get back to the same scale of units as X.

Definition of Variance and Standard Deviation

The *variance* of X, denoted $Var(X)$, is the mean squared deviation of X from its expected value $\mu = E(X)$:

$$Var(X) = E[(X - \mu)^2]$$

The *standard deviation* of X, denoted $SD(X)$, is the square root of the variance of X:

$$SD(X) = \sqrt{Var(X)}$$

Intuitively, $SD(X)$ should be understood as a measure of how spread out the distribution of X is around its mean μ. Because $Var(X)$ is a central value in the distribution of $(X - \mu)^2$, its square root $SD(X)$ gives a rough idea of the typical size of the absolute deviation $|X - \mu|$. Variance always appears as an intermediate step in the calculation of standard deviation. Variance is harder to interpret than SD, but has simpler algebraic properties. Notice that $E(X)$, $Var(X)$, and $SD(X)$ are all determined by the distribution of X. That is to say, if two random variables have the same distribution, then they have the same mean, variance, and SD. So we may speak of the mean, variance, and SD of a distribution rather than a random variable.

Parameters of a normal curve. If a histogram displaying the distribution of X follows an approximately normal curve, the curve will be centered near the mean $E(X)$, and $SD(X)$ will be approximately the distance between the center of the curve and its shoulders, where the curve switches from being concave to convex. See Figure 1 of Section 2.2. This observation is justified at the end of Section 4.1. For histograms which are approximately normal in shape, about 68% of the probability will lie in the interval within one standard deviation of the mean.

Meaning of SD when the distribution is not roughly normal. If the distribution of X is not roughly normal, there is no simple way to visualize $SD(X)$ in terms of the histogram of X. But no matter what the distribution of X, you should expect X to be around $E(X)$, plus or minus a few times $SD(X)$. This is made more precise later in this section by Chebychev's inequality. Like the mean $E(X)$, the standard deviation $SD(X)$ can be interpreted in terms of a sum $S_n = X_1 + \cdots + X_n$ of a large number n of random variables X_i with the same distribution as X. What happens is that for large n the distribution of S_n follows an approximately normal curve with parameters determined by $E(X)$, $SD(X)$, and n. This is made precise by the *central limit theorem* stated later in this section.

It is often simpler to calculate an SD using the following formula for variance rather than the definition.

Computational Formula for Variance

$$Var(X) = E(X^2) - [E(X)]^2 = \sum_{\text{all } x} x^2 P(X = x) - \left[\sum_{\text{all } x} x P(X = x)\right]^2$$

In words: Variance is the mean of the square minus the square of the mean.

Remark. The order of the two operations, squaring and taking expectation, is extremely important. Since from its original definition $Var(X)$ is non-negative, and zero if and only if $P(X = \mu) = 1$, the computational formula shows that

$$E(X^2) \geq [E(X)]^2$$

with equality if and only if X is a constant random variable.

Proof.

$$\begin{aligned}
E[(X - \mu)^2] &= E[X^2 - 2\mu X + \mu^2] \\
&= E(X^2) - 2\mu^2 + \mu^2 \qquad \text{by rules of } E \text{ using } E(X) = \mu \\
&= E(X^2) - \mu^2 \\
&= E(X^2) - [E(X)]^2 \qquad \text{because } \mu = E(X)
\end{aligned}$$

The second expression in the box comes from the formula for the expectation of a function of X, applied to $f(x) = x^2$, and the definition of $E(X)$. □

Example 1. Random sampling.

Suppose n tickets numbered $x_1, \ldots, x_n$ are put in a box and a ticket is drawn at random. Let X be the x-value on the ticket drawn. Then $E(X) = \bar{x}$, the average of the list of numbers in the box, as shown in Example 3.2.1. The corresponding formula for the standard deviation is $SD(X) = \sqrt{Var(X)}$ where

$$Var(X) = \frac{1}{n}\sum_i (x_i - \bar{x})^2 = \frac{1}{n}\sum_i x_i^2 - \bar{x}^2$$

The first formula comes from writing $X = x_I$ where I has uniform distribution on $\{1, 2, \ldots, n\}$, so $E[(X-\mu)^2] = E[(x_I - \bar{x})^2]$ is the expectation of a function of I. The second formula follows similarly from the computational formula for variance. The numbers $Var(X)$ and $SD(X)$ determined this way by a list of numbers are called the variance and standard deviation of the list. For a list of measurements on a scale of units like feet or inches, the SD of the list gives an indication of the typical magnitude of the difference between measurements in the list and their average, on the same scale of units as the measurements.

Example 2. Indicators.

Problem. Suppose X is the indicator of an event with probability p. Find $SD(X)$.

Solution. Since $0^2 = 0$ and $1^2 = 1$, we have $X^2 = X$. Therefore,

$$E(X^2) = E(X) = p$$

so the computational formula gives

$$Var(X) = E(X^2) - [E(X)]^2 = p - p^2 = p(1-p)$$
$$SD(X) = \sqrt{Var(X)} = \sqrt{p(1-p)}$$

Discussion. Since X has a binomial $(1, p)$ distribution, this agrees with the formula $\sqrt{npq}$ for the SD of the binomial (n, p) distribution given in Chapter 2. This formula for $n > 1$ is checked in a later example.

Example 3. Number on a die.

Problem. Let X be the number on a fair die. Find $SD(X)$.

Solution. By the computational formula

$$Var(X) = E(X^2) - \mu^2 = \frac{(1^2 + 2^2 + 3^2 + 4^2 + 5^2 + 6^2)}{6} - (3.5)^2 = \frac{35}{12}$$

$$SD(X) = \sqrt{35/12} = 1.71$$

Scaling and Shifting

For constants a and b, $SD(aX + b) = |a|SD(X)$

Shifting by a constant doesn't change the spread of the distribution, but multiplying by a or $-a$ spreads out the distribution by a factor of $|a|$. You can check this from the definition of SD, using properties of expectation. Compare with the corresponding formula for expectation:

$$E(aX + b) = aE(X) + b$$

Example 4. **Celsius to Fahrenheit.**

Problem. Suppose X represents a temperature in degrees Celsius, Y the same temperature in degrees Fahrenheit, so

$$Y = \frac{9}{5}X + 32$$

How are $E(Y)$ and $SD(Y)$ related to $E(X)$ and $SD(X)$?

Solution. $E(Y) = \frac{9}{5}E(X) + 32$ is $E(X)$ converted to degrees Fahrenheit. But the SD behaves differently

$$SD(Y) = \frac{9}{5}SD(X)$$

because standard deviation, as a measure of spread, is affected only by the scale factor 9/5, and not by the shift of 32.

Example 5. **Successes and failures.**

Problem. Let X be the number of successes in n trials of some kind, Y the number of failures in the same sequence of trials. Assuming that every trial results in either success or failure, how are $E(Y)$ and $SD(Y)$ related to $E(X)$ and $SD(X)$?

Solution.

$$X + Y = n \quad \text{so} \quad Y = n - X$$

$$E(Y) = n - E(X) \qquad SD(Y) = SD(X)$$

FIGURE 1. **Scaling and shifting.** The histograms display distributions of $Y = aX + b$ for various a and b. These are derived by rescaling the histogram of X shown at the center of the top row. Under the histogram of each Y are marked the points $E(Y) - SD(Y)$, $E(Y)$, and $E(Y) + SD(Y)$.

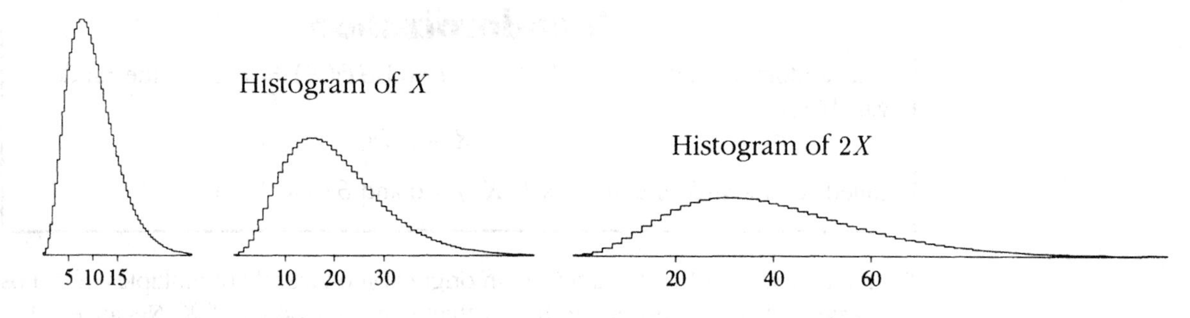

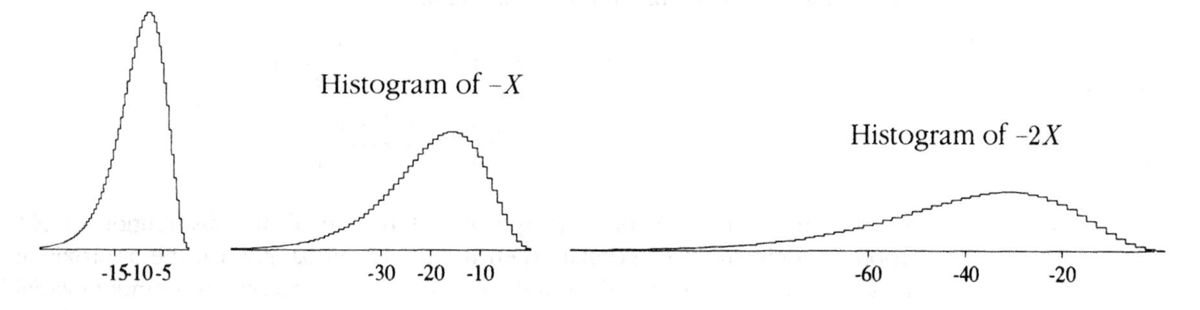

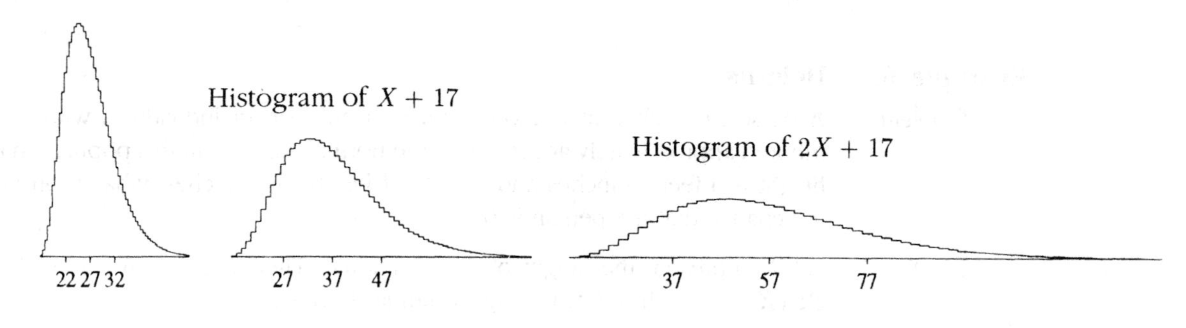

When making a normal approximation, it is convenient to transform a random variable X into a standardized variable X^*, which gives the number of SDs by which X differs from its expected value.

Standardization

If a random variable X has $E(X) = \mu$ and $SD(X) = \sigma > 0$, the random variable

$$X^* = (X - \mu)/\sigma$$

called X *in standard units,* has $E(X^*) = 0$ and $SD(X^*) = 1$.

Put another way, X^* is X relative to an origin at μ on a scale of multiples of σ. Positive values of X^* correspond to higher than expected values of X. Negative values of X^* correspond to lower than expected values of X. Any event determined by the value of X can be rewritten in terms of X^*. Usually, this is done by manipulating inequalities. For example, for any number b,

$$\begin{aligned} P(X \le b) &= P\left(\frac{X-\mu}{\sigma} \le \frac{b-\mu}{\sigma}\right) \\ &= P\left(X^* \le \frac{b-\mu}{\sigma}\right) \end{aligned}$$

In case the distribution of X is approximately normal, the distribution of X^* is approximately standard normal. Then the above probability can be approximated by $\Phi[(b-\mu)/\sigma]$, where Φ is the standard normal c.d.f. For a binomial random variable X this is the normal approximation of Chapter 2, except we are now ignoring the correction from b to $b + 1/2$ (called the *continuity correction*) which is appropriate only if the range of possible values of X is a sequence of consecutive integers.

Example 6. Heights.

Problem. A person is picked at random from a population of individuals with heights distributed approximately according to the normal curve. If in this population the mean height is 5 feet 10 inches and the SD of heights is 2 inches, what approximately is the chance that the person is over 6 feet tall?

Solution. Let X represent the height of the individual. Then $E(X) = 5$ feet 10 inches and $SD(X) = 2$ inches. Converting to standard units gives

$$\begin{aligned} P(X > 6 \text{ feet}) &= P\left(\frac{X - 5 \text{ feet } 10 \text{ inches}}{2 \text{ inches}} > 1\right) \\ &= P(X^* > 1) \approx 1 - \Phi(1) \approx 16\% \end{aligned}$$

by the normal approximation.

Tail Probabilities

Consider the event that a random variable X is more than three standard deviations from its mean. To get used to some notation, look at the following six equivalent symbolic expressions of this event, in terms of

$$E(X) = \mu, SD(X) = \sigma, \text{ and } X^* = (X - \mu)/\sigma.$$

The inequalities are manipulated by adding an arbitrary constant or multiplying by a positive constant. For example, division by σ turns (1) into (6):

$$|X - \mu| > 3\sigma \tag{1}$$

$$\text{either} \quad X - \mu < -3\sigma \quad \text{or} \quad X - \mu > 3\sigma \tag{2}$$

$$\text{either} \quad X < \mu - 3\sigma \quad \text{or} \quad X > \mu + 3\sigma \tag{3}$$

$$\text{either} \quad \frac{X - \mu}{\sigma} < -3 \quad \text{or} \quad \frac{X - \mu}{\sigma} > 3 \tag{4}$$

$$\text{either} \quad X^* < -3 \quad \text{or} \quad X^* > 3 \tag{5}$$

$$|X^*| > 3 \tag{6}$$

If the distribution of X closely follows the normal curve, the probability of this event will be very small: around $3/10$ of 1%, according to the normal table. But what if the distribution is not normal? How big could this probability be? 3%? or 30%? The answer is that it might be 3%, but not 30%. The largest this probability could possibly be, for any X whatsoever, is $1/9$, or about 11%. This is due to the following inequality, which makes precise the idea that a random variable is unlikely to be more than a few SDs away from its mean.

Chebychev's Inequality

For any random variable X, and any $k > 0$,

$$P[\,|X - E(X)| \geq k\,SD(X)\,] \leq \frac{1}{k^2}$$

In words: The probability that a random variable differs from its expected value by more than k standard deviations is at most $1/k^2$.

FIGURE 2. The probability bounded by Chebychev's inequality.

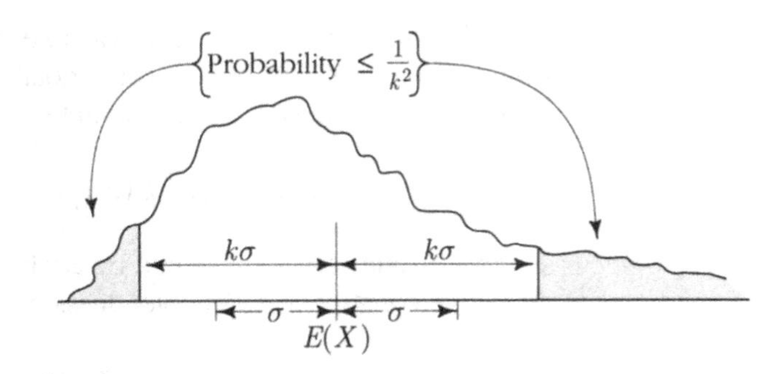

Proof. Let $\mu = E(X)$ and $\sigma = SD(X)$. The first step is yet another way of writing the event $[|X - \mu| \geq k\sigma]$, namely, $[(X - \mu)^2 \geq k^2\sigma^2]$. Now define $Y = (X - \mu)^2$, $a = k^2\sigma^2$, to see

$$P[|X - \mu| \geq k\sigma] = P(Y \geq a)$$

$$\leq \frac{E(Y)}{a} \quad \text{by Markov's inequality of Section 3.2, using } Y \geq 0,$$

$$= \frac{\sigma^2}{k^2\sigma^2} = \frac{1}{k^2} \quad \text{by definition of } Y, \sigma, \text{ and } a. \square$$

Comparison of the Chebychev bound with normal probabilities. Chebychev's inequality gives universal inequalities, satisfied by all distributions, no matter what their shape. For $k \leq 1$ the inequality is trivial, because then $1/k^2 \geq 1$. Here are the bounds for some values of $k \geq 1$ compared with corresponding probabilities for the normal distribution with parameters μ and σ.

Probability	Chebychev bound	Normal value
$P(\lvert X - \mu\rvert \geq \sigma)$	at most 1	0.3173
$P(\lvert X - \mu\rvert \geq 2\sigma)$	at most $1/2^2 = 0.25$	0.0465
$P(\lvert X - \mu\rvert \geq 3\sigma)$	at most $1/3^2 \approx 0.11$	0.00270
$P(\lvert X - \mu\rvert \geq 4\sigma)$	at most $1/4^2 \approx 0.06$	0.000063

As the table shows, Chebychev's bound will be very crude for a distribution that is approximately normal. Its importance is that it holds no matter what the shape of the distribution, so it gives some information about two-sided tail probabilities whenever the mean and standard deviation of a distribution can be calculated.

Example 7. Bounds for a list of numbers.

Problem. The average of a list of a million numbers is 10 and the average of the squares of the numbers is 101. Find an upper bound on how many of the entries in the list are 14 or more.

Solution. Let X represent a number picked at random from the list. Then $\mu = E(X) = 10$, $E(X^2) = 101$, so

$$\sigma = SD(X) = \sqrt{101 - 10^2} = 1,$$
$$P(X \geq 14) = P(X - \mu \geq 4\sigma) \leq P(|X - \mu| \geq 4\sigma) \leq 1/4^2,$$

by Chebychev's inequality. Consequently, the number of entries 14 or over is at most

$$10^6 P(X \geq 14) \leq 10^6/16 = 62,500$$

Remark. If the distribution of the list were known to be symmetric about 10, the probabilities $P(X \geq 14)$ and $P(X \leq 6)$ would be equal. Since it is the sum of these two probabilities which is at most $1/16$, the bound in this case could be reduced by a factor of 2 to $31,250$. If the distribution of the list were approximately normal, the number would be more like

$$10^6 \times [1 - \Phi(4)] \approx 32$$

Sums and Averages of Independent Random Variables

The main reason for the importance of variance is the following simple rule for the variance of a sum of two independent variables. This rule leads to the right SD to use in the normal approximation for a sum of n independent random variables for large n.

Addition Rule for Variances

$Var(X+Y) = Var(X) + Var(Y)$ if X and Y are **independent**.

$Var(X_1 + \cdots + X_n) = Var(X_1) + \cdots + Var(X_n)$ if $X_1, \ldots, X_n$ are **independent**.

The assumption of independence is important. In contrast to expectations, variances do not always add for dependent random variables. For example, if $X = Y$, then

$$Var(X+Y) = Var(2X) = [SD(2X)]^2 = [2\,SD(X)]^2 = 4\,Var(X)$$

while

$$Var(X) + Var(Y) = Var(X) + Var(X) = 2\,Var(X)$$

Proof of the addition rule for variances. Let $S = X + Y$. Then $E(S) = E(X) + E(Y)$, so

$$S - E(S) = [X - E(X)] + [Y - E(Y)]$$

Now square both sides and then take expectations to get

$$[S - E(S)]^2 = [X - E(X)]^2 + [Y - E(Y)]^2 + 2[X - E(X)][Y - E(Y)]$$
$$Var(S) = Var(X) + Var(Y) + 2E\{[X - E(X)][Y - E(Y)]\}$$

If X and Y are independent, then so are $X - E(X)$ and $Y - E(Y)$. So by the rule for the expectation of a product of independent variables, the last term above is the product of $E[X - E(X)]$ and $E[Y - E(Y)]$. This is zero times zero which equals zero, giving the addition rule for two independent variables. Apply this addition rule for two variables repeatedly to get the result for n variables. □

Sums of independent random variables with the same distribution. Suppose $X_1, \ldots, X_n$ are independent with the same distribution as X. You can think of the X_i as the results of repeated measurements of some kind. Because all the expectations and variances are determined by the same distribution,

$$E(X_k) = E(X) \qquad Var(X_k) = Var(X) \qquad (k = 1, \ldots, n)$$

So for the sum $S_n = X_1 + \cdots + X_n$

$$E(S_n) = nE(X) \qquad \text{by the addition rule for expectation}$$
$$Var(S_n) = nVar(X) \qquad \text{by the addition rule for variance.}$$

Taking square roots in the last formula gives the formula for $SD(S_n)$ in the next box. The results for the average follow by scaling the sum by the constant factor of $1/n$.

Square Root Law

Let S_n be the sum, $\bar{X}_n = S_n/n$ the average, of n independent random variables $X_1, \ldots, X_n$, each with the same distribution as X. Then

$$E(S_n) = nE(X) \qquad SD(S_n) = \sqrt{n}SD(X)$$

$$E(\bar{X}_n) = E(X) \qquad SD(\bar{X}_n) = \frac{SD(X)}{\sqrt{n}}$$

The expectation of a sum of n independent trials grows linearly with n. But the SD grows more slowly, according to a multiple of $\sqrt{n}$. This slow growth of the SD is due to the high probability of cancellation between terms which are above the expected value and terms which are below. The square root law for $SD(S_n)$ gives a precise mathematical measure of the extent to which this cancellation tends to occur.

Example 8. **Standard deviation of the binomial distribution.**

Problem. Derive the formula $\sqrt{npq}$ for the SD of the binomial (n, p) distribution.

Solution. This is the distribution of the sum $S_n = X_1 + \cdots + X_n$ of n indicators of independent events, each with probability p. So $\sqrt{npq}$ comes from the square root law for $SD(S_n)$ and the formula $\sqrt{pq}$ for the SD of an indicator, found in Example 2.

The law of averages. While as n increases $SD(S_n)$ grows as a constant times $\sqrt{n}$, dividing by n makes $SD(\bar{X}_n)$ tend to zero as a constant divided by $\sqrt{n}$. So the SD of the average of n independent trials tends to 0 as $n \to \infty$. This is an expression of the *law of averages*, which generalizes the law of large numbers stated in Section 2.2 for the proportion of successes in n Bernoulli (p) trials. Roughly speaking, the law of averages says that the average of a long sequence of independent trials $X_1, X_2, \ldots, X_n$ is likely to be close to the expected value of $X = X_1$. Here is a more precise formulation:

Law of Averages

Let $X_1, X_2, \ldots$ be a sequence of independent random variables, with the same distribution as X. Let $\mu = E(X)$ denote the common expected value of the X_i, and let

$$\bar{X}_n = (X_1 + X_2 + \cdots + X_n)/n$$

be the random variable representing the average of $X_1, \ldots, X_n$. Then for every $\epsilon > 0$, no matter how small,

$$P(|\bar{X}_n - \mu| < \epsilon) \to 1 \quad \text{as} \quad n \to \infty$$

In words: as the number of variables increases, with probability approaching 1, the average will be arbitrarily close to the expected value.

Proof. From the box for the square root law, $E(\bar{X}_n) = \mu$, $SD(\bar{X}_n) = \sigma/\sqrt{n}$, where $\sigma = SD(X_1)$. Chebychev's inequality applied to $\bar{X}_n$ now gives

$$P(|\bar{X}_n - \mu| \geq \epsilon) = P\left(|\bar{X}_n - \mu| \geq \frac{\epsilon}{SD(\bar{X}_n)} SD(\bar{X}_n)\right) \leq \left(\frac{SD(\bar{X}_n)}{\epsilon}\right)^2 = \frac{\sigma^2}{n\epsilon^2}$$

But for each fixed ϵ the right side tends to 0 as $n \to \infty$, hence so does the left side since probabilities are non-negative. Taking complements yields the result. □

Exact distribution of sums of independent variables. Suppose the X_i are independent indicator variables, with $P(X_i = 1) = p$ and $P(X_i = 0) = 1 - p$ for some $0 < p < 1$. For example, X_i could be the indicator of success on the ith trial in a sequence of independent trials. Then $S_n = X_1 + \cdots + X_n$ represents the number of successes in n trials, and S_n has the binomial (n, p) distribution studied in Chapter 2. In theory, and numerically by computer, the formula of Exercise 3.1.16 for the distribution of the sum of two random variables can be applied repeatedly to find the distribution of S_n for other distributions of X_i. But the resulting formulae are manageable only in a few other cases (e.g., the Poisson and geometric cases treated in the next section.)

Approximate distribution of sums of independent variables. Because there is no simple formula for the distribution of the sum S_n of n independent random variables with the same distribution as X, it is both surprising and useful that no matter what the distribution of X, there is a simple normal approximation for the distribution of S_n. This generalizes the normal approximation to the binomial distribution treated in Section 2.2.

The Normal Approximation (Central Limit Theorem)

Let $S_n = X_1 + \cdots + X_n$ be the sum of n independent random variables each with the same distribution over some finite set of values. For large n, the distribution of S_n is approximately normal, with mean $E(S_n) = n\mu$, and standard deviation $SD(S_n) = \sigma\sqrt{n}$, where $\mu = E(X_i)$ and $\sigma = SD(X_i)$. That is to say, for all $a \leq b$

$$P\left(a \leq \frac{S_n - n\mu}{\sigma\sqrt{n}} \leq b\right) \approx \Phi(b) - \Phi(a)$$

where Φ is the standard normal c.d.f. No matter what the distribution of the terms X_i, for every $a \leq b$ the error in using this normal approximation tends to zero as $n \to \infty$. The same result holds for X_i with an infinite range of possible values, provided the standard deviation is defined and finite.

Note that the random variable $(S_n - n\mu)/\sigma\sqrt{n}$ appearing in the normal approximation is S_n in standard units. If the possible values of the X_i form a sequence of consecutive integers, the continuity correction should be used as in Section 2.2 to obtain a better approximation. The normal approximation works just as well for averages as for sums, because the factor of n has no effect on the standardized variables. For any distribution of X_i with just two possible values, the above normal approximation follows from the normal approximation to the binomial distribution, derived in Section 2.3, by using scaling properties of the mean and standard deviation to reduce to the case when the two possible values are 0 and 1. But a full proof of the central limit theorem is beyond the scope of this text.

The pictures at the end of the section show how the distribution of the sum S_n of independent and identically distributed $X_1, X_2, \ldots, X_n$ depends on the number of terms n and the common distribution of the X_i. As a general rule, the more symmetric the distribution, and the thinner its tails, the faster the approach to normality as n increases. On each page, all histograms are scaled horizontally in standard units, and vertically to keep the total area constant.

Example 9. Random walk.

Physicists use random walks to model the process of diffusion, or random motion of particles. The position S_n of a particle at time n can be thought of as a sum of displacements $X_1, \ldots, X_n$. Assuming the displacements are independent and identically distributed, the theory of this section applies.

Problem. Suppose at each step a particle moving on sites labeled by integers is equally likely to move one step to the right, one step to the left, or stay where it is.

Find approximately the probability that after 10,000 steps the particle ends up more than 100 sites to the right of its starting point.

Solution. Let X represent a single step. Then $E(X) = 0$,

$$Var(X) = E(X^2) - 0^2 = \frac{(-1)^2}{3} + \frac{0^2}{3} + \frac{1^2}{3} = \frac{2}{3}$$

and $SD(X) = \sqrt{2/3} = 0.8165$. The problem is to find $P(S_{10,000} > 100)$, where

$$E(S_{10,000}) = 10,000E(X) = 0 \quad \text{and}$$

$$SD(S_{10,000}) = \sqrt{10,000}\, SD(X) = 100 \times 0.8165 = 81.65$$

by the square root law. The normal approximation gives

$$P(S_{10,000} > 100) = P\left(\frac{S_{10,000}}{81.65} > \frac{100}{81.65}\right) \approx 1 - \Phi(\frac{100}{81.65}) \approx 11\%$$

Skewness

Let X be a random variable with $E(X) = \mu$ and $SD(X) = \sigma$. Let $X_* = (X - \mu)/\sigma$ be X in standard units. So the first two moments of X_* are

$$E(X_*) = 0 \quad and \quad E(X_*^2) = 1$$

The *skewness* of X, (or of the distribution of X) denoted here by Skewness(X), is the third moment of X_*:

$$\text{Skewness}(X) = E(X_*^3) = E[(X - \mu)^3]/\sigma^3$$

Skewness is a measure of the degree of asymmetry in the distribution of X. For any X with finite third moment, there is the simple formula (Exercise 33):

$$\text{Skewness}(S_n) = \text{Skewness}(X)/\sqrt{n} \tag{7}$$

for S_n the sum of n independent random variables with the same distribution as X. This implies the formula $(1 - 2p)/\sqrt{npq}$ used in Section 2.2 for the skewness of binomial (n, p) distribution.

It is easy to see that if the distribution of X is symmetric about μ, then Skewness$(X) = 0$. If the normal approximation to the distribution of X is good, the distribution of X must be nearly symmetric about μ, so it is be expected that Skewness$(X) \approx 0$. In case Skewness(X) is significantly different from 0, the normal approximation to the distribution of X will usually not be very good. Formula (7) shows that no matter what the skewness of the distribution of X, the skewness of the sum S_n tends to zero as $n \to \infty$, though rather slowly. This is evidence of the central limit theorem: the distribution of S_n is asymptotically normal with skewness 0 in the limit, so has small skewness for large n. As in the binomial case studied in Section 2.2, an improvement to the normal approximation of S_n is obtained by replacing $\Phi(z)$ in the usual normal approximation by

$$\Phi(z) - \frac{1}{6\sqrt{n}}\text{Skewness}(X)\,(z^2 - 1)\,\phi(z)$$

where $\phi(z)$ is the standard normal curve. See Section 3.5 for an application to the Poisson distribution.

Figure 3. Distribution of the sum of n die rolls for $n = 1, 2, 4, 8, 16, 32$.

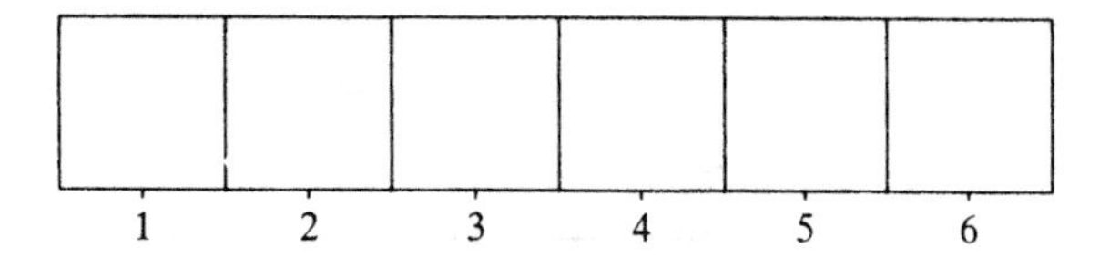

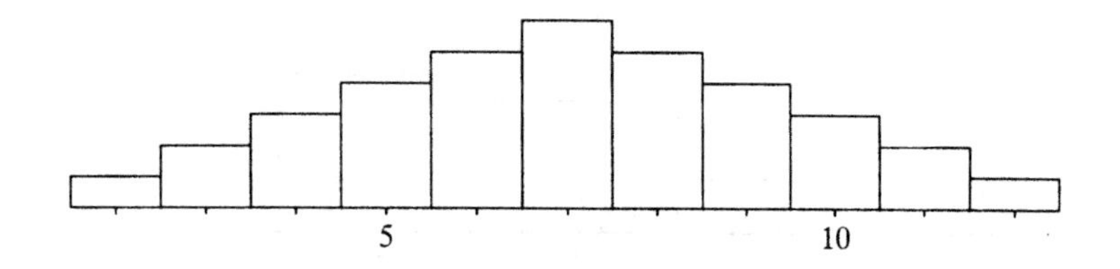

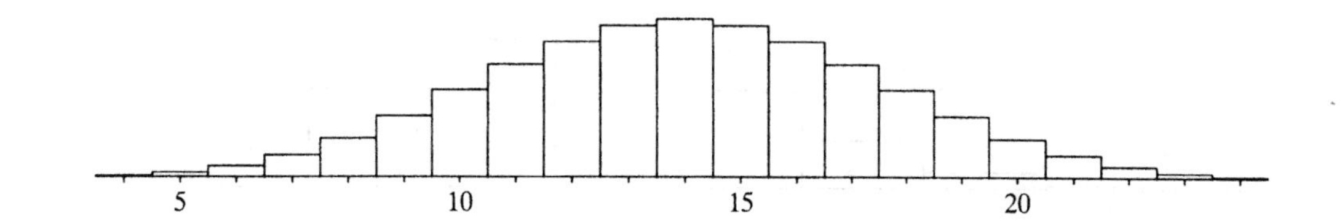

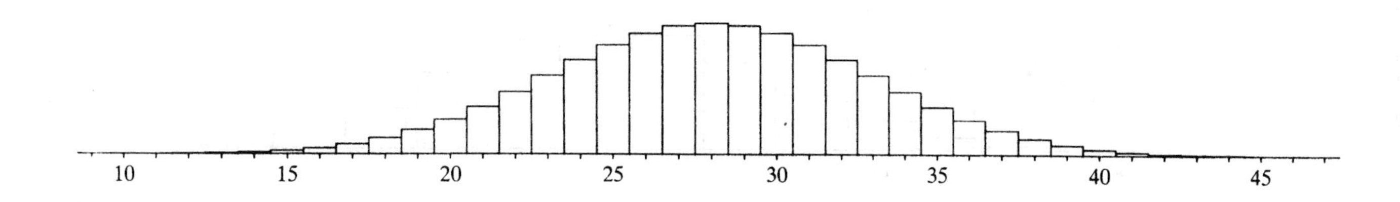

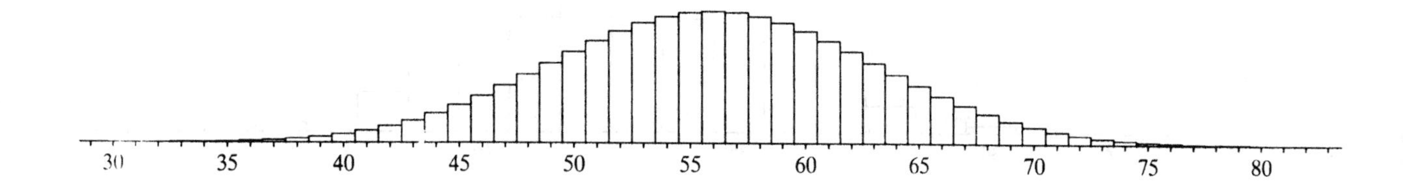

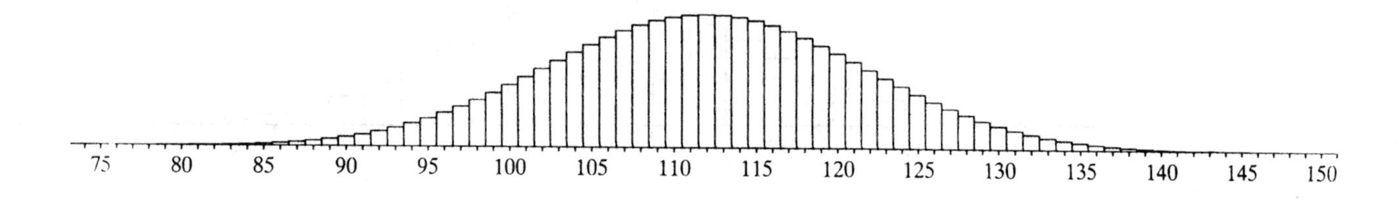

Figure 4. Distribution of S_n for $n = 1, 2, 4, 8, 16, 32$.

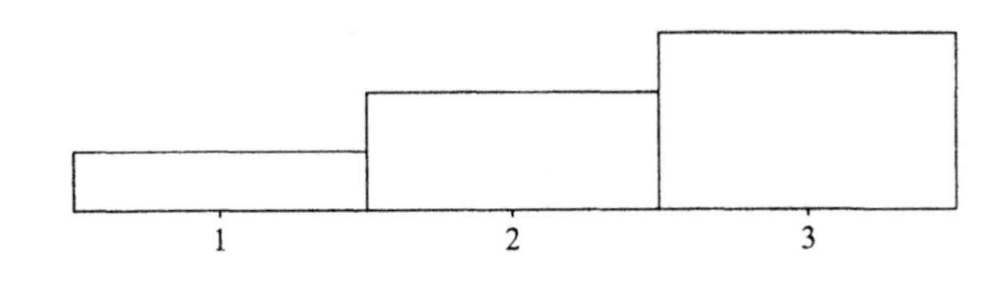

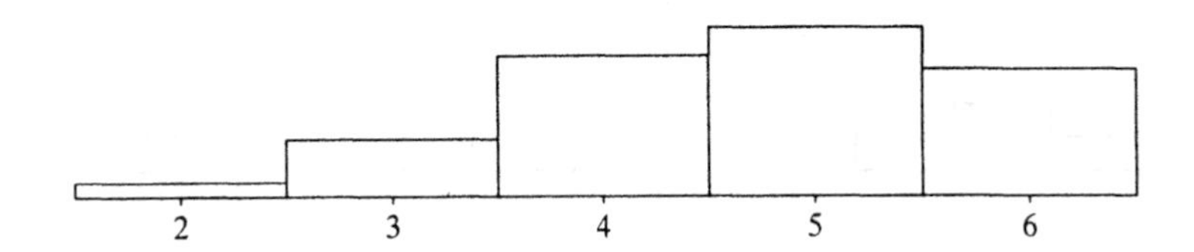

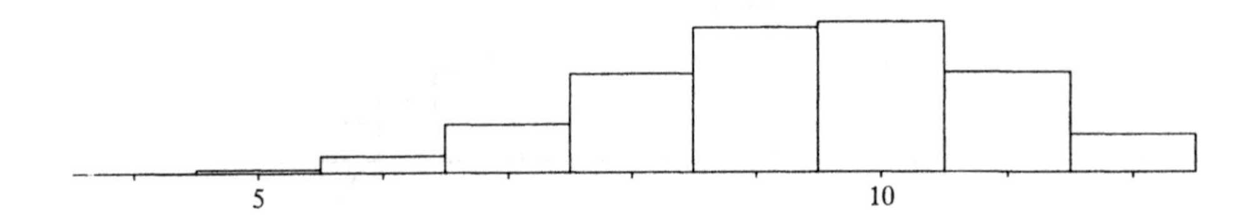

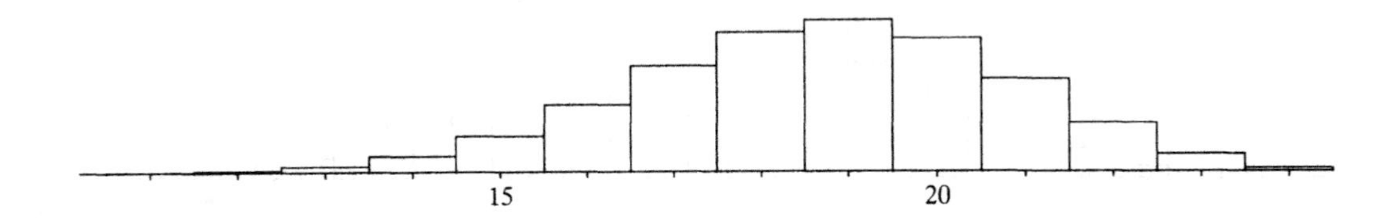

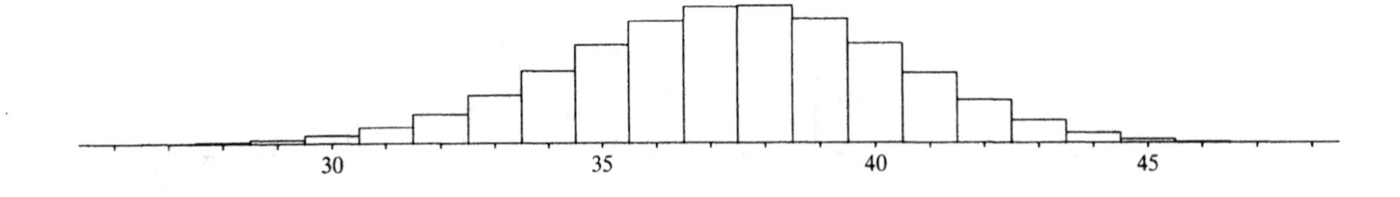

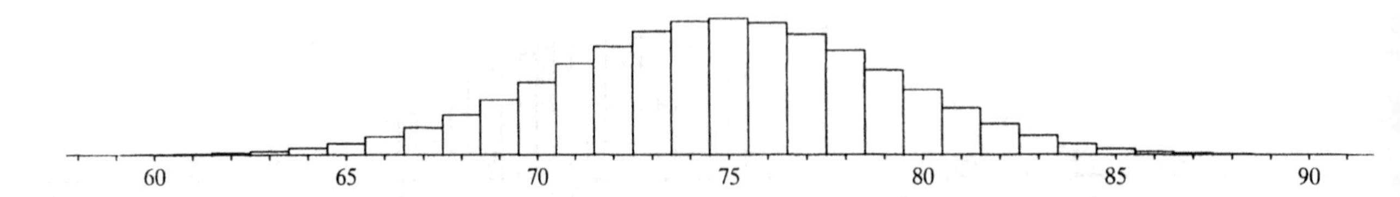

Figure 5. Distribution of S_n for $n = 1, 2, 4, 8, 16, 32$.

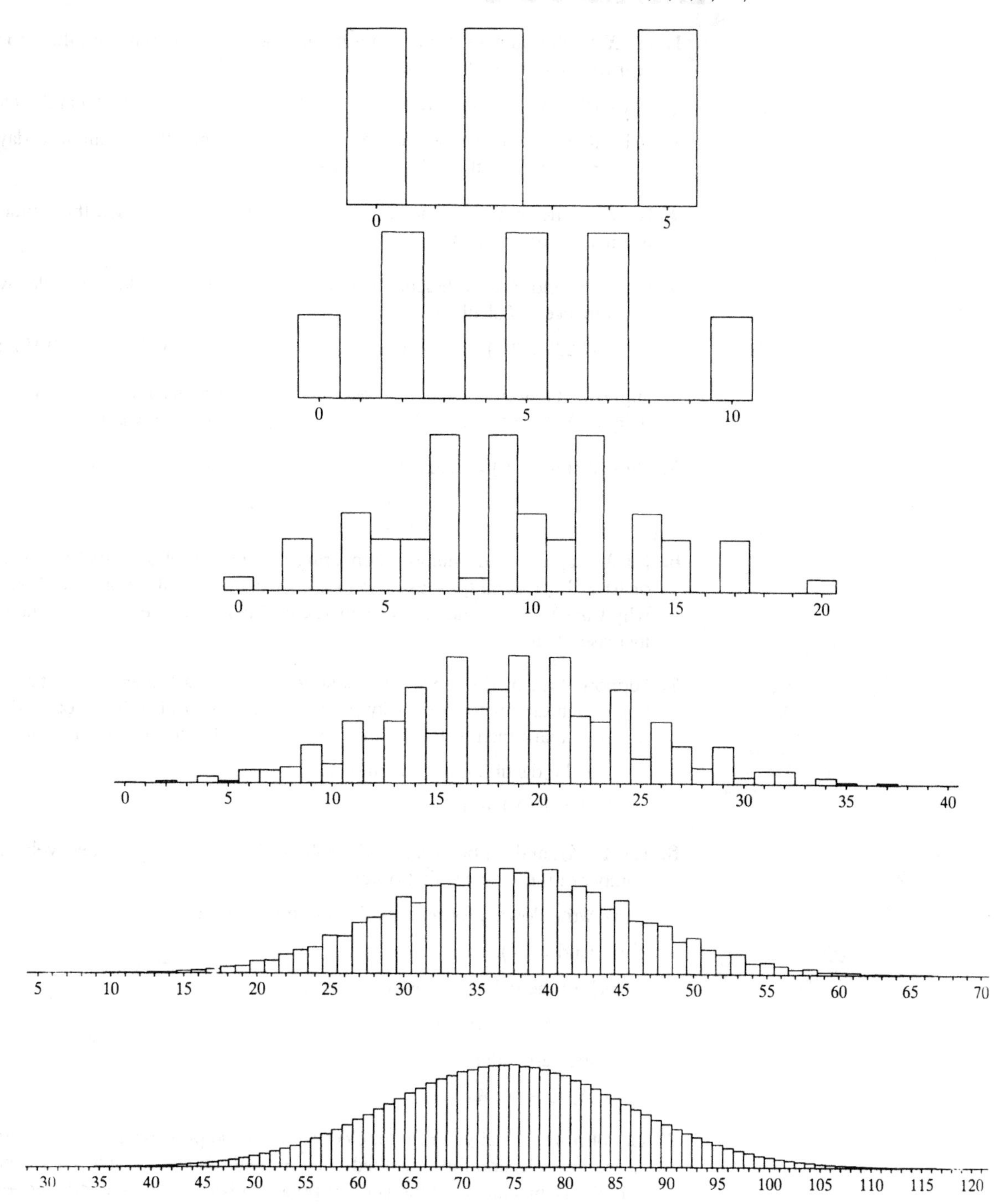

Exercises 3.3

1. Let X be the number of days in a month picked at random from the 12 months of a year (not a leap year).

a) Display the distribution of X in a table, and calculate $E(X)$ and $SD(X)$.

b) Repeat with X the number of days in the month containing a day picked at random from the 365 days of 1991.

2. Let Y be the number of heads obtained if a fair coin is tossed three times. Find the mean and variance of Y^2.

3. Let X, Y, and Z be independent identically distributed random variables with mean 1 and variance 2. Calculate:

a) $E(2X + 3Y)$; b) $Var(2X + 3Y)$; c) $E(XYZ)$; d) $Var(XYZ)$.

4. Suppose X_1 and X_2 are independent. Find a formula for $Var(X_1 X_2)$ in terms of $\mu_1 = E(X_1)$, $\sigma_1^2 = Var(X_1)$, $\mu_2 = E(X_2)$, and $\sigma_2^2 = Var(X_2)$.

5. Show that if $E(X) = \mu$ and $Var(X) = \sigma^2$, then for every constant a

$$E\left[(X - a)^2\right] = \sigma^2 + (\mu - a)^2.$$

6. Let X_p represent the number appearing on one roll of a 'shape' which lands flat (1 or 6) with probability p, as described in Example 1.3.3. Explain without calculation why $Var(X_p)$ must increase as p increases. Then compute $Var(X_p)$ and check that it increases as p increases.

7. Suppose three marksmen shoot at a target. The ith marksman fires n_i times, hitting the target each time with probability p_i, independently of his other shots and the shots of the other marksmen. Let X be the total number of times the target is hit.

a) Is the distribution of X binomial?

b) Find $E(X)$ and $Var(X)$.

8. Let A_1, A_2, and A_3 be events with probabilities $\frac{1}{5}$, $\frac{1}{4}$, and $\frac{1}{3}$, respectively. Let N be the number of these events that occur.

a) Write down a formula for N in terms of indicators.

b) Find $E(N)$.

In each of the following cases, calculate $Var(N)$:

c) A_1, A_2, A_3 are disjoint;

d) they are independent;

e) $A_1 \subset A_2 \subset A_3$.

9. Out of n individual voters at an election, r vote Republican and $n - r$ vote Democrat. At the next election the probability of a Republican switching to vote Democrat is p_1, and of a Democrat switching is p_2. Suppose individuals behave independently. Find a) the expectation and b) the variance of the number of Republican votes at the second election.

10. Moments of the uniform distribution. Let X be uniformly distributed on $\{1, 2, \ldots, n\}$. Let $s(k, n) = 1^k + 2^k + \cdots + n^k$ be the sum of the kth powers of the first n integers.

a) Show that $E(X^k) = \frac{s(k,n)}{n}$ and $E\left[(X+1)^k\right] = \frac{s(k,n+1)-1}{n}$.

b) Deduce that $E\left[kX^{k-1} + \binom{k}{2}X^{k-2} + \cdots + 1\right] = \frac{(n+1)^k - 1}{n}$.

c) Use b) for $k = 2$ to obtain $E(X) = (n+1)/2$ (also obvious by symmetry), and hence $s(1,n) = n(n+1)/2$.

d) Use b) for $k = 3$ and the above formula for $E(X)$ to deduce that $E(X^2) = \frac{1}{6}(n+1)(2n+1)$ and hence $s(2,n) = \frac{1}{6}n(n+1)(2n+1)$.

e) Show that $Var(X) = (n^2 - 1)/12$.

f) Check that your formulae c) and e) agree in the case $n = 6$ with the results obtained in Example 3 for X the number on a die.

g) Use the same method to show that $s(3,n) = [s(1,n)]^2$.

[This method can be used to obtain formulae for $s(k,n)$ for an arbitrary positive integer k. But the formulae get more complicated as k increases.]

11. Suppose that Y has uniform distribution on the n numbers $\{a, a+b, \ldots, a+(n-1)b\}$, and that X has uniform distribution on $\{1, 2, \ldots, n\}$. By writing Y as a linear function of X and using results of Exercise 10, find formulae for the mean and variance of Y in terms of a, b, and n.

12. A random variable X has expectation 10 and standard deviation 5.

a) Find the smallest upper bound you can for $P(X \geq 20)$.

b) Could X be a binomial random variable?

13. Suppose the IQ scores of a million individuals have a mean of 100 and an SD of 10.

a) Without making any further assumptions about the distribution of the scores, find an upper bound on the number of scores exceeding 130.

b) Find a smaller upper bound on the number of scores exceeding 130 assuming the distribution of scores is symmetric about 100.

c) Estimate the number of scores exceeding 130 assuming that the distribution is approximately normal.

14. Suppose the average family income in an area is \$10,000.

a) Find an upper bound for the percentage of families with incomes over \$50,000.

b) Find a better upper bound if it is known that the standard deviation of incomes is \$8000.

15. a) Show that if X and Y are independent random variables, then

$$Var(X - Y) = Var(X + Y)$$

b) Let D_1 and D_2 represent two draws at random with replacement from a population, with $E(D_1) = 10$ and $SD(D_1) = 2$. Find a number c so that

$$P(|D_1 - D_2| < c) \geq 99\%$$

16. A game consists of drawing tickets with numbers on them from a box, independently with replacement. In order to play you have to stake \$2 each time you draw a ticket. Your net gain is the number on the ticket you draw. Suppose there are 4 tickets in the box with numbers $-2, -1, 0, 3$ on them. If, for example the ticket shows \$3 then you get your stake back, plus an additional \$3.

a) Let X stand for your net gain in one game. What is the distribution of X? Find $E(X)$ and $Var(X)$.

b) If you play 100 times, what is your chance of winning \$25 or more?

17. Let X be a random variable with

$$P(X = -1) = P(X = 0) = 1/4,$$

and $P(X = 1) = 1/2$. Let S be the sum of 25 independent random variables, each with the same distribution as X. Calculate approximately
a) $P(S < 0)$, b) $P(S = 0)$, and c) $P(S > 0)$.

18. In roulette, the "house special" is a bet on the five pockets 0, 00, 1, 2 and 3. There are 5 chances in 38 to win, and the bet pays 6 to 1. That is, if you place a dollar bet on the house special and the ball lands in one of the five pockets, you get your dollar back plus 6 dollars in winnings; if the ball lands in any other pocket, you lose your dollar. If you make 300 one-dollar bets on the house special, approximately what is the chance that you come out ahead?

19. A new elevator in a large hotel is designed to carry about 30 people, with a total weight of up to 5000 lbs. More than 5000 lbs. overloads the elevator. The average weight of guests at this hotel is 150 lbs., with an SD of 55 lbs. Suppose 30 of the hotel's guests get into the elevator. Assuming the weights of these guests are independent random variables, what is the chance of overloading the elevator? Give your approximate answer as a decimal.

20. Suppose you have \$100,000 to invest in stocks. If you invest \$1000 in any particular stock your profit will be \$200, \$100, \$0 or $-$\$100 (a loss), with probability 0.25 each. There are 100 different stocks you can choose from, and they all behave independently of each other. Consider the two cases: (1) Invest \$100,000 in one stock. (2) Invest \$1000 in each of 100 stocks.

(a) For case (1) find the probability that your profit will be \$8000 or more.

(b) Do the same for case (2).

21. **Roundoff errors.** Suppose you balance your checkbook by rounding amounts to the nearest dollar. Between 0 and 49 cents, drop the cents; between 50 and 99 cents, drop the cents and add a dollar. Find approximately the probability that the accumulated error in 100 transactions is greater than 5 dollars (either way)

a) assuming the numbers of cents involved are independent and uniformly distributed between 0 and 99;

b) assuming each transaction is an exact dollar amount with probability 1/4, and given not an exact dollar amount the number of cents is uniformly distributed between 1 and 99, independently for different transactions.

22. Suppose n dice are rolled.

a) Find approximately the probability that the average number is between $3\frac{5}{12}$ and $3\frac{7}{12}$ for the following values of n: 105, 420, 1680, 6720.

b) Use these values to sketch the graph of this probability as a function of n.

c) Suppose that the numbers $3\frac{5}{12}$ and $3\frac{7}{12}$ were replaced by $3\frac{1}{2}-\epsilon$ and $3\frac{1}{2}+\epsilon$ for some other small number ϵ instead of $\epsilon = \frac{1}{12}$, say $\epsilon = \frac{1}{24}$. How would this affect the graph?

23. Suppose that in a particular application requiring a single battery, the mean lifetime of the battery is 4 weeks, with a standard deviation of 1 week. The battery is replaced by a new one when it dies, and so on. Assume lifetimes of batteries are independent. What, approximately, is the probability that more than 26 replacements will have to be made in a two-year period, starting at the time of installation of a new battery, and not counting that new battery as a replacement? [*Hint*: Use the normal approximation to the distribution of the total lifetime of n batteries for a suitable n.]

24. A box contains four tickets, numbered 0, 1, 1, and 2. Let S_n be the sum of the numbers obtained from n draws at random with replacement from the box.

a) Display the distribution of S_2 in a suitable table.

b) Find $P(S_{50} = 50)$ approximately.

c) Find an exact formula for $P(S_n = k) \quad (k = 0, 1, 2, \ldots)$.

25. Equality in Chebychev's inequality. Let μ, σ, and k be three numbers, with $\sigma > 0$ and $k \geq 1$. Let X be a random variable with the following distribution:

$$P(X = x) = \begin{cases} \dfrac{1}{2k^2} & \text{if } \; x = \mu + k\sigma \quad \text{or} \quad \mu - k\sigma \\ 1 - \dfrac{1}{k^2} & \text{if } \; x = \mu \\ 0 & \text{otherwise.} \end{cases}$$

a) Sketch the histogram of this distribution for $\mu = 0$, $\sigma = 10$, $k = 1, 2, 3$.

b) Show that $E(X) = \mu, \quad Var(X) = \sigma^2, \quad P(|X - \mu| \geq k\sigma) = 1/k^2$.

So there is equality in Chebychev's inequality for this distribution of X. This means Chebychev's inequality cannot be improved without additional hypotheses on the distribution of X.

c) Show that if Y has $E(Y) = \mu$, $Var(Y) = \sigma^2$, and $P(|Y - \mu| < \sigma) = 0$, then Y has the same distribution as X described above for $k = 1$.

26. Mean absolute deviation.

a) Calculate the *mean absolute deviation* $E(|X - \mu|)$ for X, the number on a six-sided die.

Your answer should be slightly smaller than the standard deviation found in Example 3. This is a general phenomenon, which occurs because the operation of squaring the absolute deviations before averaging them tends to put more weight on large deviations than on small ones.

b) Use the fact that $Var(|X - \mu|) \geq 0$ to show that $SD(X) \geq E(|X - \mu|)$, with equality if and only if $|X - \mu|$ is a constant.

That is to say, unless $|X - \mu|$ is a constant, the standard deviation of a random variable is always strictly larger than the mean absolute deviation. If X is a constant, then both measures of spread are zero.

27. The SD of a bounded random variable.

a) Let X be a random variable with $0 \le X \le 1$ and $E(X) = \mu$. Show that:

(i) $0 \le \mu \le 1$; (ii) $0 \le Var(X) \le \mu(1-\mu) \le \frac{1}{4}$ [*Hint*: Use $X^2 \le X$]

b) Let X be a random variable with $a \le X \le b$ and $E(X) = \mu$. Show that:

(i) $a \le \mu \le b$; (ii) $0 \le Var(X) \le (\mu - a)(b - \mu) \le \frac{1}{4}(b-a)^2$;

(iii) $0 \le SD(X) \le (b-a)/2$.

c) The standard deviation of a list of a million digits 0, 1, 2, ..., 9 is exactly $4\frac{1}{2}$. How many nines are there in the list? Or is it impossible to answer this question without more information?

28. Let S be the number of successes in n independent Bernoulli trials, with possibly different probabilities $p_1, \ldots, p_n$ on different trials. Show that for fixed $\mu = E(S)$, $Var(S)$ is largest in case the probabilities are all equal.

29. Let $\bar{D}_n$ be the average of n independent random digits from $\{0, \ldots, 9\}$.

a) Guess the first digit of $\bar{D}_n$ so as to maximize your chance of being correct.

b) Calculate the chance that your guess is correct exactly for $n = 1, 2$, and approximately for a selection of larger values of n, and show the results in a graph.

c) How large must n be for you to be 99% sure of guessing correctly?

30. Let X_i be the last digit of D_i^2, where D_i is a random digit between 0 and 9. For instance, if $D_i = 7$ then $D_i^2 = 49$ and $X_i = 9$. Let $\bar{X}_n = (X_1 + \cdots + X_n)/n$ be the average of a large number n of such last digits, obtained from independent random digits $D_1, \ldots, D_n$.

a) Predict the value of $\bar{X}_n$ for large n.

b) Find a number ϵ such that for $n = 10{,}000$ the chance that your prediction is off by more than ϵ is about 1 in 200.

c) Find approximately the least value of n such that your prediction of $\bar{X}_n$ is correct to within 0.01 with probability at least 0.99.

d) Which can be predicted more accurately for large n: the value of $\bar{X}_n$, or the value of $\bar{D}_n = (D_1 + \cdots + D_n)/n$?

e) If you just had to predict the first digit of $\bar{X}_{100}$, what digit should you choose to maximize your chance of being correct, and what is that chance?

31. Normal approximation for individual probabilities. Let X be an integer valued random variable, $S_n = X_1 + \cdots + X_n$ where the X_i are independent with the same distribution as X. If the set of possible values of X contains two consecutive integers it can be shown that there is the following normal approximation to individual probabilities in the distribution of S_n:

$$P(S_n = k) \approx \frac{1}{\sqrt{2\pi n}\sigma} e^{-\frac{1}{2}(k-n\mu)^2/(n\sigma^2)} \quad \text{where} \quad \mu = E(X) \text{ and } \sigma = SD(X)$$

This approximation holds in the sense described below formula (3) of Section 2.3, which is the special case when X has Bernoulli (p) distribution. (Note the change of notation: in formula (3), μ stands for $E(S_n)$ and σ for $SD(S_n)$.) Suppose the distribution of X is uniform on $\{0, 1, \ldots, 9\}$, as in Example 3.1.9.

a) Find μ and σ for this distribution of X.

b) Use the above normal approximation to verify the claim in the discussion of Example 3.1.9 that

$$P(S_{2m} = 9m) \sim 2/\sqrt{33\pi m} \text{ as } m \to \infty.$$

c) Let $[x]$ denote the integer part of x. Find b such that in the limit as $n \to \infty$

$$\frac{P(S_n = [(4.5)n + b\sqrt{n}\,])}{P(S_n = [(4.5)n])} \to \frac{1}{2}$$

d) For b as in part c), evaluate $\lim_{n \to \infty} P(|S_n - (4.5)n| \leq b\sqrt{n}\,)$.

32. Skewness. For a random variable X with moments $\mu_k = E(X^k)$, derive the following properties of Skewnesss $(X) = E[((X - \mu)/\sigma)^3]$, where $\mu = \mu_1$ and $\sigma = \sqrt{\mu_2 - \mu^2}$ is assumed strictly positive:

a) Skewness $(X) = (\mu_3 - 3\mu\mu_2 + 2\mu^3)/\sigma^3$

b) If the distribution of X is symmetric about some point then Skewness$(X) = 0$.

c) If $a > 0$ then Skewness$(aX + b) =$ Skewness(X). What if $a < 0$?

33. Skewness of sums. Show the following:

a) If X and Y are independent with $E(X) = E(Y) = 0$ then

$$E[(X + Y)^3] = E(X^3) + E(Y^3).$$

b) If $S_n = X_1 + \cdots + X_n$ for independent X_i with the same distribution as X, then

$$\text{Skewness}(S_n) = \text{Skewness}(X)/\sqrt{n}$$

c) If S_n has binomial (n, p) distribution,

$$\text{Skewness}(S_n) = (1 - 2p)/\sqrt{npq}.$$

3.4 Discrete Distributions

Up to now, random variables were assumed to have a finite number of possible values. Probabilities and expectations were calculated as finite sums. But already in Chapter 2 useful approximations were obtained by letting the number of trials n tend to infinity. These approximations, the normal and the Poisson, lead naturally to the study of infinite outcome spaces. This section extends the basic concepts to allow a discrete distribution over an infinite sequence of possible outcomes. Important examples are the geometric and negative binomial distributions appearing in this section, and the Poisson distribution in the next. The following chapters study random variables with continuous distributions, like the uniform and normal, with an interval of possible values.

The distribution of the number of times T that you have to roll a fair die to get a six was found in Example 2 of Section 1.6:

$$P(T = i) = q^{i-1}p \qquad (i = 1, 2, \dots)$$

where $q = 5/6$ and $p = 1/6$. This is the *geometric distribution on* $\{1, 2, 3, \dots\}$ with parameter $p = 1/6$. Here the set of possible values of T can be counted one by one, but there is no largest possible value. This is an example of a *discrete distribution* on the positive integers.

A feature of infinite outcome spaces is that individual outcomes or sets of outcomes may be assigned probability zero. Consider, for example, the event $T = \infty$ that a six never shows up in repeated rolling of a die. This is an imaginable outcome, and you might want to include it in an outcome space. To find the probability of the event $T = \infty$ notice that if $T = \infty$, then the first n rolls are not 6. So the rules of probability imply

$$0 \leq P(T = \infty) \leq P(\text{first } n \text{ rolls not } 6) = (5/6)^n$$

assuming the die is fair and the rolls are independent. But since $q^n \to 0$ as $n \to \infty$ for $|q| < 1$, in particular for $q = 5/6$, this implies $P(T = \infty) = 0$.

A *discrete distribution* on the set of non-negative integers $\{0, 1, 2, \dots\}$ is defined by a sequence of probabilities $p_0, p_1, p_2, \dots$, such that

$$p_i \geq 0 \qquad \text{for all } i \text{ and} \qquad \sum_i p_i = 1$$

where i ranges over $0, 1, 2, \dots$. By allowing p_i to be zero for all but a finite set of i, any distribution over a finite set labeled $0, 1, 2, \dots, n$ could be presented like this. Probabilities involving discrete distributions can be calculated using the familiar rules of probability, together with a natural extension of the addition rule.

Infinite Sum Rule

If event A is partitioned into $A_1, A_2, A_3, \ldots$,

$$A = A_1 \cup A_2 \cup A_3 \cup \cdots \qquad \text{where} \quad A_i \cap A_j = \emptyset \qquad i \neq j$$

then

$$P(A) = P(A_1) + P(A_2) + P(A_3) + \cdots$$

To illustrate, for a random variable X with discrete distribution on $\{0, 1, 2, \ldots\}$ given by

$$P(X = i) = p_i \qquad (i = 0, 1, \ldots)$$

$$P(X \leq 5) = \sum_{i=1}^{5} p_i$$

$$P(X > 5) = \sum_{i=6}^{\infty} p_i = 1 - \sum_{i=1}^{5} p_i$$

$$P(X \text{ is even}) = \sum_{i=0}^{\infty} p_{2i}$$

The theory of discrete distributions is mostly a straightforward extension of the theory of distributions on finite sets, treated in the previous chapters. The basic concepts of conditional probability, random variable, distribution of a random variable, joint distribution, and independence, all remain the same. All general formulae involving these concepts, in particular the rule of average conditional probabilities and Bayes' rule, remain valid simply with infinite sums of probabilities replacing finite ones. This can be proved using the infinite sum rule, which justifies familiar formulae such as

$$P(X = x) = \sum_{y} P(X = x, Y = y)$$

for discrete random variables X and Y. Here the sum over y is understood to range over the set of possible values of Y, and the infinite series can be evaluated in an arbitrary order, which is left unspecified.

Examples

Example 1. Odd or even.

Problem 1. Suppose you and I take turns at rolling a die, to see who can first roll a six. Suppose I roll first, then you roll, then I roll, and so on, until one of us has rolled a six. What is the chance that you roll the first six?

Solution. In terms of T, the number of rolls required to produce the first six, the problem is to find the probability that T is even, i.e., either 2, or 4, or 6, or .... By the infinite sum rule

$$\begin{aligned}
P(T \text{ even}) &= P(T=2) + P(T=4) + P(T=6) + \cdots \\
&= qp + q^3p + q^5p + \cdots \qquad \text{where} \quad q = \frac{5}{6}, \quad p = \frac{1}{6} \\
&= qp(1 + q^2 + q^4 + \cdots) \\
&= qp/(1-q^2) \qquad \text{(geometric series with ratio } q^2\text{)} \\
&= \frac{5}{6} \times \frac{1}{6} \Big/ \left(1 - \frac{25}{36}\right) = \frac{5}{11}
\end{aligned}$$

Problem 2. What is the chance that I roll the first six?

Solution. This is $P(T \text{ odd})$. Of course, a similar calculation could be done again. But there is no need. Since we argued earlier that T is certain to be finite, and then T must be either even or odd, so

$$P(T \text{ odd}) = 1 - \frac{5}{11} = \frac{6}{11}$$

Example 2. The craps principle.

Suppose A and B play over and over, independently, a game which each time results in a win for A, a win for B, or a draw (meaning no decision), with probabilities $P(A)$, $P(B)$, and $P(D)$. Suppose they keep playing until the first game that does not result in a draw, and call the winner of that game the overall winner.

Problem 1. Show that

$$P(\text{A wins overall}) = \frac{P(A)}{P(A)+P(B)} \quad \text{and} \quad P(\text{B wins overall}) = \frac{P(B)}{P(A)+P(B)}$$

Solution.

$$\begin{aligned}
P(\text{A wins at game } n) &= P(\text{first } n-1 \text{ games drawn, and A wins game } n) \\
&= [P(D)]^{n-1}P(A), \qquad \text{so}
\end{aligned}$$

$$P(\text{A wins}) = \sum_{n=1}^{\infty} [P(D)]^{n-1}P(A) = \frac{P(A)}{1-P(D)} = \frac{P(A)}{P(A)+P(B)}$$

Remark. Put another way, $P(\text{A wins}) = P(A \mid A \text{ or } B)$, which you may find intuitively clear without calculation. This is the basic principle behind the calculation of probabilities in the game of craps, taken up in the exercises.

Problem 2. Let G be the number of games played, X the name of the winner. Show that G has a geometric distribution, and that G and X are independent.

Solution. G is geometric with $p = 1 - P(D)$ (wait until the first nondraw)

$$\begin{aligned} P(G = n, \text{A is the winner}) &= P(n-1 \text{ games drawn, then A wins}) \\ &= [P(D)]^{n-1}P(A) \\ &= [P(D)]^{n-1} \cdot [1 - P(D)] \cdot \frac{P(A)}{1 - P(D)} \\ &= P(G = n) \cdot P(\text{A wins}) \end{aligned}$$

Similarly, $P(G = n, \text{B wins}) = P(G = n)P(\text{B wins})$. So G and X are independent.

Moments

The concept of expectation extends to most discrete distributions.

Expectation of a Discrete Random Variable

The *expectation* of a discrete random variable X is defined by

$$E(X) = \sum_x xP(X = x)$$

provided that the series is absolutely convergent, that is to say, provided

$$\sum_x |x|P(X = x) < \infty$$

Here X is allowed to have both positive and negative values. The assumption of absolute convergence is necessary to ensure that the value of $E(X)$ is the same, regardless of the order in which the terms are summed. If $X \geq 0$ then the expression for $E(X)$ at least always makes sense, provided that $E(X) = \infty$ is allowed as a possibility.

If $Y = g(X)$ is a numerical function of a discrete random variable X there is the usual formula

$$E[g(X)] = \sum_x g(x)P(X = x)$$

This formula holds in the sense that if either side is defined (possibly as ∞) then so is the other, and they are equal. The right side is regarded as defined provided either $g(x) \geq 0$, or the series is absolutely convergent. For example, taking X to be numerical and $g(x) = |x|$

$$E(|X|) = \sum_x |x| P(X = x)$$

This is the quantity that must be finite for $E(X)$ to be defined and finite.

Proof of these facts about expectation involves the theory of absolutely convergent series. But you need not worry about this. Just accept that the basic properties of expectation listed in Section 3.2 remain valid for discrete random variables provided finite sums are replaced where necessary by infinite ones, and it is assumed that the sums converge absolutely. It is still important to recognize a random variable as a sum of simpler ones and use the addition rule of expectation. Similar remarks apply to variance, which is defined for all random variables X with $E(X^2) < \infty$. In particular, Chebychev's inequality, the law of averages, and the normal approximation all hold for discrete random variables X with $E(X^2) < \infty$. In fact, the law of averages holds for independent and identically distributed random variables $X_1, X_2, \ldots$ provided that $E(X_1)$ is defined. But proof of this is beyond the scope of this course.

Example 3. Moments of the geometric distribution.

Let T be the waiting time until the first success in a sequence of *Bernoulli (p) trials*, meaning independent trials each of which results in either success with probability p, or failure with probability $q = 1 - p$. So T has geometric distribution on $\{1, 2, \ldots\}$ with parameter p.

Problem 1. Find $E(T)$.

Solution. $E(T) = \sum_{n=1}^{\infty} nP(T = n) = \sum_{n=1}^{\infty} nq^{n-1}p = p\Sigma_1$ where $\Sigma_1 = \sum_{n=1}^{\infty} nq^{n-1}$.

A simple formula for Σ_1 can be found by a method used also to obtain the formula for the sum Σ_0 of a geometric series

$$\Sigma_0 = 1 + q + q^2 + \cdots = 1/(1-q)$$

Here is the calculation of Σ_1:

$$\begin{aligned} \Sigma_1 &= 1 + 2q + 3q^2 + \cdots \\ q\Sigma_1 &= \phantom{1 + {}} q + 2q^2 + \cdots \\ (1-q)\Sigma_1 &= 1 + q + q^2 + \cdots = \Sigma_0 = 1/(1-q) \\ \Sigma_1 &= 1/(1-q)^2 \end{aligned}$$

This gives $E(T) = p/(1-q)^2 = 1/p$.

Discussion. The formula $E(T) = 1/p$ is quite intuitive if you think about long-run averages. Over the long run, the average number of successes per trial is p. And the average number of trials per success is $1/p$.

Problem 2. Find $SD(T)$.

Solution. $SD(T) = \sqrt{E(T^2) - [E(T)]^2}$ where $E(T) = 1/p$ from above, and

$$E(T^2) = \sum_{n=1}^{\infty} n^2 P(T = n) = p\Sigma_2$$

where

$$\begin{aligned}
\Sigma_2 &= 1 + 4q + 9q^2 + \cdots + n^2 q^{n-1} + \cdots \\
q\Sigma_2 &= \qquad q + 4q^2 + \cdots + (n-1)^2 q^{n-1} + \cdots \\
(1-q)\Sigma_2 &= 1 + 3q + 5q^2 + \cdots + (2n-1)q^{n-1} + \cdots = 2\Sigma_1 - \Sigma_0
\end{aligned}$$

so $\quad \Sigma_2 = (1+q)/(1-q)^3$

Substituting these expressions gives $SD(T) = \sqrt{q}/p$.

Example 4. Waiting until the rth success (negative binomial distribution).

Let T_r denote the number of trials until the rth success in Bernoulli (p) trials. To illustrate the definition, for the following sequence of results, with $1 =$ success, $0 =$ failure,

$$0001000000100100000010000000\ldots$$

$$T_1 = 4; \quad T_2 = 11; \quad T_3 = 14; \quad T_4 = 21; \quad T_5 = ??$$

Problem 1. What is the distribution of T_r?

Solution. The possible values of T_r are $r, r+1, r+2, \ldots$. For t in this range

$$\begin{aligned}
P(T_r = t) &= P(r-1 \text{ successes in first } t-1 \text{ trials, and trial } t \text{ success}) \\
&= \binom{t-1}{r-1} p^{r-1}(1-p)^{t-r} p = \binom{t-1}{r-1} p^r (1-p)^{t-r}
\end{aligned}$$

Problem 2. Find $E(T_r)$ and $SD(T_r)$.

Solution. Direct calculation from the formula for the distribution is tedious. The key to a quick solution is to notice that

$$T_r = W_1 + W_2 + \cdots + W_r$$

where W_i is the waiting time after the $(i-1)$th success till the ith success. It is intuitively clear, and not hard to check, that

$$W_1, W_2, W_3, \ldots$$

FIGURE 1. Geometric and negative binomial histograms. The histogram in row r and column p shows the negative binomial (r, p) distribution of $T_r - r$ the number of failures before the rth success in Bernoulli (p) trials, for $r = 1, 2, 3, 4, 5$ and $p = 0.75, 0.5$, and 0.25. Note how as either p decreases or r increases, the distributions shift to the right and flatten out.

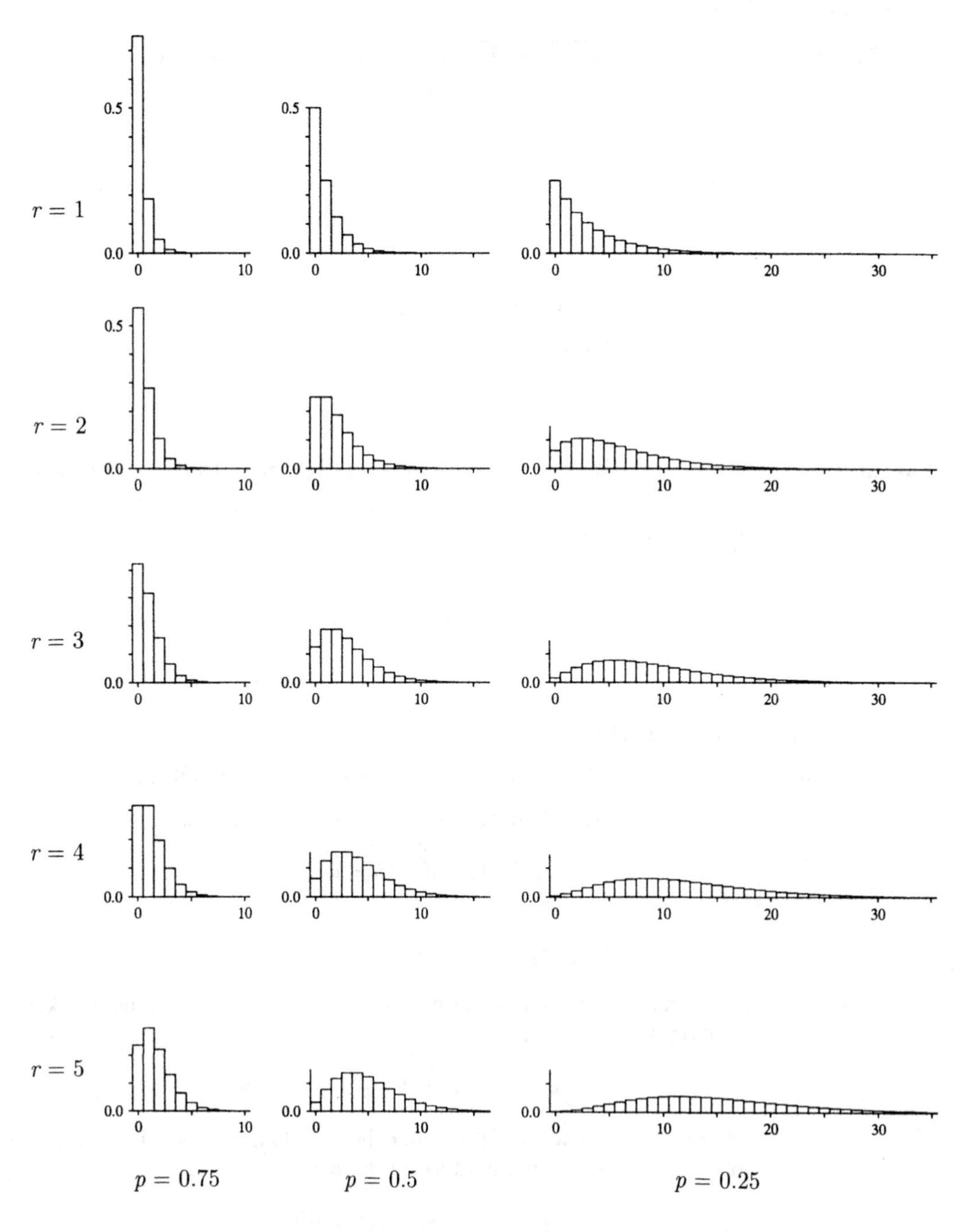

are independent, each with geometric (p) distribution. So by the results of the last example, the addition rule for expectation, and the square root law,

$$E(T_r) = r/p \qquad SD(T_r) = \sqrt{rq}/p$$

Remarks. (i) As $r \to \infty$ the distribution of T_r becomes asymptotically normal, another example of the central limit theorem. But due to the skewness of the geometric distribution of the terms being added, the approach to normality is rather slow. Particularly for p near 0.5, better approximations are obtained using the relation $P(T_r > n) = P(S_n < r)$, where S_n is the number of successes in the first n trials, and the normal approximation to the binomial (n, p) distribution of S_n.

(ii) The distribution of $T_r - r$, the number of failures before the rth success, in independent Bernoulli (p) trials, is called *negative binomial* with parameters r and p. This is just the distribution of T_r, shifted from $\{r, r+1, r+2, \ldots\}$ to $\{0, 1, 2, \ldots\}$

$$P(T_r - r = n) = P(T_r = n + r) = \binom{n+r-1}{r-1} p^r (1-p)^n \qquad (n = 0, 1, \ldots)$$

Example 5. The collector's problem.

Each box of a particular brand of cereal contains one out of a set of n different plastic animals. Suppose that the animal in each box is equally likely to be any one of the set of n, independently of what animals are in other boxes.

Problem. What is the expected number of cereal boxes a collector must buy in order to obtain the complete set of animals?

Solution. The collector gets one of the n animals in the first box. Each subsequent box contains an animal that is different from this first one with probability $(n-1)/n$, and the same with probability $1/n$. Using the independence assumption, the additional number of boxes required to get two different animals is a geometric random variable with parameter $p = (n-1)/n$ and mean

$$\frac{1}{p} = \frac{n}{n-1}$$

So the number of boxes required to get two different animals has mean

$$1 + \frac{n}{n-1}$$

Once two different animals are obtained, each box contains a new animal with probability $(n-2)/n$, and one of the old ones with probability $2/n$. So the additional time to get three different animals once two have been obtained is a geometric random variable with parameter $p = (n-2)/n$, and mean

$$\frac{1}{p} = \frac{n}{n-2}$$

So the number of boxes required to get three different animals has mean

$$1 + \frac{n}{n-1} + \frac{n}{n-2}$$

Continuing in this way, the mean μ_n of the overall waiting time for the set of all n animals is the sum of n terms

$$\begin{aligned}\mu_n &= 1 + \frac{n}{n-1} + \frac{n}{n-2} + \cdots + \frac{n}{2} + \frac{n}{1} \\ &= n\left(\frac{1}{n} + \frac{1}{n-1} + \frac{1}{n-2} + \cdots + \frac{1}{2} + 1\right) \\ &= n\left(1 + \frac{1}{2} + \frac{1}{3} + \cdots + \frac{1}{n}\right)\end{aligned}$$

by reversing the order of the terms.

Discussion. To illustrate, for $n = 6$ animals, the expected number of boxes required is

$$\mu_6 = 6\left(1 + \frac{1}{2} + \frac{1}{3} + \frac{1}{4} + \frac{1}{5} + \frac{1}{6}\right) = 14.7$$

As a variation of the problem, this is the long-run average number of times you have to roll a die in order to see every one of its faces. Similarly, the long-run average number of places you must inspect in a table of random digits, before seeing every one of the digits 0 through 9, is

$$\mu_{10} = 10\left(1 + \frac{1}{2} + \cdots + \frac{1}{10}\right) = 29.29$$

For large n, approximate values of μ_n can be obtained using Euler's approximation for the harmonic series

$$1 + \frac{1}{2} + \frac{1}{3} + \cdots + \frac{1}{n} \approx \log(n) + \gamma + \frac{1}{2n}$$

where $\gamma = 0.57721\ldots$ is Euler's constant. So

$$\mu_n \approx n\log(n) + \gamma n + \frac{1}{2}$$

This approximation is good even for small n, as you can check on a calculator.

Technical remarks. The infinite sum rule looks natural enough, but there is more to it than meets the eye. Consider, for example, a sequence of mutually exclusive events A_j, each determined by a finite number of independent trials, for example, $A_j = (T = 2j)$ that a die first shows six on roll number $2j$. As j increases, so may the number of trials required to determine whether or not A_j occurs, so the event $A = A_1 \cup A_2 \cup \cdots$ may involve an unlimited number of trials, like the event $A = (T \text{ even})$ in the die example. It seems natural to *define* $P(A)$ as the sum of the infinite series

$$P(A_1) + P(A_2) + \cdots = \sum_{j=1}^{\infty} P(A_j) = \lim_{n\to\infty} \sum_{j=1}^{n} P(A_j)$$

to use three common notations for the same thing. This limit exists and is a number between 0 and 1 because the rules of probability for a finite number of trials imply that the partial sums are non-negative, increasing and bounded above by 1. That much is fairly straightforward. The hard thing to show is that this definition is consistent, because a given event A might be split up in lots of different ways, and it is not obvious that the infinite sum rule gives the same result no matter how the event A is split up. Still, mathematicians have shown that it does. So the infinite sum rule gives a consistent way of extending the definition of probability from events for a finite number of trials to events for an infinite number of trials. Mathematically, the infinite sum rule is usually taken to be an *axiom*. It is then a nontrivial theorem that the various distributions studied in this book can be defined over suitable classes of subsets so as to satisfy this axiom. Proof of this goes beyond the scope of this course; see, for example, Billingsley's book, *Probability and Measure*.

Exercises 3.4

Note: Geometric series should *not* be left unsimplified. Use

$$1 + x + x^2 + x^3 + \cdots = \frac{1}{1-x} \qquad (|x| < 1)$$

1. A coin which lands heads with probability p is tossed repeatedly. Assuming independence of the tosses, find formulae for

a) P(exactly 5 heads appear in the first 9 tosses);

b) P(the first head appears on the 7th toss);

c) P(the fifth head appears on the 12th toss);

d) P(the same number of heads appear in the first 8 tosses as in the next 5 tosses).

2. An urn contains 10 red balls and 10 black balls. Balls are drawn out at random with replacement until at least one ball of each color has been drawn out. Let D be the number of draws. Find: a) the distribution of D; b) $E(D)$; c) $SD(D)$.

3. Suppose you pick people at random and ask them what month of the year they were born. Let X be the number of people you have to question until you find a person who was born in December. What is $E(X)$, approximately?

4. In the game of "odd one out" three people each toss a fair coin to see if one of their coins shows a different face from the other two.

a) After one play, what is the probability of some person being the "odd one out"?

b) Suppose play continues until there is an "odd one out". What is the probability that the duration is r plays?

c) What is the expected duration of play?

5. Bill, Mary, and Tom have coins with respective probabilities p_1, p_2, p_3 of turning up heads. They toss their coins independently at the same times.

a) What is the probability it takes Mary more than n tosses to get a head?

b) What is the probability that the first person to get a head has to toss more than n times?

c) What is the probability that the first person to get a head has to toss exactly n times?

d) What is the probability that neither Bill nor Tom get a head before Mary?

6. The geometric (p) distribution on { 0, 1, 2, ... }. The geometric (p) distribution is often defined as a distribution on $\{0, 1, 2, ...\}$ instead of $\{1, 2, 3, ...\}$. A random variable W has geometric (p) distribution on $\{0, 1, 2, ...\}$ if

$$P(W = k) = q^k p \qquad (k = 0, 1, \ldots)$$

a) Show that this is the distribution of the number of failures before the first success in Bernoulli (p) trials.

b) Find $P(W > k)$ $\quad (k = 0, 1, \ldots)$ c) Find $E(W)$. d) Find $Var(W)$.

7. Suppose that A and B take turns in tossing a biased coin which lands heads with probability p. Suppose that A tosses first.

a) What is the probability that A tosses the first head?

b) What is the probability that B tosses the first head? For both a) and b) above, find formulae in terms of p and sketch graphs.

No matter what the value of p, A is more likely to toss the first head than B. To try to compensate for this, let A toss once, then B twice, then A once, B twice, and so on.

c) Repeat a) and b) with this scheme. Give formulae and graphs.

d) For what value of p do A and B have the same chance of tossing the first head?

e) What, approximately, is B's chance of winning for very small values of p? Give both an intuitive explanation and an evaluation of the limit as $p \to 0$ by calculus.

8. Craps. In this game a player throws two dice and observes the sum. A throw of 7 or 11 is an immediate win. A throw of 2, 3, or 12 is an immediate loss. A throw of 4, 5, 6, 8, 9, or 10 becomes the player's *point*. In order to win the game now, the player must continue to throw the dice, and obtain the point before throwing a 7. The problem is to calculate the probability of winning at craps. Let X_0 represent the first sum thrown. The basic idea of the calculation is first to calculate $P(\text{Win} \mid X_0 = x)$ for every possible value x of X_0, then use the law of average conditional probabilities to obtain $P(\text{Win})$.

a) Show that for $x = 4, 5, 6, 8, 9, 10$,

$$P(\text{Win} \mid X_0 = x) = P(x)/[P(x) + P(7)]$$

where $P(x) = P(X_i = x)$ is the probability of rolling a sum of x. (Refer to Example 2).

b) Write down $P(\text{Win} \mid X_0 = x)$ for the other possible values x of X_0.

c) Deduce that the probability of winning at craps is

$$P(\text{Win}) = \frac{1952}{36 \times 11 \times 10} = 0.493\ldots$$

9. Suppose we play the following game based on tosses of a fair coin. You pay me \$10, and I agree to pay you $\$n^2$ if heads comes up first on the nth toss. If we play this game repeatedly, how much money do you expect to win or lose per game over the long run?

10. Let X be the number of Bernoulli (p) trials required to produce at least one success and at least one failure. Find:

a) the distribution of X; b) $E(X)$; c) $Var(X)$.

11. Suppose that A tosses a coin which lands heads with probability p_A, and B tosses one which lands heads with probability p_B. They toss their coins simultaneously over and over again, in a competition to see who gets the first head. The one to get the first head is the winner, except that a draw results if they get their first heads together. Calculate:

a) $P(\text{A wins})$; b) $P(\text{B wins})$; c) $P(\text{draw})$;

d) the distribution of the number of times A and B must toss.

12. Let W_1 and W_2 be independent geometric random variables with parameters p_1 and p_2. Find:

a) $P(W_1 = W_2)$; b) $P(W_1 < W_2)$; c) $P(W_1 > W_2)$;

d) the distribution of $\min(W_1, W_2)$;

e) the distribution of $\max(W_1, W_2)$.

13. Consider the following gambling game for two players, Black and White. Black puts b black balls and White puts w white balls in a box. Black and White take turns at drawing at random from the box, with replacement between draws until either Black wins by drawing a black ball or White wins by drawing a white ball. Suppose Black gets to draw first.

a) Calculate $P(\text{Black wins})$ and $P(\text{White wins})$ in terms of $p = b/(b + w)$.

b) What value of p would make the game fair (equal chances of winning)?

c) Is the game ever fair?

d) What is the least total number of balls in the game, $(b + w)$, such that neither player has more than a 51% chance of winning?

14. In Bernoulli (p) trials let V_n be the number of trials required to produce either n successes or n failures, whichever comes first. Find the distribution of V_n.

15. The memoryless property. Suppose F has geometric distribution on $\{0, 1, 2, ...\}$ as in Exercise 6.

a) Show that for every $k \geq 0$,

$$P(F - k = m \mid F \geq k) = P(F = m), \quad m = 0, 1, ...$$

b) Show the geometric distribution is the only discrete distribution on $\{0, 1, 2, ...\}$ with this property.

c) What is the corresponding characterization of the geometric (p) distribution on $\{1, 2, \ldots\}$?

16. Fix r and p and let $P(k)$, $k = 0, 1, \ldots$, denote the probabilities in the negative binomial (r, p) distribution.

a) Show that the consecutive odds ratios are

$$P(k)/P(k-1) = (r + k - 1)q/k \qquad (k = 1, 2, \ldots)$$

b) Find a formula for the mode m of the negative binomial distribution.

c) For what values of r and p does the distribution have a double maximum? Which values k attain it?

17. Suppose the probability that a family has exactly n children is $(1 - p)p^n$, $n \geq 0$. Assuming each child is equally likely to be a boy or a girl, independently of previous children, find a formula for the probability that a family contains exactly k boys.

18. Suppose two teams play a series of games, each producing a winner and a loser, until one team has won two more games than the other. Let G be the total number of games played. Assuming your favorite team wins each game with probability p, independently of the results of all previous games, find:

a) $P(G = n)$ for $n = 2, 3, ...$;

b) $E(G)$;

c) $Var(G)$.

19. Let T_r be the number of fair coin tosses required to produce r heads. Show that:

a) $E(T_r) = 2r$;

b) $P(T_r < 2r) = 1/2$;

c) for every non-negative integer n, $\displaystyle\sum_{i=0}^{n} \binom{n+i}{n} 2^{-i} = 2^n$

20. Tail sums. Show that for a random variable X with possible values 0, 1, 2,...

a) $E(X) = \sum_{n=1}^{\infty} P(X \geq n)$;

b) $E[\frac{1}{2}X(X+1)] = \sum_{n=1}^{\infty} nP(X \geq n)$;

c) Call the first sum above Σ_1 and the second Σ_2. Find a formula for $Var(X)$ in terms of Σ_1 and Σ_2, assuming Σ_2 is finite.

21. Section 2.4 shows that the binomial (n,p) distribution approaches the Poisson (μ) distribution as $n \to \infty$, and $p \to 0$ with $np = \mu$ held fixed. Consider the negative binomial distribution with parameters r and $p = 1 - q$. Let $r \to \infty$, and let $p \to 1$ so that $rq = \mu$ is held fixed.

a) What does the mean become in the limit?

b) What does the variance become in the limit?

c) Show the distribution approaches the Poisson (μ) distribution in the limit.

22. Factorial moments and the probability generating function. The kth factorial moment of X is $f_k = E[(X)_k]$ where $(X)_k = X(X-1)\cdots(X-k+1)$. For many distributions of X with range $\{0,1,\ldots\}$ it is easier to compute the factorial moments than the ordinary moments $\mu_k = E[X^k]$. Note that $x^n = \sum_1^n S_{n,k}(x)_k$ for some integer coefficients $S_{n,k}$. These $S_{n,k}$ are known as *Stirling numbers of the second kind.*

a) Find $S_{n,k}$ for $1 \le n \le 3$ and $1 \le k \le n$.

b) Find a formula for μ_n in terms of $f_k, 1 \le k \le n$.

c) Assuming X has non-negative integer values, let $P(X = i) = p_i$ for $i = 0, 1, \ldots$. Let $G(z) = \sum_{i=0}^{\infty} p_i z^i$, known as the *probability generating function* of X. Assume $G(r) < \infty$ for some $r > 1$. Show by switching the order of summation and differentiation k times, (which can be justified, but you need not show this) that the kth derivative $G^{(k)}(z)$ of the function $G(z)$ is $G^{(k)}(z) = \sum_{i=k}^{\infty} p_i(i)_k z^{i-k}$. Deduce that $f_k = G^{(k)}(1)$.

23. Geometric generating function and moments. Using the notation and results of Exercise 22:

a) Find the generating function of the geometric (p) distribution on $\{0,1,2,\ldots\}$.

b) Find the first three factorial moments of the geometric (p) distribution on the integer set $\{0,1,2,\ldots\}$ by differentiation of the generating function. Check the first two factorial moments yield the mean and variance as given in the text.

c) Referring to Exercise 3.3.33 for properties of skewness, use the result of b) to find the skewness of the geometric (p) distribution on $\{0,1,2,\ldots\}$. Without further calculation, find the skewness of the geometric (p) distribution $\{1,2,\ldots\}$ and of the negative binomial (r,p) distribution.

24. The collector's problem. In the setting of Example 5, let T_n denote the number of boxes to get a complete set of animals.

a) Find a formula for $\sigma_n = SD(T_n)$.

b) Show that $\sigma_n < cn$ for a constant $c > 0$.

c) Deduce from Chebychev's inequality that T_n will most likely differ from $n \log n$ by only a small multiple of n.

d) (*Hard.*) Find the asymptotic distribution as $n \to \infty$ of $(T_n - n \log n)/n$. (It's not normal.)

3.5 The Poisson Distribution

The Poisson distribution is an approximation to the distribution of the number N of occurrences of events of some kind, when the events all have small probabilities, and are independent or nearly so. For example, N might be one of the following counting variables:

N_{wins}**:** the number of wins in n games of roulette for a gambler who bets on a single number each game.

N_{drops}**:** the number of raindrops which fall on a particular square inch of roof during a one-second interval of time.

$N_{\text{particles}}$**:** the number of radioactive particles emitted by a piece of radioactive material during an interval of time.

In case there are n independent events with equal probability p, the exact distribution of the number N that occurs is binomial (n, p). As shown in Section 2.4, if p is small this distribution is closely approximated by the Poisson distribution with parameter $\mu = E(N) = np$:

$$P(N = k) \approx e^{-\mu}\mu^k/k! \qquad (k = 0, 1, \ldots)$$

This justifies the use of the Poisson distribution in each case above. For instance, in the raindrops example, think of the square inch as divided into 100 hundredths of a square inch, each of which might or might not be hit by a raindrop. Suppose each hundredth of a square inch has the same small chance of being hit by a raindrop in the given second, independently of what happens elsewhere on the roof, and ignore the extremely small probability of the same hundredth of a square inch being hit more than once. Then N_{drops} is the number of successes in 100 independent trials, with small probability of success on each trial. You can think of $N_{\text{particles}}$ in a similar way, by dividing time into small units. By passing to a limit in which the raindrops are regarded as hitting random points in the plane, or the particles arrive at random instants on the time line, a mathematical model is obtained in which the distribution of the count is *exactly* Poisson. This is the idea of a *Poisson random scatter*, or *Poisson process*, discussed later in this section.

Features

Features of the Poisson (μ) distribution come from corresponding features of the binomial (n, p) distribution, by the passage to the limit as $n \to \infty$ and $p \to 0$ with $np = \mu$ kept fixed. It was shown in Section 2.4 that in this limit the probabilities of individual values converge

$$\binom{n}{k} p^k (1-p)^{n-k} \to e^{-\mu}\mu^k/k! \quad \text{as} \quad n \to \infty \quad \text{and} \quad p \to 0 \quad \text{with} \quad np = \mu$$

Since the binomial (n, p) distribution has mean $np = \mu$, it is natural that the Poisson (μ) limit should also have mean μ. And the SD of the binomial (n, p) distribution is $\sqrt{npq}$, which tends to $\sqrt{\mu}$ as $n \to \infty$ and $p \to 0$ with $np = \mu$.

Poisson Mean and Standard Deviation

If N has Poisson (μ) distribution,

$$E(N) = \mu \qquad SD(N) = \sqrt{\mu}$$

These formulae, made plausible by passage to the limit from binomial, will now be verified using the Poisson probability formula and the definitions of mean and SD for a discrete distribution.

Derivation of the mean.

$$\begin{aligned} E(N) &= \sum_{k=0}^{\infty} kP(N = k) \\ &= \sum_{k=1}^{\infty} k e^{-\mu} \frac{\mu^k}{k!} \\ &= e^{-\mu} \mu \sum_{k=1}^{\infty} \frac{\mu^{k-1}}{(k-1)!} \\ &= e^{-\mu} \mu \sum_{j=0}^{\infty} \frac{\mu^j}{j!} \\ &= e^{-\mu} \mu e^{\mu} = \mu \end{aligned}$$

Derivation of the SD. A direct attempt to find $E(N^2)$ would be to try to repeat the last calculation with $k^2 P(N = k)$ instead of $kP(N = k)$. This gives terms of a constant times $\mu^k k^2/k!$ which are not easy to sum. But $\mu^k k(k-1)/k!$ can easily be summed, and this solves the problem:

$$\begin{aligned} E(N(N-1)) &= \sum_{k=0}^{\infty} k(k-1) e^{-\mu} \frac{\mu^k}{k!} \\ &= e^{-\mu} \sum_{k=2}^{\infty} k(k-1) \frac{\mu^k}{k!} \\ &= e^{-\mu} \mu^2 \sum_{k=2}^{\infty} \frac{\mu^{k-2}}{(k-2)!} \\ &= e^{-\mu} \mu^2 e^{\mu} = \mu^2 \end{aligned}$$

$$\text{so} \quad E[N^2] = E[N(N-1)+N] = E[N(N-1)] + E(N) = \mu^2 + \mu$$
$$\text{and} \quad Var(N) = E(N^2) - [E(N)]^2 = \mu^2 + \mu - \mu^2 = \mu$$
$$SD(N) = \sqrt{\mu}$$

How μ affects the shape of the distribution. Let N_μ have Poisson (μ) distribution. For example, think of N_μ as the number of raindrops which hit a portion of a roof of area μ in a given length of time, assuming one raindrop is expected per unit area. Since N_μ has mean μ and SD $\sqrt{\mu}$, you should expect N_μ to be around μ plus or minus a small multiple of $\sqrt{\mu}$.

If μ is so close to 0 that μ^2 is negligible in comparison to μ (for example, when $\mu = 0.01$, $\mu^2 = 0.0001$), terms of order μ^2 and higher can be neglected in the expansion

$$e^{-\mu} = 1 - \mu + \mu^2/2 + \cdots$$

so

$$P(N_\mu = 0) = e^{-\mu} \approx 1 - \mu$$
$$P(N_\mu = 1) = \mu e^{-\mu} \approx \mu$$
$$P(N_\mu \geq 2) \approx 0$$

where $\approx$ means an approximation for small μ with an error of at most about μ^2. In the raindrops example, with one drop expected per unit area, this means that for a small area $\mu \ll 1$ the chance of being hit by one drop is about μ, and the chance of being hit by more than one drop is negligible in comparison.

Look again at the histograms of Poisson distributions at the end of Section 2.4. For $0 < \mu < 1$ the Poisson (μ) distribution has most probability at 0, and strictly decreasing probabilities for higher counts. As μ increases, the distribution shifts toward larger values and slowly flattens out, consistent with the formulae μ and $\sqrt{\mu}$ for the mean and SD.

Normal approximation. For μ large enough that the standard deviation $\sqrt{\mu}$ of the Poisson distribution is small in comparison to its mean μ, the distribution starts to become normal in shape. The distribution of the standardized Poisson variable $(N_\mu - \mu)/\sqrt{\mu}$ approaches standard normal as $\mu \to \infty$. This can be shown by study of consecutive odds ratios as in the binomial case treated in Section 2.3. It is yet another instance of the central limit theorem, due to the fact, discussed below, that sums of independent Poisson variables are Poisson.

Skewness. The Poisson(μ) distribution has skewness $1/\sqrt{\mu}$ (Exercise 20). Because this skewness tends to zero very slowly as $\mu \to \infty$ the approach of the Poisson distribution to normality is rather slow. Numerical calculations shown in Table 1 confirm what is apparent in Figure 1: for moderate values of μ the Poisson histogram follows a skew-normal curve much more closely than it does the normal curve.

FIGURE 1. **Normal and skew-normal approximation to the Poisson (9) distribution** Both the normal curve $y = \phi(z)$ and the skew–normal curve $y = \phi(z) - (1/18)\phi'''(z)$ are shown. The skew–normal curve follows the histogram much more closely.

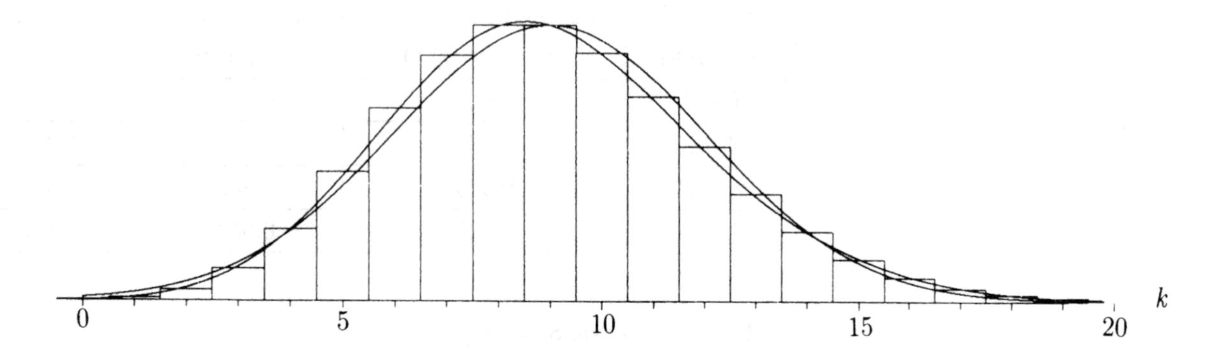

Skew-normal Approximation to the Poisson Distribution

If N_μ has Poisson (μ) distribution, then for $b = 0, 1, \ldots$

$$P(N_\mu \le b) \approx \Phi(z) - \frac{1}{6\sqrt{\mu}}(z^2 - 1)\phi(z) \quad \text{where } z = (b + \tfrac{1}{2} - \mu)/\sqrt{\mu}.$$

Here $\Phi(z)$ is the standard normal c.d.f. and $\phi(z)$ is the standard normal curve.

It can be shown that if this skew-normal approximation is used twice to approximate interval probabilities, the worst error is less than $1/(20\mu)$ for all μ. If the skewness correction term is ignored, the resulting normal approximation with continuity but not skewness correction gives interval probabilities with much larger errors up to about $1/(10\sqrt{\mu})$ for the worst cases $a \approx \mu - \sqrt{3\mu}, b \approx \mu$ and $a \approx \mu, b \approx \mu + \sqrt{3\mu}$. If μ is sufficiently large such errors can be ignored.

The following table shows some numerical results for $\mu = 9$. The numbers are correct to three decimal places. Compare with the very similar behavior of the binomial $(100, 1/10)$ distribution displayed in Table 2 at the end of Section 2.2. As in that table, the ranges selected are the ranges over which the normal approximation is first too high, then too low, too high, and too low again. The normal approximation to the Poisson (9) distribution is very rough, but the skew-normal approximation is excellent.

TABLE 1. **Approximations to the Poisson (9) distribution.** The interval probability $P(a \leq N_9 \leq b)$ is shown for a Poisson (9) random variable N_9 along with approximations using the normal and skew–normal curves.

range of values a to b	Poisson (9) probability $P(a \leq N_9 \leq b)$	skew-normal approximation	normal approximation
$0 - 3$	0.021	0.024	0.033
$4 - 8$	0.434	0.431	0.400
$9 - 14$	0.503	0.502	0.533
$15 - \infty$	0.041	0.043	0.033

Law of large numbers. Since $E(N_\mu/\mu) = \mu/\mu = 1$ and

$$SD(N_\mu/\mu) = \sqrt{\mu}/\mu = 1/\sqrt{\mu} \to 0 \quad \text{as} \quad \mu \to \infty$$

$$N_\mu/\mu \approx 1 \quad \text{for large } \mu$$

in the probabilistic sense that N_μ/μ will most likely be very close to 1. This is the law of large numbers in the Poisson context. In terms of the raindrops example, with one drop expected per unit area, this law of large numbers says that over a large area μ the average number of drops per unit area is nearly certain to be close to its expected value of 1. Both the normal approximation and the law of large numbers for the Poisson distribution are instances of more general results for sums of independent random variables, due to the result of the next paragraph.

Sums. If a big area is broken up into, say, j small areas, the number of raindrops hitting the big area is the sum of the numbers of drops in the j small areas. So the following result is very natural:

Sums of Independent Poisson Variables are Poisson

If $N_1, \ldots, N_j$ are independent Poisson random variables with parameters $\mu_1, \ldots, \mu_j$, then $N_1 + \cdots + N_j$ is a Poisson random variable with parameter $\mu_1 + \cdots + \mu_j$.

To see this via the approximation to binomial, first consider two separate blocks of Bernoulli trials of lengths n_1 and n_2 to see the following:

If N_1 and N_2 are independent with binomial (n_1, p) and binomial (n_2, p) distributions, then $N_1 + N_2$ has binomial $(n_1 + n_2, p)$ distribution.

Now let n_1 and n_2 both tend to ∞, and $p \to 0$, with $n_1 p \to \mu_1$ and $n_2 p \to \mu_2$. Then $(n_1 + n_2)p \to \mu_1 + \mu_2$. So N_1 and N_2 approach independent Poisson variables with means μ_1 and μ_2, while $N_1 + N_2$ approaches Poisson $(\mu_1 + \mu_2)$.

Here is an alternative derivation. To simplify notation, let $\alpha = \mu_1$ and $\beta = \mu_2$.

$$\begin{aligned}
P(N_1 + N_2 = k) &= \sum_{j=0}^{k} P(N_1 = j)P(N_2 = k - j) \\
&= \sum_{j=0}^{k} e^{-\alpha}\frac{\alpha^j}{j!}e^{-\beta}\frac{\beta^{k-j}}{(k-j)!} \\
&= e^{-(\alpha+\beta)}\frac{(\alpha+\beta)^k}{k!}\sum_{j=0}^{k}\frac{k!}{j!(k-j)!}\left(\frac{\alpha}{\alpha+\beta}\right)^j\left(\frac{\beta}{\alpha+\beta}\right)^{k-j} \\
&= e^{-(\alpha+\beta)}\frac{(\alpha+\beta)^k}{k!}
\end{aligned}$$

because the terms in the previous sum are all the terms in a binomial distribution, with sum 1. Thus $N_1 + N_2$ has Poisson $(\alpha + \beta)$ distribution. Repeated application of this result for two terms gives the result for any number of terms.

Example 1. Number of wins.

Problem. Suppose a gambler bets ten times on events of probability $1/10$, then twenty times on events of probability $1/20$, then thirty times on events of probability $1/30$, then forty times on events of probability $1/40$. Assuming the events are independent, what is the approximate distribution of the number of times the gambler wins?

Solution. Let N_1 be the number of wins on the first 10 events of probability $1/10$, N_2 the number of wins on the next 20, N_3 the number of wins on the next 30, and N_4 the number of wins on the next 40. The exact distribution of the gambler's winnings is the distribution of

$$N = N_1 + N_2 + N_3 + N_4$$

The random variables N_i, $i = 1, 2, 3, 4$, are independent, and each N_i is binomial $(10i, 1/10i)$, hence approximately Poisson (1). Thus the distribution of N must be approximately Poisson (4), by the Poisson sums theorem.

Remark. As the example suggests, the Poisson approximation to the binomial distribution extends to the case of independent trials with possibly different probabilities of success. It can be shown that if N is the number of events which occur among n independent events with probabilities $p_1, \ldots, p_n$, then provided all the probabilities p_i are small, the distribution of N is approximately Poisson (μ), where

$$\mu = E(N) = p_1 + p_2 + \cdots + p_n.$$

Random Scatter

It has already been argued informally that it would be reasonable to assume a Poisson distribution for a random variable like the number of raindrops to hit a given area in a given period of time. This idea will now be developed further to give a mathematical model for a random scatter of points in a plane such as in the diagram below.

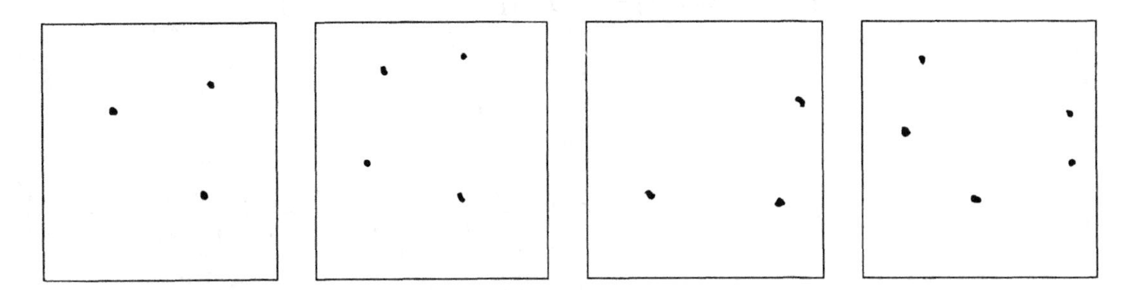

The points might indicate, for example:

(i) points on a surface hit by particles of some kind, for example, raindrops, dust particles, atomic particles, or photons;

(ii) positions of cells of some kind on a microscopic slide;

(iii) positions of stars on a photographic plate.

The model is based on simple intuitive assumptions which turn out to imply that the number of points in a fixed area will have a Poisson distribution. The same idea of a random scatter makes sense in any number of dimensions, with length or volume instead of area. For example, a mist of raindrops is a three-dimensional scatter. And a process of random arrivals, like calls coming into a telephone exchange, can be thought of as defining a scatter of points on a time line. The basic ideas will be set out here for a scatter in two dimensions. But similar assumptions in any number of dimensions lead to the same conclusion of Poisson distributed counts.

A random scatter has both a discrete and a continuous aspect. Counting the number of points in a given region or interval gives a discrete variable. If you know enough counts for different regions you can say more or less where the points are. And the probabilities of events determined by the scatter can be derived from assumptions about the counting variables. This is the approach taken here, with assumptions which imply the counts are Poisson distributed. On the other hand, the positions in space or time of points in a scatter are typical continuous variables. Section 4.2 shows how the continuous distributions of these variables are related to the discrete Poisson distribution of counts.

Assumptions. Consider a scatter of a finite number of points in a square. To distinguish points in the scatter from other points in the square, call the points in the scatter *hits*. These are the places hit by the raindrops, particles or whatever, idealized as points in the square.

Assumption 1: No Multiple Hits

That is to say, distinct hits define distinct points in the square.

To state the next assumption, suppose that for each $n = 4, 16, 64, \ldots$, the square is divided into n subsquares of equal area $1/n$, as in the following diagrams. Say a subsquare is *hit* if it contains one or more hits of the scatter, and *missed* if it contains no hits. Hit squares are black and missed squares white in the diagrams. For each n, the pattern of hit squares provides some information about the scatter. This pattern gives a digital representation of the scatter, with some loss of information. As the number of subsquares n increases the pattern of hit subsquares becomes more and more sharply focused on the scatter. This can be seen in the following diagram, which shows patterns derived from a scatter of 5 points in the square.

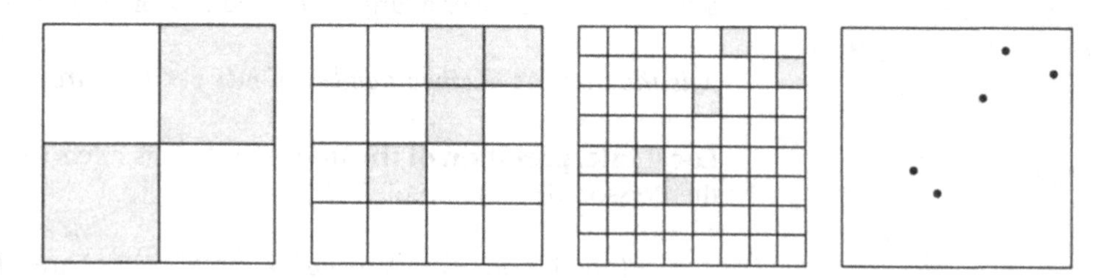

Assumption 2: Randomness of Hits on Subsquares

For each n, any one of the n subsquares is hit with the same probability, say p_n, independently of hits on the other $n - 1$ subsquares.

Note that the randomness assumption refers separately to each digital representation. The digital representations of a random scatter for different values of n are, in fact, highly dependent. If you know the digital representation for some value of n, the representation for smaller values of n is completely determined.

Poisson Scatter Theorem

The assumptions of no multiple hits and randomness imply there is a positive constant λ such that:

(i) for each subset B of the square, the number $N(B)$ of hits in B is a Poisson random variable with mean $\lambda \times \text{area}(B)$;

(ii) for disjoint subsets $B_1, \ldots, B_j$, the numbers of hits $N(B_1), \ldots, N(B_j)$ are mutually independent.

The random scatter is then called a *Poisson scatter with intensity* λ. The intensity is the expected number of hits per unit area. Conversely, (i) and (ii) imply the assumptions of no multiple hits and randomness.

A proof of the Poisson scatter theorem is sketched at the end of the section.

Global interpretation of the intensity λ. If the scatter in the square is just part of a Poisson scatter over a larger area, the law of large numbers shows that

λ is the limiting average number of hits per unit area over a large area.

Local interpretation of the intensity λ. This refers to sets B with small area. From the Poisson distribution of $N(B)$,

$$P(\text{one hit on } B) = \lambda \, \text{area}(B) \, e^{-\lambda \, \text{area}(B)} \sim \lambda \, \text{area}(B) \quad \text{as} \quad \text{area}(B) \to 0$$

and the probability of two or more hits on B is negligible in comparison. So

λ is the probability of a hit per unit area, as the area tends to zero.

Sums again. The fact that sums of independent Poisson variables are again Poisson is built into the concept of a Poisson scatter. For if $B_1, \ldots, B_j$ is a partition of a unit square into sets with areas $p_1, \ldots, p_j$, where $\sum_i p_i = 1$, then the total number of hits is $N = \sum_i N(B_i)$. If the scatter is Poisson with intensity λ, then N is Poisson (λ), while the $N(B_i)$, $1 \leq i \leq j$, are independent Poisson variables with means λp_i, which could be any positive numbers with sum λ.

Scatters over other sets. The theorem extends to scatters over other subsets of the plane than a square, and scatters on the line or in higher dimensions. Then length or volume replaces area.

Example 2. Particle hits.

Problem. Suppose particles hit a square at random according to a Poisson random scatter, with 8 particles expected in the whole square. What is the probability that the four equal subsquares in the diagram are hit by exactly 0, 1, 2, and 3 particles, respectively?

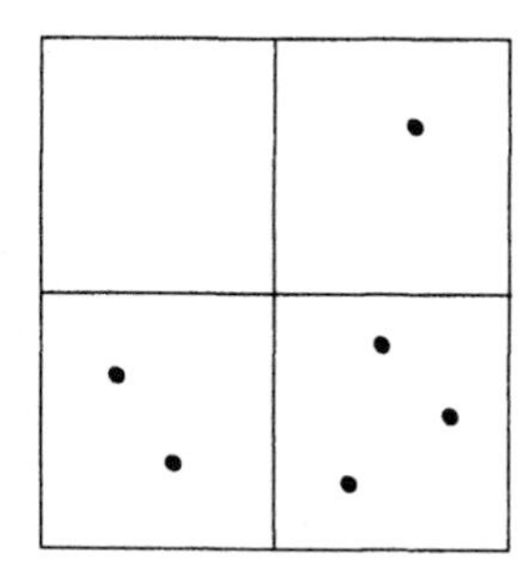

Solution. Since the numbers of hits on the four squares are independent Poisson random variables, all with parameter $8 \times 1/4 = 2$, the probability in question is

$$\frac{e^{-2}2^0}{0!} \times \frac{e^{-2}2^1}{1!} \times \frac{e^{-2}2^2}{2!} \times \frac{e^{-2}2^3}{3!} = \frac{e^{-8}2^6}{12}$$

Example 3. Bacterial colonies.

Problem 1. Suppose a volume of 1000 drops of water contains 2000 bacteria, separate from each other and thoroughly mixed in the water. A single drop is smeared uniformly over the surface of a dish. The dish contains nutrients on which the bacteria feed and multiply. After a few days, wherever a bacterium was deposited on the dish a visible colony of bacteria appears. Find the distribution of the number of colonies that appear: a) over the whole plate, b) over an area of half the plate.

Solution. It seems reasonable to suppose that the positions of bacterial colonies over the plate form a Poisson random scatter. Since 1000 drops contain 2000 bacteria, the expected number of bacteria per drop may be estimated as $2000/1000 = 2$. So the distribution of the number of bacteria on the whole plate is Poisson with mean 2. And the distribution of the number in half the plate is Poisson with mean 1.

Remark. Instead of thinking of the scatter over the plate to justify the Poisson distribution, you might think that each of the 2000 bacteria was present in the drop smeared on the plate with probability $1/1000$, independently of the others. Then the number of bacteria on the plate would have binomial $(2000, 1/1000)$ distribution, which is Poisson (2) for all practical purposes. Similarly, for the number on half the plate, you get binomial $(2000, 1/2000)$, which is approximately Poisson (1). But the assumption of random scatter implies that the numbers in the two halves of the plate are independent, something not so obvious by the second method.

Problem 2. Suppose now it is not certain that a bacterium will survive and produce a visible colony, but that this happens with probability p for each bacterium on the plate, independently of the others. What now is the distribution for the number of colonies?

Solution. It is intuitively clear that the scatter of colonies must still satisfy the hypotheses of a Poisson scatter. The intensity of the colonies can be calculated from its local interpretation. Take the area of the whole plate to be 1, so by the previous example the intensity for the scatter of all bacteria landing on the plate is 2 per unit area, and take a region B so small that

$$P(\text{one bacterium in } B) \approx 2\,\text{area}(B) \qquad P(2 \text{ or more bacteria in } B) \approx 0$$

where $\approx$ allows an error of order area(B) squared. Then

$$\begin{aligned} P(\text{one colony in } B) &= P(\text{one bacterium in } B \text{ and colony}) \\ &\quad + P(2 \text{ or more bacteria in } B \text{ and colony}) \\ &\approx 2\,\text{area}(B)\,p = 2p\,\text{area}(B) \end{aligned}$$

So the scatter of colonies has intensity $2p$ per unit area. The number of colonies on the whole plate therefore has Poisson $(2p)$ distribution.

Remark. Again, the same conclusion can be obtained another way. Think of the number of colonies as the sum of 2000 independent indicator random variables, indicating whether or not each of the 2000 bacteria gets deposited on the plate and then produces a colony. The chance of a bacterium getting on to the plate is $1/1000$, and the chance of it producing a colony, given that it gets on the plate, is p. So the overall probability of being deposited on the plate and then surviving is $p/1000$. This makes the number of colonies have binomial $(2000, p/1000)$ distribution, which is Poisson $(2p)$ for all practical purposes.

The last example illustrates a useful property of Poisson scatters, which can be derived in general by the same argument:

Thinning a Poisson Scatter

Suppose that in a Poisson scatter with intensity λ, each point of the scatter is kept with probability p, and erased (or *thinned*) with probability $1 - p$, independently both of the positions of points in the scatter and of all other thinnings. Then the scatter of points that are kept is a Poisson scatter with intensity λp.

Similarly, the scatter of points that are thinned is a Poisson process with intensity λq, where $q = 1 - p$. It can be shown, moreover, that the two scatters, one of points that are kept, and the other of points that are thinned, are independent. This means that any event determined by the numbers and positions of points in one scatter is independent of any such event determined by the other. In the example with the bacterial colonies, the numbers and positions on the plate of the bacteria that survive to produce colonies are independent of the numbers and positions of those that do not.

If you combine or *superpose* these two independent Poisson scatters, with intensities, say, $\alpha = \lambda p$ and $\beta = \lambda q$, you get back the original Poisson scatter with intensity $\lambda = \alpha + \beta$. So thinning can be understood as a kind of inverse to the more obvious operation of superposition of two independent Poisson scatters, which gives a new Poisson scatter whose intensity is the sum of the intensities of the component scatters.

Sketch Proof of the Poisson Scatter Theorem

Step 1. Poisson distribution for the total number of hits. Let N be the total number of hits in the whole square, assumed to be of unit area. Let N_n be the number of subsquares hit when the unit square is divided into n subsquares. Then N_n increases as n increases, because each hit on one of the n subsquares must contribute one or more hits to all counts with more subsquares. And $N_n = N$ for all n large enough that the distance across one of the n subsquares is shorter than the smallest distance between two of the hits in the scatter, since then the N hits must fall in N different subsquares. (This is where the assumption of no multiple hits is essential.) Just how large n must be before $N_n = N$ depends on the scatter. But whatever the scatter, N_n eventually equals N. So the distribution of N can be found as the limit as $n \to \infty$ of the distribution of N_n. (Technically, this uses the infinite sum rule for probabilities, taken here as an axiom.) By the randomness assumption, N_n has binomial (n, p_n) distribution, where p_n is the probability that one of the subsquares of area $1/n$ is occupied. Since N_n increases with n, so does its expectation np_n. Therefore np_n converges to a limit λ as $n \to \infty$, and you can show that λ must be finite (exercise). Consequently, the limit distribution of N_n is Poisson (λ). This is the distribution of N.

Step 2. Poisson distribution for the number of hits on a subset B. Assuming B is a *simple* subset of the unit square, meaning a finite union of subsquares at some level, this is similar to the argument above, with N replaced by $N(B)$ and N_n replaced by $N_n(B)$, the number of hit squares of area $1/n$ within B. For large enough n, the simple set B is the union of some number n_B of subsquares of area $1/n$. In fact, $n_B = n \,\text{area}(B)$, since we assume the whole square has unit area, so $\text{area}(B) = n_B/n$. Now $N_n(B)$ has binomial (n_B, p_n) distribution, where

$$n_B p_n = np_n \,\text{area}(B) \to \lambda \,\text{area}(B) \quad \text{as} \quad n \to \infty$$

So in the limit the distribution of $N(B)$ is Poisson with mean $\lambda \,\text{area}(B)$. The same conclusion for more general subsets B is justified by approximation arguments or measure theory.

Step 3. Independence of counts in disjoint subsets. This comes from the assumed independence of hits in different subsquares, by letting the number of subsquares tend to infinity. □

Exercises 3.5

1. Suppose 1% of people in a large population are over 6 feet 3 inches tall. Approximately what is the chance that from a group of 200 people picked at random from this population, at least four people will be over 6 feet 3 inches tall?

2. How many raisins must cookies contain on average for the chance of a cookie containing at least one raisin to be at least 99%?

3. The cookie dough used by a bakery to make 2-ounce cookies contains an average of 32 raisins per pound of dough. The bakery sells cookies in bags of a dozen.

a) Suppose that customers complain if one or more of the cookies in a bag contains no raisins. Over the long run, about what proportion of bags of cookies give rise to complaints?

b) Approximately what average number of raisins per pound would ensure that only 5% of the bags give rise to complaints?

4. Books from a certain publisher contain an average of 1 misprint per page. What is the probability that on at least one page in a 300-page book from this publisher there will be at least 5 misprints?

5. Microbes are smeared over a plate at an average density of 5000 per square inch. The viewing field of a microscope is 10^{-4} square inches of this plate. What is the chance that at least one microbe is in the viewing field? What assumptions are you making?

6. Suppose rain is falling at an average rate of 30 drops per square inch per minute. What is the chance that a particular square inch is not hit by any drops during a given 10-second period? What assumptions are you making?

7. Suppose raisin muffins from the recycling bakery have an average of 3 fresh raisins and 2 rotten raisins per muffin.

a) What is an appropriate distribution for the number of each kind of raisin, and for the total?

b) If you bite off 20% of a muffin, what is the probability you get no raisins?

8. A Geiger counter receives pulses at an average rate of 10 per minute. What is the probability of three pulses appearing in a given half-minute period? What assumptions are you making?

9. Suppose that X and Y are independent Poisson random variables with parameters 1 and 2, respectively. Find:

a) $P(X = 1 \text{ and } Y = 2)$;

b) $P\left(\frac{X+Y}{2} \geq 1\right)$;

c) $P\left(X = 1 \,|\, \frac{X+Y}{2} = 2\right)$

10. Let X have Poisson (λ) distribution. Calculate:

a) $E(3X + 5)$; b) $Var(3X + 5)$; c) $E\left[\frac{1}{1+X}\right]$.

11. Suppose X, Y, and Z are independent Poisson random variables, each with mean 1. Find

a) $P(X + Y = 4)$; b) $E\left[(X + Y)^2\right]$; c) $P(X + Y + Z = 4)$.

12. Radioactive substances emit α-particles. The number of such particles reaching a counter over a given time period follows the Poisson distribution. Suppose two substances emit α-particles independently of each other. The first substance gives out α-particles which reach the counter according to the Poisson (3.87) distribution, while the second substance emits α-particles which reach the counter according to the Poisson (5.41) distribution. Find the chance that the counter is hit by at most 4 particles.

13. Regard the positions of molecules in a room as the points of a Poisson random scatter in 3 dimensions. According to physics, there are about 6.023×10^{23} molecules in every 22.4 liters of air at normal temperature and pressure. (A liter is 1000 cubic centimeters.) Let $N(x)$ be the random number of molecules in a particular cube of air with sides of length x centimeters.

a) Calculate the mean $\mu(x)$ and standard deviation $\sigma(x)$ of $N(x)$.

b) How small does x have to be in order that $\sigma(x)$ be 1% of $\mu(x)$, so fluctuations in density of around 1% over a cube of length x are likely to occur?

14. Assume that each of 2000 individuals living near a nuclear power plant is exposed to particles of a certain kind of radiation at an average rate of one per week. Suppose that each hit by a particle is harmless with probability $1 - 10^{-5}$, and produces a tumor with probability 10^{-5}. Find the approximate distribution of:

a) the total number of tumors produced in the whole population over a one-year period by this kind of radiation;

b) the total number of individuals acquiring at least one tumor over a year from this radiation.

Sketch the histograms of each distribution, and find the means and SD's.

15. A book has 200 pages. The number of mistakes on each page is a Poisson random variable with mean 0.01, and is independent of the number of mistakes on all other pages.

(a) What is the expected number of pages with no mistakes? What is the variance of the number of pages with no mistakes?

(b) A person proofreading the book finds a given mistake with probability 0.9. What is the expected number of pages where this person will find a mistake?

(c) What, approximately, is the probability that the book has two or more pages with mistakes?

16. On average, one cubic inch of Granma's cookie dough contains 2 chocolate chips and 1 marshmallow.

a) Granma makes a cookie using three cubic inches of her dough. Find the chance that the cookie contains at most four chocolate chips. State your assumptions.

b) Assume the number of marshmallows in Granma's dough is independent of the number of chocolate chips. I take three cookies, one of which is made with two cubic inches of dough, the other two with three cubic inches each. What is the chance that at most 1 of my cookies contains neither chocolate chips nor marshmallows?

17. Raindrops are falling at an average rate of 30 drops per square inch per minute.

a) What is the chance that a particular square inch is not hit by any drops during a given 10-second period?

b) If each drop is a big drop with probability 2/3 and a small drop with probability 1/3, independently of the other drops, what is the chance that during 10 seconds a particular square inch gets hit by precisely four big drops and five small ones?

18. A population comprises X_n individuals at time $n = 0, 1, 2, \ldots$. Suppose that X_0 has Poisson (μ) distribution. Between time n and time $n+1$ each of the X_n individuals dies with probability p, independently of the others. The population at time $n+1$ is formed from the survivors together with a random number of immigrants who arrive independently according to a Poisson (μ) distribution.

a) What is the distribution of X_n?

b) What happens to this distribution as $n \to \infty$?

19. Poisson generating function and moments. Suppose X has Poisson(μ) distribution. Using the notation and results of Exercise 3.4.22,

a) Show that $G(z) = e^{-\mu+\mu z}$.

b) Find the first three factorial moments X.

c) Deduce the values of the first three ordinary moments of X.

d) Show that $E(X-\mu)^3 = \mu$ and Skewness$(X) = 1/\sqrt{\mu}$.

20. Skewness of the Poisson(μ) distribution. Derive the formula $1/\sqrt{\mu}$ for the skewness of the Poisson(μ) distribution from the Poisson approximation to binomial distribution (you can assume the required switches of sums and limits are justified).

21. Skew-normal approximation to the Poisson distribution. Derive the skew-normal approximation to the Poisson (μ) distribution stated in this section:

a) from the skew-normal approximation to the binomial (n, p) distribution (in Section 2.2) by passage to the Poisson limit as $n \to \infty$ and $p \to 0$ with $np = \mu$;

b) from the skew-normal approximation for the sum of n independent random variables stated at the end of Section 3.3.

c) For N_{10} with Poisson (10) distribution, find $P(N_{10} \leq 10)$ correct to three significant figures.

d) Find the normal approximation to $P(N_{10} \leq 10)$ with continuity but not skewness correction, "correct" to three significant figures. Observe that the last two figures are useless: the error of approximation exceeds 0.02.

e) Find the normal approximation to $P(N_{10} \leq 10)$ with continuity and skewness correction, correct to three significant figures. [All three figures should be correct. The actual error of approximation is about 2×10^{-5}.]

3.6 Symmetry (Optional)

This section studies a symmetry property for joint distributions, and illustrates it by applications to sampling without replacement. Let (X, Y) be a pair of random variables with joint distribution defined by

$$P(x, y) = P(X = x, Y = y)$$

The joint distribution is called *symmetric* if $P(x, y)$ is a symmetric function of x and y. That is to say,

$$P(x, y) = P(y, x) \qquad \text{for all } (x, y)$$

Graphically, this means that the distribution in the plane is symmetric with respect to a flip about the upward sloping diagonal line $y = x$. A glance at the figure on page 148 shows that a symmetric joint distribution is obtained for X and Y derived by sampling either with or without replacement from the set $\{1, 2, 3\}$. A symmetric joint distribution is obtained more generally whenever X and Y are two values picked by random sampling from some arbitrary list of values, either with or without replacement. This is obvious for sampling with replacement, and verified below for sampling without replacement.

In terms of random variables, the joint distribution of (X, Y) is symmetric if and only if (X, Y) has the same joint distribution as (Y, X). Then X and Y are called *exchangeable*. If X and Y are exchangeable then X and Y have the same distribution. This is true by the change of variable principle: X is a function (the first coordinate) of (X, Y), and Y is the same function of (Y, X).

The joint distribution of three random variables X, Y, and Z is called *symmetric* if

$$P(x, y, z) = P(X = x, Y = y, Z = z)$$

is a symmetric function of (x, y, z). That is to say, for all (x, y, z)

$$P(x, y, z) = P(x, z, y) = P(y, x, z) = P(y, z, x) = P(z, x, y) = P(z, y, x)$$

(all 3! = 6 possible orders of x,y and z). Equivalently, the 6 possible orderings of the random variables,

$$(X, Y, Z),\ (X, Z, Y),\ (Y, X, Z),\ (Y, Z, X),\ (Z, X, Y),\ (Z, Y, X)$$

all have the same joint distribution. Then X, Y, and Z have the same distribution, and each of the three pairs (X, Y), (X, Z), and (Y, Z) has the same (exchangeable) joint distribution, by the change of variable principle again.

A function of n variables, say $f(x_1, \ldots, x_n)$, is called *symmetric* if the value of f remains unchanged for all of the $n!$ possible permutations of the variables. Examples of symmetric functions are the sum $g(x_1) + g(x_2) + \cdots + g(x_n)$ and the product $g(x_1)g(x_2) \cdots g(x_n)$ for any numerical function $g(x)$.

Symmetry of a Joint Distribution

Let $X_1, \ldots, X_n$ be random variables with joint distribution defined by

$$P(x_1, \ldots, x_n) = P(X_1 = x_1, \ldots, X_n = x_n)$$

The joint distribution is *symmetric* if $P(x_1, \ldots, x_n)$ is a symmetric function of $(x_1, \ldots, x_n)$. Equivalently, all $n!$ possible orderings of the random variables $X_1, \ldots, X_n$ have the same joint distribution. Then $X_1, \ldots, X_n$ are called *exchangeable*. Exchangeable random variables have the same distribution. For $2 \leq m \leq n$, every subset of m out of n exchangeable random variables has the same symmetric joint distribution of m variables.

The simplest example of an exchangeable sequence of random variables is n independent trials $X_1, \ldots, X_n$. Then

$$P(x_1, x_2, \ldots, x_n) = p(x_1)p(x_2)\cdots p(x_n)$$

where $p(x) = P(X_i = x)$ defines the common distribution of the X_i. This a symmetric function of $(x_1, x_2, \ldots, x_n)$ because the product is the same evaluated in any order. Sampling with replacement is a special case of independent trials. Here is a more interesting example:

Sampling Without Replacement

The basic setup for sampling without replacement was described in Section 2.5. Suppose there is some population of N individuals. Suppose the ith individual in the population has some attribute b_i, for example the color of the ith ball in a box, or the height of the ith individual in a human population. Suppose n items are drawn one by one without replacement from the population. Let X_j be the attribute of the jth individual in the sample. So $X_1, \ldots, X_n$ might represent the random sequence of colors of n balls drawn at random without replacement from a box, or the random sequence of heights in a sample without replacement from a human population.

Symmetry in Sampling Without Replacement

Let $X_1, \ldots, X_n$ be a sample of size n without replacement from a list of values $\{b_1, \ldots, b_N\}$, where $2 \leq n \leq N$. Then $X_1, \ldots, X_n$ are exchangeable. In particular, for $1 \leq m \leq n$ the joint distribution of any subset of m of the X_i has the same distribution as a random sample of size m without replacement from the list $\{b_1, \ldots, b_N\}$.

This is proved in three stages as follows:

Proof for $n = N$ and $b_i = i$, $1 \le i \le n$. In this case $(X_1, \ldots, X_n)$ is an exhaustive random sample without replacement from the list $1, 2, \ldots n$, that is, a random permutation of $\{1, 2, \ldots, n\}$, as in Example 3.1.6. The joint probability function was calculated in that example and found to be symmetric. So $(X_1, \ldots, X_n)$ is exchangeable. □

Remark. The exchangeability of a random permutation is quite intuitive if you think of generating $(X_1, \ldots, X_n)$ by shuffling and then dealing out in order a deck of n cards labeled $1, 2, \ldots, n$. Any particular rearrangement of the variables $(X_1, \ldots, X_n)$ then corresponds to a particular deterministic shuffle before the deal. And it is intuitively clear that any particular additional deterministic shuffle of a perfectly shuffled deck must keep the deck perfectly shuffled. The exchangeability of a random permutation $(X_1, \ldots, X_n)$ is not so intuitive, but still true, for $X_1, \ldots, X_n$ generated by drawing balls at random one by one from an urn containing n balls labeled $1, 2, \ldots, n$.

Proof for $n = N$ and a general list $\{b_1, \ldots, b_n\}$. Now $\{b_1, \ldots, b_n\}$ can be any list of values whatever, allowing repetitions of values. The values need not be numerical. For example, for $n = N = 6$, $b_1 = b_2 = b_3 = b$, $b_4 = b_5 = r$, and $b_6 = w$, might represent a listing of the colors of balls in a box of 3 black balls, 2 red balls, and 1 white ball. A typical result of 6 draws from the box without replacement would then be the event

$$(X_1, X_2, X_3, X_4, X_5) = (b, r, w, b, b, r)$$

Think of a general list $\{b_1, \ldots, b_n\}$ listing the contents of a box. The result $(X_1, \ldots, X_n)$ of exhaustive sampling without replacement is a random permutation of the values in the list, with all $n!$ possible permutations of the indices equally likely. Write $b(k) = b_k$. Then, $X_i = b(Y_i)$ where $(Y_1, \ldots, Y_n)$ is random permutation of $1, 2, \ldots, n$. So

$$X_i = b(Y_i) \text{ where } Y_1, \ldots, Y_n \text{ are exchangeable}$$

But it is intuitively clear (and a consequence of the change of variable principle), that a function b applied to all variables in an exchangeable sequence yields another exchangeable sequence. □

Proof for $2 \le n \le N$ and a general list $\{b_1, \ldots, b_N\}$. For a sample of size n without replacement from a list of N values, the exchangeability follows by viewing the sample of size n as the first n variables in an exhaustive sample, which is exchangeable by the previous case, and appealing to the general fact that subsets of exchangeable variables are exchangeable. □

Examples. The symmetry of sampling without replacement appeared already in Section 2.5, in the derivation of the probability of getting g good elements and b bad elements in a sample of size n without replacement from a population of G good and B bad elements. That calculation used the fact that the probability of getting g

good elements and b bad elements in a particular order is the same for all possible orders. Other consequences of the symmetry appear in Example 1.4.7 and Example 3.1.6. Here are two more examples.

Example 1. Dealing cards.

Five cards are dealt from a standard deck of 52 cards.

Problem 1. What is the probability that the fifth card is a king?

Solution. It is confusing in this problem to think about which of the first four cards are kings. Rather, ignore the first four cards. The fifth card is a card drawn at random from the deck, just like the first card. So the probability that the fifth card is a king is the same as the probability that the first card is a king, that is $1/13$.

Problem 2. What is the chance that the third and fifth cards are black?

Solution. Ignore the first, second, and fourth cards. By the symmetry of sampling without replacement, the third and fifth cards are two cards drawn at random without replacement from the deck, just like the first two cards. So the probability that the third and fifth cards are black is the same as the probability that the first and second cards are black, that is $\frac{26}{52} \times \frac{25}{51}$.

Discussion. This kind of intuitive argument is precisely what is justified by the symmetry of sampling without replacement. Particular problems like these can be solved quickly "by symmetry" without using random variable notation. But the theoretical justification is symmetry of the joint distribution involved.

Example 2. Red and black balls.

Suppose 20 balls are drawn at random without replacement from a box containing 50 red balls and 50 black balls.

Problem 1. What is the probability that the 10th ball is red given that the 18th and 19th balls are red?

Solution. Let X_i be the color of the ith ball drawn. Then $(X_1, X_2, \ldots, X_{20})$ represents a random sample of size 20 without replacement from the population of 100 red and black balls. The problem is to calculate

$$P(X_{10} = red \mid X_{18} = red \text{ and } X_{19} = red) = \frac{P(X_{10} = red \text{ and } X_{18} = red \text{ and } X_{19} = red)}{P(X_{18} = red \text{ and } X_{19} = red)}$$

This conditional probability is determined by the joint distribution of X_{10}, X_{18}, and X_{19}, which is the same as the joint distribution of X_3, X_2 and X_1 by the symmetry of sampling without replacement. So the required probability is the same as

$$P(X_3 = red \mid X_2 = red \text{ and } X_1 = red) = \frac{48}{98}$$

since after drawing two red balls on the first two draws there are 48 red balls remaining out of 98 balls total.

Mean and Variance of the Hypergeometric Distribution

Recall from Section 2.5 the distribution of the number of good elements S_n in a sample of size n for a population of size N containing G good elements:

$$P(S_n = g) = \binom{G}{g}\binom{B}{b}\Big/\binom{N}{n} \quad 0 \le g \le n$$

where $b = n - g$, $B = N - G$ represent numbers of bad elements. The mean and standard deviation of S_n are as follows:

$$E(S_n) = np \text{ and } SD(S_n) = \sqrt{\frac{N-n}{N-1}}\sqrt{npq}$$

where $p = G/N$ is the proportion of good elements in the population, $q = B/N$ the proportion of bad elements in the population. Note that the mean is the same as if the sampling were done with replacement, when the distribution of S_n is binomial(n, p). And the standard deviation is just the familiar binomial standard deviation of $\sqrt{npq}$ multiplied by the factor $\sqrt{\frac{N-n}{N-1}}$, called the *finite population correction factor.*

Proof. Write

$$S_n = I_1 + I_2 + \cdots + I_n,$$

where for each $j = 1, 2, \ldots, n$, I_j is the indicator of the event that the jth draw yields a good element. By the symmetry of sampling without replacement just discussed, the distribution of I_j is the same Bernoulli (G/N) distribution for every j. Thus the expectation of S_n can be computed as

$$E(S_n) = E(I_1) + E(I_2) + \cdots + E(I_n) = nE(I_1) = n\frac{G}{N}$$

The variance can now be computed, starting from a calculation of

$$\begin{aligned} E(S_n^2) &= E[(\sum_j I_j)^2] \\ &= E[\sum_j I_j^2 + 2\sum_{j<k} I_j I_k] \\ &= \sum_j E(I_j^2) + 2\sum_{j<k} E(I_j I_k) \end{aligned}$$

FIGURE 1. Normal approximation for sampling with and without replacement. The top histogram shows the binomial $(100, 0.5)$ distribution of the number of good elements in a sample of size $n = 100$ with replacement from a population of size $N = 200$ containing $G = 100$ good elements and $B = 100$ bad ones. The approximating normal curve is superimposed. The bottom histogram shows the corresponding hypergeometric distribution for sampling *without* replacement from this population, together with its normal approximation. Note how the two means are the same, but the standard deviation is noticeably smaller for sampling without replacement because the sample size is a significant fraction (half) of the population size.

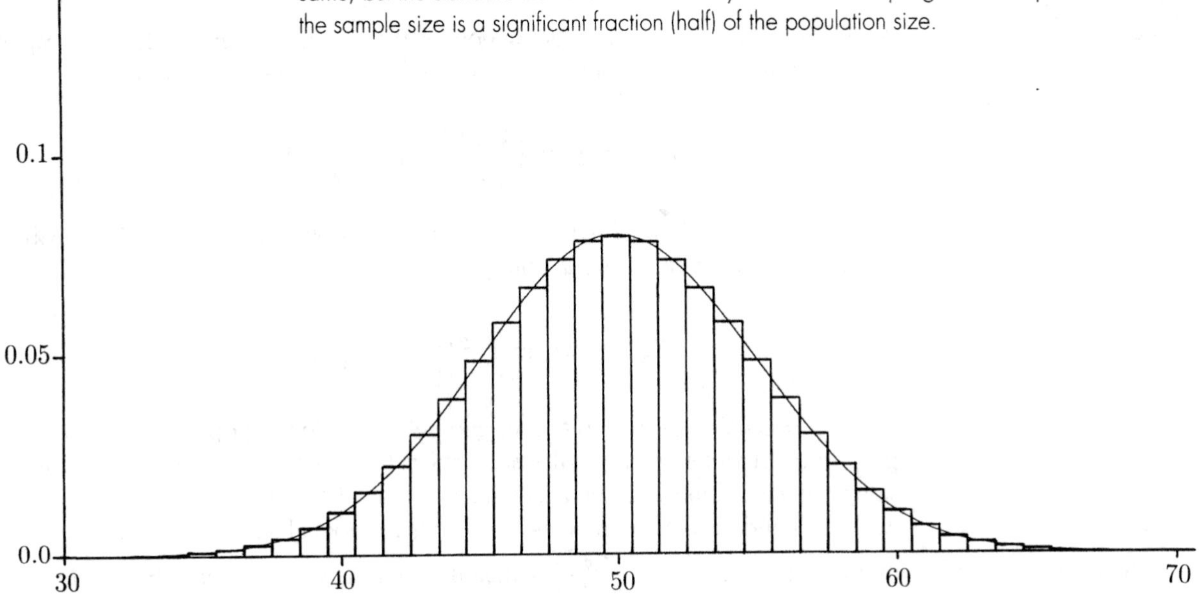

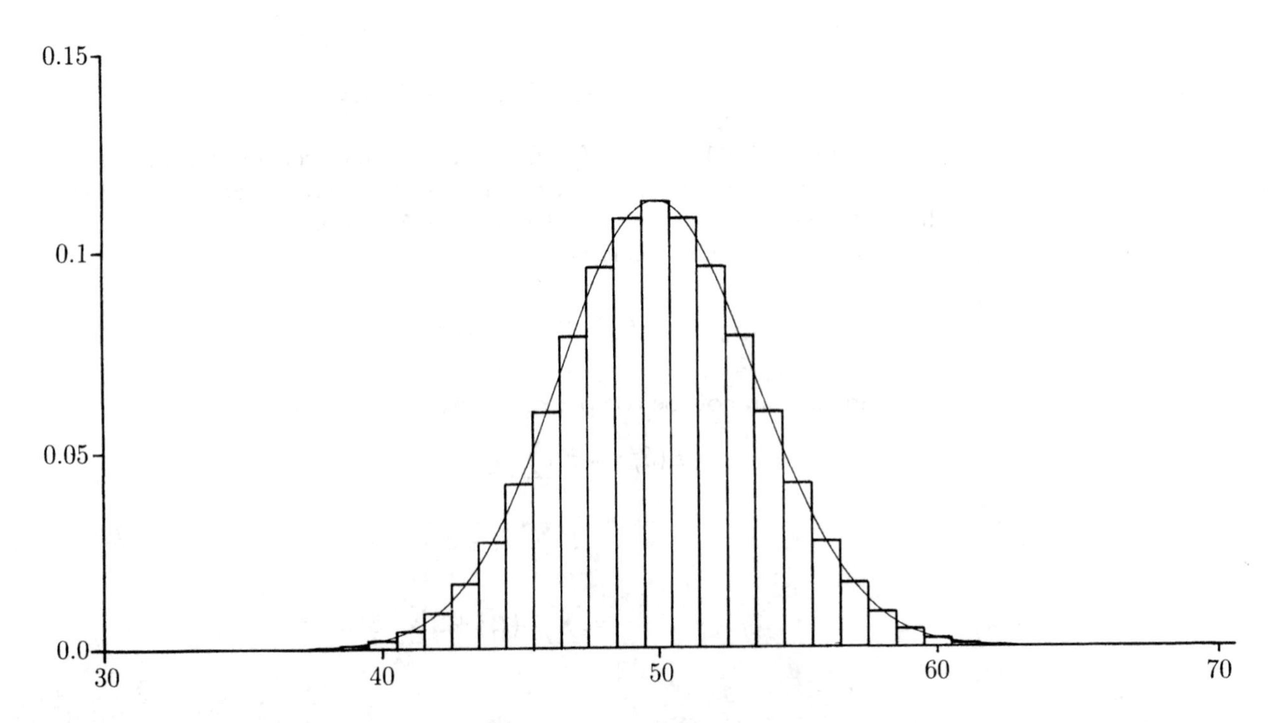

$$= n\frac{G}{N} + 2\binom{n}{2}\frac{G}{N}\frac{(G-1)}{(N-1)}$$

because in the first sum there are n identical terms of

$$E(I_j^2) = E(I_j) = \frac{G}{N}$$

(I_j is an indicator variable with value 0 or 1, so $I_j^2 = I_j$) and in the second sum there are $\binom{n}{2}$ identical terms with value

$$E(I_j I_k) = E(I_1 I_2) = \frac{G}{N} \cdot \frac{(G-1)}{(N-1)}$$

the probability of getting good elements on two consecutive draws, since $I_1 I_2$ is one if both I_1 and I_2 are 1, and 0 otherwise. Now use

$$Var(S_n) = E(S_n^2) - [E(S_n)]^2$$

and simplify to obtain the expression for $SD(S_n) = \sqrt{Var(S_n)}$. □

Remark. A similar argument shows that the same finite population correction factor applies for sums or averages of other kinds of variables in sampling without replacement, not just indicator variables. See Example 6.4.7.

The normal approximation. This can be used for sampling without replacement exactly as in the binomial case for sampling with replacement, provided the finite population correction factor is used for the standard deviation. The approximation is good provided the standard deviation is sufficiently large. This can be shown by consideration of consecutive odds ratios, just as in the binomial case. See Figure 1 for an illustration.

Exercises 3.6

1. Five cards are dealt from a standard deck of 52. Find

a) the probability that the third card is an ace;

b) the probability that the third card is an ace given the last two cards are not aces;

c) the probability that all cards are of the same suit;

d) the probability of two or more aces.

2. Cards. A deck of 52 cards is shuffled and dealt. Find the probabilities of the following events:

a) the tenth card is a queen;

b) the twentieth card is a spade;

c) the last five cards are spades;

d) The last king appears on the 48th card.

3. Conditional probabilities. In the setting of Exercise 2, denote by A, B, C, and D the events defined in parts a), b), c) and d) of that exercise. Find:

a) $P(B|C)$; b) $P(C|B)$; c) $P(B|A)$; d) $P(A|B)$; e) $P(D|C)$; f) $P(C|D)$;

4. Testing for defectives. Suppose a lot of 5 items contains two defective items. The items are tested one by one in random order. Let T_1 be the number of the test on which the first defective item is discovered, and T_2 the number of the test on which the second is discovered.

a) Display the distribution table of T_1.

b) Without further calculation, display the distribution table of $6 - T_2$.

c) Without further calculation, display the distribution table of T_2.

d) Display the joint distribution table of T_1 and T_2.

e) Are the random variables $T_1, T_2 - T_1, 6 - T_2$ exchangeable? Prove your answer.

f) Find the distribution of $T_2 - T_1$.

5. Suppose n balls are thrown independently at random into b boxes. Let X be the number of boxes left empty. Use the method of indicators to find expressions for $E(X)$ and $Var(X)$.

6. Mean and SD of the number of matches. There are n balls labeled 1 through n, and n boxes labeled 1 through n. The balls are distributed randomly into the boxes, one in each box, so that all $n!$ permutations are equally likely. Say that a match occurs at place i if the ball labeled i happens to fall in the box labeled i. Let M be the total number of matches.

a) Find $E(M)$. b) Find $SD(M)$.

c) For very large n, what do you think is the approximate distribution of M? Give an intuitive explanation for your answer. Check that your answer makes sense in view of your answers to a) and b) and the answer to Exercise 28 from the Chapter 2 Review Exercises.

7. Suppose n cards are dealt from a standard deck of 52 cards. Calculate a) the expectation and b) the variance of the number of red cards among the n cards dealt.

8. A deck of 52 cards is shuffled and split into two halves. Let X be the number of red cards in the first half. Find: a) a formula for $P(X = k)$;

b) $E(X)$; c) $SD(X)$; d) $P(X \geq 15)$, approximately, using the normal curve.

9. A population contains G good and B bad elements, $G + B = N$. Elements are drawn one by one at random without replacement. Suppose the first good element appears on draw number X. Find simple formulae, not involving any summation from 1 to N, for:

a) $E(X)$; b) $SD(X)$.

[*Hint:* Write $X - 1$ as a sum of B indicators.]

10. Success runs in sampling without replacement. Repeat Exercise 3.2.22 for the random sequence of successes and failures obtained by a sampling n times without replacement from a population of G good and $N - G$ bad elements, where each draw of a good element is called success, and each draw of a bad element a failure.

11. Sampling without replacement. Let X_j be the indicator of the event that a good element appears at place j in a random ordering of n elements consisting of g good elements and $n - g$ bad ones.

a) Find a formula for $P(x_1, \ldots, x_n) = P(X_1 = x_1, \ldots, X_n = x_n)$.

b) Are the random variables $X_1, \ldots, X_n$ independent? Prove your answer.

c) Are they exchangeable? Prove your answer.

12. Discrete order statistics. In an exhaustive random sample without replacement of a population of N elements, containing n good and $N - n$ bad elements, let $1 \leq T_1 < T_2 < \cdots < T_n \leq N$ denote when the good elements appear. Part d) of this exercise explains why the random variables $T_1, \ldots, T_n$ with possible values in $\{1, \ldots, N\}$ are discrete analogs of the order statistics of n independent uniform $(0, 1)$ variables, studied in Section 4.6.

a) Show that $\{T_1, \ldots, T_n\}$, the random set of times when good elements appear, is uniformly distributed over all subsets of n elements of $\{1, \ldots, N\}$. That is to say, the set of times when good elements appear is a simple unordered random sample of size n from $\{1, \ldots, N\}$.

b) Find a formula for $P(T_1 = t_1, \ldots, T_n = t_n)$ for $1 \leq t_1 < t_2 < \cdots < t_n \leq N$.

c) Use a counting argument to find a formula for $P(T_i = t)$ for each $i = 1, \ldots, n$ and $t = 1, \ldots, N$.

d) Let $U_{(1)} \leq U_{(2)} \leq \ldots \leq U_{(n)}$ denote the order statistics, that is, the values in increasing order, of n independent trials $U_1, \ldots, U_n$ with uniform distribution on $\{1, \ldots, N\}$. Let D denote the event that the $U_i, 1 \leq i \leq n$ are all distinct. Show that the conditional joint distribution of $U_{(1)}, \ldots, U_{(n)}$ given D is identical to the joint distribution of $T_1, \ldots, T_n$ found in part b). What is $P(D)$? Show that $P(D) \to 1$ as $N \to \infty$ for fixed n.

[It follows that for fixed n, as $N \to \infty$, the limiting joint distribution of $(T_1, \ldots, T_n)/N$ is the joint distribution of the order statistics of n independent uniform $(0, 1)$ random variables. In particular, part c) implies the asymptotic distribution of T_i/N is the beta $(i, n - i + 1)$ distribution, as obtained directly from the continuous model in Section 4.6. A number of interesting results for continuous uniform order statistics can be derived via this passage to the limit. See Chapter 6 Review Exercises 31, 32, and 33.

13. Discrete spacings. As in Exercise 12, let $T_1 < \ldots < T_n$ be the places that good elements appear in a random ordering of n good and $N - n$ bad elements. (In terms of a shuffled deck of N cards with n aces, T_i represents the place in the deck where the ith ace lies.) Let $W_1 = T_1 - 1$, the number of bad elements before the first good one. For $2 \leq i \leq n$, let $W_i = T_i - T_{i-1} - 1$, the number of bad elements between the $(i - 1)$th and ith good ones. Let $W_{n+1} = N - T_n$, the number of bad elements after the last good one. Think of the W_i as spacings between the good elements.

a) Find the joint distribution of $W_1, \ldots, W_{n+1}$.

b) Show that the $n+1$ random variables $W_1, \ldots, W_{n+1}$ are exchangeable, hence identically distributed, but not independent.

c) Find a formula for $P(W_i = w)$ for $0 \le w \le N$.

d) Find $E(W_i)$ for $1 \le i \le n+1$ and $E(T_i)$ for $1 \le i \le n$. [*Hint*: Use the symmetry.] Evaluate in the case $N = 52$ and $n = 4$ to find the mean number of cards between any two aces, and the mean position in the deck of the ith ace. (See Chapter 6 Review Exercise 29 for the variance.)

e) Show that for $1 \le i < j \le n+1$ the random variable $W_i + W_j$ has the same distribution as $T_2 - 2$. Deduce from Exercise 12c) a formula for $P(W_i + W_j = t)$ for $0 \le t \le N$.

f) Let $D_n = T_n - T_1 - 1$, the number of elements between the first and last good elements (including the other $n-2$ good ones). Use the result of e) to find a formula for $P(D_n = d)$, $0 \le d \le N$, and find $E(D_n)$.

14. Consecutive pairs. Consider a well-shuffled deck of N cards, with n aces and $N-n$ non-aces.

a) Show by a counting argument that the probability that there are at least two consecutive aces somewhere in the deck is $1 - \binom{N-n+1}{n} \Big/ \binom{N}{n}$ [*Hint*: Look for a one-to-one correspondence].

b) Check the above formula by more direct counting arguments in each of the following three special cases: $n = 2, N = 2n-1$, and $N = 2n$.

For the following parts, assume a standard deck of 52 cards, and evaluate the probabilities of the events as decimals:

c) The ace of spades is next to the ace of clubs.

d) There are at least two consecutive aces somewhere in the deck.

e) There are at least two consecutive spades somewhere in the deck.

f) There is no pair of adjacent black cards anywhere in the deck.

15. Runs and Spacings. As in Exercise 13 let $W_1, W_2, \ldots, W_{n+1}$ be the exchangeable sequence of spacings defined by a random ordering of n aces and $N-n$ non-aces.

a) Explain why the probability evaluated in Exercise 14, that there are at least two consecutive aces somewhere in the deck, is

$$1 - P(W_i \ge 1 \text{ for every } 2 \le i \le n)$$

b) Show that for any sequence of $n+1$ non-negative integers $t_1, \ldots, t_{n+1}$ with $t_1 + \cdots + t_{n+1} = t$,

$$P(W_i \ge t_i \text{ for every } 1 \le i \le n+1) = \binom{N-t}{n} \Big/ \binom{N}{n}$$

c) What special case of b) yields the result of Exercise 14?

16. Distribution of the longest run. As in Exercises 13 and 15, let $W_1, W_2, \ldots, W_{n+1}$ be the exchangeable sequence of spacings defined by a random ordering of n aces and $N-n$ non-aces. Let $W_{\max} = \max_i W_i$ where the max is over $1 \le i \le n+1$. So $W_{\max}$ is the length of the longest run of non-aces in the deck.

a) Show by using the result of Exercise 15, and the inclusion–exclusion formula of Exercise 1.3.12 that

$$P(W_{\max} \geq r) = \sum_{i=1}^{n+1} (-1)^{i-1} \binom{n+1}{i} \binom{N-ir}{n} \Big/ \binom{N}{n}$$

b) Denote the above expression for $P(W_{\max} \geq r)$, which depends on $N, n,$ and r, by $P(N, n, r)$. Let S_N be the number of successes in N Bernoulli (p) trials and R_N be the longest run of successes in the N trials. Explain why

$$P(R_N \geq r | S_N = k) = P(N, N-k, r)$$

and why this conditional probability does not depend on p.

c) Show that the probability that there is a run of at least r consecutive successes in N Bernoulli (p) trials is

$$P(R_N \geq r) = \sum_{k=0}^{N} \binom{N}{k} p^k (1-p)^{N-k} P(N, N-k, r)$$

d) Find as a decimal the probability that the longest run of heads in 10 fair coin tosses is exactly r for each $0 \leq r \leq 10$. What is the most likely length of the longest run? What is the expected length of the longest run?

e) What is the probability that there is a run of either at least 5 heads or at least 5 tails in 10 fair coin tosses?

Random Variables: Summary

Random variable X: symbol representing an outcome.

Range of X: set of all possible values of X.

Distribution of X: The probability distribution over the range of X defined by probabilities $P(X = x)$ for x in the range of X.
$P(X \in B) = \sum_{x \in B} P(X = x)$ for B a subset of the range of X.

Change of variable formula: $P(f(X) = y) = \sum_{x: f(x)=y} P(X = x)$ gives the distribution of a function $f(X)$ in terms of the distribution of X.

Joint outcome (X, Y): $P(x, y) = P(X = x, Y = y)$

$$P(X = x) = \sum_{\text{all } y} P(x, y) \qquad P(X < Y) = \sum_{x} \sum_{y > x} P(x, y)$$

Equality of random variables: $X = Y$ means $P(X = Y) = 1$.

Equality in distribution: X and Y have the same distribution if $P(X = x) = P(Y = x)$ for all x in the range of X (= range of Y). If $X = Y$ then X and Y have the same distribution, but not conversely.

Independence: For n random variables

$$P(X_1{=}x_1, X_2{=}x_2, \ldots, X_n{=}x_n) = P(X_1{=}x_1)P(X_2{=}x_2) \cdots P(X_n{=}x_n)$$

for all possible values x_i of X_i, $i = 1, \ldots, n$,
- functions of disjoint blocks of independent random variables are independent.

Expectation: $E(X) = \sum_x x P(X = x)$

- average value of X weighted by probabilities;
- long-run average value of independent variables with same distribution as X;
- center of mass of distribution of X
- properties: generalize properties of averages: see summary on pages 180 – 181

Variance: $Var(X) = E(X-\mu)^2 = E(X^2) - \mu^2$ where $\mu = E(X)$.

Standard deviation: $SD(X) = \sqrt{Var(X)}$: measure of spread in the distribution of X.

Scaling: $Var(aX+b) = a^2 Var(X)$, $SD(aX+b) = |a| SD(X)$.

Chebychev's inequality: $P\left[|X - E(X)| > kSD(X)\right] \le \dfrac{1}{k^2}$

Sums: For independent random variables $X_1, \ldots, X_n$, if $S_n = X_1 + \cdots + X_n$,

$$\begin{aligned} Var(S_n) &= Var(X_1) + \cdots + Var(X_n) \\ &= nVar(X_1) \qquad \text{if the } X_i \text{ all have same distribution.} \end{aligned}$$

Compare

$$\begin{aligned} E(S_n) &= E(X_1) + \cdots + E(X_n) \qquad \text{(true even if dependent)} \\ &= nE(X_1) \qquad \text{if the } X_i \text{ all have same distribution.} \end{aligned}$$

Square root law: For independent X_i with same distribution, S_n as above, and $\bar{X}_n = S_n/n$ the average

$$SD(S_n) = SD(X_1)\sqrt{n} \qquad SD(\bar{X}_n) = SD(X_1)/\sqrt{n}$$

Law of averages: $\bar{X}_n$ is nearly certain to be close to $E(X_1)$ for large n.

Normal approximation: For S_n as above, with $E(X_i) = \mu, \mathrm{SD}(X_i) = \sigma$,

$$\frac{S_n - n\mu}{\sigma\sqrt{n}} = \frac{(\bar{X}_n - \mu)\sqrt{n}}{\sigma}$$

has distribution which approaches standard normal as $n \to \infty$, no matter what the common distribution of the X_i.

Infinite sum rule. If event A splits into an infinite sequence of mutually exclusive cases $A_1, A_2, A_3, \ldots$, so $A = A_1 \cup A_2 \cup A_3 \cup \cdots$, where $A_i \cap A_j = \emptyset$, $i \ne j$, then

$$P(A) = P(A_1) + P(A_2) + P(A_3) + \cdots$$

Discrete distribution on $\{0, 1, 2, \ldots\}$: defined by a sequence of probabilities $p_0, p_1, p_2, \ldots$ such that $p_i \ge 0$ for all i, and $\sum_i p_i = 1$.

Geometric, negative binomial, and Poisson distributions.
See Distribution Summaries on pages 476 – 488.

Review Exercises

1. A fair die is rolled ten times. Write down numerical expressions for:

a) the probability of at least one six in the ten rolls;

b) the expected number of sixes in the ten rolls;

c) the expected sum of the numbers in the ten rolls;

d) the probability of 2 sixes in the first five rolls given 4 sixes in the ten rolls;

e) the probability of getting strictly more sixes in the second five rolls than in the first five.

2. A fair die is rolled repeatedly. Calculate, correct to at least two decimal places:

a) the chance that the first 6 appears before the tenth roll;

b) the chance that the third 6 appears on the tenth roll;

c) the chance of seeing three 6's among the first ten rolls, given that there were six 6's among the first twenty rolls;

d) the expected number of rolls until six 6's appear;

e) the expected number of rolls until all six faces appear.

3. Two fair dice are rolled independently. Let X be the maximum of the two rolls, and Y the minimum.

a) What is $P(X = x)$ for $x = 1, \ldots, 6$?

b) What is $P(Y = y | X = 3)$ for $y = 1, \ldots, 6$?

c) What is the joint distribution of X and Y?

d) What is $E(X + Y)$?

4. Let X and Y be independent, each uniform on $\{0, 1, \ldots, 100\}$. Let $S = X + Y$. For $n = 0, \ldots, 200$, find:

a) $P(S = n)$; b) $P(S \leq n)$. c) Sketch graphs of these functions of n.

5. Someone plays roulette the following way: before each spin he rolls a die, and then he bets on red as many dollars as there were spots on the die. For example, if there were 4 spots he bets \$4.
If red comes up he gets the stake back plus an amount equal to the stake. If red does not come up he loses the stake. In the example above, if red comes up he gets the stake of \$4 back plus an additional \$4. If red does not come up he loses his stake of \$4. The probability of red coming up is $18/38$.

a) What is his expected gain on one spin?

b) What is the expected number of spins it will take until red comes up for the first time?

c) What is the expected number of spins it will take until the first time the person bets exactly \$4 on one spin and wins.

6. A gambler repeatedly bets 10 dollars on red at a roulette table, winning 10 dollars with probability $18/38$, losing 10 dollars with probability $20/38$. He starts with capital of 100 dollars, and can borrow money if necessary to keep in the game.

a) Find exact expressions for the probabilities that after 50 plays the gambler is:

i) ahead; ii) not in debt.

b) Find the mean and variance of the gambler's capital after 50 plays.

c) Use the normal approximation to estimate the probabilities in a) above.

7. Suppose an airline accepted 12 reservations for a commuter plane with 10 seats. They know that 7 reservations went to regular commuters who will show up for sure. The other 5 passengers will show up with a 50% chance, independently of each other.

a) Find the probability that the flight will be overbooked, i.e., more passengers will show up than seats are available.

b) Find the probability that there will be empty seats.

c) Let X be the number of passengers turned away. Find $E(X)$.

8. A box contains w white balls and b black balls. Balls are drawn one by one at random from the box, until b black balls have been drawn. Let X be the number of draws made. Find the distribution of X,

a) if the draws are made with replacement;

b) if the draws are made without replacement.

9. The doubling cube. A doubling cube is a die with faces marked $2, 4, 8, 16, 32$, and 64. Suppose two doubling cubes are rolled. Let XY be the product of the two numbers. Find a) $P(XY < 100)$; b) $P(XY < 200)$; c) $E(XY)$; d) $SD(XY)$.

10. Matching. Suppose each of n balls labeled 1 to n is placed in one of n boxes labeled 1 to n. Assume the n placements are made independently and uniformly at random (so each box can contain more than one ball). A match occurs at place k if ball number k falls in box k. Find:

a) the probability of a match at i and no match at j;

b) the expected number of matches.

11. Data for performances of a particular surgical operation show that two operations per thousand have resulted in the death of the patient. Let X be the number of deaths due to the next thousand operations of this kind. Which of these three numbers is the smallest and which is the largest

$$P(X < 2), \quad P(X = 2), \quad P(X > 2)?$$

Explain carefully the assumptions of your answer.

12. Consider an unlimited sequence of independent trials resulting in success with probability p, failure with probability q. For $s = 1, 2, ..., f = 1, 2, ...$ calculate the probability that s successes in a row occur before f failures in a row. [*Hint:* Let A be the event in question, $P_1 = P(A \mid \text{first trial a success})$, and $P_0 = P(A \mid \text{first trial a failure})$. Given the first trial is a success, for A to occur, either the next $s - 1$ trials must be successes, or the first failure must come at the tth trial for some $2 \le t \le s$, then subsequently the event A must occur starting from a failure. This gives one equation relating P_1 to P_0. Find another by conditioning on the first trial being a failure, then solve for P_0 and P_1, hence $P(A)$.]

13. Let X and Y be independent random variables with $E(X) = E(Y) = \mu$, $Var(X) = Var(Y) = \sigma^2$. Show that $Var(XY) = \sigma^2(2\mu^2 + \sigma^2)$.

14. A circuit contains 10 switches, arranged as in the figure below. Assume switches perform independently of each other, and are closed with probabilities indicated in the figure. Current flows through a switch if and only if it is closed.

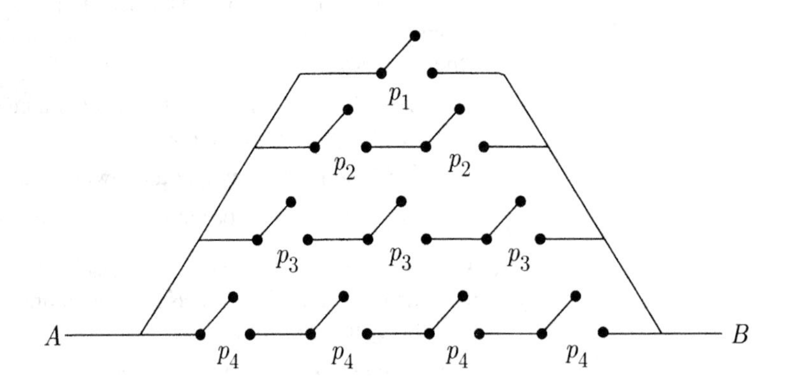

a) What is the probability that current flows between points A and B?

b) Find the mean and standard deviation of the number of closed switches.

15. A roulette wheel is spun independently many times. On each spin the chance of a seven appearing is $1/38$.

a) What is the exact distribution of the number of sevens in the first 100 spins?

b) Give a simple approximation for this distribution.

c) What is the distribution of the number Z of spins required to produce three sevens?

d) What is $E(Z)$?

16. Random products mod 10. Pick two successive digits from a table of random digits from $\{0, 1, \ldots, 9\}$. Multiply them together, and let D be the last digit of this random product. For example,

$$(3, 9) \to 27 \to 7$$

$$(2, 4) \to 8 \to 8$$

Find the distribution of D, and calculate its mean.

17. Suppose N dice are rolled, where $1 \le N \le 6$.

a) Given that no two of the N dice show the same face, what is the probability that one of the dice shows a six? Give a formula in terms of N.

b) In a) the number of dice N was fixed, but now repeat assuming instead that N is random, determined as the value of another die roll. Your answer now should be simply a number, not involving N.

18. Expected number of records. Suppose 100 cards numbered 1 to 100 are shuffled and dealt one by one.

a) What is the fair price to pay in advance if you receive one cent for the first card and then one cent for each card dealt whose number is greater than those of all previous cards dealt?

b) If you paid 10 cents for each play of this game, and played 25 times (meaning you paid a total of 250 cents for 25 separate deals of the 100 card deck) what, approximately, is the chance that you would come out ahead?

19. Suppose that X has Poisson (μ) distribution, and that Y has geometric (p) distribution on $\{0, 1, 2, \ldots\}$ independently of X.

a) Find a formula for $P(Y \geq X)$ in terms of p and μ.

b) Evaluate numerically for $p = 1/2$ and $\mu = 1$.

20. a) Show that for all p between 0 and 1: $p(1-p) \leq 1/4$.

b) A certain university has about 12,000 students. To estimate the percentage of students who have part-time jobs, someone takes a random sample from a list of all students in the university. How big does the sample need to be so that the margin of error in the estimate (i.e., the standard deviation of the percentage in the sample) is at most 5%?

21. Suppose X and Y are independent with $P(X = j) = p(1-p)^j$ for $j = 0, 1, \ldots$ and $P(Y = k) = (k+1)p^2(1-p)^k$ for $k = 0, 1, \ldots$. Find the distribution of $Z = X + Y$. [*Hint*: Represent X and Y in terms of a biased coin-tossing sequence.]

22. The newsboy problem. A newsboy buys papers at 10 cents a copy and sells them on the street corner at 25 cents a copy. He must buy all his papers at once, but he can sell only as many as are demanded on the street. Left-over papers are a dead loss. Over the last few years, demand has been fluctuating at around 100 papers per day. He has been buying 100 papers and selling them all about half the time. Assuming that the demand for papers has an approximately Poisson distribution, find:

a) the newsboy's long-run average profit per day;

b) how many papers the newsboy should buy each day to maximize his long-run average profit.

23. Suppose you economize your use of toothpicks by breaking whole toothpicks in half and only using half at a time. Starting from a full box of n toothpicks, you draw repeatedly at random from the box. In case you draw a whole toothpick, you use half and throw it away, and replace the other half. In case you draw half a toothpick, you use it and throw it away. So the box will be empty after exactly $2n$ draws. Suppose that on any draw, each whole toothpick in the box has the same chance of being drawn, and so does each half toothpick, but the halves have half the chance of the wholes. Let H be the random number of half toothpicks remaining in the box after the last whole toothpick has been drawn and half of it replaced. So H has possible values between 1 (e.g., if you draw alternately whole–half–whole–half ...) and n (e.g., if you draw n wholes in a row, followed by n halves).

a) Find a formula for $P(H = k)$, $k = 1, 2, \ldots, n$.

b) What happens to the distribution of H as $n \to \infty$?

c) Find an asymptotic formula for $E(H)$ as $n \to \infty$.

d) If you start with $n = 100$ toothpicks, about how many halves do you expect to be left with?

e) For $n = 100$, find a and b so that $P(a \le H \le b) \approx 95\%$ with $b - a$ as small as possible.

24. The voter paradox.

a) Can random variables X, Y, and Z be such that each of the three probabilities $P(X > Y)$, $P(Y > Z)$, and $P(Z > X)$, is strictly greater than $\frac{1}{2}$? [*Hint*: Try a joint distribution of X, Y, and Z which is uniform on some of the 6 permutations of $(1, 2, 3)$.]

b) What is the largest that the minimum of the above three probabilities can possibly be? Prove your answer. [*Hint*: The sum of the probabilities is an expectation.]

c) A survey is conducted to determine the popularity of three candidates A, B, and C. Each voter is asked to rank the candidates in order of preference. When the results are analyzed, it is found that more than 50% of the voters prefer A to B, more than 50% prefer B to C, and more than 50% prefer C to A. How is this possible? Explain carefully the connection to previous parts.

d) Generalize a) and b) to $n \ge 3$ random variables instead of $n = 3$.

e) Repeat a) for independent X, Y, and Z. [*Hint*: Try $P(X = 5) = p_1$, $P(X = 2) = 1 - p_1$, $P(Y = 4) = p_2$, $P(Y = 1) = 1 - p_2$, and $P(Z = 3) = 1$. Deduce that the three probabilities can all be as large as the golden mean $(-1 + \sqrt{5})/2$. This is known to be the largest possible for independent variables, but I don't know the proof.]

25. Let Y_1 and Y_2 be independent random variables each with probability distribution defined by the following table:

value	0	1	2
probability	1/2	1/3	1/6

a) Display the probability distribution of $Y_1 + Y_2$ in a table. Express all probabilities as multiples of $1/36$.

b) Calculate $E(3Y_1 + 2Y_2)$.

c) Let X_1 and X_2 be the numbers on two rolls of a fair die. Define a function f so that $(f(X_1), f(X_2))$ has the same distribution as (Y_1, Y_2).

26. The horn on an auto operates on demand 99% of the time. Assume that each time you hit the horn, it works or fails independently of all other times.

a) How many times would you expect to be able to honk the horn with a 50% probability of not having any failures?

b) What is the expected number of times you hit the horn before the fourth failure?

27. A certain test is going to be repeated until done satisfactorily. Assume that repetitions of the test are independent and that each has probability 0.25 of being satisfactory. The first 5 tests cost \$100 each to perform and thereafter cost \$40 each, regardless of the outcomes. Find the expected cost of running the tests until a satisfactory result is obtained.

28. Let $X_1, X_2, \ldots$ be a sequence of independent trials, and suppose that each X_i has distribution P_1 over some range space Ω_1. Let $W_1, W_2, \ldots$ be the successive waiting times between trials s such that X_s is in A, where A is some subset Ω_1, and let $Y_1, Y_2, \ldots$ be the successive values in A which appear at trials W_1, $W_1 + W_2$, $W_1 + W_2 + W_3, \ldots$.

a) Show that $W_1, W_2, \ldots,\quad Y_1, Y_2, \ldots$ are independent random variables, the W's all having geometric distribution on $\{1, 2, \ldots\}$ with parameter $P_1(A)$, and the Y's all having the distribution P_1 conditioned on A.

b) Deduce from the law of large numbers the long run frequency interpretation of $P_1(B|A)$ as the limiting proportion of those trials which are A's that turn out also to be B's.

29. Polya's urn scheme. (Continuation of Exercise 1.5.2). An urn contains w white and b black balls. A ball is drawn from the urn, then replaced along with d more balls of the same color. So after n such draws with multiple replacement, the urn contains $w+b+nd$ balls. Let $X_i = 1$ if the ith ball drawn is black and $X_i = 0$ if the ith ball drawn is white.

a) Find a formula for the probability $P(X_1 = x_1, \ldots, X_n = x_n)$ in terms of w, b, d, n and k, where $k = x_1 + \cdots + x_n$ is the number of 1's in the sequence $(x_1, \ldots, x_n)$.

b) Let $S_n = X_1 + \cdots + X_n$. What does S_n represent? Find a formula for $P(S_n = k)$ for $0 \le k \le n$.

c) What is the distribution of S_n in the special case $b = w = d = 1$?

d) Are $X_1, \ldots, X_n$ independent? Are they exchangeable? (Refer to Section 3.6.)

e) Find a formula for $P(X_n = 1)$, the probability of a black ball on draw n, in terms of b,w,d, and n. [*Hint*: The probability does not depend on all of the parameters.]

f) Find the probability that the fifth ball drawn is black given that the tenth ball drawn is black.

30. Diagonal neighbor random walk. Let (S_n, T_n) denote the position after n steps of a random walk on the lattice of points in the plane with integer coordinates, starting from $(S_0, T_0) = (0, 0)$. Suppose that $S_{n+1} = S_n \pm 1$ and $T_{n+1} = T_n \pm 1$ where the signs are picked by two independent tosses of a fair coin, independently at each step.

a) For $c > 0$, find the limit as $n \to \infty$ of the probability that (S_n, T_n) is inside the square with corners at $(\pm c\sqrt{n}, \pm c\sqrt{n})$.

b) Let $R_n = \sqrt{S_n^2 + T_n^2}$, the distance from the origin. Find $E(R_n^2)$.

c) Find b, as small as you can, such that $E(R_n) \le \sqrt{bn}$ for every n.

d) Let p_n denote the probability that the random walk is at $(0, 0)$ after n steps. Find p_4 as a decimal.

e) Show that $p_{2m} \sim c/m$ as $m \to \infty$ for a constant c. What is c?

31. Nearest neighbor random walk. Let (S_n, T_n) be the position after n steps of a random walk as in the previous exercise, but now instead of diagonal moves, suppose at each step the move is made with equal probability up, down, left or right, to one of the four nearest neighbors in the lattice. For $c > 0$, find the limit as $n \to \infty$ of the probability that $|S_n| < c\sqrt{n}$. The events $|S_n| < c\sqrt{n}$ and $|T_n| < c\sqrt{n}$ are clearly not independent for this random walk, but they turn out to be approximately independent for large n. Assuming the error of this approximation tends to zero as $n \to \infty$ (something not obvious, but true: see Chapter 5 Review Exercise 31 for an explanation), repeat part a) of the previous exercise for this random walk. Now repeat the rest of the previous exercise for this random walk.

32. King's random walk. Same as Exercise 30, but now make each move like a king on an infinite chessboard, with equal probabilites to the 8 nearest or diagonal neighbors. [The two components are still asymptotically independent. This can be proved for any step distribution with mean zero and uncorrelated components, that is to say $E(S_1T_1) = 0$.]

33. From a very large collection of red and black balls, half of them red and half black, I pick n balls at random and put these n balls in a bag. Suppose you now draw k balls from the bag, with replacement and mixing of the balls between draws.

a) Show that given that all k balls you pick are red, the chance that the n balls in the bag are all red is

$$P(n \text{ red in bag} \mid \text{pick } k \text{ red}) = \frac{n^k}{2^n E(X^k)}$$

where X is a binomial $(n, 1/2)$ random variable.

b) Simplify this expression further in the cases $k = 1$ and $k = 2$.

c) Find a similar formula assuming instead that the sample of size k is drawn from the bag *without* replacement. Deduce by calculating the same quantity in a different way that

$$E(X)_k = (n)_k/2^k,$$

where $(X)_k = X(X-1)\cdots(X-k+1)$.

d) Use the identity of c) to simplify the answer to a) in case $k = 3$.

e) Show by a variation of the above calculations that for a binomial (n, p) random variable X,

$$E(X)_k = (n)_k\, p^k.$$

Check that for $k = 1$ and 2 this agrees with the formulae for $E(X)$ and $Var(X)$.

34. Probability generating functions. For a random variable X with non-negative integer values, let $G_X(z) = \sum_{i=0}^{\infty} P(X = i)z^i$, be the probability generating function of X, defined for $|z| < 1$. (Refer to Exercises 3.4.22, 3.4.23 and 3.5.19.) Show that:

a) $G_X(z) = E(z^X)$.

b) If X and Y are independent, then $G_{X+Y}(z) = G_X(z)G_Y(z)$. That is to say, $P(X+Y=k)$ is the coefficient of z^k in $G_X(z)G_Y(z)$.

Generalize the above result to obtain the probability generating function of $S_n = X_1 + \cdots + X_n$ for independent X_i. Now identify the generating function and hence the distribution of S_n in case the distribution of the X_i is c) binomial (n_i, p);

d) Poisson (μ_i); e) geometric (p); f) negative binomial (r_i, p);

35. Binomial moments and the inclusion–exclusion formula. Let X be the number of events that occur in some collection of events $A_1, \ldots, A_n$. So $X = \sum_j I_j$ where I_j is the indicator of A_j.

a) Explain the identity of random variables $\binom{X}{2} = \sum_{i<j} I_i I_j$. [*Hint*: Think in terms of a gambler who for every $i < j$ bets that both A_i and A_j will occur. If the number of events that occurs is, say x, how many bets has the gambler won?]

b) For $k = 0, 1, \ldots, n$ the kth *binomial moment* of X is $b_k = E[\binom{X}{k}]$. Show:

$$b_1 = \sum_i P(A_i); \qquad b_2 = \sum_{i<j} P(A_i A_j); \qquad b_3 = \sum_{i<j<k} P(A_i A_j A_k) \quad \text{and so on.}$$

c) Notice that these are the sums of probabilities that appear in the inclusion–exclusion formula from Exercise 1.3.12. Note also that $b_0 = 1$. Deduce that

$$P(X = 0) = \sum_{k=0}^{n} (-1)^k b_k$$

d) **Sieve formula.** [*Hard.*] Show that for every $m = 1, 2, \ldots n$

$$P(X = m) = \sum_{k=m}^{n} \binom{k}{m} (-1)^{m-k} b_k \text{ and } P(X \geq m) = \sum_{k=m}^{n} (-1)^{k-m} \binom{k-1}{m-1} b_k$$

[*Hint*: $P(X = m)$ is the coefficient of z^m in the probability generating function $G_X(z)$ (see Exercise 3.4.22). Consider the Taylor series of $G_X(z)$ about 1, and use the fact that $G_X(z)$ is a polynomial.]

36. Moments of the binomial distribution. Let S_n be the number of successes in n Bernoulli (p) trials.

a) Use the formula for binomial moments in Exercise 35 to find a simple formula for the kth binomial moment of S_n.

b) Check that your formula implies the usual formulae for the mean and variance, and the formula of Exercise 3.3.33 for the skewness of the binomial (n, p) distribution of S_n.

37. Binomial moments of the hypergeometric distribution. Let S_n be the number of good elements in a sample of size n without replacement from a population of G good and $N - G$ bad elements.

a) Use the formula for binomial moments in Exercise 35 to find a formula for the kth binomial moment of S_n for $k = 1, 2, 3$.

b) Check that your formula implies the formulae of this section for the mean and variance.

c) Find the skewness of the distribution of S_n.

38. Limit distribution for the number of matches. Let M_n denote the number of matches in the matching problem of Chapter 2 Review Exercise 28, for a random permutation of n items.

a) Use the method of Exercise 35 to find the kth factorial moment of M_n.

b) Show that for $1 \leq k \leq n$ this kth factorial moment is identical to the kth factorial moment of the Poisson (1) distribution.

c) Show that for $1 \le k \le n$ the ordinary kth moment of M_n equals the ordinary kth moment of the Poisson (1) distribution. Deduce that for every k, as $n \to \infty$, the kth moment of the distribution of M_n converges to the kth moment of the Poisson (1) distribution.

d) It is known (though not easy to prove) that if all the moments of a sequence of distributions P_n on $\{0, 1, \ldots\}$ converge to those of a Poisson (λ) distribution, then for every $k = 1, 2, \ldots$, $P_n(k)$ converges to the Poisson (λ) probability of k. In the present problem, this implies that as $n \to \infty$, the limiting distribution of M_n is Poisson (1): $P(M_n = k) \to e^{-1}/k!$. Deduce this result another way by applying part a) and the sieve formula of Exercise 35.

39. Recovering a distribution over $\{0,1,\ldots,n\}$ from its moments. Let X be a random variable with possible values $\{0, 1, \ldots, n\}$. Assuming the results of Exercise 35, show

a) For some coefficients $c_{n,k}$ not depending on the distribution of X, (which you need not determine explicitly)

$$P(X = 0) = \sum_{k=0}^{n} c_{n,k} E[X^k]$$

b) Find the values of $c_{n,k}$ for $0 \le k \le n \le 3$.

c) Show that for every $m = 1, \ldots n$, the probability $P(X = m)$ can be expressed as a linear combination (which you need not determine explicitly) of the first n ordinary moments of X. [Exercise 40 gives a generalization.]

40. Recovering a distribution on n values from its moments. For a random variable X and $k = 1, 2, \ldots$, let $\mu_k = E(X^k)$, the kth moment of X. Suppose X has n possible values $x_1, \ldots, x_n$. Show that the n probabilities

$$p_i = P(X = x_i) \qquad (i = 1, \ldots, n)$$

are determined by the first $n - 1$ moments. [*Hint:* The vector $\mu = (1, \mu_1, \ldots, \mu_{n-1})$ is determined from the vector $p = (p_1, \ldots, p_n)$ as $\mu = pM$ for a suitable matrix M. Show that M has rank n, because if there were a linear combination of its columns which was identically zero, there would be a polynomial of degree $n - 1$ with n roots. Deduce that M has an inverse M^{-1}, so that $p = \mu M^{-1}$.]

41. (*Hard.*) Suppose you toss a coin ten times and record the exact sequence of outcomes, e.g.,

H T H H T T H H T H .

Of course, many other sequences are possible. About how many times n would you have to repeat this ten toss experiment

a) to be 90% sure of seeing this particular sequence again in these n repetitions?

b) to be 90% sure of seeing at least one of the possible sequences twice in the n repetitions?

c) to be 90% sure of seeing every possible sequence at least once in the n repetitions?

d) to be 90% sure of seeing at least once every sequence in a set comprising exactly half of all possible outcomes, where the set is specified in advance.

e) Same as d), but for a set not specified in advance.

4 Continuous Distributions

The basic ideas of previous sections were the notions of a random variable, its probability distribution, expectation, and standard deviation. These ideas will now be extended from discrete distributions to continuous distributions on a line, in a plane, or in higher dimensions. This chapter concerns continuous probability distributions over an interval of real numbers. One example is the normal distribution, seen already as an approximation to various discrete distributions. A simpler example is the uniform distribution on an interval, defined by relative lengths. Another example, the exponential distribution, treated in Section 4.2, is the continuous analog of the geometric distribution. Each of these distributions is defined by a ***probability density function***, like the familiar normal curve associated with the normal distribution. The way a continuous distribution can be specified by such a density function is the subject of Section 4.1. Change of variable for distributions defined by densities is the subject of Section 4.4.

The concept of a continuously distributed random variable is an idealization which allows probabilities to be computed by calculus. This gives models for chance phenomena involving continuous variables. Such models arise both:

(i) as limits from discrete models (e.g., the normal distribution as an approximation to the binomial, or the exponential approximation to the geometric discussed in Section 4.2), and

(ii) directly from physical phenomena most naturally modeled by continuous variables (e.g., the normal distribution as a model for measurement error, or the exponential distribution as a model for the lifetime of an atom).

4.1 Probability Densities

In Chapters 2 and 3 the normal distribution was used as an approximation to the distribution of a sum or average of a large number of independent random variables. The idea there was to approximate a discrete distribution of many small individual probabilities by scaling the histogram to make it follow a continuous curve. The function defining such a curve is called a ***probability density***, denoted $f(x)$ here. This function determines probabilities over an infinite continuous range of possible values.

The basic idea is that probabilities are defined by areas under the graph of $f(x)$. That is, a random variable X has density $f(x)$ if for all $a \leq b$

$$P(a \leq X \leq b) = \int_a^b f(x)dx,$$

which is the area shaded in the following diagram:

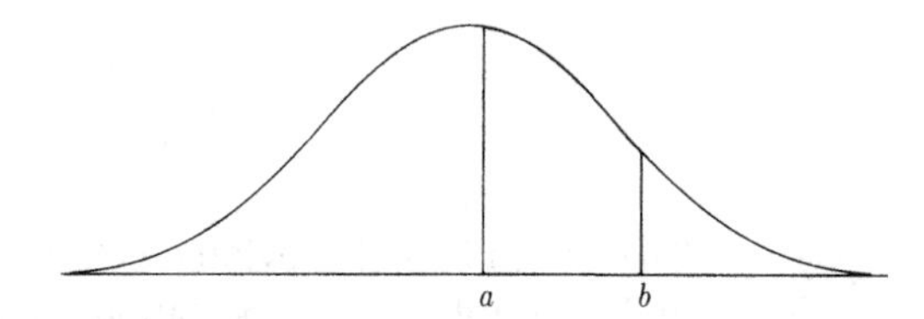

The boxes on pages 262 and 263 show the analogy between a discrete distribution of a random variable X defined by the probabilities $P(x) = P(X = x)$ of individual values x, and a continuous distribution defined by a probability density $f(x)$. In the density case, it is of no use to consider $P(X = x)$. This probability is zero for every x for a distribution with a density, so it gives no information about the distribution. Rather, everything is determined by the density $f(x)$, which gives the probability per unit length for values near x. The individual probability $P(x)$ of the event $(X = x)$ is replaced everywhere by the infinitesimal probability $f(x)dx$ of the event $(X \in dx)$, and sums are replaced by integrals. Here $(X \in dx)$ stands for the event that X falls in an infinitesimal interval of length dx near x, for example, $(x \leq X \leq x + dx)$, or $(x - dx \leq X \leq x)$.

Assuming f is continuous at x, the area representing $P(X \in dx)$ is essentially a rectangle of sides $f(x)$ and dx, hence area $f(x)dx$. Note well that it is $f(x)dx$, not just $f(x)$, which is the analog of $P(x)$. It may well be that $f(x) > 1$ for some values of x. Thus $f(x)$ is *not* a probability, but a probability ***density***. When multiplied by small lengths, $f(x)$ gives approximate probabilities of small intervals near x. If you cut the interval $[a, b]$ into lots of tiny intervals between a and b, add the probabilities of all the tiny intervals, and pass to the limit as the interval widths tend to zero, you get the integral formula for $P(a \leq X \leq b)$. So when integrated over an interval, $f(x)$ gives the exact probability of the interval. A probability density $f(x)$ thus describes a continuous distribution of probability over a number line.

Mean, variance and standard deviation. These are defined just as before in terms of expectations.

$$E(X) = \int_{-\infty}^{\infty} x f(x) dx$$

$$E(X^2) = \int_{-\infty}^{\infty} x^2 f(x) dx$$

If the second integral is finite, then so is the first, and then $E(X^2)$ and $E(X)$ can be used to calculate $Var(X)$ and $SD(X)$ in the usual way

$$Var(X) = E(X^2) - [E(X)]^2 \qquad SD(X) = \sqrt{Var(X)}$$

The basic properties of expectation, variance, and standard deviation are the same as in the discrete case. For example, Chebychev's inequality holds just as well for X with a density as for a discrete random variable X. Proofs of such things parallel the discrete case, using properties of integrals instead of properties of sums.

Independence. Numerical random variables X and Y are called *independent* if the events $(X \in A)$ and $(Y \in B)$ are independent for any choice of two intervals A and B, or more generally any choice of subsets A and B of the line for which the probabilities of these events are defined. That is to say

$$P(X \in A, Y \in B) = P(X \in A)P(Y \in B)$$

Only for discrete random variables can this definition be reduced to the case $A = [x, x]$ and $B = [y, y]$, when the rule becomes simply

$$P(X = x, Y = y) = P(X = x)P(Y = y).$$

If X has a distribution with a density, then $P(X = x) = 0$ for every x, which implies $P(X = x, Y = y) = 0 = P(X = x)P(Y = y)$ for all x and y for any random variable Y whatever. See Section 5.2 for a more careful treatment of independence of X and Y with densities in terms of their joint distribution. Independence of several variables is defined by a similar product rule. The basic properties of independent random variables are the same in the density case as in the discrete case. In particular, if X and Y are independent and both $E(X)$ and $E(Y)$ are defined and finite, then $E(XY) = E(X)E(Y)$. The addition rule for the variance of a sum of independent random variables follows from this.

Discrete Distributions

Point Probability:

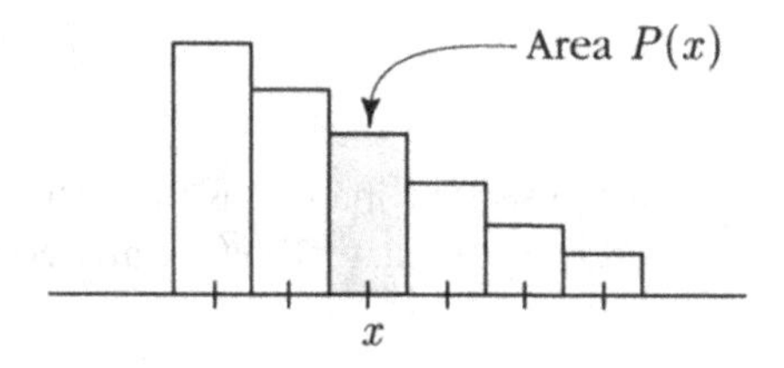

$$P(X = x) = P(x)$$

So $P(x)$ is the probability that X has integer value x.

Interval Probability:

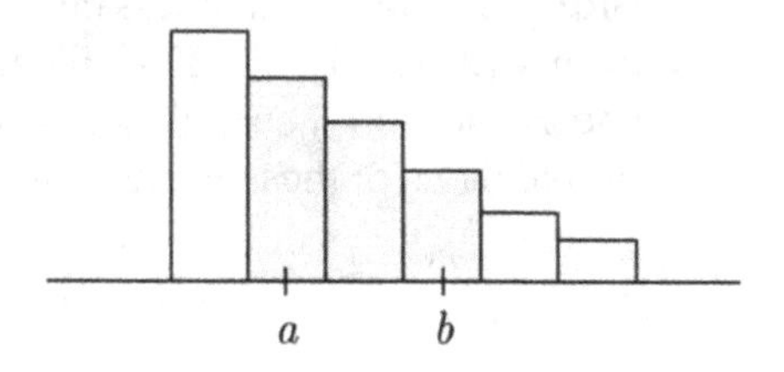

$$P(a \le X \le b) = \sum_{a \le x \le b} P(x)$$

the relative area under a histogram between $a - 1/2$ and $b + 1/2$.

Constraints: Non-negative with Total Sum 1

$$P(x) \ge 0 \quad \text{for all } x \qquad \text{and} \qquad \sum_{\text{all } x} P(x) = 1$$

Expectation of a Function g of X, e.g., X, X^2:

$$E\left(g(X)\right) = \sum_{\text{all } x} g(x)P(x)$$

provided the sum converges absolutely.

Distributions Defined by a Density

Infinitesimal Probability:

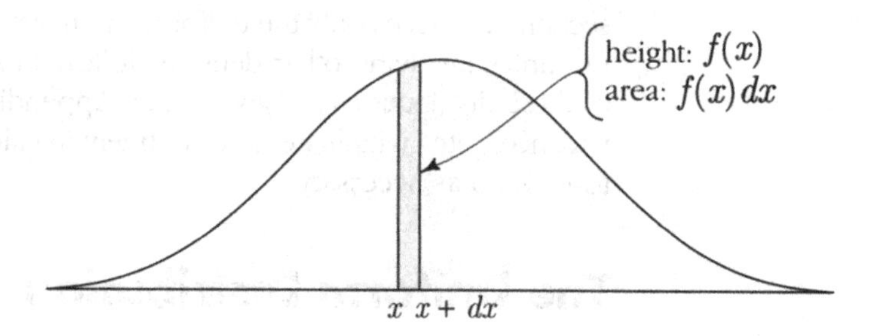

$$P(X \in dx) = f(x)dx$$

The *density* $f(x)$ gives the probability per unit length for values near x.

Interval Probability:

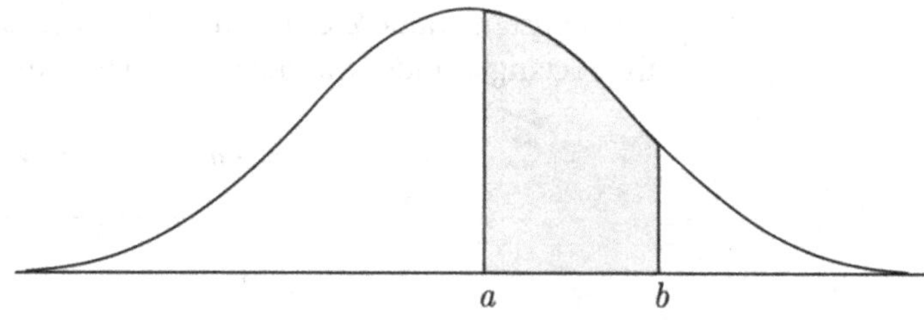

$$P(a \leq X \leq b) = \int_a^b f(x)dx$$

the area under the graph of $f(x)$ between a and b.

Constraints: Non-negative with Total Integral 1

$$f(x) \geq 0 \quad \text{for all } x \qquad \text{and} \qquad \int_{-\infty}^{\infty} f(x)dx = 1$$

Expectation of a Function g of x, e.g., X, X^2:

$$E(g(X)) = \int_{-\infty}^{\infty} g(x)f(x)dx$$

provided the integral converges absolutely.

Special densities. There are a few particularly important probability densities which appear over and over again, both in theory and applications. Most notable are the uniform, normal, exponential, gamma, and beta densities. Why these few should be so important is not at first obvious, but emerges gradually after study of their properties and relationships, both with each other and with other discrete distributions. This section introduces only the uniform and normal densities. Further developments and examples involving other densities follow in subsequent sections. Also, summaries of these distributions are given in an Appendix. These include formulae for means, variances, etc., which are used routinely in calculations and which you are expected to look up as necessary.

The Uniform Distribution

A random variable X has *uniform distribution* on the interval (a, b), if X has density $f(x)$ which is constant on (a, b), and 0 elsewhere. The uniform (a, b) density is

$$f(x) = \begin{cases} 1/(b-a) & \text{if } \ a < x < b \\ 0 & \text{otherwise} \end{cases}$$

The constant value c of the density on (a, b) is $1/(b-a)$, because the total area of the rectangle under the density function must be 1:

$$(b-a)c = 1 \quad \Longrightarrow \quad c = 1/(b-a)$$

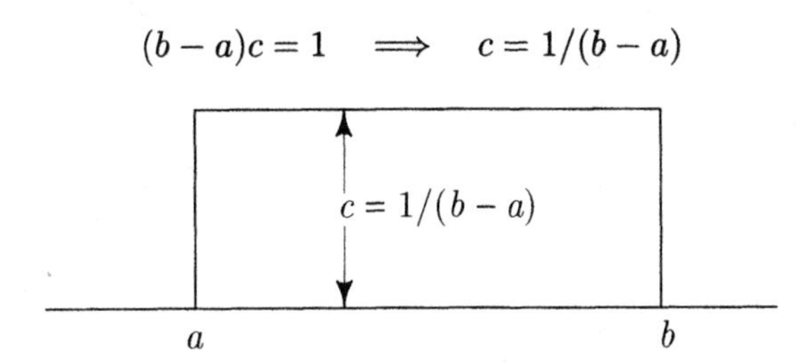

As suggested by the verticals at $x = a$ and $x = b$, the values of $f(x)$ at these endpoints do not affect the probabilities defined by areas under the graph. The area of a line is zero, and so is the probability that any continuously distributed random variable X takes any particular real value. This is an idealization based on the idea that a real number is specified with infinite precision. In practice, it would only ever be possible to know that X was equal to x to some finite number of decimal places. For X distributed uniformly on (a, b), and $a < x < b$, this event would always have strictly positive probability.

For a uniform distribution, probabilities reduce to relative lengths. So if X has uniform (a, b) distribution, then for $a < x < y < b$,

$$P(x < X < y) = \frac{\text{length } (x, y)}{\text{length } (a, b)} = \frac{y - x}{b - a}$$

as is obvious from the diagram.

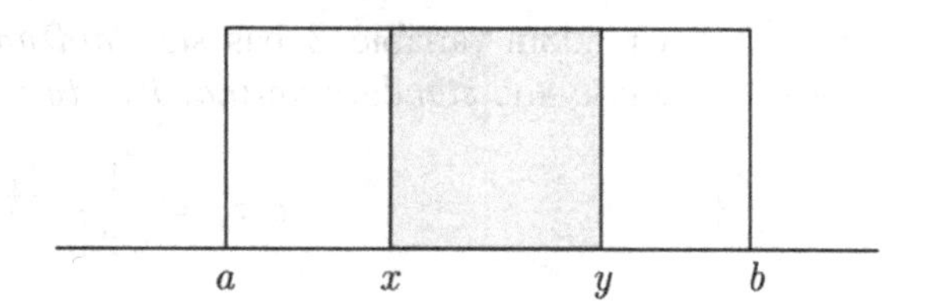

For example, if X has uniform $(0, 2)$ distribution, the probability that X is 1.23 correct to two decimal places is

$$P(1.225 < X < 1.235) = \frac{1.235 - 1.225}{2} = 0.01/2 = 0.5\%$$

A simple rescaling transforms the interval (a, b) into $(0, 1)$. The uniform (a, b) distribution then transforms into the uniform $(0, 1)$ distribution, whose density is simply 1 on $(0, 1)$, and 0 elsewhere. In terms of random variables, any problem involving a uniform (a, b) random variable X reduces easily to one involving a uniform $(0, 1)$ random variable U defined by

$$U = (X - a)/(b - a) \qquad \text{so} \quad X = a + (b - a)U$$

This kind of *scaling* or *linear change of variable*, is a basic technique for reducing problems to the simplest case to avoid unnecessary calculation. To illustrate, the expected value of X is

$$\begin{aligned} E(X) &= E(a + (b - a)U) \\ &= a + (b - a)E(U) \\ &= a + (b - a)\frac{1}{2} = (a + b)/2 \end{aligned}$$

This is obvious anyway by symmetry, since $(a + b)/2$ is the midpoint of (a, b). The variance of X is

$$\begin{aligned} Var(X) &= Var(a + (b - a)U) \\ &= (b - a)^2 Var(U) \\ &= (b - a)^2[E(U^2) - (E(U))^2] = (b - a)^2[1/3 - (1/2)^2] = (b - a)^2/12 \end{aligned}$$

Here $E(U) = 1/2$ without calculation, but $E(U^2)$ requires an integral:

$$\begin{aligned} E(U^2) &= \int_{-\infty}^{\infty} u^2 f(u)du \\ &= \int_0^1 u^2 du \qquad \text{since } U \text{ has density } f(u) = 1 \text{ for } 0 < u < 1,\ 0 \text{ otherwise} \\ &= \frac{1}{3}u^3\Big|_0^1 = \frac{1}{3} \end{aligned}$$

The Normal Distribution

A random variable Z has *standard normal distribution* if Z has as its probability density the *standard normal density*

$$\phi(z) = \frac{1}{\sqrt{2\pi}} e^{-\frac{1}{2}z^2} \qquad (-\infty < z < \infty)$$

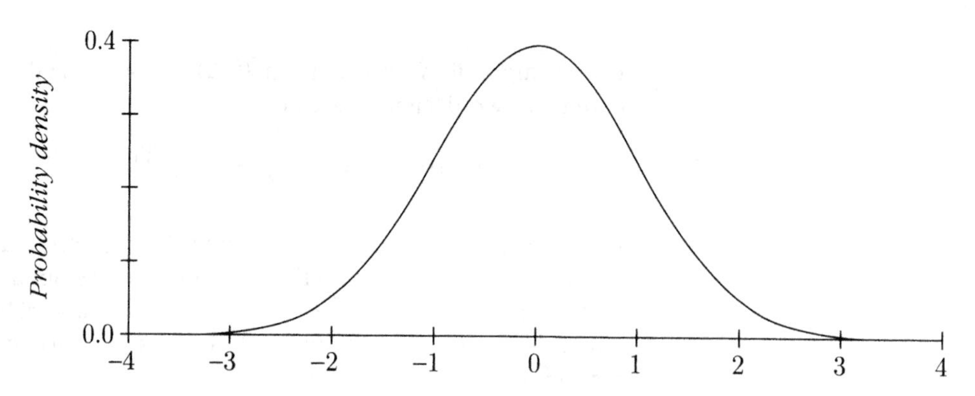

The constant $1/\sqrt{2\pi}$ is put in the definition of the standard normal density so the total area under the standard normal curve $y = \phi(z)$ is 1. This is the first integral in the following box:

Standard Normal Integrals

$$\int_{-\infty}^{\infty} \phi(z)\, dz = 1; \qquad \int_{-\infty}^{\infty} z\phi(z)\, dz = 0; \qquad \int_{-\infty}^{\infty} z^2\phi(z)\, dz = 1.$$

The first and third of these integrals are evaluated in Section 5.3. The second and third integrals show that the standard normal distribution has mean 0 and second moment 1, hence variance 1. The mean of this distribution is zero, because of the symmetry about zero of the standard normal curve. The third integral in the box can be reduced to the first integral by integration by parts.

There is no simple formula for the standard normal probability of an interval

$$\Phi(a, b) = P(a < Z < b) = \int_a^b \phi(z)dz$$

Instead, this probability is found, as in Section 2.2, using a table of the standard normal c.d.f.

$$\Phi(b) = \Phi(-\infty, b) = P(Z \le b) = \int_{-\infty}^{b} \phi(z)dz$$

Normal (μ, σ^2) Distribution

If Z has standard normal distribution and μ and σ are constants with $\sigma \geq 0$, then

$$X = \mu + \sigma Z$$

has mean μ, standard deviation σ, and variance σ^2. The distribution of X is called the *normal distribution with mean μ and variance σ^2*, abbreviated normal (μ, σ^2). So X has normal (μ, σ^2) distribution if and only if the standardized variable

$$Z = (X - \mu)/\sigma$$

has normal $(0, 1)$ or *standard normal* distribution. To find $P(c < X < d)$, change to standard units and use the standard normal table

$$P(c < X < d) = P(a < Z < b) = \Phi(b) - \Phi(a)$$

where $\quad a = (c - \mu)/\sigma \qquad Z = (X - \mu)/\sigma \qquad b = (d - \mu)/\sigma$

Formula for the normal (μ, σ^2) density. For $\sigma > 0$, the formula is

$$\frac{1}{\sigma}\phi((x - \mu)/\sigma) = \frac{1}{\sqrt{2\pi}\sigma} e^{-\frac{1}{2}(x-\mu)^2/\sigma^2} \qquad (-\infty < x < \infty).$$

This is the transformation of the standard normal density $\phi(z)$ corresponding to the linear change of variable from Z to $X = \mu + \sigma Z$. See Section 4.4 for details of this kind of transformation. This formula is rarely used in calculations. It is always simpler to transform to standard units as in Example 1 below. If $\sigma^2 = 0$ the normal (μ, σ^2) distribution is just the distribution of the constant random variable with value μ, with probability one at μ. For $\sigma^2 > 0$, the normal (μ, σ^2) distribution piles up around μ for small values of σ^2, and become more and more spread out as σ^2 increases. See Figure 1 on the next page.

Normal approximation to an empirical distribution. The normal distribution is often fitted to an empirical distribution of observations. The parameters μ and σ are usually estimated by the mean and standard deviation of the list of observations. This is justified by the integral approximation for averages discussed later in this section. How well such an approximation works depends on the source of the data and the measurement technique. Examples of the kinds of observations where the normal approximation has been found to be good are weighings on a chemical balance, and measurements of the angular position of a star.

FIGURE 1. Some normal (μ, σ^2) densities.

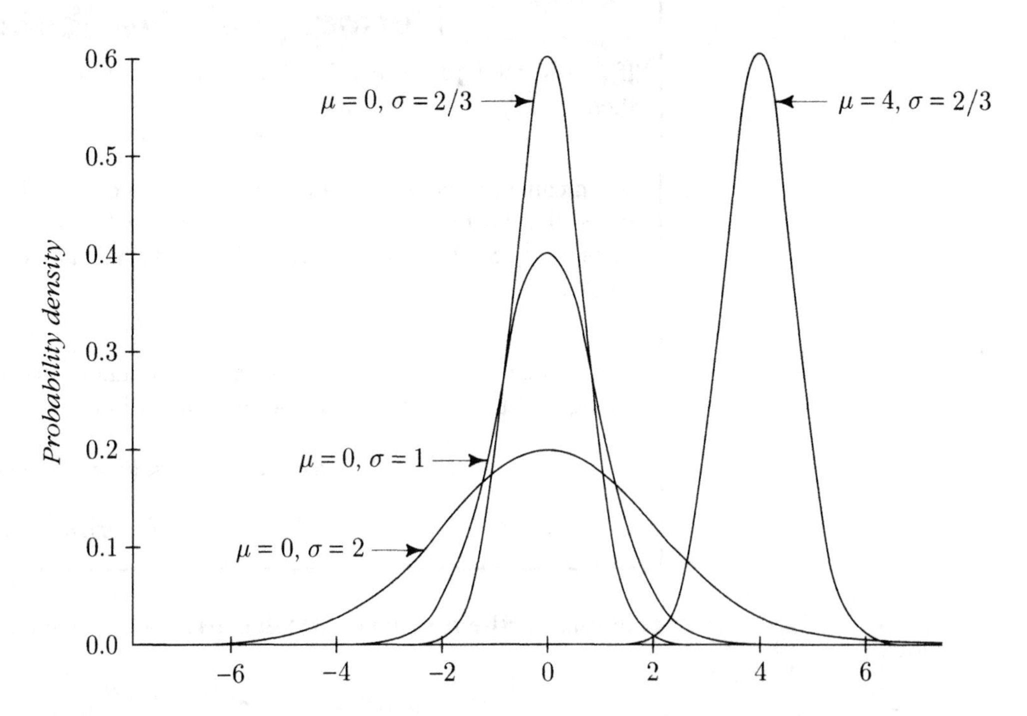

The central limit theorem. The appearance of the normal distribution in many contexts is explained by the central limit theorem, stated in Section 3.3. According to this result, for independent random variables with the same distribution and finite variance, as $n \to \infty$, the distribution of the standardized sum (or average) of n variables approaches the standard normal distribution. It can be shown that this happens no matter what the common distribution of the random variables summed or averaged, discrete or continuous, provided the distribution has finite variance. In particular, the central limit theorem implies that the distribution of the sum or average of a large number of independent measurements will typically tend to follow the normal curve, even if the distribution of the individual measurements does not. This mathematical fact is the basis for most statistical applications of the normal distribution.

History. The normal distribution is also known as the Gaussian distribution, and in France as Laplace's distribution. Gauss (1777–1855) and Laplace (1749–1827) brought out the central role of the normal distribution in the theory of errors of observation. Quetelet (1796–1874) and Galton (1822–1911) fitted the normal distribution to empirical data such as heights and weights in human and animal populations. But the normal distribution was actually first discovered around 1720 by Abraham De Moivre (1667–1754), as the approximation to the binomial (n, p) distribution for large n described in Section 2.2.

Example 1. **Repeated measurements.**

Suppose a long series of repeated measurements of the weight of a standard kilogram yield results that are normally distributed with a mean of one kilogram and an SD of 20 micrograms.

Problem 1. About what proportion of measurements are correct to within 10 micrograms?

Solution. By converting to standard units, this is $P(-0.5 \le Z \le 0.5) = 2\Phi(0.5) - 1 = 38.29\%$.

Problem 2. In 100 measurements, what is the probability that more than 45 measurements will be correct to within 10 micrograms?

Solution. It seems reasonable to assume that each measurement is correct to within 10 micrograms with chance 38.29%, independently of all others. Out of 100 measurements, the number correct to within 10 micrograms has the binomial (100, 0.3829) distribution. This is approximately normal, with

$$\mu = 38.29 \qquad \sigma = \sqrt{100 \times 0.3829 \times (1 - 0.3829)} = 4.86$$

The probability that more than 45 measurements are correct to within 10 micrograms is approximately

$$1 - \Phi\left(\frac{45.5 - 38.29}{4.86}\right) = 1 - \Phi(1.48) = 6.94\%$$

Problem 3. In the long series of measurements, some errors are positive and some are negative. What is the approximate average absolute size of these errors?

Solution. Here X = observed weight $-$ 1 kilogram, in micrograms, and has normal $(0, 20^2)$ distribution. We want $E|X|$. In terms of a standard normal variable Z, $X = 20Z$, so

$$\begin{aligned}
E|X| = 20E|Z| &= 20 \int_{-\infty}^{\infty} |z|\phi(z)dz \\
&= 40 \int_0^{\infty} z\phi(z)\, dz \quad \text{by symmetry} \\
&= 40 \int_0^{\infty} z \frac{1}{\sqrt{2\pi}} e^{-\frac{1}{2}z^2} dz \\
&= -\frac{40}{\sqrt{2\pi}} e^{-\frac{1}{2}z^2}\Big|_0^{\infty} = \frac{40}{\sqrt{2\pi}} = 15.96 \text{ micrograms.}
\end{aligned}$$

Further Examples

Example 2. Radial distance.

Suppose a bacterial colony appears at a point uniformly distributed at random on a circular plate of radius 1. Let R be the distance of the point from the center of the plate.

Problem 1. Find the probability density of R.

Solution. The basic assumption is that the probability of the colony appearing in any particular region of the plate is proportional to the area of the region. From the diagram, for $0 < r < 1$,

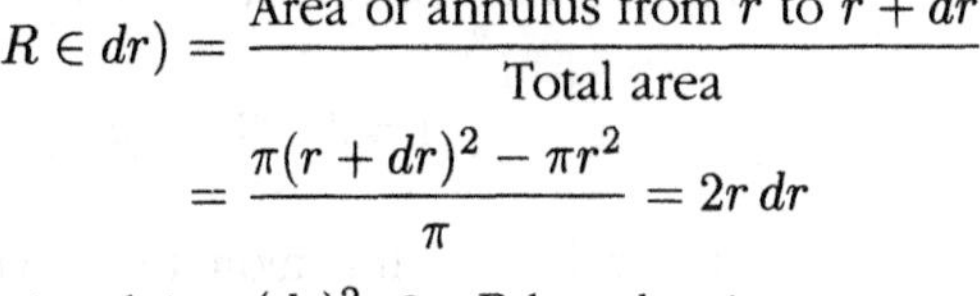

$$P(R \in dr) = \frac{\text{Area of annulus from } r \text{ to } r + dr}{\text{Total area}} = \frac{\pi(r+dr)^2 - \pi r^2}{\pi} = 2r\,dr$$

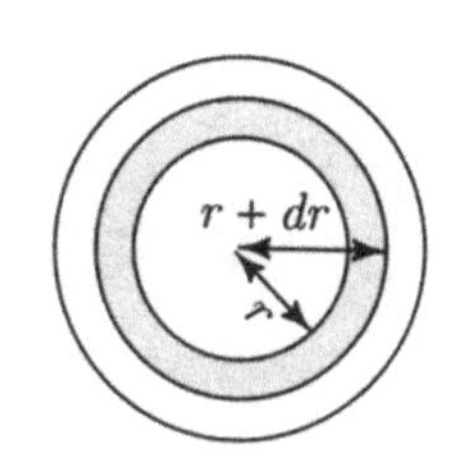

by ignoring the term involving $(dr)^2$. So R has density

$$f(r) = \begin{cases} 2r & 0 < r < 1 \\ 0 & \text{otherwise} \end{cases}$$

Problem 2. Find $P(a \leq R \leq b)$ for $0 < a < b < 1$.

Solution.

$$P(a \leq R \leq b) = \int_a^b 2r dr = r^2 \Big|_a^b = b^2 - a^2$$

(This can also be done using areas in the plane.)

Problem 3. Find the mean and variance of R.

Solution.

$$E(R) = \int_{-\infty}^{\infty} r f(r) dr = \int_0^1 2r^2 dr = \frac{2}{3} r^3 \Big|_0^1 = \frac{2}{3}$$

$$E(R^2) = \int_{-\infty}^{\infty} r^2 f(r) dr = \int_0^1 2r^3 dr = \frac{2}{4} r^4 \Big|_0^1 = \frac{1}{2}$$

$$Var(R) = E(R^2) - (E(R))^2 = \frac{1}{2} - \frac{4}{9} = \frac{1}{18}$$

Problem 4. Suppose 100 bacterial colonies are distributed independently and uniformly at random on a circular plate of radius 1. What is the probability that the mean distance of the colonies from the center of the plate is at least 0.7?

Solution. The problem is to find $P(A_{100} > 0.7)$ where

$$A_{100} = (R_1 + R_2 + \cdots + R_{100})/100$$

and the R_i are independent random variables with the same distribution as that of R calculated in Problem 1. Basic formulae for means and SDs derived in Chapter 3 still apply to give $E(A_{100}) = E(R) = 0.667$

$$SD(A_{100}) = SD(R)/\sqrt{100} = \sqrt{\frac{1}{18} \cdot \frac{1}{100}} \approx 0.0236$$

Using the normal approximation, the required probability is approximately

$$1 - \Phi\left(\frac{0.7 - 0.667}{0.0236}\right) = 1 - \Phi(1.40) = 8.7\%$$

Example 3. A distribution with infinite mean.

Suppose that X has probability density

$$f(x) = \begin{cases} 1/(1+x)^2 & \text{if } x > 0 \\ 0 & \text{otherwise} \end{cases}$$

Problem 1. Find $P(X > 3)$.

Solution. $P(X > 3) = \displaystyle\int_3^\infty \frac{1}{(1+x)^2}\,dx = -\frac{1}{1+x}\Big|_3^\infty = \frac{1}{4}.$

Problem 2. Let X_1, X_2, X_3, X_4 be independent random variables with the same distribution as X. Find the chance that exactly two of these variables are greater than 3.

Solution. Since $P(X_i > 3) = P(X > 3) = 1/4$, and the random variables X_i are independent, the events $(X_i > 3)$, $i = 1, 2, 3, 4$, are four independent events, each with probability $1/4$. The number of these events which occur is therefore a binomial $(4, 1/4)$ random variable. Call this random variable N. The required probability is then

$$P(N = 2) = \binom{4}{2}\left(\frac{1}{4}\right)^2\left(\frac{3}{4}\right)^2 = \frac{27}{128}$$

Problem 3. Find $E(X)$.

Solution.

$$E(X) = \int_0^\infty \frac{x}{(1+x)^2}\,dx = \int_0^\infty \left(\frac{1}{1+x} - \frac{1}{(1+x)^2}\right)dx$$
$$= \int_0^\infty \frac{1}{1+x}\,dx - 1 = \log(1+x)\Big|_0^\infty - 1 = \infty$$

Remark. The long-run interpretation is that the average $(X_1 + \cdots + X_n)/n$ of independent random variables chosen according to this distribution will, with overwhelming probability, tend to increase beyond all finite bounds as $n \to \infty$.

Fitting a Curve to an Empirical Distribution

The empirical distribution of a data list $(x_1, \ldots, x_n)$ can be displayed in a histogram, as in Figure 4 at the end of Section 1.3. This histogram smoothes out the data to display the general shape of the empirical distribution. Such a histogram often follows a smooth curve, say $y = f(x)$, as shown in Figure 2. Since histograms are non-negative it is natural to assume that $f(x) \geq 0$ for every x.

FIGURE 2. A smooth curve fitted to a data histogram.

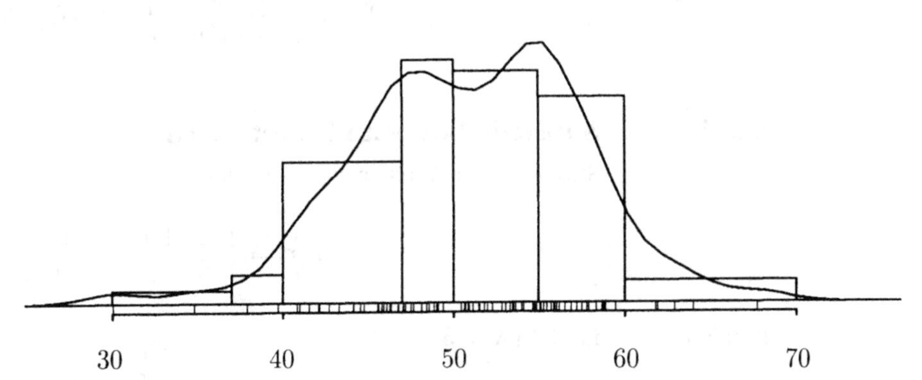

The basic idea is that if (a, b) is a bin interval, then the area of the bar over (a, b) should approximately equal the area under the curve from a to b. Summing such approximations over bins, and interpolating between the cut points, suggests a more general approximation: for any interval (a, b) the proportion of data in the interval should be approximately the area under the curve from a to b. Since the area under the curve can be evaluated as an integral, this amounts to the following:

Integral Approximation for Empirical Proportions

If a histogram of an empirical distribution follows the curve $y = f(x)$, then the proportion $P_n(a, b)$ of observations between a and b is approximated by

$$P_n(a, b) \approx \int_a^b f(x)\, dx$$

Since $P_n(-\infty, \infty) = 1$, whatever the empirical distribution, any reasonable approximation $f(x)$ to a data histogram must satisfy

$$\int_{-\infty}^{\infty} f(x)\, dx = 1 \tag{1}$$

Then $f(x)$ is a probability density function, and the empirical distribution of the data is approximated by the theoretical probability distribution with density $f(x)$.

Averages and Integrals

Given a data list $(x_1, \ldots, x_n)$ and an interval (a, b), the method of indicators provides a useful way to express the proportion of values in (a, b) as an average. Define the *indicator function* of (a, b) by

$$I_{(a,b)}(x) = \begin{cases} 1 & \text{if } x \in (a, b) \\ 0 & \text{otherwise} \end{cases}$$

Given a list $(x_1, \ldots, x_n)$, the number of i such that $x_i \in (a, b)$ can be calculated by going through the list and for each i adding 1 if $x_i \in (a, b)$ and adding 0 otherwise. The term added for the ith element of the list is $I_{(a,b)}(x_i)$. The empirical proportion of values in (a, b) is therefore

$$P_n(a, b) = \frac{1}{n} \sum_{i=1}^{n} I_{(a,b)}(x_i)$$

In words: the proportion of x-values in (a, b) is the average of $I_{(a,b)}(x)$ as x ranges over the n values in the list. Suppose now that the empirical distribution is well approximated by a theoretical distribution with density $f(x)$. The integral approximation for empirical proportions becomes an integral approximation for an empirical average:

$$\frac{1}{n} \sum_{i=1}^{n} I_{(a,b)}(x_i) = P_n(a, b) \approx \int_a^b f(x)dx = \int_{-\infty}^{\infty} I_{(a,b)}(x) f(x)\, dx$$

where the last equality holds because $I_{(a,b)}(x) = 0$ for x outside (a, b). The point of writing the integral approximation this way is that it suggests a very useful generalization for other functions $g(x)$ besides $g(x) = I_{(a,b)}(x)$.

Integral Approximation for Averages

If the empirical distribution of a list $(x_1, \ldots, x_n)$ is well approximated by the theoretical distribution with density $f(x)$, then the average of a function $g(x)$ over the n values in the list is approximated by the integral of $g(x)$ times the density $f(x)$ over all values of x:

$$\frac{1}{n} \sum_{i=1}^{n} g(x_i) \approx \int_{-\infty}^{\infty} g(x) f(x)\, dx$$

Notice that the left-hand average is $E[g(X)]$ for X picked at random from the list of n values $(x_1, \ldots, x_n)$. The right-hand integral is $E[g(X)]$ for a random variable X with density $f(x)$.

Apart from indicator functions $g(x)$, the integral approximation is most commonly applied to the powers $g(x) = x^k$:

$$\frac{1}{n}\sum_{i=1}^{n} x_i^k \approx \int_{-\infty}^{\infty} x^k f(x)\,dx$$

The left side is the average value of x^k as x ranges over values in the data list, and is called the *kth moment of the empirical distribution.* The right side is called the *kth moment of the theoretical distribution* with density $f(x)$. The cases $k = 1$ and $k = 2$ together imply that the mean and variance of the empirical distribution are close to the mean and variance of the theoretical distribution. Thus if a data histogram looks like a normal curve, then the mean and variance of the data can be used to estimate the parameters of the normal curve.

Heuristic derivation of the integral approximation for averages. For $g(x)$ the indicator of an interval, this is just the integral approximation for proportions. A step function $g(x)$ that has a finite number of different values on a finite number of disjoint intervals can be written as a finite linear combination

$$g(x) = c_1 I_{(a_1,b_1)}(x) + \cdots + c_m I_{(a_m,b_m)}(x)$$

of indicator functions of intervals. So for a step function $g(x)$ the integral approximation for the average follows by combining the integral approximation for the proportions $P_n(a_i, b_i)$, using the linearity properties of sums and integrals. The approximation for a more general function $g(x)$ is obtained by approximating $g(x)$ by a step function, much as in the usual approximation of integrals by Riemann sums. □

How good is the integral approximation for an average? This depends both on how closely the empirical distribution conforms to the theoretical density $f(x)$, and on how rapidly $g(x)$ varies as a function of x. (If $g(x)$ grows too rapidly for large absolute values of x the integral $\int_{-\infty}^{\infty} g(x)f(x)\,dx$ might not even be defined.) Provided a data histogram follows the density curve closely, and $g(x)$ is a fairly smooth function of x that does not grow too rapidly for large $|x|$, the data average $\frac{1}{n}\sum_1^n g(x_i)$ will be well approximated by $\int_{-\infty}^{\infty} g(x)f(x)\,dx$.

The law of averages. This is a probabilistic way to make the statement of the previous paragraph more precise. If the data list $(x_1, \ldots, x_n)$ is obtained by a process of repeated measurements of some kind, it may be reasonable to assume that $(x_1, \ldots, x_n)$ is the result of independent random sampling of points from the theoretical distribution with density $f(x)$. More formally, $(x_1, \ldots, x_n)$ is regarded as the observed result of $(X_1, \ldots, X_n)$ for a sequence of independent random variables X_i, each distributed like X with density $f(x)$. According to the law of averages of

Section 3.3, which holds just as well for X with a density as for discrete X, provided the integral that defines $E[g(X)]$ is absolutely convergent, for large n it is highly probable that

$$\frac{1}{n}\sum_{i=1}^{n} g(X_i) \approx E[g(X)] = \int_{-\infty}^{\infty} g(x)f(x)dx$$

Assuming $Var[g(X)] < \infty$, Chebychev's inequality gives for any $\epsilon > 0$

$$P\left(\left|\frac{1}{n}\sum_{i=1}^{n} g(X_i) - \int_{-\infty}^{\infty} g(x)f(x)dx\right| > \epsilon\right) \leq \frac{Var[g(X)]}{n\epsilon^2}$$

Provided n is large enough that $Var[g(X)]/n\epsilon^2$ is small, the integral approximation for the average of n values of $g(X_i)$ will probably be correct to within ϵ. Note that the variance of $g(X)$ will tend to be small provided $g(x)$ does not vary too rapidly over the typical range of values x of X, and provided $g(x)$ does not grow too rapidly for less typical values x in the tails of the distribution of X. So the factor $Var(g[X])$ in the above probability estimate captures nicely the idea of the previous paragraph that the integral approximation for averages will tend to work better for smoother functions $g(x)$. The estimate given by Chebychev's inequality is very conservative. More realistic approximations to the probability of errors of various sizes in the integral approximation for averages are provided by the normal approximation.

The Monte-Carlo method. It may be that the integral $\int_{-\infty}^{\infty} g(x)f(x)dx$ is difficult to evaluate by calculus or numerical integration, but it is easy to generate pseudo-random numbers X_i distributed according to density $f(x)$. The value of the integral can then be estimated by the average value of $g(X_i)$ for a large number of such X_i. For instance, the value of $\int_0^1 g(x)dx$ can be estimated this way using X_i with uniform $(0, 1)$ distribution. Assuming that some bound on $Var[g(X)]$ is available (e.g., if $g(x)$ is a bounded function of x), error probabilities can be estimated using Chebychev's inequality or a normal approximation. The same method can be applied in higher dimensions to approximate multiple integrals.

Exercises 4.1

1. What is the probability that a standard normal random variable has value

a) between 0 and 0.001? b) between 1 and 1.001?

2. Suppose X has density $f(x) = c/x^4$ for $x > 1$, and $f(x) = 0$ otherwise, where c is a constant. Find a) c; b) $E(X)$; c) $Var(X)$.

3. Suppose X is a random variable whose density is $f(x) = cx(1-x)$ for $0 < x < 1$, and $f(x) = 0$ otherwise. Find:

a) the value of c; b) $P(X \leq 1/2)$; c) $P(X \leq 1/3)$;

d) $P(1/3 < X \leq 1/2)$; e) the mean and variance of X.

4. Suppose X with values in $(0, 1)$ has density $f(x) = cx^2(1-x)^2$ for $0 < x < 1$. Find:

a) the constant c; b) $E(X)$; c) $Var(X)$.

5. Suppose that X is a random variable whose density is

$$f(x) = \frac{1}{2(1+|x|)^2} \qquad (-\infty < x < \infty)$$

a) Draw the graph of $f(x)$. b) Find $P(-1 < X < 2)$.

c) Find $P(|X| > 1)$. d) Is $E(X)$ defined?

6. Suppose X has normal (μ, σ^2) distribution, and $P(X \leq 0) = 1/3$, $P(X \leq 1) = 2/3$.

a) What are the values of μ and σ? b) What if instead $P(X \leq 1) = 3/4$?

7. Suppose the distribution of height over a large population of individuals is approximately normal. Ten percent of individuals in the population are over 6 feet tall, while the average height is 5 feet 10 inches. What, approximately, is the probability that in a group of 100 people picked at random from this population there will be two or more individuals over 6 feet 2 inches tall?

8. Measurements on the weight of a lump of metal are believed to be independent and identically distributed; each measurement has mean 12 grams and SD 1.1 gram.

a) Find the chance that a single measurement is between 11.8 and 12.2 grams, assuming that individual measurements are normally distributed.

b) Estimate the chance that the average of 100 measurements is between 11.8 and 12.2 grams. For this calculation, is it necessary to assume that individual measurements are normally distributed? Explain.

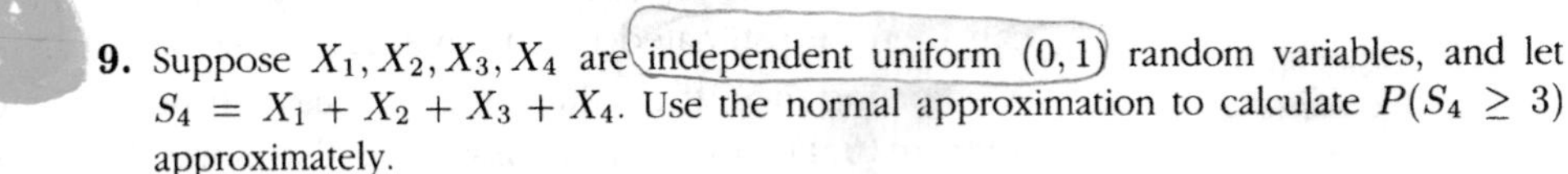

9. Suppose X_1, X_2, X_3, X_4 are independent uniform $(0, 1)$ random variables, and let $S_4 = X_1 + X_2 + X_3 + X_4$. Use the normal approximation to calculate $P(S_4 \geq 3)$ approximately.

10. The distribution of repeated measurements of the weight of an object is approximately normal with a mean of 9.7800 gm and a standard deviation of 0.0031 gm. Calculate:

a) the chance that the next measurement will be between 9.7840 and 9.8000 gm;

b) the proportion of measurements smaller than 9.7794 gm;

c) the weight that the next measurement has a 10% chance of exceeding.

11. A large lot of marbles have diameters which are approximately normally distributed with a mean of 1 cm. One third have diameters greater than 1.1 cm. Find:

a) the standard deviation of the distribution;

b) the proportion whose diameters are within 0.2 cm of the mean;

c) the diameter that is exceeded by 75% of the marbles.

12. Consider a point picked uniformly at random from the area inside one of the following shapes:

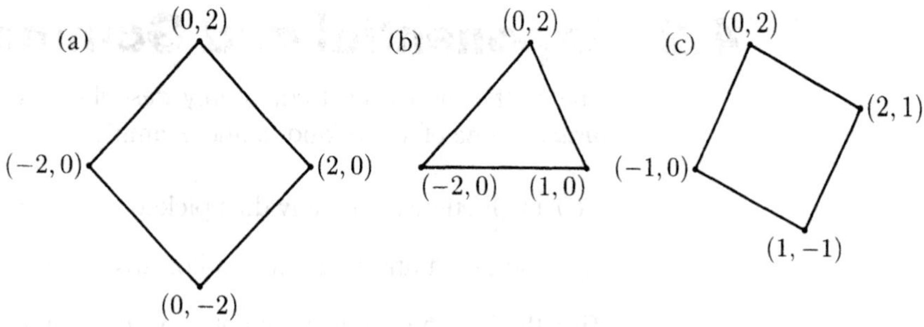

In each case find the density function of the x coordinate.

13. Suppose a manufacturing process designed to produce rods of length 1 inch exactly, in fact produces rods with length distributed according to the density graphed below.

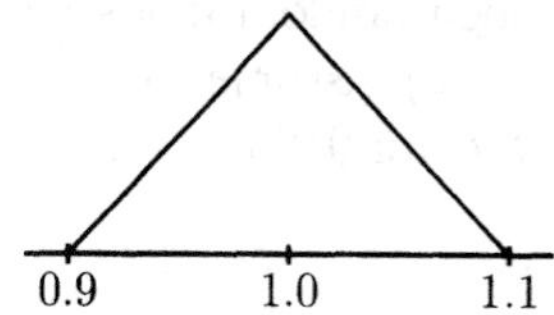

For quality control, the manufacturer scraps all rods except those with length between 0.925 and 1.075 inches before he offers them to buyers.

a) What proportion of output is scrapped?

b) A particular customer wants 100 rods with length between 0.95 and 1.05 inches. Assuming lengths of successive rods produced by the process are independent, how many rods must this customer buy to be 95% sure of getting at least 100 of the prescribed quality?

14. Another manufacturer produces similar rods by a process that produces lengths with the same mean and standard deviation as in Exercise 13, but with a distribution following the normal curve. This manufacturer uses the same quality control procedure of scrapping rods not within 0.075 inches of 1 inch in length.

a) What proportion of output is scrapped by this manufacturer?

b) If you were the customer with requirements as in part b) in Exercise 13, which manufacturer would you prefer? Explain.

15. Standard normal c.d.f. in terms of the error function. Many calculators and computer languages have built in the *error function* $\text{erf}(x) = (2/\sqrt{\pi}) \int_0^x e^{-t^2} dt$.

a) Find μ and σ^2 so that $P(|X| \leq x) = \text{erf}(x)$ if X has normal (μ, σ^2) distribution.

b) Express $\text{erf}(x)$ in terms of the standard normal c.d.f. $\Phi(z)$.

c) Express $\Phi(z)$ in terms of $\text{erf}(x)$.

4.2 Exponential and Gamma Distributions

One of the things most commonly described by a distribution with a density is a *random time* of some kind. Some examples are:

(i) the lifetime of an individual picked at random from some biological population;

(ii) the time until decay of a radioactive atom;

(iii) the length of time a patient survives after an operation of some kind;

(iv) the time it takes a computer to process a job of some kind.

Such random times will be regarded as random variables with range the interval $[0, \infty)$. Assume the distribution of a random time T is defined by a probability density $f(t)$ for $0 \leq t < \infty$, so for $0 \leq a < b < \infty$

$$P(a < T \leq b) = \int_a^b f(t)\, dt$$

If T is interpreted as the lifetime of something, the probability of the thing surviving past time s is

$$P(T > s) = \int_s^\infty f(t)\, dt$$

This is a decreasing function of s, called the *survival function*. By the difference rule for probabilities

$$P(a < T \leq b) = P(T > a) - P(T > b)$$

So the probability of the random time falling in any interval can be found from the survival function.

The simplest model for a random time with no upper bound on its range is the *exponential distribution*. This distribution fits the lifetimes of a variety of inanimate objects that experience no aging effect. More importantly, many models for systems that evolve randomly over time, called *stochastic processes*, are built up from some combination of independent exponential random times. A case in point is the *Poisson process* on a time line, which models the times of successive arrivals of some kind, such as the times customers arrive at a store. In this model, the successive interarrival times are independent exponential random variables. And the time of the rth arrival has a *gamma* distribution. These exponential and gamma distributions, studied in this section, are the continuous analogs of the geometric and negative binomial distributions of Section 3.4.

The following section introduces the concept of a *death* or *hazard rate* associated with a random time. For the exponential distribution this is constant over time, but for more general distributions the death rate varies over time, indicating an aging effect.

Exponential Distribution

A random time T has *exponential distribution with rate* λ, denoted exponential (λ), where λ is a positive parameter, if T has probability density

$$f(t) = \lambda e^{-\lambda t} \qquad (t \geq 0)$$

Equivalently, for $0 \leq a < b < \infty$

$$P(a < T \leq b) = \int_a^b \lambda e^{-\lambda t} dt = -e^{-\lambda t}\Big|_a^b = e^{-\lambda a} - e^{-\lambda b}$$

To see that $f(t)$ is a probability density on $[0, \infty)$, let $a = 0$, and let $b \to \infty$ to find the total probability of 1 on $[0, \infty)$. Set $a = t$ and let $b \to \infty$ to get the next formula for the survival function. Calculation of the mean and SD are left as an exercise.

Exponential Survival Function

A random time T has exponential distribution with rate λ if and only if T has survival function

$$P(T > t) = e^{-\lambda t} \qquad (t \geq 0)$$

Mean and SD: $E(T) = SD(T) = \frac{1}{\lambda}$

Note that the rate λ is the inverse of the mean, so an exponential random time with a large rate is likely to be small, and one with a small rate is likely to be large. A better interpretation of λ as a hazard rate will be given shortly.

Memoryless Property of the Exponential Distribution

A positive random variable T has exponential (λ) distribution for some $\lambda > 0$ if and only if T has the *memoryless property*

$$P(T > t + s \mid T > t) = P(T > s) \qquad (s \geq 0, \quad t \geq 0)$$

In words: Given survival to time t, the chance of surviving a further time s is the same as the chance of surviving to time s in the first place.

FIGURE 1. Exponential densities for $\lambda = 0.5, 1, 2$.

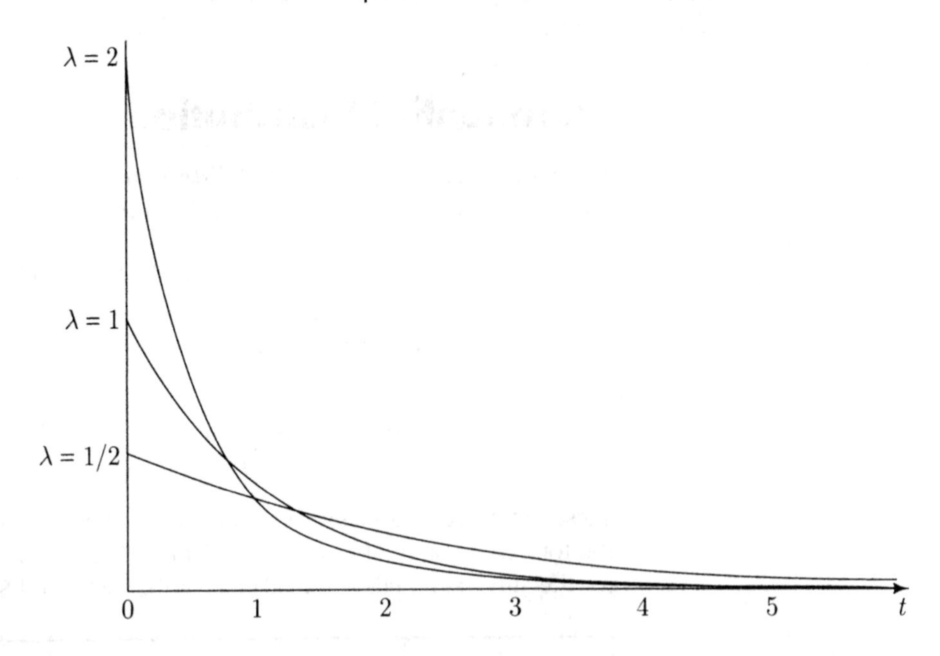

The memoryless property follows immediately from the formula for the survival function, as you should check. The converse hinges on the fact that if T has the memoryless property then the survival function $G(t) = P(T > t)$ must be a solution of the functional equation

$$G(t+s) = G(t)G(s) \qquad (t > 0, \quad s > 0)$$

with $G(t)$ decreasing and bounded between 0 and 1. It can be shown that every such function $G(t)$ is of the form $e^{-\lambda t}$ for some λ.

Thinking of T as the lifetime of something, the memoryless property is this: Whatever the current age of the thing, the distribution of the remaining lifetime is the same as the original lifetime distribution. Some things, such as atoms or electrical components, have this property, hence exponential lifetime distribution. But most forms of life do not have exponential lifetime distribution because they experience an aging process.

Interpretation of the rate λ. For something with an exponentially distributed lifetime, λ is the constant value of the instantaneous *death rate* or *hazard rate*. That is to say, λ measures the probability of death per unit time just after time t, given survival up to time t. To see why, for a time t and a further length of time Δ, calculate

$$\begin{aligned} P(T \le t + \Delta \mid T > t) &= 1 - P(T > t + \Delta \mid T > t) \\ &= 1 - P(T > \Delta) \quad \text{by the memoryless property} \\ &= 1 - e^{-\lambda\Delta} \end{aligned}$$

$$= 1 - [1 - \lambda\Delta + \frac{1}{2}\lambda^2\Delta^2 - \cdots]$$
$$\approx \lambda\Delta \qquad \text{for small } \Delta$$

where $\approx$ is an approximation with error negligible in comparison to Δ as $\Delta \to 0$. Less formally, for an infinitesimal time increment dt, the result of this calculation is that

$$P(T \le t + dt \,|\, T > t) = \lambda\, dt \qquad \text{or}$$
$$P(t < T \le t + dt)/dt = \lambda P(T > t)$$

Since the left side is the density of T at time t, this explains why the exponential (λ) density at t is the death rate λ times the probability $e^{-\lambda t}$ of survival to time t. The characteristic feature of exponentially distributed lifetimes is that the death rate is constant, not depending on t. Other continuous distributions on $(0, \infty)$ correspond to a time-dependent death rate $\lambda(t)$: see Section 4.3.

Example 1. Reliability.

Under suitably constant conditions of use, some kinds of electrical components, for example, fuses and transistors, have a lifetime distribution well fitted by an exponential distribution. Such a component does not wear out gradually. Rather, it stops functioning suddenly and unpredictably. No matter how long the component has been in use, the chance that it survives a further time interval of length Δ is always the same. This probability must then be $e^{-\lambda\Delta}$ for some rate λ, called the *failure rate* in this context. The lifetime distribution is then exponential with rate λ. Roughly speaking, so long as it is still functioning, such a component is as good as new.

Problem 1. Suppose the average lifetime of a particular kind of transistor is 100 working hours, and that the lifetime distribution is approximately exponential. Estimate the probability that the transistor will work for at least 50 hours.

Solution. Since the mean of the exponential distribution is $1/\lambda$, put

$$1/\lambda = 100 \qquad \text{so} \quad \lambda = 0.01$$

and calculate $\quad P(T > 50) = e^{-\lambda 50} = e^{-0.5} = 0.606\ldots$

Problem 2. Given that the transistor has functioned for 50 hours, what is the chance that it fails in the next minute of use?

Solution. From the interpretation of $\lambda = 0.01$, as the instantaneous rate of failure per hour given survival so far, the chance is about $0.01 \times 1/60 \approx 0.00017$.

Example 2. Radioactive decay.

Atoms of radioactive isotopes like Carbon 14, Uranium 235, or Strontium 90 remain intact up to a random instant of time when they suddenly decay, meaning that they split or turn into some other kind of atom, and emit a pulse of radiation or particles of some kind. This radioactive decay can be detected by a Geiger counter. Let T be the random lifetime, or time until decay, of such an atom, starting at some arbitrary time when the atom is intact. It is reasonable to assume that the distribution of T must have the memoryless property. Consequently, there is a rate $\lambda > 0$, the *rate of decay* for the isotope in question, such that T has exponential (λ) distribution: $P(T > t) = e^{-\lambda t}$.

Probabilities here have a clear interpretation due to the large numbers of atoms typically involved (for example, a few grams of a substance will consist of around 10^{23} atoms). Assume a large number N of such atoms decay independently of each other. Then, by the law of large numbers, the proportion of these N atoms that survives up to time t is bound to be close to $e^{-\lambda t}$, the survival probability for each individual atom. This exponential decay over time of the mass of radioactive substance has been experimentally verified, confirming the hypothesis that lifetimes of individual atoms are exponentially distributed. The decay rates λ for individual isotopes can be measured with great accuracy, using this exponential decay of mass. These rates λ show no apparent dependence on physical conditions such as temperature and pressure.

A common way to indicate the rate of decay of a radioactive isotope is by the *half life* h. This is the time it takes for half of a substantial amount of the isotope to disintegrate. So

$$e^{-\lambda h} = 1/2 \quad \text{or} \quad h = \log(2)/\lambda$$

In other words, the half life h is the *median* of the atomic lifetime distribution

$$P(T \le h) = P(T > h) = 1/2$$

The median lifetime is smaller than the mean lifetime $1/\lambda$, by the factor of $\log(2) = 0.693147\ldots$. This is due to the very skewed shape of the exponential distribution.

Numerical illustration. Strontium 90 is a particularly dangerous component of fallout from nuclear explosions. The substance is toxic, easily absorbed into bones when eaten, and has a long half-life of about 28 years. Assuming this value for the half-life h, let us calculate:

a) *The decay rate* λ: From above, this is

$$\lambda = \frac{\log(2)}{h} = 0.693147.../28 = 0.0248 \text{ per year}$$

b) *The mean lifetime of a Strontium* 90 *atom*: This is

$$\frac{1}{\lambda} = \frac{h}{\log(2)} = \frac{28}{0.693147...} = 40.4 \text{ years}$$

c) *The probability that a Strontium 90 atom survives at least 50 years*: This is

$$P(T > 50) = e^{-\lambda 50} = e^{-0.0248 \times 50} = 0.29$$

d) *The proportion of one gram of Strontium 90 that remains after 50 years.* This proportion is the same as the above probability, by the law of large numbers.

e) *The number of years after a nuclear explosion before 99% of the Strontium 90 produced by the explosion has decayed.* Let y be the number of years. Then

$$e^{-0.0248y} = 1/100 \qquad \text{so} \qquad y = \log(100)/0.0248 \approx 186 \text{ years}$$

Relation to the geometric distribution. The exponential distribution on $(0, \infty)$ is the continuous analog of the geometric distribution on $\{1, 2, 3, \ldots\}$. For instance, in the formulation of the memoryless property it was assumed that s and t range over all non-negative real numbers. This property for integers s and t, and an integer-valued random variable T, is a characterization of the geometric distribution. An exponential distribution is the limit of rescaled geometric (p) distributions as the parameter p tends to 0. More precisely, if G has geometric (p) distribution, so that $P(G > n) = (1-p)^n$, and p is small so that $E(G) = 1/p$ is large, then the rescaled variable $G/E(G) = pG$ has approximately exponential distribution with rate $\lambda = 1$:

$$P(pG > t) = P(G > t/p) \approx (1-p)^{t/p} \text{ (only} \approx \text{because } t/p \text{ may not be an integer)}$$
$$\approx e^{-t}$$

by the usual exponential approximation $(1-p) \approx e^{-p}$ for small p. This approximation has been used already in the gambler's rule example in Section 1.6. The factor of $\log(2)$ which appeared there was the median of the exponential distribution with rate 1.

Relation to a Poisson process. A sequence of independent Bernoulli trials, with probability p of success on each trial, can be characterized in two different ways as follows:

I. Counts of successes. The distribution of the number of successes in n trials is binomial (n, p), and numbers of successes in disjoint blocks of trials are independent.

II. Times between successes. The distribution of the waiting time until the first success is geometric (p), and the waiting times between each success and the next are independent with the same geometric distribution.

After a passage to the limit by discrete approximations, as in Section 3.5, these characterizations of Bernoulli trials lead to the two descriptions in the next box of a *Poisson arrival process with rate* λ. This means a Poisson random scatter of points, as in Section 3.5, for points now called *arrivals* on the interval $(0, \infty)$ interpreted as a time line, instead of hits on a region in the plane. In the diagram inside the box, arrivals are at times marked × on the time line. Think of arrivals representing something like calls coming into a telephone exchange, particles arriving at a counter, or customers entering a store.

Two Descriptions of a Poisson Arrival Process

I. Counts of arrivals. The distribution of the number of arrivals $N(I)$ in a fixed time interval I of length t is Poisson (λt), and numbers of arrivals in disjoint time intervals are independent.

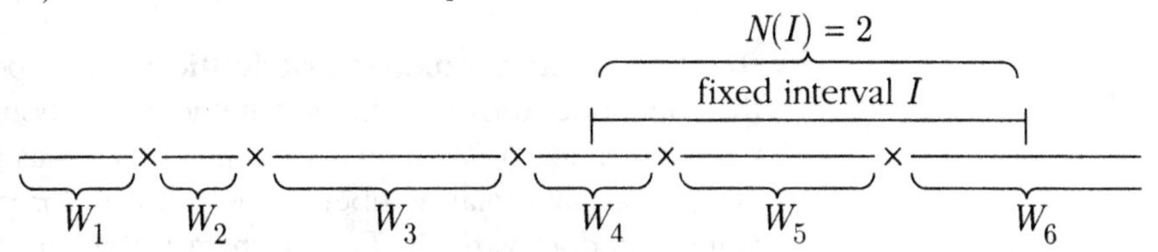

II. Times between arrivals. The distribution of the waiting time W_1 until the first arrival is exponential (λ), and W_1 and the subsequent waiting times W_2, $W_3, \ldots$ between each arrival and the next are independent, all with the same exponential distribution.

These two descriptions of a random arrival process are equivalent.

Probabilities of events defined by a Poisson arrival process can be calculated from whichever of these two descriptions is more convenient.

Example 3. Telephone calls.

Suppose calls are coming into a telephone exchange at an average rate of 3 per minute, according to a Poisson arrival process. So, for instance, $N(2,4)$, the number of calls coming in between $t = 2$ and $t = 4$, has Poisson distribution with mean $\lambda(4-2) = 3 \times 2 = 6$; and W_3, the waiting time between the second and third calls, has exponential (3) distribution. Let us calculate:

a) *The probability that no calls arrive between* $t = 0$ *and* $t = 2$: Since $N(0,2]$, the number of calls arriving in this interval has Poisson (6) distribution, this is

$$P(N(0,2] = 0) = e^{-6} = 0.0025$$

b) *The probability that the first call after* $t = 0$ *takes more than 2 minutes to arrive.* From the exponential (3) distribution of W_1 this is

$$P(W_1 > 2) = e^{-3\times 2}$$

The answer is the same as in a) because the events are, in fact, identical.

c) *The probability that no calls arrive between* $t = 0$ *and* $t = 2$ *and at most four calls arrive between* $t = 2$ *and* $t = 3$. By independence of $N(0, 2]$ and $N(2, 3]$, this is

$$P(N(0,2] = 0) \cdot P(N(2,3] \leq 4) = e^{-6} \cdot e^{-3}(1 + 3 + \frac{3^2}{2!} + \frac{3^3}{3!} + \frac{3^4}{4!}) = 0.0020$$

d) *The probability that the fourth call arrives within* 30 *seconds of the third.* This is

$$P(W_4 \leq 0.5) = 1 - P(W_4 > 0.5) = 1 - e^{-3 \times 0.5} = 0.7769$$

e) *The probability that the first call after* $t = 0$ *takes less than* 20 *seconds to arrive, and the waiting time between the first and second calls is more than* 3 *minutes.* By independence of W_1 and W_2, this is

$$P(W_1 < 1/3) \cdot P(W_2 > 3) = (1 - e^{-3 \times 20/60})e^{-3 \times 3}$$

f) *The probability that the fifth call takes more than* 2 *minutes to arrive.* Since the arrival time of the fifth call is the sum of the first five interarrival times, the problem is to find $P(W_1 + W_2 + W_3 + W_4 + W_5 > 2)$ where the W_i are independent, all with exponential (3) distribution. The general technique for finding the distribution of a sum of continuously distributed random variables is not discussed until Section 5.4. But this particular problem is solved easily by recoding it in terms of the Poisson distributed counts. The fifth call takes more than 2 minutes to arrive if and only if at most four calls arrive between $t = 0$ and $t = 2$. So the required probability is

$$\begin{aligned} P(W_1 + W_2 + W_3 + W_4 + W_5 > 2) &= P(N(0,2] \leq 4) \\ &= e^{-6}(1 + 6 + \frac{6^2}{2!} + \frac{6^3}{3!} + \frac{6^4}{4!}) = 0.2851 \end{aligned}$$

Gamma Distribution

As in the previous example, let $W_1, W_2, \ldots$ be independent exponential (λ) variables, and interpret the W_i as the waiting times between arrivals in a Poisson process with rate λ. The method used in the last part f) of the example can be used to find the distribution of the time T_r of the rth arrival, for any $r = 1, 2, \ldots$. Here is a general statement of the result:

Poisson Arrival Times (Gamma Distribution)

If T_r is the time of the rth arrival after time 0 in a Poisson process with rate λ, or if $T_r = W_1 + W_2 + \cdots + W_r$ where the W_i are independent with exponential (λ) distribution, then T_r has the *gamma* (r, λ) distribution defined by either (1) or (2) for all $t \geq 0$:

(1) **Density:** $$P(T_r \in dt)/dt = P(N_t = r-1)\lambda = e^{-\lambda t}\frac{(\lambda t)^{r-1}}{(r-1)!}\lambda$$

where N_t, the number of arrivals by time t in the Poisson process with rate λ, has Poisson (λt) distribution. In words, the probability per unit time that the rth arrival comes around time t is the probability of exactly $r-1$ arrivals by time t multiplied by the arrival rate.

(2) **Right tail probability:** $$P(T_r > t) = P(N_t \leq r-1) = \sum_{k=0}^{r-1} e^{-\lambda t}\frac{(\lambda t)^k}{k!}$$

because $T_r > t$ if and only if there are at most $r-1$ arrivals in the interval $(0, t]$.

(3) **Mean and SD:** $E(T_r) = r/\lambda \qquad SD(T_r) = \sqrt{r}/\lambda$

Formula (2) is the extension of the numerical example f) above from the case $r = 5, \lambda = 3, t = 2$ to general r, λ, and t. Formula (1) for the density can be derived from (2) by calculus. But here is a neater way. For the rth arrival to come in an infinitesimal interval of time of length dt just after time t, it must be that:

A: there is an arrival in the time dt,
where $P(A) = \lambda dt$, by the local interpretation of the arrival rate λ;

and (since the possibility of more than one arrival in the infinitesimal interval can be safely ignored), that:

B: there were exactly $r-1$ arrivals in the preceding time t,
where $P(B) = P(N_t = r-1) = e^{-\lambda t}(\lambda t)^{r-1}/(r-1)!$

These events A and B are defined by arrivals in disjoint time intervals, so they are independent by the basic assumptions of a Poisson process. Multiplying their probabilities gives formula (1) for $P(AB) = P(T_r \in dt)$. The formulae (3) for the mean and SD are immediate from the representation of T_r as a sum of r independent exponential (λ) variables, and the formulae for the case $r = 1$, when the gamma $(1, \lambda)$ distribution is just exponential (λ).

The full extent of the analogy between Bernoulli trials and a Poisson process is brought out in the display on pages 288 and 289. In this analogy the continuous gamma (r, λ) distribution of the time until the rth arrival corresponds to the discrete negative binomial (r, p) distribution of the number of trials until the rth success, as derived in Section 3.4. As the display shows, the formulae relating the gamma to the Poisson distribution are like similar formulae relating the negative binomial to the binomial distribution.

FIGURE 2. Gamma density of the rth arrival for $r = 1$ to 10. Note how the distributions shift to the right and flatten out as r increases, in keeping with the formulae r/λ and $\sqrt{r}/\lambda$ for the mean and SD. Due to the central limit theorem, the gamma (r, λ) distribution becomes asymptotically normal as $r \to \infty$.

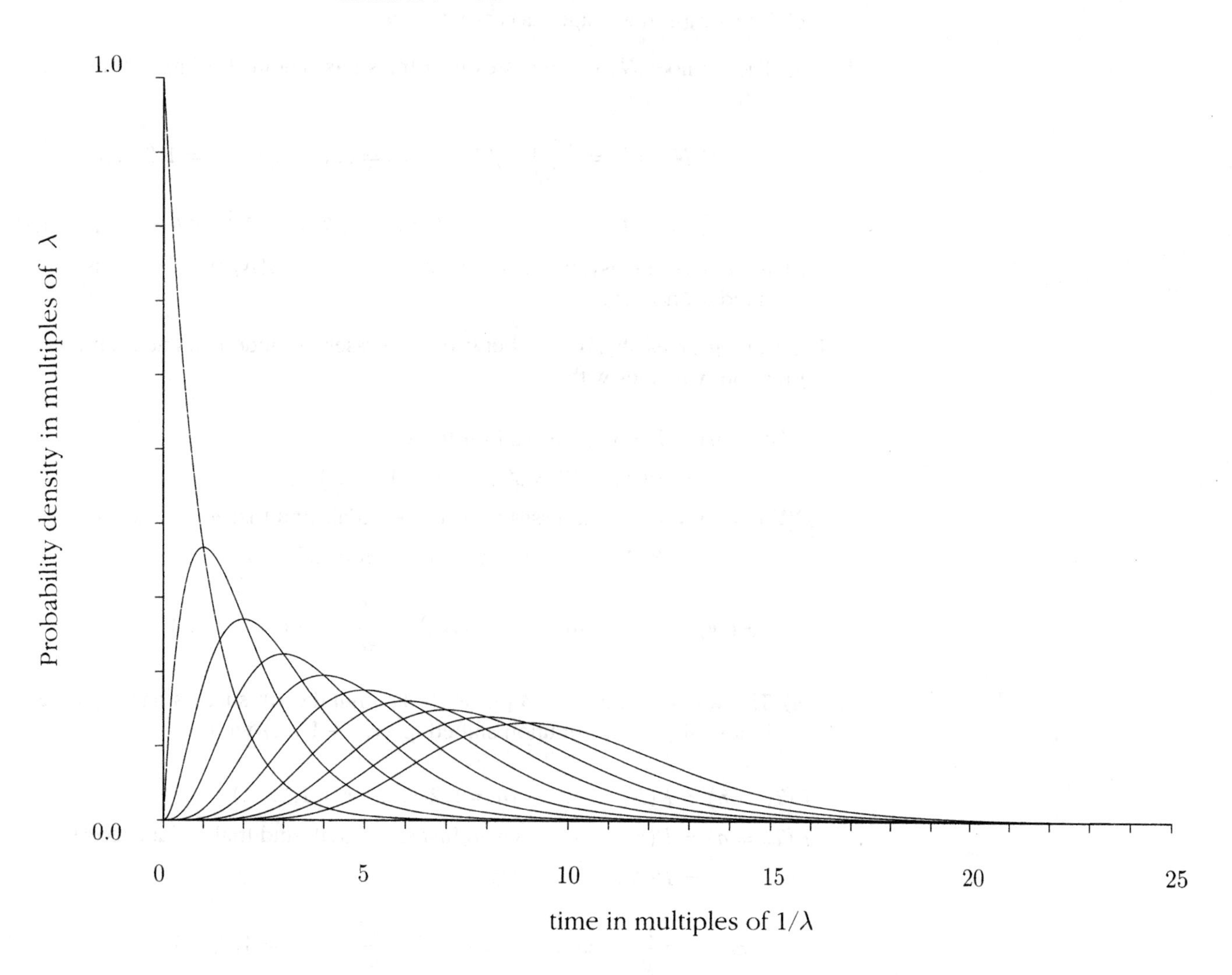

Summary of Properties of a Bernoulli (p) Trials Process

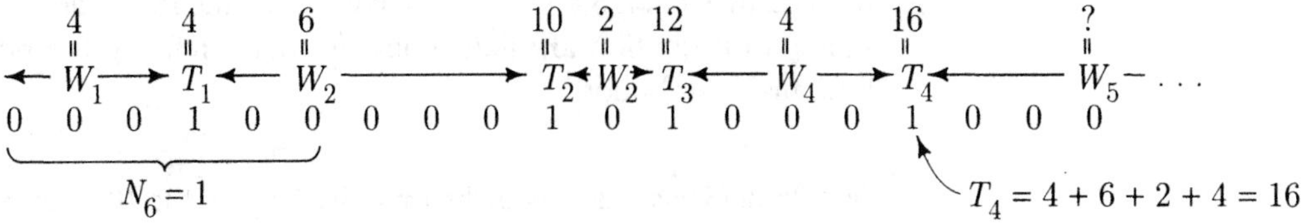

$$\begin{array}{ccccccccc} 4 & 4 & 6 & 10 & 2 & 12 & 4 & 16 & ? \\ \| & \| & \| & \| & \| & \| & \| & \| & \| \\ W_1 & T_1 & W_2 & T_2 & W_2 & T_3 & W_4 & T_4 & W_5 \end{array} \ \ldots$$

$$0\ 0\ 0\ 1\ 0\ 0\ 0\ 0\ 0\ 1\ 0\ 1\ 0\ 0\ 0\ 1\ 0\ 0\ 0$$

$N_6 = 1$ $\qquad$ $T_4 = 4 + 6 + 2 + 4 = 16$

1. a) The probability of success per trial is p.

   b) The events of successes on different trials are independent.

   c) The long-run average success rate is p.

2. a) The number N_n of successes in n trials has binomial (n, p) distribution, with

$$P(N_n = k) = \binom{n}{k} p^k q^{n-k} \qquad (k = 0, 1, \ldots, n, \quad n = 1, 2, \ldots)$$

$$E(N_n) = np \qquad \text{and} \qquad SD(N_n) = \sqrt{npq} \qquad \text{where } q = 1 - p$$

   b) As $n \to \infty$ the asymptotic distribution of $(N_n - E(N_n))/SD(N_n)$ is standard normal.

3. The waiting times $W_1, W_2, \ldots$ between successes are independent geometric (p) random variables with

$$\begin{aligned} P(W_k > n) &= P(\text{no successes in } n \text{ trials}) \\ &= P(N_n = 0) = q^n \qquad (n = 1, 2, \ldots) \\ P(W_k = n) &= P(\text{no successes in first } n - 1 \text{ trials and trial } n \text{ is a success}) \\ &= P(N_{n-1} = 0)p = q^{n-1}p \qquad (n = 1, 2, \ldots) \end{aligned}$$

$$E(W_k) = \frac{1}{p} \qquad \text{and} \qquad SD(W_k) = \frac{\sqrt{q}}{p} \qquad (k = 1, 2, \ldots)$$

4. a) The waiting time $T_r = W_1 + \cdots + W_r$ until the rth success has negative binomial (r, p) distribution shifted to $\{r, r+1, \ldots\}$ with

$$\begin{aligned} P(T_r > n) &= P(N_n < r) \qquad (n = 1, 2, \ldots, \quad r = 1, 2, \ldots) \\ P(T_r = n) &= P(r - 1 \text{ successes in first } n - 1 \text{ trials and trial } n \text{ is a success}) \\ &= P(N_{n-1} = r - 1)p \end{aligned}$$

$$E(T_r) = \frac{r}{p} \qquad \text{and} \qquad SD(T_r) = \frac{\sqrt{rq}}{p} \qquad (r = 1, 2, \ldots)$$

   b) As $r \to \infty$ the distribution of $(T_r - E(T_r))/SD(T_r)$ converges to standard normal.

Summary of Properties of a Poisson (λ) Arrival Process

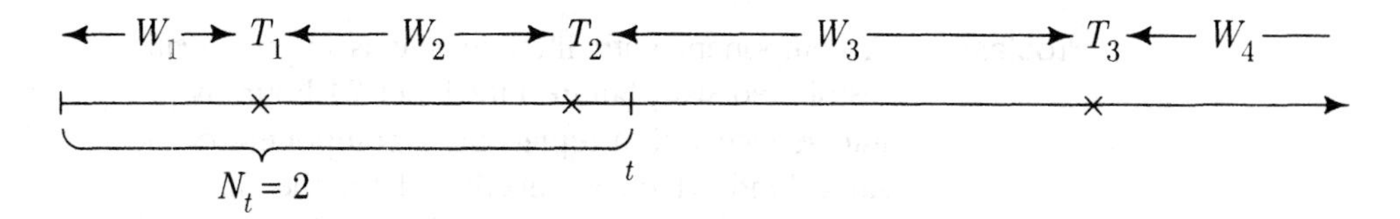

1. a) $P(\text{arrival in interval } \Delta t) \approx \lambda \Delta t$ as $\Delta t \to 0$.

   b) The events of arrivals in disjoint intervals are independent.

   c) The long-run average rate of arrivals per unit time is λ.

2. a) The number N_t of arrivals in time t has Poisson (λt) distribution with

$$P(N_t = k) = e^{-\lambda t}(\lambda t)^k / k! \qquad (k = 0, 1, \ldots, \quad t \geq 0)$$

$$E(N_t) = \lambda t \qquad \text{and} \qquad SD(N_t) = \sqrt{\lambda t}$$

   b) As $t \to \infty$ the asymptotic distribution of $(N_t - E(N_t)) / SD(N_t)$ is standard normal.

3. The waiting times $W_1, W_2, \ldots$ between arrivals are independent exponential (λ) random variables with

$$\begin{aligned} P(W_k > t) &= P(\text{no arrivals in time } t) \\ &= P(N_t = 0) = e^{-\lambda t} \qquad (t \geq 0) \\ P(W_k \in dt) &= P(\text{no arrivals in time } t, \text{ arrival in time } dt) \\ &= P(N_t = 0) P(\text{arrival in time } dt) = e^{-\lambda t} \lambda dt \qquad (t \geq 0) \end{aligned}$$

$$E(W_k) = \frac{1}{\lambda} \qquad \text{and} \qquad SD(W_k) = \frac{1}{\lambda} \qquad (k = 1, 2, \ldots)$$

4. a) The waiting time $T_r = W_1 + \cdots + W_r$ until the rth arrival has gamma (r, λ) distribution with

$$\begin{aligned} P(T_r > t) &= P(N_t < r) \qquad (t \geq 0, \quad r = 1, 2, \ldots) \\ P(T_r \in dt) &= P(r - 1 \text{ arrivals in time } t \text{ and arrival in time } dt) \\ &= P(N_t = r - 1) \lambda dt \end{aligned}$$

$$E(T_r) = \frac{r}{\lambda} \qquad \text{and} \qquad SD(T_r) = \frac{\sqrt{r}}{\lambda} \qquad (r = 1, 2, \ldots)$$

   b) As $r \to \infty$ the distribution of $(T_r - E(T_r)) / SD(T_r)$ converges to standard normal.

Example 4. Sum of two lifetimes.

Problem. A component with lifetime that is exponentially distributed with failure rate 1 per 24 hours is put into service with a replacement component of the same kind which is substituted for the first one when it fails. What is the median of the total time to failure of both components?

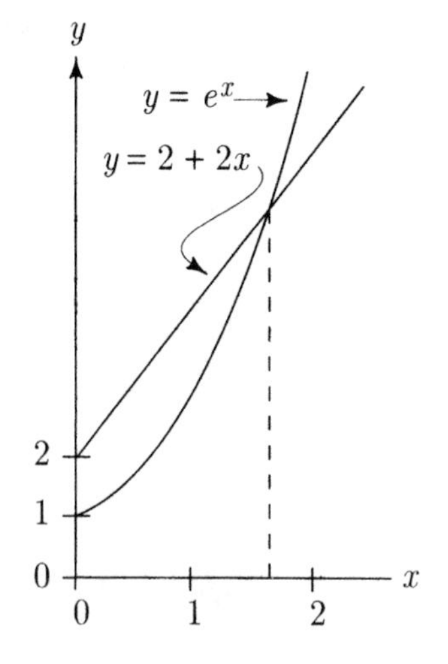

Solution. The problem is to find m such that $P(T_2 \geq m) = 1/2$, where $T_2 = W_1 + W_2$ is the sum of two independent exponential lifetimes with rate $\lambda = 1/24$ per hour. But from formula (2) on page 286

$$P(T_2 \geq m) = P(N_m \leq 1) = e^{-\lambda m}(1+\lambda m)$$

where N_m has Poisson (λm) distribution. Put $x = \lambda m$. Then $m = x/\lambda$ where x solves

$$1/2 = e^{-x}(1+x)$$
$$e^x = 2 + 2x$$

Some trial and error with a calculator gives $x \approx 1.675$. So the median is about $1.675/(1/24) \approx 40.3$ hours.

Discussion. Note how the Poisson formula for $P(T_2 \geq t)$ can be used here for the gamma $(2, \lambda)$ distribution of the sum $T_2 = W_1 + W_2$ of two independent exponential (λ) variables, even though these exponential random variables are not originally defined as inter-arrival times for a Poisson process. Technically, this is because the distribution of a sum of independent random variables is determined by the distributions of the individual variables. Section 5.4 goes into this in more detail. Intuitively, you may as well suppose the two lifetimes W_1 and W_2 are just the first two in an infinite sequence of independent exponentially distributed lifetimes of components replaced one after another. In that case the times of replacements would make a Poisson process, with N_t representing the total number of replacements by time t.

Gamma Distribution for Non-Integer Shape Parameter

A gamma distribution is defined for all positive values of the parameters r and λ by a variation of the density formula (1) on page 286 for integer r. A random variable T has *gamma distribution with parameters r and λ*, or *gamma (r, λ) distribution*, if

T has probability density

$$f_{r,\lambda}(t) = \begin{cases} [\Gamma(r)]^{-1}\lambda^r t^{r-1}e^{-\lambda t} & t \geq 0 \\ 0 & t < 0 \end{cases} \qquad \text{where} \quad \Gamma(r) = \int_0^\infty t^{r-1}e^{-t}dt$$

is a constant of integration, depending on r, called the *gamma function*. The parameter r is called the *index* or *shape* parameter. And $1/\lambda$ is a scale parameter. Comparison with formula (1) on page 286 shows that

$$\Gamma(r) = (r-1)! \qquad (r = 1, 2, \ldots)$$

You should think of the gamma function $\Gamma(r)$ as a continuous interpolation of the factorial function $(r-1)!$ for non-integer r. Integration by parts gives the following:

Recursion formula for the gamma function: $\Gamma(r+1) = r\Gamma(r) \quad (r > 0)$

Since it is easy to see that $\Gamma(1) = 1$, the recursion formula implies $\Gamma(r) = (r-1)!$ for integer r by mathematical induction.

But there is no explicit formula for $\Gamma(r)$ except in case r is a positive integer, or a positive half-integer, starting from $\Gamma(1/2) = \sqrt{\pi}$. See Exercise 5.3.15. Section 5.3 shows that for half integer r the gamma distributions arise from sums of squares of independent normal variables.

As will be shown in Section 5.4, several algebraic functions of gamma random variables have distributions which are easy to compute. See the gamma distribution summary for a survey. In applications, the distribution of a random variable may be unknown, but reasonably well approximated by some gamma distribution. Then results obtained assuming a gamma distribution might provide useful approximations.

For non-integer values of r the gamma (r, λ) distribution has a shape which varies continuously between the shapes for integers r, as illustrated by the following diagrams:

In Figures 3, 4, and 5, both horizontal and vertical scales change from one figure to the next. Figure 3 shows how the gamma (r, λ) density is unbounded near zero for $0 < r < 1$. As $r \to 0$ the distribution piles up more and more near zero, approaching the distribution of a constant random variable with value 0. This is a discrete distribution, which does not have a probability density, but assigns probability one to the point zero, and may be thought of as the gamma $(0, \lambda)$ distribution.

FIGURE 3. Gamma (r, λ) densities for $\lambda = 1$ and r a multiple of $1/4$, $0 < r \leq 1$.

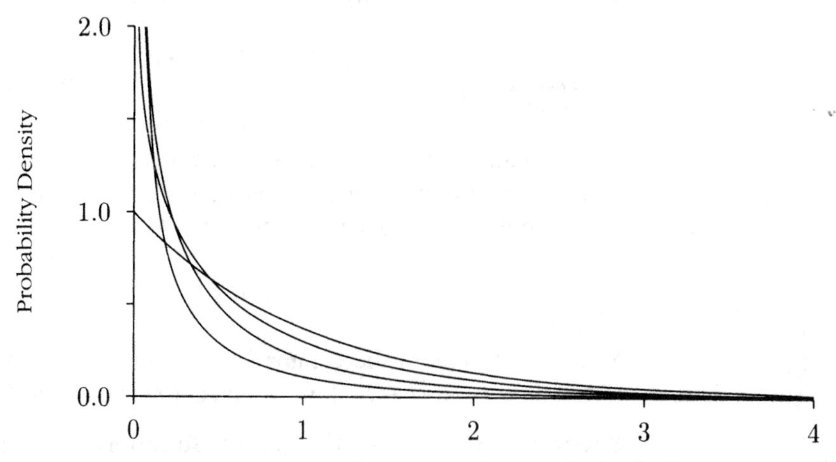

FIGURE 4. Gamma (r, λ) densities for $\lambda = 1$ and r a multiple of $1/4$, $1 \leq r \leq 2$.

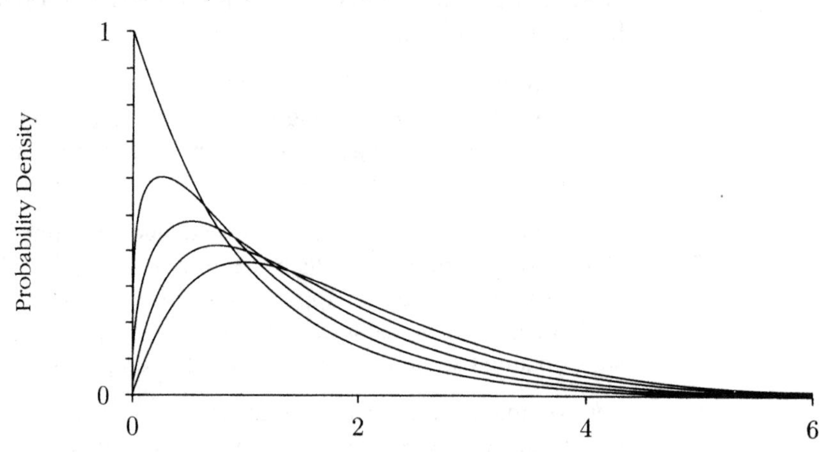

FIGURE 5. Gamma (r, λ) densities for $\lambda = 1$ and r a multiple of $1/4$, $2 \leq r \leq 3$.

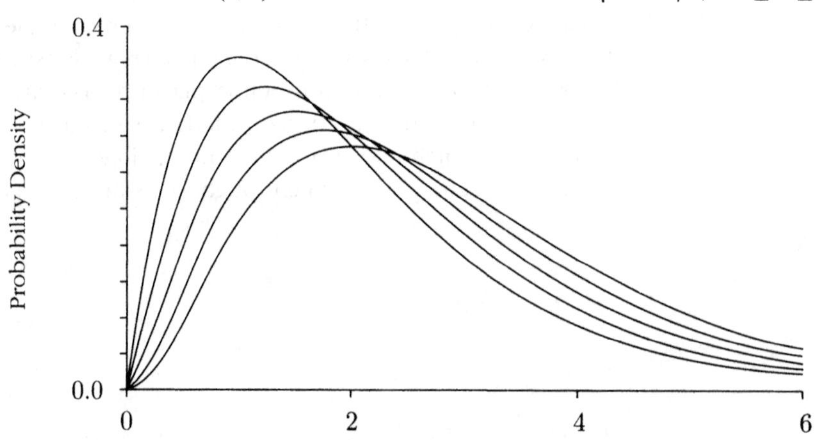

Exercises 4.2

1. Suppose a particular kind of atom has a half-life of 1 year. Find:

a) the probability that an atom of this type survives at least 5 years;

b) the time at which the expected number of atoms is 10% of the original;

c) if there are 1024 atoms present initially, the time at which the expected number of atoms remaining is one;

d) the chance that in fact none of the 1024 original atoms remains after the time calculated in c).

2. A piece of rock contains 10^{20} atoms of a particular substance. Each atom has an exponentially distributed lifetime with a half-life of one century. How many centuries must pass before

a) it is most likely that about 100 atoms remain;

b) there is about a 50% chance that at least one atom remains. What assumptions are you making?

3. Suppose the time until the next earthquake in a particular place is exponentially distributed with rate 1 per year. Find the probability that the next earthquake happens within

a) one year; b) six months; c) two years; d) 10 years.

4. Suppose component lifetimes are exponentially distributed with mean 10 hours. Find:

a) the probability that a component survives 20 hours;

b) the median component lifetime;

c) the SD of component lifetime;

d) the probability that the average lifetime of 100 independent components exceeds 11 hours;

e) the probability that the average lifetime of 2 independent components exceeds 11 hours.

5. Suppose calls are arriving at a telephone exchange at an average rate of one per second, according to a Poisson arrival process. Find:

a) the probability that the fourth call after time $t = 0$ arrives within 2 seconds of the third call;

b) the probability that the fourth call arrives by time $t = 5$ seconds;

c) the expected time at which the fourth call arrives.

6. A Geiger counter is recording background radiation at an average rate of one hit per minute. Let T_3 be the time in minutes when the third hit occurs after the counter is switched on. Find $P(2 \le T_3 \le 4)$.

7. Let $0 < p < 1$. For the exponential distribution with rate λ, find a formula for the $100p$th percentile point t_p such that $P(T \le t_p) = 100p\%$.

8. Transistors produced by one machine have a lifetime which is exponentially distributed with mean 100 hours. Those produced by a second machine have an exponentially distributed lifetime with mean 200 hours. A package of 12 transistors contains 4 produced by the first machine and 8 produced by the second. Let X be the lifetime of a transistor picked at random from this package. Find:

a) $P(X \geq 200 \text{ hours})$; b) $E(X)$; c) $Var(X)$.

9. Gamma function and moments of the exponential distribution. Consider the gamma function $\Gamma(r) = \int_0^\infty x^{r-1}e^{-x}dx \quad (r > 0)$

a) Use integration by parts to show that $\Gamma(r+1) = r\Gamma(r) \quad (r > 0)$

b) Deduce from a) that $\Gamma(r) = (r-1)! \quad (r = 1, 2, \ldots)$

c) If T has exponential distribution with rate 1, then

$$E(T^n) = n! \qquad (n = 0, 1, 2, \ldots) \qquad \text{and} \quad SD(T) = 1$$

d) If T has exponential distribution with rate λ, then show λT has exponential distribution with rate 1, hence

$$E(T^n) = n!/\lambda^n \qquad (n = 0, 1, 2, \ldots) \qquad \text{and} \quad SD(T) = 1/\lambda$$

10. Geometric from exponential.

a) Show that if T has exponential distribution with rate λ, then int(T), the greatest integer less than or equal to T, has a geometric (p) distribution on $\{0, 1, 2, \ldots\}$, and find p in terms of λ.

b) Let $T_m = \text{int}(mT)/m$, the greatest multiple of $1/m$ less than or equal to T. Show that T has exponential distribution on $(0, \infty)$ for some λ, if and only if for every m there is some p_m such that mT_m has geometric (p_m) distribution on $\{0, 1, 2, \ldots\}$. Find p_m in terms of λ.

c) Use b) and $T_m \leq T \leq T_m + 1/m$ to calculate $E(T)$ and $SD(T)$, from the formulae for the mean and standard deviation of a geometric random variable.

11. Suppose the probability that a given kind of atom disintegrates in any particular microsecond, given that it was alive at the beginning of the microsecond, is $\lambda \times 10^{-6}$ where $\lambda > 0$ is a constant. Let T be the random lifetime of the atom in seconds.

a) Show that the distribution of T is approximately exponential with parameter λ. [*Hint*: Consider $P(T \geq t)$ for t a multiple of 10^{-6}.]

b) What is the chance that the atom has a lifetime of between 1 and 2 seconds?

12. Gamma distribution. Derive the following features of the gamma (r, λ) distribution for all positive r:

a) For $r \geq 1$ the mode (i.e., the value that maximizes the density) is $(r-1)/\lambda$. What if $0 < r < 1$?

b) For $k > 0$, the kth moment of T with gamma (r, λ) distribution is

$$E(T^k) = \frac{1}{\lambda^k}\frac{\Gamma(r+k)}{\Gamma(r)}$$

In particular $E(T) = r/\lambda$.

c) $SD(T) = \sqrt{r}/\lambda$ and Skewness$(T) = 2/\sqrt{r}$.

13. Suppose that under normal operating conditions the operating time until failure of a certain type of component has exponential (λ) distribution for some $\lambda > 0$. And suppose that the random variables representing lifetimes of different components of this type may be regarded as independent.

a) The average lifetime of 10,000 components is found to be 20 days. Estimate the value of λ based on this information.

b) Assuming the exponential lifetime model with $\lambda = 5\%$ per day, let N_d be the number of components among 10,000 components which survive more than d days. Find $E(N_d)$ and $SD(N_d)$ for $d = 10, 20, 30$.

14. Interpretation of the rate. In Exercise 13, the exponential model with $\lambda = 5\%$ per day implies the probability of a component failing in the first day of its use is:
a) exactly 5%; b) approximately 5%, but slightly less;
c) approximately 5%, but slightly more. Without doing any numerical calculations, pick out which of a), b), or c) is true, and explain your choice. Confirm your choice by numerical calculation of the exact probability.

15. Satellite problem. Suppose that a system using one of the components described in Exercise 13, with failure rate 5% per day, is sent up in a satellite together with three spare components of the same type. Assume that as soon as the original component fails, it is replaced by one of the spares, and when that component fails it is replaced by a second spare, and so on. The total operating time of the component plus three spares is then $T_{\text{total}} = T_1 + T_2 + T_3 + T_4$ where T_1 is the operating time of the first component, T_2 is the operating time of the first spare, and so on. Assuming that the satellite launch is successful, and normal operating conditions obtain once the satellite is in orbit, calculate:

a) $E(T_{\text{total}}])$; b) $SD(T_{\text{total}})$; c) $P(T_{\text{total}} \geq 60 \text{ days})$.

16. In the satellite problem of Exercise 15, how many spares would have to be provided to make $P(T_{\text{total}} \geq 60 \text{ days})$ at least 90%?

17. Another type of component has lifetime distribution which is approximately gamma $(2, \lambda)$ with $\lambda = 10\%$ per day.

a) Redo Exercise 15 for this type of component, making similar independence assumptions. After calculating the answers to a) and b), guess without calculation whether the answer to c) should be larger or smaller than under the original assumptions of the satellite problem. Confirm your guess by calculation.

b) Redo Exercise 16 for this type of component.

4.3 Hazard Rates (Optional)

Let T be a positive random variable with probability density $f(t)$, where t ranges over $(0, \infty)$. Think of T as the lifetime of some kind of component. The *hazard rate* $\lambda(t)$ is the probability per unit time that the component will fail just after time t, *given* that the component has survived up to time t. Thus

$$P(T \in dt | T > t) = \lambda(t)dt$$

where $(T \in dt)$ stands for the event $(t < T \leq t + dt)$ that the component fails in an infinitesimal time interval of length dt just after time t. As usual, this is shorthand for a limit statement:

$$\lambda(t) = \lim_{\Delta t \to 0} \frac{P(T \in (t, t + \Delta t) | T > t)}{\Delta t}$$

Depending on what lifetime T represents in an application, the hazard rate $\lambda(t)$ may also be called a *death rate* or *failure rate*. For example, T might represent the lifetime of some kind of component. Then $\lambda(t)$ would represent the failure rate for components that have been in use for time t, estimated, for example, by the number of failures per hour among similar components in use for time t.

In practice, failure rates can be estimated empirically as suggested above. Often it is found that empirically estimated hazard rates based on large amounts of data tend to follow a smooth curve. It is then reasonable to fit an ideal model in which $\lambda(t)$ would usually be a continuous function of t. The exponential distribution of the previous section is the simplest possible model corresponding to *constant* failure rate $\lambda(t) = \lambda$ for some $\lambda > 0$. Other distributions with densities on $(0, \infty)$ correspond to time-varying failure rates. The following box summarizes the basic terminology and analytic relationships between the probability density, survival function, and hazard rate.

Formulae (1), (2), and (3) in the box are simply definitions, and (4) is the usual integral for the probability of an interval. Formulae (4) and (5) are equivalent by the fundamental theorem of calculus. Informally, (5) results from

$$\begin{aligned} f(t)dt &= P(T \in dt) && \text{by (1)} \\ &= P(T > t) - P(T > t + dt) && \text{by the difference rule} \\ &= G(t) - G(t + dt) && \text{by (2)} \\ &= -dG(t) \end{aligned}$$

Random Lifetimes

Probability density: $$P(T \in dt) = f(t)dt \tag{1}$$

Survival function: $$P(T > t) = G(t) \tag{2}$$

Hazard rate: $$P(T \in dt \,|\, T > t) = \lambda(t)dt \tag{3}$$

Survival from density: $$G(t) = \int_t^\infty f(u)du \tag{4}$$

Density from survival: $$f(t) = -\frac{dG(t)}{dt} \tag{5}$$

Hazard from density and survival: $$\lambda(t) = \frac{f(t)}{G(t)} \tag{6}$$

Survival from hazard: $$G(t) = \exp\left(-\int_0^t \lambda(u)du\right) \tag{7}$$

To obtain (6), use $P(A|B) = P(AB)/P(B)$, with $A = (T \in dt)$, $B = (T > t)$. Since $A \subset B$, $AB = A$,

$$\lambda(t)dt = P(T \in dt \,|\, T > t) = \frac{P(T \in dt)}{P(T > t)} = \frac{f(t)dt}{G(t)}$$

by (1) and (2).

The most interesting formula is (7). To illustrate, in case $\lambda(t) = \lambda$ is constant,

$$\int_0^t \lambda(u)du = \lambda t$$

so (7) becomes the familiar exponential survival probability

$$P(T > t) = e^{-\lambda t} \qquad \text{if } T \text{ has exponential } (\lambda) \text{ distribution.}$$

In general, the exponential of the integral in (7) represents a kind of continuous product obtained as a limit of discrete products of conditional probabilities. This is explained at the end of the section. Formula (7) follows also from (5) and (6) by calculus as you can check as an exercise.

Example 1. Linear failure rate.

Problem. Suppose that a component has linear increasing failure rate, such that after 10 hours the failure rate is 5% per hour, and after 20 hours 10% per hour.

(a) Find the probability that the component survives 20 hours.

(b) Calculate the density of the lifetime distribution.

(c) Find the mean lifetime.

Solution. By assumption,

$$\lambda(t) = (t/2)\% = t/200$$

(a) The required probability is by (7)

$$P(\text{survive 20 hours}) = G(20) = \exp\left(-\int_0^{20} \lambda(u)du\right)$$

The integral inside the exponent is

$$\int_0^{20} \frac{u\,du}{200} = \frac{1}{400}u^2\Big|_0^{20} = 1$$

Thus $P(\text{survive 20 hours}) = e^{-1} \approx 0.368$

(b) Put t instead of 20 above to get

$$G(t) = \exp(-t^2/400)$$

Now by (5)

$$f(t) = -\frac{d}{dt}G(t)$$
$$= \frac{t}{200}\exp\left(-\frac{t^2}{400}\right)$$

You can sketch the density by calculating a few points, as in the following table and graph:

t	0	5	10	15	20
$f(t)$	0	0.023	0.039	0.043	0.037

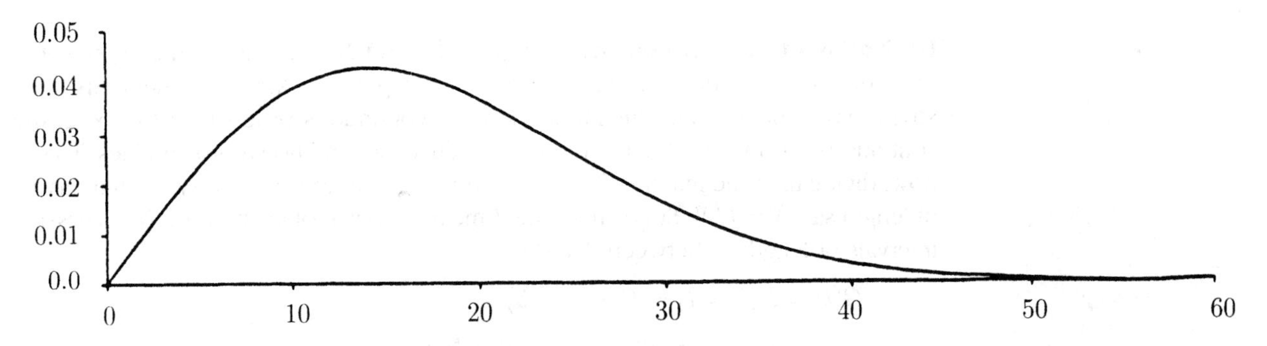

(c) The mean can be calculated from

$$E(T) = \int_0^\infty t f(t)\, dt$$

but there is a shortcut for examples like this where the survival function $G(t)$ is simpler than the density $f(t)$. This is to use the following formula:

Mean Lifetime from Survival Function

$$E(T) = \int_0^\infty G(t)\, dt \tag{8}$$

This follows by integration by parts from the previous formula for $E(T)$, using $\dfrac{dG(t)}{dt} = -f(t)$. It is the continuous analog of the formula

$$E(T) = \sum_{n=1}^{\infty} P(T \geq n)$$

valid for a random variable T with possible values $0, 1, 2, \ldots$. In the present example, (8) gives

$$E(T) = \int_0^\infty \exp(-t^2/400)\, dt \tag{9}$$

Now the problem is that you cannot integrate the function $\exp(-t^2/400)$ in closed form. But you should recognize this integral as similar to the standard Gaussian integral

$$\int_0^\infty e^{-z^2/2}\, dz = \frac{1}{2}\int_{-\infty}^\infty e^{-z^2/2}\, dz = \frac{1}{2}\sqrt{2\pi} = \sqrt{\frac{\pi}{2}}$$

Since $t^2/400 = \dfrac{1}{2}\left(\dfrac{t}{10\sqrt{2}}\right)^2$, make the change of variable $z = t/10\sqrt{2}$, $dz = dt/10\sqrt{2}$, $dt = 10\sqrt{2}dz$ in (9) to get

$$E(T) = 10\sqrt{2}\int_0^\infty e^{-z^2/2}\, dz = 10\sqrt{2}\sqrt{\frac{\pi}{2}} \approx 17.72$$

Derivation of the formula $G(t) = \exp\left(-\int_0^t \lambda(u)\,du\right)$. Recall that the exponential of a sum is the product of exponentials. An integral is a kind of continuous sum, so an exponential of an integral is a kind of continuous product. In this case, the continuous product is a limit of discrete products of conditional probabilities. To see how, divide the time interval $[0, t]$ into a very large number N of very small intervals of length say $\Delta = t/N$. Survival to time t means survival of each of the N successive intervals of length Δ between 0 and t

$$
\begin{aligned}
G(t) &= P(T > t) = P(T > N\Delta) \\
&= P(T > \Delta, T > 2\Delta, \cdots, T > N\Delta) \\
&= P(T > \Delta)P(T > 2\Delta \mid T > \Delta) \cdots P(T > N\Delta \mid T > (N-1)\Delta) \\
&= [1 - P(T \leq \Delta)]\,[1 - P(\Delta \leq T \leq 2\Delta \mid T > \Delta)] \cdots \\
&\approx [1 - \Delta\lambda(0)]\,[1 - \Delta\lambda(\Delta)]\,[1 - \Delta\lambda(2\Delta)] \cdots [1 - \Delta\lambda((N-1)\Delta)] \\
&\qquad \text{for small } \Delta \text{, by the definition of } \lambda(t) \\
&\approx e^{-\Delta\lambda(0)} e^{-\Delta\lambda(\Delta)} \cdots e^{-\Delta\lambda((N-1)\Delta)} \\
&\qquad \text{for small } \Delta \text{, by the approximation } 1 - x \approx e^{-x} \text{ for small } x \\
&= \exp\left[-\Delta \sum_{i=0}^{N-1} \lambda(i\Delta)\right] \\
&\approx \exp\left[-\int_0^t \lambda(u)\,du\right] \\
&\qquad \text{for small } \Delta \text{, by a Riemann sum approximation of the integral.}
\end{aligned}
$$

As $\Delta \to 0$, the errors in each of the three approximations $\approx$ above tend to zero. So the approximate equality between the first and last expressions not involving Δ must in fact be an exact equality. This is (7).

Note how the exponential appears here, as always, as the limit of a product of more and more factors all approaching 1 in the limit.

Exercises 4.3

1. For T with survival function $G(t) = P(T > t)$, find:
   a) $P(T \leq b)$; b) $P(a \leq T \leq b)$.

2. Use the formulae of this section to show that the hazard rate $\lambda(t)$ is constant if and only if the distribution is exponential (λ) for some λ.

3. Business enterprises have the feature that the longer an enterprise has been in business, the less likely it is to fail in the next month. This indicates a decreasing failure rate. One that has been successfully fitted to empirical data of lifetimes of businesses is $\lambda(t) = a/(b + t)$, where a, b, and t are greater than 0. For this $\lambda(t)$:
   a) find a formula for $G(t)$; b) find a formula for $f(t)$.

4. **Weibull distribution.** Show that the following are equivalent:

   (i) $\lambda(t) = \lambda\alpha t^{\alpha-1}$ for constants $\lambda > 0$ and $\alpha > 0$

   (ii) $G(t) = e^{-\lambda t^{\alpha}}$

   (iii) $f(t) = \lambda\alpha t^{\alpha-1}e^{-\lambda t^{\alpha}}$

   This is called the *Weibull* distribution with parameters λ and α. This family of distributions is widely used in engineering practice. It can be verified both theoretically and practically that the distribution of the lifetime of a component which consists of many parts, and fails when the first of these parts fails, can be well approximated by a Weibull distribution.

5. **Moments of the Weibull distribution.** Let T have the Weibull distribution described in Exercise 4. a) Show that $E(T^k) = \Gamma(1 + \frac{k}{\alpha})\,\lambda^{-\frac{k}{\alpha}}$ b) Find $E(T)$ and $Var(T)$.

6. Suppose that a component is subject to failure at constant rate 5% per hour for the first 10 hours in use. After 10 hours the component is subject to additional stress producing a failure rate of 10% per hour.

   a) Find the probability that the component survives 15 hours.

   b) Calculate and sketch the survival probability function.

   c) Calculate and sketch the probability density function.

   d) Find the mean lifetime.

7. **Second moment from survival function.**

   a) Show that $E(T^2) = 2\int_0^{\infty} tG(t)\,dt$

   b) Use this formula to calculate the SD of the component in Example 1.

   c) If 100 components of this type operate independently, what approximately is the probability that the average lifetime of these components exceeds 20 hours?

8. Suppose the failure rate is $\lambda(t) = at + b$ for $t \geq 0$.

   a) For what parameter values a and b does this make sense?

   b) Find the formula for $G(t)$. c) Find the formula for $f(t)$.

   d) Find the mean lifetime. e) Find the SD of the lifetime.

9. **Calculus derivation of** $G(t) = \exp\{-\int_0^t \lambda(u)du\}$ (Formula (7)).

   a) Use (5) and (6) to show $\lambda(t) = -\frac{d}{dt}\log G(t)$.

   b) Now derive (7) by integration from 0 to t.

10. Suppose a component has failure rate $\lambda(t)$ which is an increasing function of t.

    a) For $s, t > 0$, is $P(T > s + t|T > s)$ larger or smaller than $P(T > t)$?

    b) Prove your answer.

    c) Repeat a) and b) for $\lambda(t)$ which is decreasing.

4.4 Change of Variable

Many problems require finding the distribution of some function of X, say $Y = g(X)$, from the distribution of X. Suppose X has density $f_X(x)$, where a subscript is now used to distinguish densities of different random variables. Then provided the function $y = g(x)$ has a derivative dy/dx which does not equal zero on any interval in the range of X, the random variable $Y = g(X)$ has a density $f_Y(y)$ which can be calculated in terms of $f_X(x)$ and the derivative dy/dx. How to do this calculation is the subject of this section.

Linear Functions

To see why the derivative comes in, look first at what happens if you make a *linear* change of variable. For a linear function $y = ax + b$, the derivative is the constant $dy/dx = a$. The function stretches or shrinks the length of every interval by the same factor of $|a|$.

Example 1. **Uniform distributions.**

Suppose X has the uniform $(0, 1)$ distribution, with density

$$f_X(x) = \begin{cases} 1 & 0 < x < 1 \\ 0 & \text{otherwise} \end{cases}$$

Then for $a > 0$, you can see that $Y = aX + b$ has the uniform $(b, b + a)$ density

$$f_Y(y) = \begin{cases} 1/a, & b < y < b + a \\ 0 & \text{otherwise} \end{cases}$$

Similarly, if $a < 0$, then $Y = aX + b$ has the uniform $(b + a, b)$ distribution

$$f_Y(y) = \begin{cases} 1/|a|, & b + a < y < b \\ 0 & \text{otherwise} \end{cases}$$

You might guess the density of $Y = aX + b$ at y was the density of X at the corresponding point $x = (y - b)/a$. But this must be divided by $|a|$, because the probability density gives probability per unit length, and the transformation from x to $ax + b$ multiplies lengths by a factor of $|a|$:

Linear Change of Variable for Densities

$$f_{aX+b}(y) = \frac{1}{|a|} f_X\left(\frac{y - b}{a}\right)$$

FIGURE 1. **Linear change of variable for uniform densities.** The graphs show the densities of $Y = aX + b$ for various a and b, where X has uniform $(0, 1)$ distribution. Notice how if $a > 1$ the range is spread out and the density decreased. And if $0 < a < 1$ the range is shrunk and the density increased. Adding $b > 0$ shifts to the right by b, and adding $b < 0$ shifts to the left by $-b$.

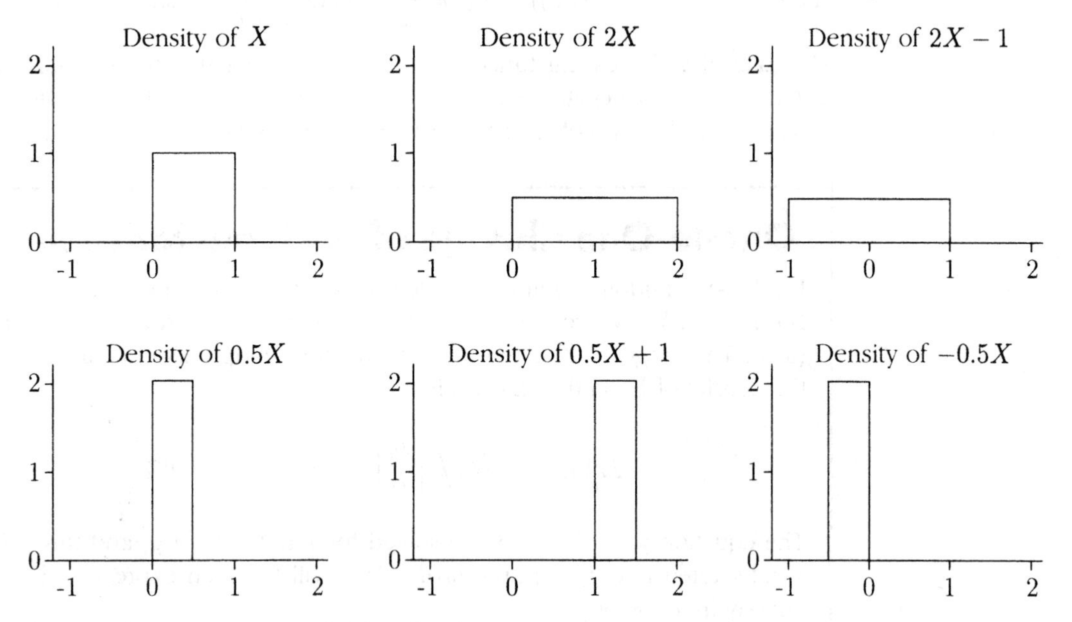

Example 2. **Normal distributions.**

Take X with standard normal density $\phi(x)$, $a = \sigma > 0$, and $b = \mu$. The linear change of variable formula then gives the density of the normal (μ, σ^2) distribution, displayed on page 267.

One-to-One Differentiable Functions

Let X be a random variable with density $f_X(x)$ on the range (a, b). Let $Y = g(X)$ where g is either strictly increasing or strictly decreasing on (a, b). For example, X might have an exponential distribution on $(0, \infty)$, and Y might be X^2, $\sqrt{X}$, or $1/X$. The range of Y is then an interval with endpoints $g(a)$ and $g(b)$.

The aim now is to calculate the probability density function $f_Y(y)$ for y in the range of Y. For an infinitesimal interval dy near y, the event $(Y \in dy)$ is identical to the event $(X \in dx)$, where dx is an infinitesimal interval near the unique x such that $y = g(x)$. See Figure 2, where each of the two shaded areas represents the probability of the same event

$$P(Y \in dy) = P(X \in dx) \quad \text{where} \quad y = g(x)$$

This identity $P(Y \in dy) = P(X \in dx)$, where $y = g(x)$, makes

$$f_Y(y)dy = f_X(x)dx$$

and so

$$f_Y(y) = f_X(x)\frac{dx}{dy} = f_X(x)\Big/\frac{dy}{dx} \quad \text{where} \quad y = g(x)$$

The case of a decreasing function g is similar except that the calculus derivative dy/dx now has a negative sign. This sign must be ignored because it is only the *magnitude* of the ratio of lengths of small intervals which is relevant. To summarize:

One-to-One Change of Variable for Densities

Let X be a random variable with density $f_X(x)$ on the range (a, b).
Let $Y = g(X)$ where g is either strictly increasing or strictly decreasing on (a, b). The range of Y is then an interval with endpoints $g(a)$ and $g(b)$. And the density of Y on this interval is

$$f_Y(y) = f_X(x)\Big/\left|\frac{dy}{dx}\right| \qquad \text{where} \quad y = g(x)$$

The equation $y = g(x)$ must be solved for x in terms of y, and this value of x substituted into $f_X(x)$ and dy/dx. This will leave an expression for $f_Y(y)$ entirely in terms of y.

Example 3. Square root of an exponential variable (illustrated by Figure 2)

Problem. Let X have the exponential density, $f_X(x) = e^{-x} \qquad (x > 0)$
Find the density of $Y = \sqrt{X}$.

Solution. **Step 1.** Find the range of y: here $0 < x < \infty$, $y = \sqrt{x}$, so $0 < y < \infty$.

Step 2. Check the function is one-to-one by solving for x in terms of y: here $x = y^2$

Step 3. Calculate $\frac{dy}{dx}$: here $\frac{dy}{dx} = \frac{d}{dx}\sqrt{x} = \frac{1}{2\sqrt{x}}$

Step 4. Plug density of X and the result of Step 3 into $f_Y(y) = f_X(x)\Big/\left|\frac{dy}{dx}\right|$:

$$f_Y(y) = e^{-x}\Big/\frac{1}{2\sqrt{x}}$$

Step 5. Use result of Step 2 to eliminate x from the right side

$$f_Y(y) = e^{-y^2}\Big/\frac{1}{2\sqrt{y^2}} = 2ye^{-y^2} \qquad (y > 0)$$

FIGURE 2. Change of variable formula for densities. The diagram shows the graph of $y = g(x)$ for the increasing function $g(x) = \sqrt{x},\ x > 0$. Density $f_X(x)$ is graphed upside down below the x-axis. Density $f_Y(y)$ is graphed on the side of the y-axis. The densities are as in Example 3.

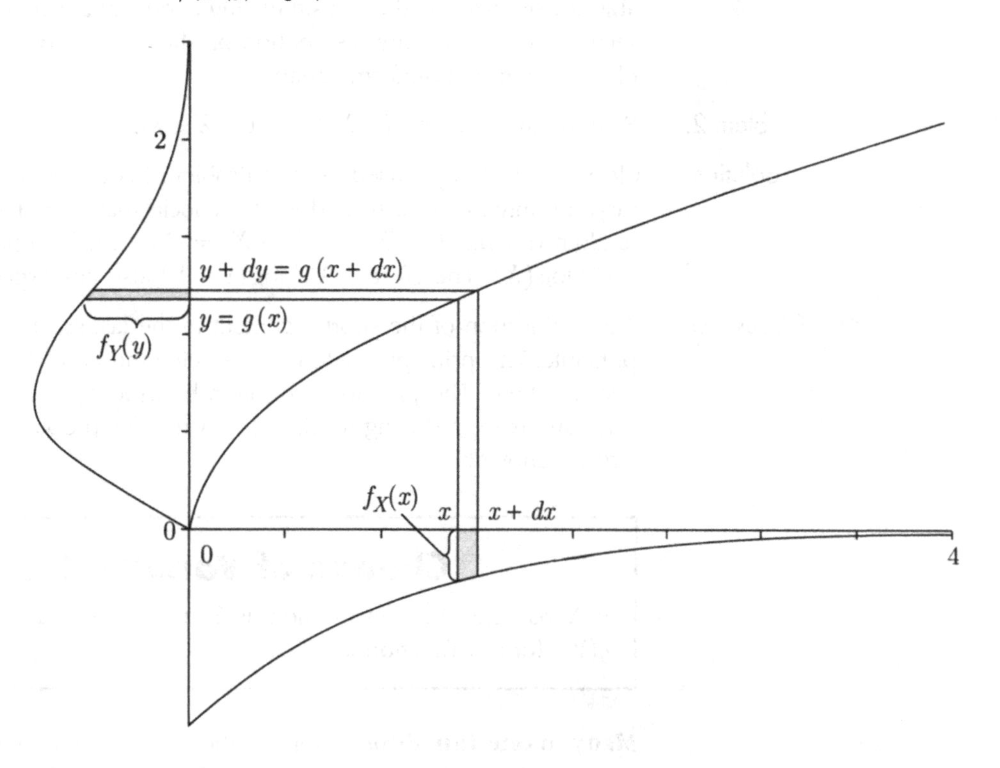

Example 4. **Log of uniform.**

Let X have uniform $(0, 1)$ distribution.

Problem 1. Find the distribution of $Y = -\lambda^{-1}\log(X)$, where $\lambda > 0$.

Solution. This follows the steps of the previous example in a slightly different order. Here $y = -\lambda^{-1}\log x$ has

$$\frac{dy}{dx} = -\frac{1}{\lambda x} < 0 \quad \text{for} \quad 0 < x < 1$$

so y decreases from ∞ to 0 as x increases from 0 to 1. The density of Y is then

$$f_Y(y) = f_X(x) \Big/ \left|\frac{dy}{dx}\right| = 1 \Big/ \frac{1}{\lambda x} = \lambda x$$

where $\quad -\lambda^{-1}\log x = y, \quad$ or $\quad x = e^{-\lambda y}$, so

$$f_Y(y) = \lambda e^{-\lambda y} \qquad (y > 0)$$

Conclusion: Y is exponentially distributed with rate λ.

Discussion. This way of obtaining an exponential variable as a function of a uniform $(0,1)$ variable is a standard method of simulating exponential variables by computer. The next section shows how any distribution on the line can be obtained as the distribution of a function of a uniform variable.

Problem 2. Find the distribution of $-\lambda^{-1}\log(1-X)$, where $\lambda > 0$.

Solution. Clearly the technique used to solve Problem 1 could be repeated. But this is unnecessary. It is intuitively clear (and easy to check) that $X' = 1-X$ is also a uniform $(0,1)$ random variable, so $-\lambda^{-1}\log(1-X) = -\lambda^{-1}\log(X')$ has the same distribution as $-\lambda^{-1}\log(X)$. Therefore, $-\lambda^{-1}\log(1-X)$ also has exponential (λ) distribution.

Discussion. The justification of the short argument in the last solution is the change of variable principle. This principle, stated for discrete random variables in Section 3.1, is worth restating here. The principle can often be used as in the last example to eliminate calculations by reducing a change of variables problem to one whose solution is already known:

Change of Variable Principle

If X has the same distribution as Y, then $g(X)$ has the same distribution as $g(Y)$, for any function g.

Many-to-one functions. Suppose the function $y = g(x)$ has a derivative that is zero at only a finite number of points. Now some values of y may come from more than one value of x. Consider $Y = g(X)$ for a random variable X. As shown in the diagram, Y will be in an infinitesimal interval dy near y when X is in one of possibly several infinitesimal intervals dx near points x such that $g(x) = y$.

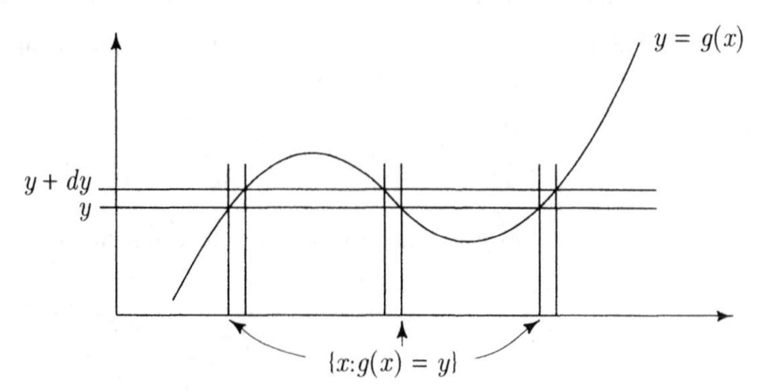

Now

$$P(Y \in dy) = \sum_{\{x:g(x)=y\}} P(X \in dx)$$

This gives

$$f_Y(y) = \sum_{\{x:\, g(x)=y\}} f_X(x) \Big/ \left|\frac{dy}{dx}\right|$$

Example 5. Density of the square of a random variable.

Problem. Suppose X has density $f_X(x)$. Find a formula for the density of $Y = X^2$.

Solution. Here, for $y > 0$, there are two values x such that $x^2 = y$, namely, $x = \sqrt{y}$ and $x = -\sqrt{y}$. Since $dy/dx = 2x$,

$$f_Y(y) = \sum_{\{x=\pm\sqrt{y}\}} f_X(x)/|2x|$$
$$= [f_X(\sqrt{y}) + f_X(-\sqrt{y})]/2\sqrt{y}.$$

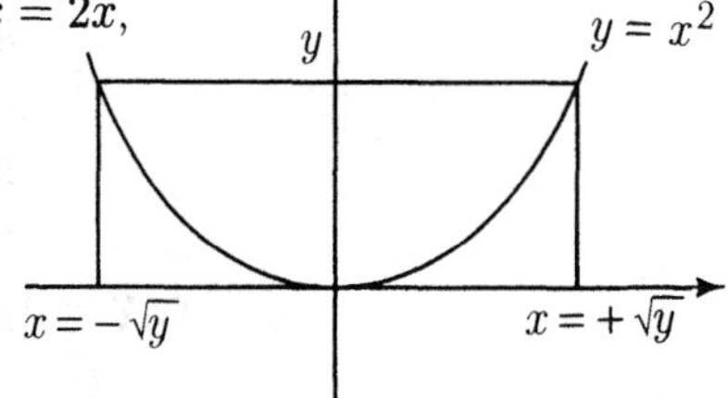

Expectation of a function of X. If you just want to calculate the expectation of $Y = g(X)$, it is not necessary to calculate the density of Y, and usually simpler not to. For instance, there is no need to use the linear change of variable formula for densities to calculate $E(Y)$ or $SD(Y)$ for $Y = aX+b$. Instead use the simple scaling rules

$$E(aX + b) = aE(X) + b \qquad \text{and} \qquad SD(aX + b) = |a|SD(X)$$

whenever $E(X)$ or $SD(X)$ are defined. More generally, if $Y = g(X)$, where both X and Y have densities, then

$$E(Y) = \int_{-\infty}^{\infty} y f_Y(y)\, dy = \int_{-\infty}^{\infty} g(x) f_X(x)\, dx$$

Often the second integral is easier to evaluate than the first. The equality of the two integrals is the density analog of the basic discrete formula for the expectation of a function of X that was derived in Section 4.1. The equality of integrals can also be checked by the calculus technique of substitution

$$y = g(x), \qquad dy = g'(x)dx.$$

Further Examples

Here are some more geometric problems solved by the same basic technique of finding the probability in an infinitesimal interval by calculus.

Example 6. Projection of a uniform random variable on a circle.

A point is picked uniformly at random from the perimeter of a unit circle.

Problem 1. Find the probability density of X, the x-coordinate of the point.

Solution. From the diagram, since two places on the circle map to one x-value,

$$P(X \in dx) = 2|d\theta|/2\pi = |d\theta|/\pi$$

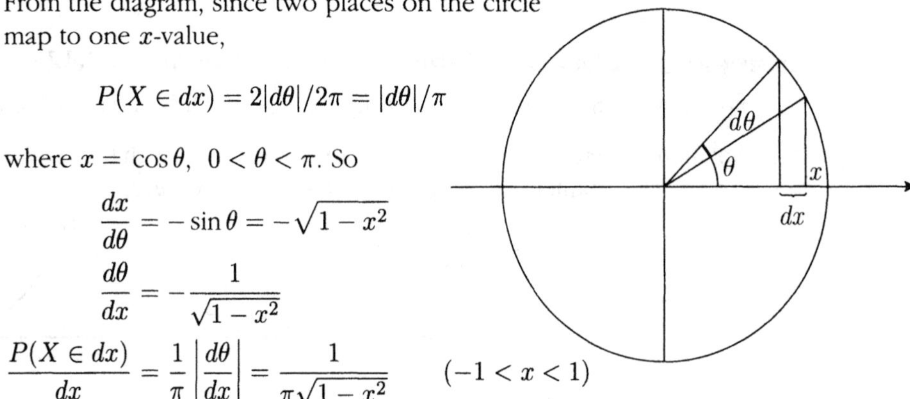

where $x = \cos\theta$, $0 < \theta < \pi$. So

$$\frac{dx}{d\theta} = -\sin\theta = -\sqrt{1-x^2}$$

$$\frac{d\theta}{dx} = -\frac{1}{\sqrt{1-x^2}}$$

$$\frac{P(X \in dx)}{dx} = \frac{1}{\pi}\left|\frac{d\theta}{dx}\right| = \frac{1}{\pi\sqrt{1-x^2}} \qquad (-1 < x < 1)$$

Problem 2. Find $E(X)$.

Solution. Easily, $E(X) = 0$, since the density of X is symmetric about 0.

Problem 3. Find the probability density of $Y = |X|$, the absolute value of X.

Solution. Since two x values $+y$ and $-y$, with the same probability density, map to any given value of y with $0 < y < 1$, $P(Y \in dy) = 2 \times P(X \in dy)$, and so

$$f_Y(y) = \frac{2}{\pi\sqrt{1-y^2}} \qquad (0 < y < 1)$$

Problem 4. Find $E(Y)$.

Solution. $E(Y) = \frac{2}{\pi}\int_0^1 \frac{y}{\sqrt{1-y^2}}dy = -\frac{2}{\pi}\sqrt{1-y^2}\Big|_0^1 = \frac{2}{\pi}$

Example 7. Projection of a uniform random variable on a sphere.

Let Θ be the latitude, between $-\pi/2$ and $\pi/2$, of a point chosen uniformly at random on the surface of a unit sphere.

Problem 1. Find the probability density of Θ.

Solution. From the diagram:

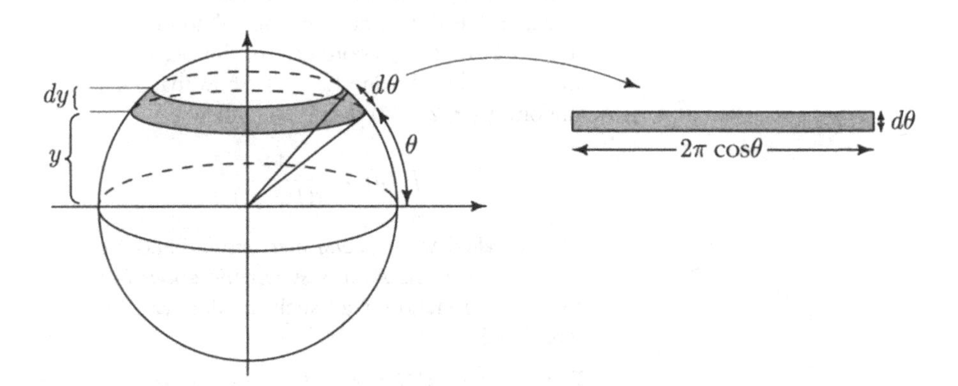

$$P(\Theta \in d\theta) = \frac{\text{Indicated Area}}{\text{Total Surface Area}} = \frac{2\pi\cos\theta d\theta}{4\pi}$$

$$f_\Theta(\theta) = \frac{\cos\theta}{2} \qquad (-\frac{\pi}{2} < \theta < \frac{\pi}{2})$$

Problem 2. Let Y be the vertical coordinate of the point on the sphere, between -1 and 1. Find the probability density of Y.

Solution. $P(Y \in dy) = P(\Theta \in d\theta)$ with $y = \sin\theta$, which implies that $dy = \cos\theta\, d\theta$ and

$$P(Y \in dy) = P(\Theta \in d\theta) = f_\Theta(\theta)d\theta = \frac{\cos\theta\, d\theta}{2} = \frac{dy}{2} \qquad (-1 < y < 1)$$

Conclusion: Y has uniform $(-1, 1)$ distribution.

Discussion. This calculation shows that the surface area of the sphere between two parallel planes cutting the sphere depends only on the distance between the planes, and not on exactly how they cut the sphere. This fact was discovered by Archimedes. The formula $4\pi r^2$ for the total surface area, used in Problem 1, is a consequence.

Exercises 4.4

1. Suppose X has an exponential (λ) distribution. What is the distribution of cX for a constant $c > 0$?

2. **Scaling of gamma distributions.** Show that a random variable T has gamma (r, λ) distribution, if and only if $T = T_1/\lambda$, where T_1 has gamma $(r, 1)$ distribution.

3. Suppose U has uniform $(0, 1)$ distribution. Find the density of U^2.

4. Suppose X has uniform distribution on $(-1, 1)$. Find the density of $Y = X^2$.

5. Suppose X has uniform $[-1, 2]$ distribution. Find the density of X^2.

6. Cauchy distribution. Suppose that a particle is fired from the origin in the (x, y)-plane in a straight line in a direction at random angle Φ to the x-axis, and let Y be the y-coordinate of the place where the particle hits the line $\{x = 1\}$. Show that if Φ has uniform $(-\pi/2, \pi/2)$ distribution, then

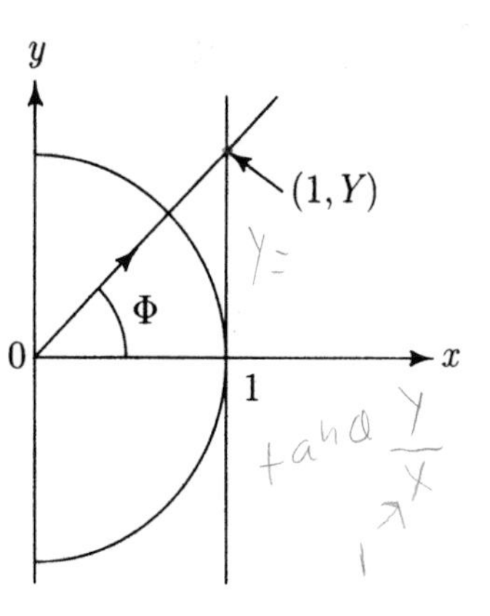

$$f_Y(y) = \frac{1}{\pi(1 + y^2)}$$

This is called the *Cauchy distribution*. Show that the Cauchy distribution is symmetric about 0, but that the expectation of a Cauchy random variable is undefined.

7. Show that if U has uniform $(0, 1)$ distribution, then $\tan(\pi U - \frac{\pi}{2})$ has the Cauchy distribution, as in Exercise 6.

8. Arcsine distribution. Suppose that Y has the Cauchy distribution as in Exercise 6. Let $Z = 1/(1 + Y^2)$.

a) Show Z has density

$$f_Z(z) = \frac{1}{\pi\sqrt{z(1 - z)}} \qquad (0 < z < 1)$$

b) Show $P(Z \le x) = (2/\pi)\arcsin(\sqrt{x}) \quad (0 < x < 1)$.

c) Find $E(Z)$. d) Find $Var(Z)$.

[This *arcsine distribution* of Z is the special case $r = s = 1/2$ of the beta(r, s) distribution. This distribution arises naturally in the context of random walks. If $S_n = X_1 + \cdots + X_n$ for X_i with values ± 1 determined by tosses of a fair coin, and L_n is the last time $k \le n$ such that $S_k = 0$, then the limit distribution of L_n/n as $n \to \infty$ is the arcsine distribution. See Feller, *An Introduction to Probability Theory and Its Applications*, Vol. I.]

9. Weibull distribution.

a) Show that if T has the Weibull (λ, α) distribution, with density

$$f(t) = \lambda \alpha t^{\alpha - 1} e^{-\lambda t^\alpha} \qquad (t > 0)$$

where $\lambda > 0$ and $\alpha > 0$, then T^α has an exponential (λ) distribution. (Note the special case when $\alpha = 1$.)

b) Show that if U is a uniform $(0, 1)$ random variable, then $T = (-\lambda^{-1} \log U)^{\frac{1}{\alpha}}$ has a Weibull (λ, α) distribution.

10. Let Z be a standard normal random variable. Find formulae for the densities of each of the following random variables:

a) $|Z|$; b) Z^2; c) $1/Z$; d) $1/Z^2$.

11. Explain how the calculations of Example 7 imply the formula $4\pi r^2$ for the surface area of a sphere of radius r.

4.5 Cumulative Distribution Functions

One way to specify a probability distribution on the line is to say how much probability is at or to the left of each point x. In terms of a random variable X with the given distribution, this probability is a function of x,

$$F(x) = P(X \leq x)$$

called the *cumulative distribution function (c.d.f.)* of X. For example, the standard normal c.d.f. is the function $F(x) = \Phi(x)$ used in calculations with the normal distribution. But the cumulative distribution function can be defined for any distribution of a random variable X over the line, whether continuous, discrete, or neither.

If you can define or calculate the c.d.f. of X then, by using the rules of probability, you can find the probability of any event determined by X, for example, the probability that X falls in an interval, or the probability that X is an even integer. To clarify terminology, the *distribution* of X refers broadly to the assignment of probabilities to all such events determined by X. Technically, this means probabilities defined for a collection of subsets of the line, satisfying the rules of probability, now including the infinite sum rule of Section 3.4. The c.d.f. just gives the probabilities of the intervals $(-\infty, x]$ as a function of the point x.

Interval probabilities. The formula $P(a < X \leq b) = F(b) - F(a)$, a consequence of the difference rule for probabilities, is familiar from the special case of the standard normal c.d.f. Because probabilities must be non-negative, this shows that a c.d.f. $F(x)$ must be a nondecreasing function of x

FIGURE 1. Graph of a continuous c.d.f

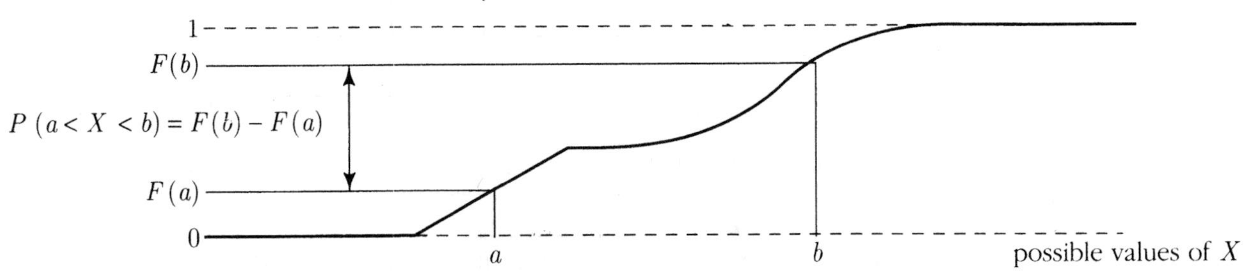

The distribution is called *continuous* if the c.d.f. is a continuous function. Then it can be shown that

$$P(X = x) = 0 \quad \text{for all } x$$

so it makes no difference in formulae involving the c.d.f. whether inequalities are strict or weak. For example, using the rule of complements,

$$P(X > x) = 1 - F(x) \qquad \text{whatever the distribution of } X$$

$$P(X \geq x) = 1 - F(x) \qquad \text{if the distribution of } X \text{ is continuous}$$

More generally, it can be shown that the c.d.f. determines the probability of every interval, and also the probability of more complicated sets by the addition rule. To summarize:

> A probability distribution over the line is completely determined by its c.d.f.

Most distributions of practical interest are either discrete or defined by densities. These two cases will now be discussed in more detail.

Discrete Case

Here is an illustration:

FIGURE 2. Individual probabilities and the c.d.f. for an indicator variable. Consider the c.d.f. of an indicator variable X which is 0 with probability 0.3 and 1 with probability 0.7. The value of $F(x)$ is 0 for $x < 0$ because there is no chance for $X \leq x$ for a negative x. The value of $F(x)$ is 0.3 for $0 \leq x < 1$, because for such an x the event $(X \leq x)$ is the same as the event $(X = 0)$, which has probability 0.3. And the value of $F(x)$ is 1 for $1 \leq x < \infty$, because for these x the event $(X \leq x)$ is certain. Thus $F(x)$ jumps by $0.3 = P(0)$ at $x = 0$ and by $1 - 0.3 = 0.7 = P(1)$ at $x = 1$.

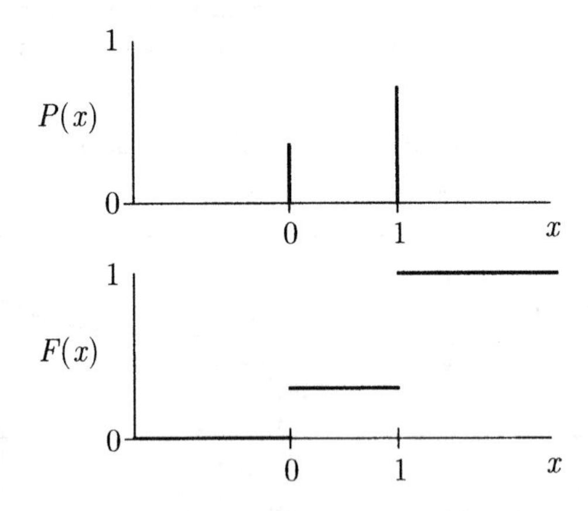

In general, the c.d.f. of a discrete random variable X looks like a staircase with a rise of $P(x) = P(X = x)$ at each possible value x of X:

$$F(x) = \sum_{y \leq x} P(y)$$

and $P(x)$ is the jump of the c.d.f. at x:

$$P(x) = F(x) - F(x-)$$

where $F(x-) = P(X < x)$ is the limit of values of F approaching x from the left. Figure 3 gives a more interesting example.

FIGURE 3. **The c.d.f. and individual probabilities for the binomial $(100, 0.5)$ distribution.** Here $F(x)$ is the probability of getting x or less heads in 100 fair coin tosses, $P(x)$ is the probability of exactly x heads. The value of $F(x)$ is simply the sum of values $P(y)$ over all integers y less than or equal to x. Each integer x introduces a new term $P(x)$ into the sum. Thus the graph of F jumps by $P(x)$ at each integer x, and is flat between. Put another way, the probability $P(x)$ of an individual value x shows the *difference* between $F(x)$ and $F(x-)$, where $F(x-) = F(x-1)$ is the value of $F(y)$ for any y in the interval $[x-1, x)$.

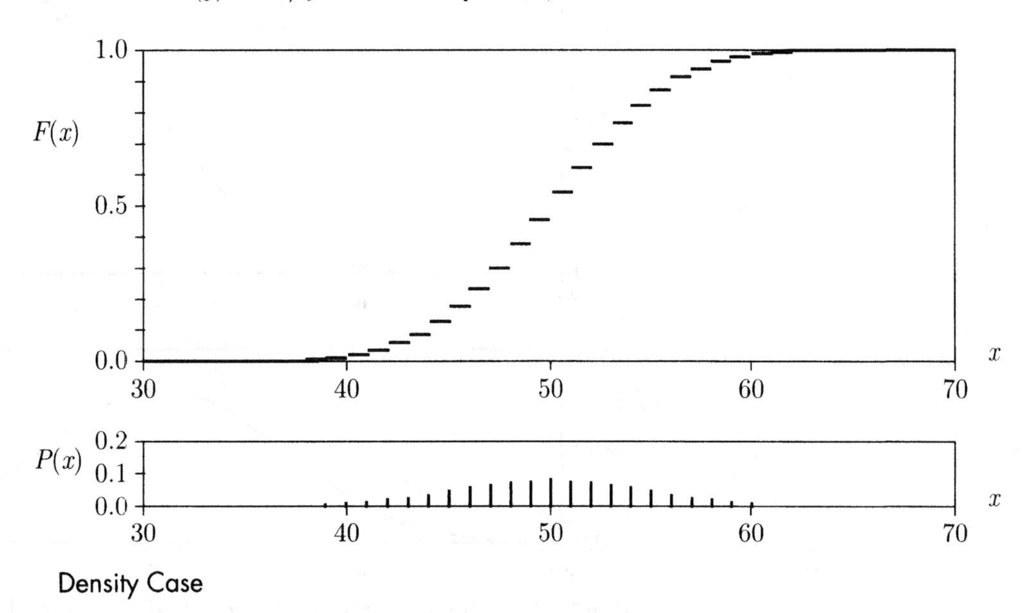

Density Case

As usual in this case, sums become integrals. So if X has density $f(x)$, then $F(x)$ is the area under the density function to the left of x

$$F(x) = P(X \leq x) = \int_{-\infty}^{x} f(y)dy$$

Similarly, discrete differences become derivatives,

$$dF(x) = F(x + dx) - F(x) = P(X \in dx) = f(x)dx$$

$$\text{so} \qquad f(x) = \frac{dF(x)}{dx} = F'(x)$$

That is to say, the density $f(x)$ is the slope at x of the c.d.f. This is an instance of the fundamental theorem of calculus. Conversely, it can be shown that if the c.d.f. is everywhere continuous, and differentiable at all except at perhaps a finite number of points, then the corresponding distribution has density $f(x) = F'(x)$. In this density case, $F(x)$ is a particular choice of an indefinite integral of $f(x)$, namely, the one which vanishes at $-\infty$.

FIGURE 4. The c.d.f. and density for the normal $(50, 25)$ distribution. This distribution, with mean 50 and variance 25, is the usual normal approximation to the preceding binomial distribution. Its c.d.f. and density are just scale changes of the standard normal ones plotted in Section 2.2.

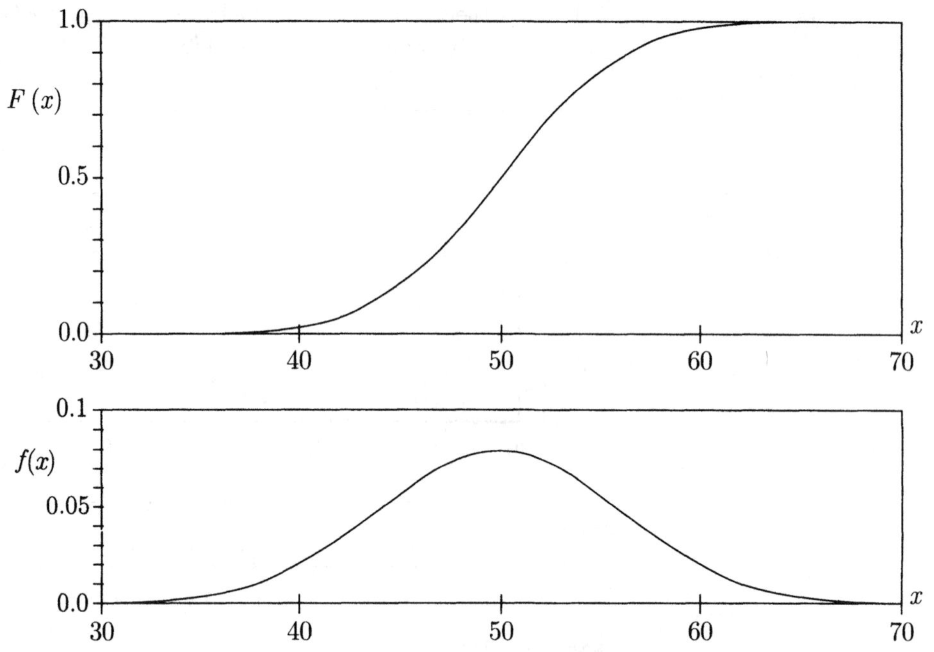

A distribution with a density can be specified by a formula for the density $f(x)$, or by a formula for the c.d.f. $F(x)$. Either of these functions can be obtained from the other by calculus.

You might think that every continuous distribution has a density, but this turns out not to be so. Still, you don't have to worry about continuous distributions without densities in this course. The famous mathematician Poincare thought such distributions "were invented by mathematicians to confound their ancestors". For a nice picture of one, see Mandelbrot's book, *The Fractal Geometry of Nature*.

Example 1. **The uniform (0, 1) distribution.**

The density is

$$f(x) = \begin{cases} 1 & \text{for } 0 < x < 1 \\ 0 & \text{otherwise} \end{cases}$$

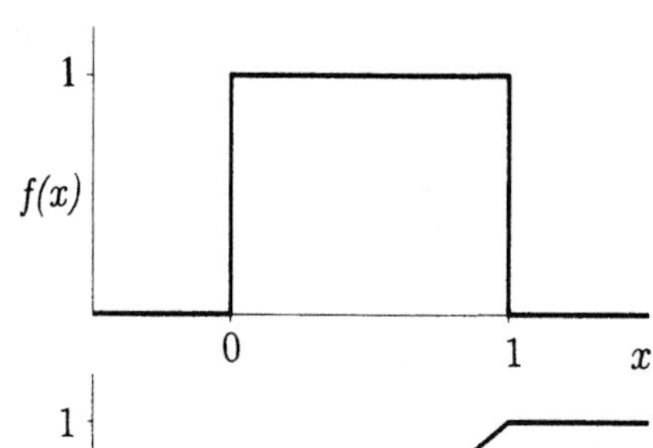

and the c.d.f. is

$$F(x) = \begin{cases} x & \text{for } 0 \le x \le 1 \\ 0 & \text{for } x < 0 \\ 1 & \text{for } x > 1 \end{cases}$$

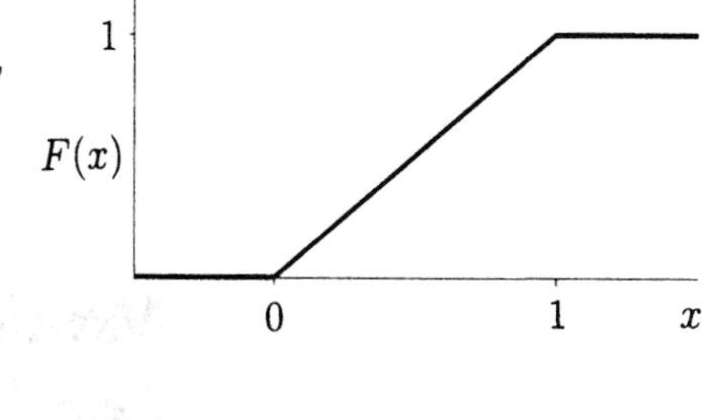

Here is an application: If U is uniform $(0, 1)$, then so is $X = 2\left|U - \frac{1}{2}\right|$, because

$$\begin{aligned} P(X \le x) &= P\left(2\left|U - \frac{1}{2}\right| \le x\right) \\ &= P\left(\frac{1}{2} - \frac{x}{2} \le U \le \frac{1}{2} + \frac{x}{2}\right) \\ &= F(x) \end{aligned}$$

as defined above. This technique is an alternative to the method of the previous section for calculating the distribution of a function of a random variable.

Example 2. **Uniform on a disc.**

Let (X, Y) be a point chosen uniformly at random from the unit disc $\{(x, y) : x^2 + y^2 \le 1\}$. Calculate the c.d.f. and density function of X.

Solution. It is easiest to find the density function first. Suppose $|x| \le 1$. The event $(X \in dx)$ is shaded in the diagram. For small dx the event in question is approximately a rectangle with height $2\sqrt{1 - x^2}$ and width dx. Dividing by the total area π gives its probability, then dividing by dx gives the density

$$f(x) = \begin{cases} \frac{2}{\pi}\sqrt{1 - x^2} & |x| \le 1 \\ 0 & \text{otherwise} \end{cases}$$

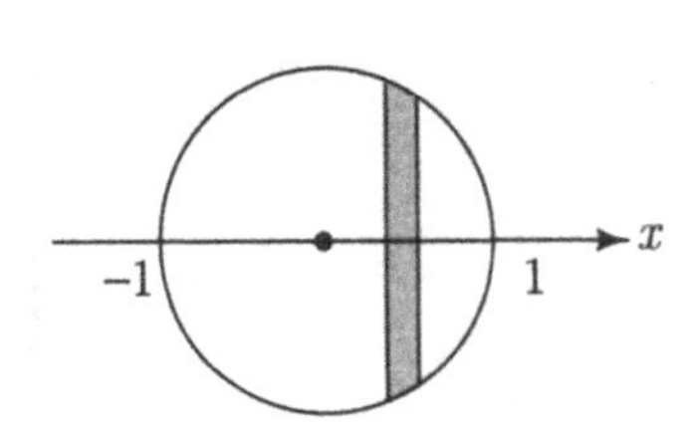

as graphed on the right. This is half an ellipse obtained by rescaling the upper semi-circle. The c.d.f. $F(x)$, which represents the relative area of the disc to the left of x, is now obtained by calculus

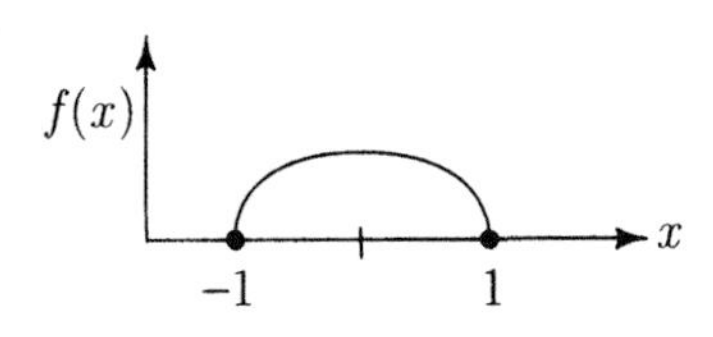

$$F(x) = \int_{-1}^{x} f(z)\,dz = \frac{1}{\pi}\int_{-1}^{x} 2\sqrt{1 - z^2}\,dz$$

This is not a very easy integral. Still, because $F(x)$ has derivative $f(x)$ which you know, and $F(x)$ is 0 for $x \leq -1$ and 1 for $x \geq 1$, you should be able to sketch the graph of $F(x)$ and see it must have the shape shown below. Some more calculus (or consulting a table of integrals) gives

$$F(x) = \frac{1}{2} + \frac{1}{\pi}\left[x\sqrt{(1-x^2)} + \arcsin x\right] \qquad (|x| \leq 1).$$

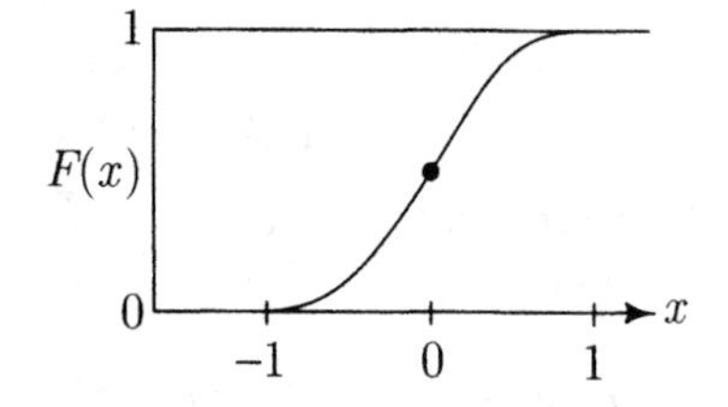

Maximum and Minimum of Independent Random Variables

Cumulative distribution functions make it easy to find the distribution of the maximum and minimum

$$X_{\text{max}} = \max(X_1, \ldots, X_n) \qquad \text{and} \qquad X_{\text{min}} = \min(X_1, \ldots, X_n)$$

of a collection of independent random variables $X_1, X_2, \ldots, X_n$. Let F_i denote the c.d.f. of X_i, $i = 1, \ldots, n$. The c.d.f. of either the maximum or the minimum of the X's can be written in terms of the individual distribution functions F_i, once you notice the following key facts:

> For any number x:
>
> (a) X_{max} is less than or equal to x if and only if all the X's are less than or equal to x;
>
> (b) X_{min} is greater than x if and only if all the X's are greater than x.

The c.d.f. of the maximum is then

$$\begin{aligned}
F_{\text{max}}(x) &= P(X_{\text{max}} \leq x) \qquad (-\infty < x < \infty) && \text{by definition} \\
&= P(X_1 \leq x, X_2 \leq x, \ldots, X_n \leq x) && \text{by (a)} \\
&= P(X_1 \leq x)P(X_2 \leq x) \cdots P(X_n \leq x) && \text{by independence} \\
&= F_1(x)F_2(x) \cdots F_n(x)
\end{aligned}$$

The c.d.f. of the minimum is

$$\begin{aligned} F_{\text{min}}(x) &= P(X_{\text{min}} \le x) \quad (-\infty < x < \infty) \\ &= 1 - P(X_{\text{min}} > x) \\ &= 1 - P(X_1 > x, X_2 > x, \ldots, X_n > x) \qquad \text{by (b)} \\ &= 1 - (1 - F_1(x))(1 - F_2(x)) \cdots (1 - F_n(x)). \end{aligned}$$

It is best not to try and memorize these formulae. Just remember (a) and (b), and derive the formulae when you need them.

Example 3. **Minimum of independent exponential variables is exponential.**

Let $X_1, X_2, \ldots, X_n$ be independent random variables, and suppose X_i has exponential distribution with rate λ_i, $i = 1, \ldots, n$.

Problem. Find the distribution of X_{min} the minimum of $X_1, \ldots, X_n$.

Solution. For $i = 1, \ldots, n$, the c.d.f. of X_i is

$$F_i(x) = \begin{cases} 0 & \text{if } x < 0 \\ 1 - e^{-\lambda_i x} & \text{if } x \ge 0 \end{cases}$$

Since the X's are non-negative, so is their minimum. So X_{min} has c.d.f.

$$F_{\text{min}}(x) = 0 \qquad (x < 0)$$

For $x \ge 0$,

$$\begin{aligned} F_{\text{min}}(x) &= 1 - e^{-\lambda_1 x} e^{-\lambda_2 x} \cdots e^{-\lambda_n x} \\ &= 1 - e^{-(\lambda_1 + \lambda_2 + \cdots + \lambda_n)x} \end{aligned}$$

This is the c.d.f. of the exponential distribution with rate $\lambda_1 + \lambda_2 + \cdots + \lambda_n$. So the minimum of independent exponential variables with rates λ_i is simply a new exponential variable with rate the sum of the rates λ_i.

Example 4. **Expected lifetime of a circuit.**

An electrical circuit consists of five components, connected as in the following diagram. The lifetimes of the components, measured in days, have independent exponential distributions with rates indicated in the diagram.

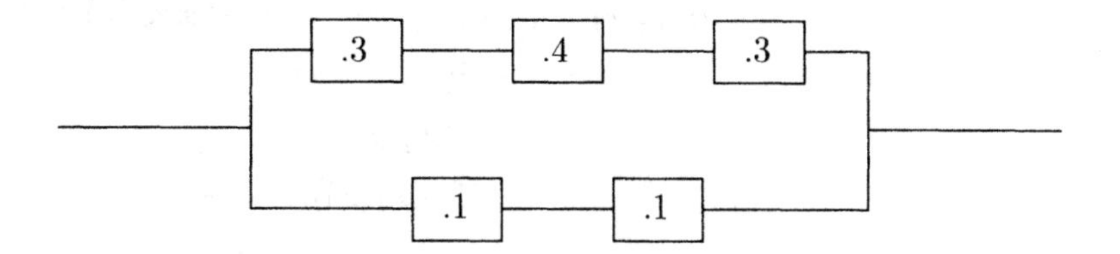

Problem. What is the expected lifetime of the circuit?

Solution. We want $E(L)$, where L denotes the lifetime of the circuit. Let L_{top} and L_{bottom} denote the lifetimes of the top and bottom parts of the circuit. Then L_{top} and L_{bottom} are independent, and

$$L = \max(L_{\text{top}}, L_{\text{bottom}})$$

since the top and bottom parts are linked in parallel.

Now L_{top} is the minimum of three independent exponential lifetimes, since the top consists of three components linked in series. By Example 3, L_{top} has exponential distribution with rate $0.3 + 0.4 + 0.3 = 1$. So the top is expected to last about 1 day. By a similar argument, L_{bottom} has exponential (0.2) distribution, so the bottom is expected to last about $1/0.2 = 5$ days.

Since L is the maximum of L_{top} and L_{bottom}, its c.d.f. is

$$F_L(x) = \begin{cases} 0 & x < 0 \\ (1 - e^{-x})(1 - e^{-0.2x}) & x \geq 0 \end{cases}$$

Since L is a positive random variable

$$E(L) = \int_0^\infty (1 - F_L(x))dx$$

(See Exercise 9 .) For $x \geq 0$,

$$F_L(x) = 1 - e^{-x} - e^{-0.2x} + e^{-1.2x}$$
$$1 - F_L(x) = e^{-x} + e^{-0.2x} - e^{-1.2x}$$

so

$$E(L) = \int_0^\infty (e^{-x} + e^{-0.2x} - e^{-1.2x})dx = 1 + (1/0.2) - (1/1.2) = 5.17$$

So the circuit is expected to last about 5.17 days.

Note. Once you have the c.d.f. of L, you can, of course, compute its expectation by first differentiating to find the density, then using the density to find the expectation by integration. But that involves more work than the method used here.

Suppose now that in addition to being independent, the X's are continuous random variables with the same density. For example, the X's could be a sequence of random numbers produced by a uniform random number generator. Let f denote the common density function of the X's, and F the common c.d.f. The maximum X_{max} and minimum X_{min} are also continuous random variables, whose densities can be obtained by differentiating their c.d.f.'s

$$F_{\text{max}}(x) = (F(x))^n \qquad (-\infty < x < \infty)$$

$$f_{\text{max}}(x) = \frac{d}{dx}(F(x))^n = n\,(F(x))^{n-1}\,f(x) \qquad (-\infty < x < \infty)$$

by the chain rule of calculus. Similarly,

$$F_{\text{min}}(x) = 1 - (1 - F(x))^n \qquad (-\infty < x < \infty)$$

$$f_{\text{min}}(x) = n\,(1 - F(x))^{n-1}\,f(x) \qquad (-\infty < x < \infty)$$

These densities can also be found more directly by a differential calculation explained in the next section.

Percentiles and the Inverse Distribution Function

Given a distribution of X and a value x, the c.d.f. $F(x)$ gives the probability that X is less than or equal to x. Often the question gets turned around. For instance: For what value of x is there probability $1/2$ that X is less than or equal to x? Such an x is a *median* of the distribution. More generally, given a probability p, for what x is $P(X \leq x) = p$? By definition of the c.d.f. this x must solve the equation

$$F(x) = p$$

In the case of $F(x)$ given by a formula, the formula can usually be rearranged to express x in terms of p. In general, assuming this equation has a unique solution, as it does for most continuous distributions of interest and $0 < p < 1$, the solution of this equation defines the *inverse c.d.f.*

$$x = F^{-1}(p)$$

FIGURE 5. Relation between a c.d.f. and its inverse.

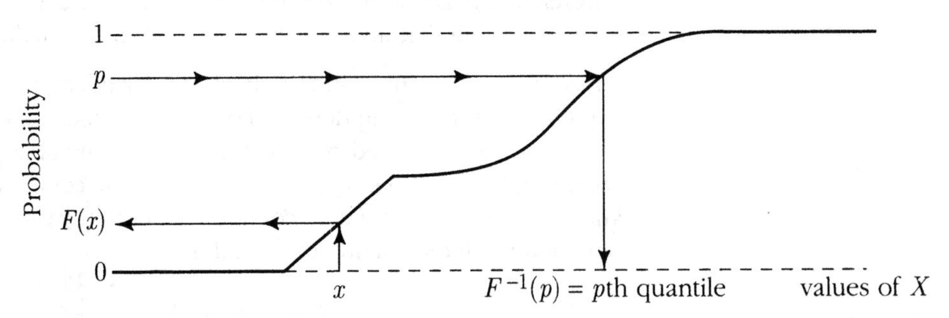

See Figure 5. This point x, such that $P(X \leq x) = p$, is called the pth *quantile* of the distribution of X. This term is a generalization of the more common *quartile, decile,* and *percentile* in case p is expressed as a multiple of $1/4$, $1/10$, or $1/100$.

Example 5. Finding percentiles.

Problem 1. For the exponential (λ) distribution, find a formula for the pth quantile, $0 < p < 1$.

Solution. Since the c.d.f. is $F(x) = 1 - e^{-\lambda x}$ for $x > 0$, the required point x is found from

$$1 - e^{-\lambda x} = p \qquad \text{so} \qquad x = -\frac{1}{\lambda}\log(1-p)$$

Problem 2. Find the 75th percentile point of the standard normal distribution.

Solution. This is $\Phi^{-1}(0.75)$ where Φ is the standard normal c.d.f. Just as there is no simple formula for Φ, there is none for Φ^{-1}. But numerical values of Φ^{-1} are easily found by backwards lookup in the table of values of Φ. Inspection of the table gives $\Phi(0.67) = 0.7486$ and $\Phi(0.68) = 0.7517$, so $\Phi^{-1}(0.75) \approx 0.675$.

Simulation via Inverse Distribution Function

Given a distribution on the line, how can you create random variables with this distribution? This problem arises in computer simulation of random variables. The random number generator on a computer provides a sequence of numbers between 0 and 1, say $U_1, U_2, \ldots$, which behaves in most respects like a sequence of independent uniform $(0,1)$ random variables. For example, the long-run proportion of values U_i in any subinterval of $[0,1]$ will be very close to the length of the subinterval. How can these variables be transformed into a sequence simulating independent random variables with some other distribution? The problem is to find a function g such that if U has uniform $(0,1)$ distribution, then $X = g(U)$ has a prescribed c.d.f., say $F(x)$:

$$P(g(U) \leq x) = F(x) \quad \text{for all } x$$

There are many ways to solve this problem by tricks depending on the desired distribution. Which method is best depends on considerations such as computational efficiency, not discussed here. One method will now be described which works no matter what the required distribution. Here is a simple example to illustrate the method.

Example 6. Simulating a binomial (2, 0.5) random variable.

The left graph shows the required c.d.f. The right graph shows a function g from $(0, 1)$ to $\{0, 1, 2\}$. This graph should be read on its side as a kind of inversion of the graph of the c.d.f. The staircase is the same in both graphs. Imagine U picked at random from the vertical unit interval. Then $g(U) \in \{0, 1, 2\}$ has the required distribution.

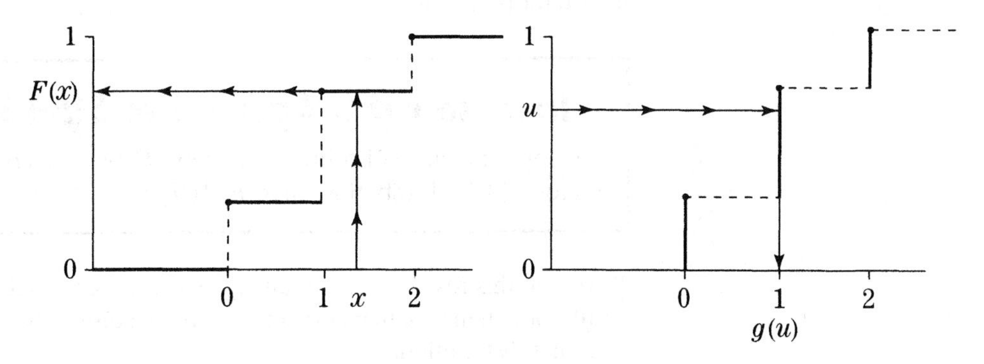

In detail, as it would be programmed on a computer, the rule for getting from the uniform $(0, 1)$ variable U to the binomial $(2, 0.5)$ variable $g(U)$ is

$$\begin{array}{lll} \text{if } 0 \leq U \leq 0.25 & \text{then} & g(U) = 0 \\ \text{if } 0.25 < U \leq 0.75 & \text{then} & g(U) = 1 \\ \text{if } 0.75 < U \leq 1.0 & \text{then} & g(U) = 2 \end{array}$$

This g does the job because by construction the intervals on which g takes the values 0, 1, and 2 have lengths 0.25, 0.5, and 0.25, respectively, as required by the binomial $(2, 0.5)$ distribution.

Simulation of a discrete distribution. The method of the previous example generalizes easily to any discrete distribution. For example, to get a random variable with discrete distribution on $1, 2, \ldots$ defined by probabilities $p_1, p_2, \ldots$ define

$$g(u) = k \quad \text{if} \quad p_1 + \cdots + p_{k-1} < u \leq p_1 + \cdots + p_{k-1} + p_k$$

Then if U has uniform $(0, 1)$ distribution

$$P(g(U) = k) = P(p_1 + \cdots + p_{k-1} < U \leq p_1 + \cdots + p_{k-1} + p_k) = p_k$$

since this is the length of the interval of U-values that make $g(U) = k$. This means $g(U)$ has the given discrete distribution.

The inverse distribution function. The function $g(u)$ defined in the discrete case above is always a kind of inverse of the c.d.f. $F(x)$, in the sense that

$$g(F(x)) = x \quad \text{for all possible values } x$$

Check this inverse relation in the example above for $x = 0, 1, 2$. Given any c.d.f. F, not necessarily continuous or strictly increasing, a function g satisfying the above inverse relation can be defined. Because of the inverse relation, $g(u)$ is usually denoted $F^{-1}(u)$, and called the *inverse c.d.f.* In general, the inverse c.d.f. $F^{-1}(u)$ can be defined as the least value x such that $F(x) \geq u$. This function has the following important property:

Inverse c.d.f. Applied to Standard Uniform

For any cumulative distribution function F, with inverse function F^{-1}, if U has uniform $(0, 1)$ distribution, then $F^{-1}(U)$ has c.d.f. F.

To restate this result more intuitively, if you pick a percentage uniformly at random on $(0, 100)$, then take that percentile point in a distribution, you get a random variable with that distribution.

Proof. The discrete case has already been treated. The continuous case is more interesting. Assume, for simplicity, that $F(x)$ is a continuous and strictly increasing function of x. Then $F^{-1}(u)$ is the usual inverse function of $F(x)$, as discussed earlier, and

$$w \leq x \iff F(w) \leq F(x)$$

The events $(F^{-1}(U) \leq x)$ and $(F(F^{-1}(U)) \leq F(x))$ are therefore identical. But since $F(F^{-1}(u)) = u$ for every u in $(0, 1)$, by definition of the inverse function, we can calculate

$$\begin{aligned} P(F^{-1}(U) \leq x) &= P(F(F^{-1}(U)) \leq F(x)) \\ &= P(U \leq F(x)) \\ &= F(x) \quad \text{from the c.d.f. of } U \end{aligned}$$

Thus the random variable $F^{-1}(U)$ has c.d.f. F. □

The method of generating random variables via F^{-1} is efficient computationally in simulations only if F^{-1} turns out to be a fairly simple function to compute, as it is for the uniform distribution on (c, d) for any $c < d$, or the exponential distribution. But F^{-1} is laborious to compute for the normal distribution. In this case it is quicker and nearly as accurate to approximate using the central limit theorem, using, for instance, a standardized sum of 12 independent uniform $(0, 1)$ variables. See also Exercise 5.3.13 for another method of generating normal variables from uniform ones.

Exercises 4.5

1. For the exponential (λ) distribution:

a) Show the c.d.f. is $F(x) = 1 - e^{-\lambda x}$ for $x \geq 0$. b) Sketch this c.d.f. for $\lambda = 1$.

2. Find and sketch the cumulative distribution functions of:

a) the binomial $(3, 1/2)$ distribution;

b) the geometric $(1/2)$ distribution on $\{1, 2, \ldots\}$.

3. Let (X, Y) be as in Example 2.

a) Find f_Y and F_Y. [*Hint*: No calculations required!]

b) Let $R = \sqrt{X^2 + Y^2}$. Sketch the event $\{R \leq r\}$ as a subset of the circle. Deduce a formula for the c.d.f. of R, and check by differentiating that you get the same density for R as in Example 4.1.2.

4. Let X be a random variable with c.d.f. $F(x)$. Find the c.d.f. of $aX + b$ first for $a > 0$, then for $a < 0$.

5. Find the c.d.f. of X with density function $f_X(x) = \frac{1}{2}e^{-|x|}$ $(-\infty < x < \infty)$.

6. Let X be a random variable with c.d.f. $F(x) = x^3$ for $0 \leq x \leq 1$. Find:

a) $P(X \geq \frac{1}{2})$; b) the density function $f(x)$; c) $E(X)$.

d) Let Y_1, Y_2, Y_3 be three points chosen independently and uniformly on the unit interval, and let X be the rightmost point. Show that X has the distribution described above.

7. Let T have the exponential distribution with parameter λ, and let $Y = \sqrt{T}$.

a) Find the density of Y.

b) Find the expectation of Y, correct to two decimal places, for $\lambda = 3$.

c) A random number generator produces uniform $[0, 1]$ random numbers. How could you use these to generate random numbers which have the distribution of Y?

8. Components in the following series-parallel systems have independent exponentially distributed lifetimes. Component i has mean lifetime μ_i. In each case, find a formula for the probability that the system operates for at least t units of time, and sketch the graph of this function of t in case $\mu_i = i$ for each i.

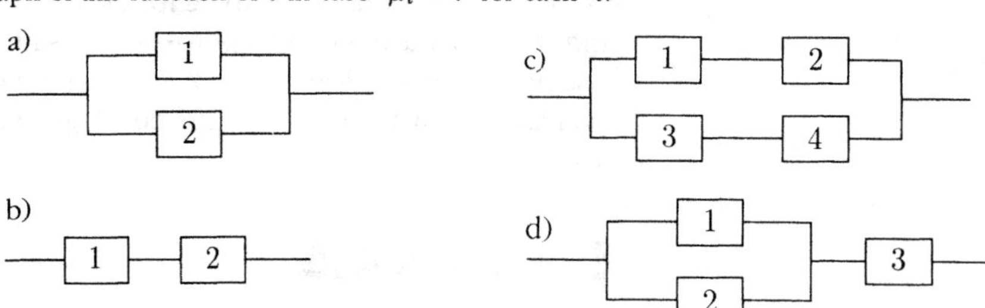

9. Expectation from c.d.f. Let X be a positive random variable, with c.d.f. F, as in the following diagram for example:

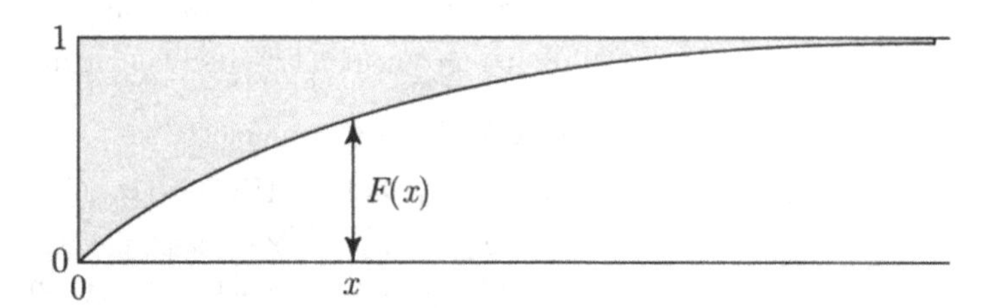

a) Show, using the representation $X = F^{-1}(U)$ for a uniform $[0, 1]$ random variable U, that $E(X)$ can be interpreted as the shaded area above the c.d.f. of X, both for X with a density, and for discrete X. Deduce that

$$E(X) = \int_0^\infty [1 - F(x)]\, dx = \int_0^\infty P(X > x)\, dx$$

b) Deduce that if X has possible values $0, 1, 2, \ldots$, then $E(X) = \sum_{n=1}^{\infty} P(X \geq n)$.

c) Use these formulae to rederive the means of the exponential and geometric distributions.

d) Show that for a random variable X with both positive and negative values (either discrete or with a density), $E(X) = E(X_+) - E(X_-)$ where $X_+ = XI(X > 0)$, and $X_- = (-X)I(X < 0)$, so $E(X)$ is area $(+)$ minus area $(-)$ defined in terms of the c.d.f. as indicated below:

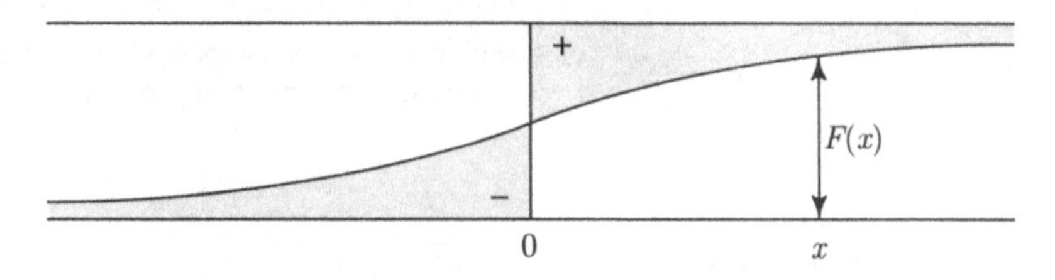

4.6 Order Statistics (Optional)

Let $X_1, X_2, \ldots, X_n$ be random variables. Let $X_{(1)}$ denote the smallest of the X's, $X_{(2)}$ the next smallest, and so on, so that

$$X_{(1)} \leq X_{(2)} \leq \cdots \leq X_{(n)}$$

This relabeling of the X's corresponds to arranging them in increasing order, as shown below, for one particular ordering of five values $X_1, \ldots, X_5$.

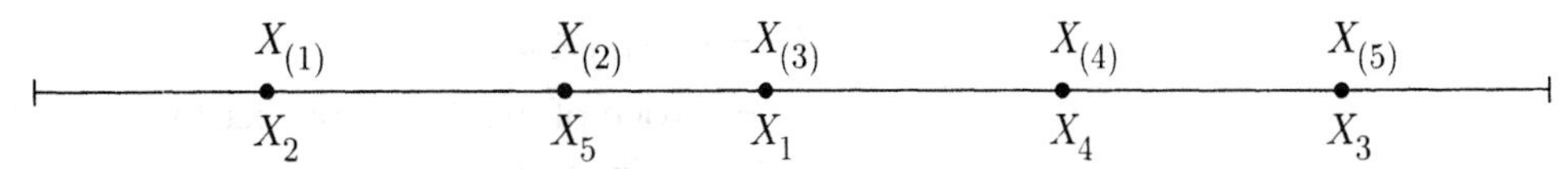

Notice that

$$X_{(1)} = \min(X_1, \ldots, X_n)$$

$$X_{(n)} = \max(X_1, \ldots, X_n)$$

In general, $X_{(k)}$ is called the kth *order statistic* of $X_1, \ldots, X_n$.

This section deals with properties of order statistics of independent and identically distributed random variables. Beta distributions appear as the distributions of order statistics of independent uniform $(0, 1)$ random variables.

Let $X_1, X_2, \ldots, X_n$ be independent random variables, all with the same density function f and cumulative distribution function F. For example, the X's could be a sequence of random numbers produced by a uniform random number generator. The object is to find a formula for the density of the kth order statistic $X_{(k)}$. This has been done already in Section 4.5 in the case of the maximum $X_{(n)}$ and minimum $X_{(1)}$ by first finding the c.d.f., then differentiating. But here is another argument in these special cases which generalizes more easily. First of all, it can be shown that in a sequence $X_1, \ldots, X_n$ of independent continuous random variables, all n values are distinct with probability 1. Taking this for granted, here is a calculation of the density of the maximum $X_{(n)}$

$$\begin{aligned}
f_{(n)}(x)dx &= P(X_{(n)} \in dx) \\
&= P(\text{one of the } X\text{'s} \in dx, \text{ all others} < x) \\
&= P(X_1 \in dx, \text{ all others} < x) + P(X_2 \in dx, \text{ all others} < x) \\
&\qquad + \cdots + P(X_n \in dx, \text{ all others} < x) \\
&= nP(X_1 \in dx, \text{ all others} < x) \qquad \text{by symmetry} \\
&= nP(X_1 \in dx)P(\text{all others} < x) \qquad \text{by independence} \\
&= nf(x)dx\,(F(x))^{n-1}
\end{aligned}$$

in agreement with the previous calculation in Section 4.5. Similarly,

$$\begin{aligned} f_{(1)}(x)\,dx &= P(X_{(1)} \in dx) \\ &= P(\text{one of the } X\text{'s} \in dx, \text{ all others} > x) \\ &= nf(x)\,dx\,(1 - F(x))^{n-1} \end{aligned}$$

The same method can be used to derive a formula for the density of the kth order statistic of $X_1, \ldots, X_n$. Recall that $X_{(k)}$ is the kth smallest of $X_1, \ldots, X_n$. The density $f_{(k)}(x)$ of $X_{(k)}$ is found as follows. For $-\infty < x < \infty$

$$\begin{aligned} f_{(k)}(x)dx &= P(X_{(k)} \in dx) \\ &= P(\text{one of the } X\text{'s} \in dx, \text{ exactly } k-1 \text{ of the others} < x) \\ &= nP(X_1 \in dx, \text{ exactly } k-1 \text{ of the others} < x) \\ &= nP(X_1 \in dx)P(\text{exactly } k-1 \text{ of the others} < x) \\ &= nf(x)dx\binom{n-1}{k-1}(F(x))^{k-1}(1-F(x))^{n-k} \end{aligned}$$

using the binomial formula. To summarize:

Density of the kth Order Statistic

Let $X_{(k)}$ denote the kth order statistic of $X_1, X_2, \ldots, X_n$, where $X_1, \ldots, X_n$ are independent, identically distributed random variables with common density f and c.d.f. F. The density of $X_{(k)}$ is given by

$$f_{(k)}(x) = nf(x)\binom{n-1}{k-1}(F(x))^{k-1}(1-F(x))^{n-k} \qquad (-\infty < x < \infty)$$

It is best not to memorize the formula, but to remember how it is derived.

Order Statistics of Uniform Random Variables

Let $X_1, \ldots, X_n$ be independent random variables each with uniform distribution on $(0, 1)$. The common density of the X's is

$$f(x) = \begin{cases} 1 & 0 < x < 1 \\ 0 & \text{otherwise} \end{cases}$$

Their common c.d.f. is

$$F(x) = \begin{cases} 0 & x < 0 \\ x & 0 \le x \le 1 \\ 1 & x > 1 \end{cases}$$

By the boxed formula above, the density of the kth order statistic of the n uniform random variables is

$$f_{(k)}(x) = \begin{cases} n\binom{n-1}{k-1}x^{k-1}(1-x)^{n-k} & 0 < x < 1 \\ 0 & \text{otherwise} \end{cases}$$

Some of these densities are graphed in Figure 1 on the next page.

Notice how as n increases, the density for the minimum gets more concentrated near 0, the density for the maximum gets more concentrated near 1, and the density for the middle value of the X's gets more concentrated near $1/2$. This is what you would expect intuitively.

Notice also the functional form of the density: a constant, times x raised to a power, times $1 - x$ raised to a power. This simple form for a density on $(0, 1)$ appears in many settings. Here is a general definition:

Beta (r,s) Distribution

For $r, s > 0$, the *beta* (r, s) distribution on $(0, 1)$ is defined by the density

$$\frac{1}{B(r,s)}x^{r-1}(1-x)^{s-1} \qquad (0 < x < 1)$$

where

$$B(r,s) = \int_0^1 x^{r-1}(1-x)^{s-1}dx$$

is the normalizing constant which makes the density integrate to 1.

Viewed as a function of r and s, $B(r, s)$ is called the *beta function*.

A comparison of the last two boxes shows the following:

Beta Distribution of Uniform Order Statistics

The kth order statistic of n independent uniform $(0, 1)$ random variables has beta $(k, n - k + 1)$ distribution.

A nice corollary of the formula for the density of $X_{(k)}$ derived above is that for integers r and s, the beta function $B(r, s)$ is evaluated. Since $f_{(k)}$ is a density it must integrate to 1 over $[0, 1]$. So

$$\int_0^1 x^{k-1}(1-x)^{n-k}dx = \frac{1}{n\binom{n-1}{k-1}} = \frac{(k-1)!(n-k)!}{n!}$$

FIGURE 1. Densities of order statistics of independent uniform variables. For $n = 1, 2, \ldots, 6$ and $k = 1, 2, \ldots, n$, the density of the kth order statistic of n independent uniform $(0, 1)$ random variables, which is the beta density with parameters k and $n - k + 1$, is plotted as the kth graph in the nth row of the diagram.

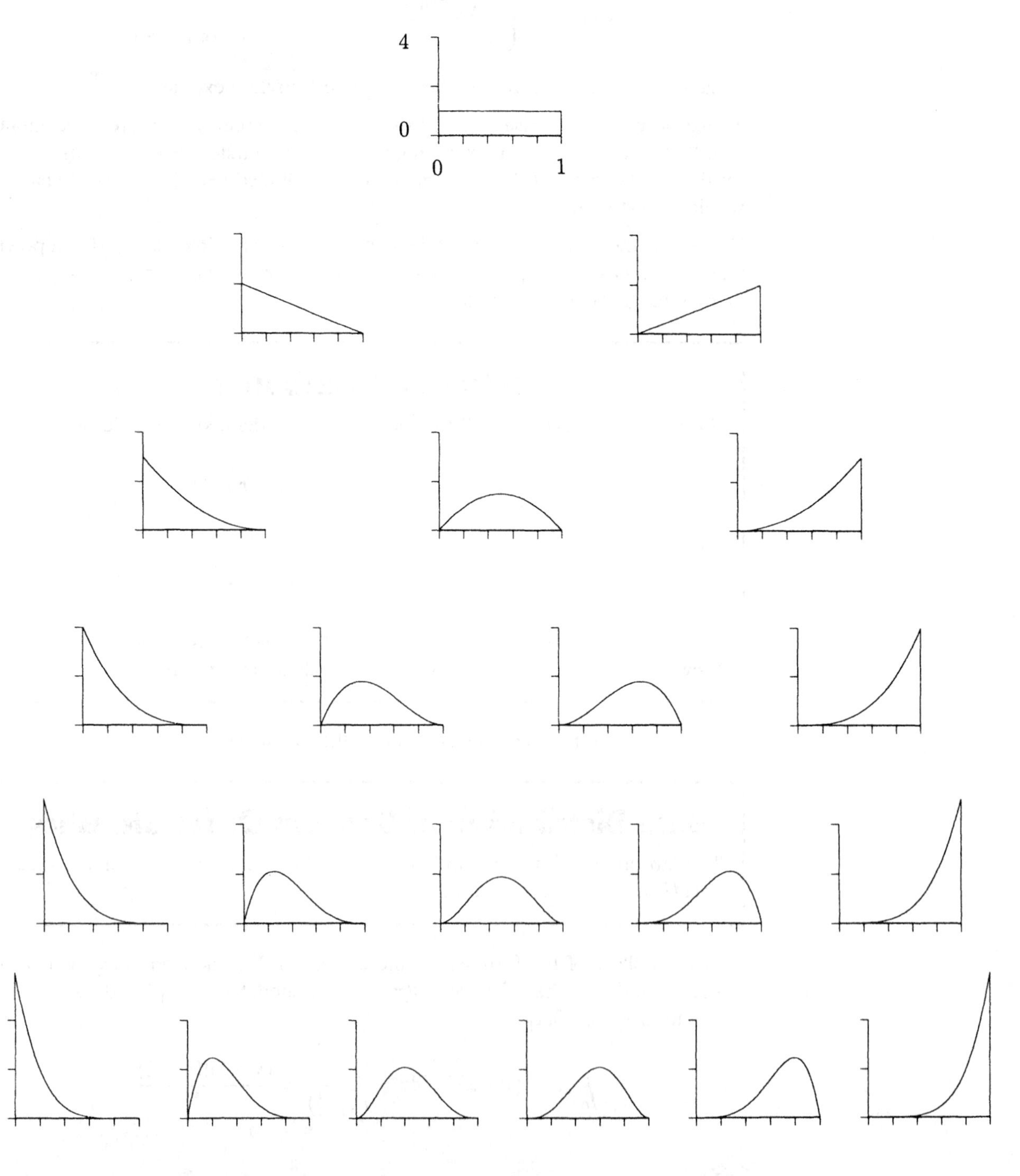

Substitute $r = k$ and $s = n-k+1$ and recall that $\Gamma(r) = (r-1)!$ for positive integers r to get the following result for integers r and s:

Evaluation of the Beta Integral

For positive r and s

$$B(r,s) = \int_0^1 x^{r-1}(1-x)^{s-1}dx = \frac{\Gamma(r)\Gamma(s)}{\Gamma(r+s)}$$

The beta (r,s) distribution is defined, and the above evaluation of the beta integral is valid, for all positive r and s, not necessarily integers. See Section 5.4, especially Exercise 5.4.19 for a proof of this and explanation of the connection between the beta and gamma distributions.

Moments of the beta distribution. The expectation and variance of a beta random variable with integer parameters are now easy to calculate. If X has beta distribution with positive integer parameters r and s,

$$\begin{aligned}
E(X) &= \int_0^1 x \cdot \frac{1}{B(r,s)} x^{r-1}(1-x)^{s-1}dx \\
&= \frac{1}{B(r,s)} \int_0^1 x^{(r+1)-1}(1-x)^{s-1}dx \\
&= \frac{B(r+1,s)}{B(r,s)} \\
&= \frac{r!(s-1)!}{(r+s)!} \cdot \frac{(r+s-1)!}{(r-1)!(s-1)!} \\
&= \frac{r}{r+s}
\end{aligned}$$

$E(X^2)$ can be calculated in the same way, and used to find a formula for the variance of X. This is left as an exercise.

The kth order statistic of n independent uniform $(0,1)$ random variables has beta distribution with parameters k and $n-k+1$, so

$$E(X_{(k)}) = \frac{k}{n+1}$$

Thus the smallest of four uniform random numbers is expected to be around $1/5$, the next smallest around $2/5$, the third smallest around $3/5$, and the largest around

4/5. In other words, if you think of picking four points at random from $[0, 1]$ as cutting the interval into five pieces

all the pieces are expected to have the same length. In fact, more is true: It can be shown that when an interval is split at random like this by any number of independent uniform random points, the length of each piece has the same beta distribution as the length of the first piece. See Chapter 6 Review Exercise 32.

Exercises 4.6

1. Four people agree to meet at a cafe at noon. Suppose each person arrives at a time normally distributed with mean 12 noon and SD 5 minutes, independently of all the others.

a) What is the chance that the first person to arrive at the cafe gets there before 11:50?

b) What is the chance that some of the four have still not arrived at 12:15?

c) Approximately what is the chance that the second person to arrive gets there within ten seconds of noon?

2. Let X have beta (r, s) distribution.

a) Find $E(X^2)$, and use the formula for $E(X)$ given in this section to find $Var(X)$.

b) Find a formula for $E(X^k)$, for integers $k \geq 1$.

3. Let $U_{(1)}, \ldots, U_{(n)}$ be the values of n independent uniform $(0, 1)$ variables arranged in increasing order. Let $0 \leq x < y \leq 1$. Find simple formulae for:

a) $P(U_{(1)} > x \text{ and } U_{(n)} < y)$; b) $P(U_{(1)} > x \text{ and } U_{(n)} > y)$;

c) $P(U_{(1)} < x \text{ and } U_{(n)} < y)$; d) $P(U_{(1)} < x \text{ and } U_{(n)} > y)$;

e) $P(U_{(k)} < x \text{ and } U_{(k+1)} > y)$ for $1 \leq k \leq n - 1$;

f) $P(U_{(k)} < x \text{ and } U_{(k+2)} > y)$ for $1 \leq k \leq n - 2$.

4. Let $X = \min(S, T)$ and $Y = \max(S, T)$ for independent random variables S and T with a common density f. Let Z denote the indicator of the event $S < T$.

a) What is the distribution of Z?

b) Are X and Z independent? Are Y and Z independent? Are (X, Y) and Z independent?

c) How can these conclusions be extended to the order statistics of three or more independent random variables with the same distribution?

5. C.d.f. of the beta distribution for integer parameters.

a) Let $X_1, X_2, \ldots, X_n$ be independent uniform $(0, 1)$ random variables, and let $X_{(k)}$ be the kth order statistic of the X's. Find the c.d.f. of $X_{(k)}$ by expressing the event $X_{(k)} \leq x$ in terms of the number of X_i that are $\leq x$.

b) Use a) to show that for positive integers r and s, the c.d.f. of the beta (r, s) distribution is given by

$$\sum_{i=r}^{r+s-1} \binom{r+s-1}{i} x^i (1-x)^{r+s-i-1} \qquad (0 \leq x \leq 1)$$

c) Expand the power of $(1-x)$ in the beta density using the binomial theorem, and then integrate, to obtain the following alternative formula for the c.d.f. of the beta (r, s) distribution:

$$\frac{x^r}{B(r,s)} \sum_{i=0}^{s-1} \binom{s-1}{i} (-1)^i x^i / (r+i) \qquad (0 \leq x \leq 1)$$

[Equating the results of these two calculations yields an algebraic identity that is not easy to prove directly.]

Continuous Distributions: Summary

For a random variable X with probability density $f(x)$:

Differential formula: $P(X \in dx) = f(x)dx$.

Integral formula: $P(a \leq X \leq b) = \int_a^b f(x)dx$.

Interpretation: $f(x)$ is the chance per unit length for values of X near x.

Properties of $f(x)$: Non-negative, total integral 1.

Expectation of a function g of X

$$E((g(X)) = \int_{-\infty}^{\infty} g(x)f(x)dx \quad \text{provided} \quad \int_{-\infty}^{\infty} |g(x)|f(x)dx < \infty$$

Uniform, exponential, normal distributions: See Distribution Summaries.

Hazard rates

Let T be a positive random variable with probability density f. Think of T as the lifetime of a component. The *hazard rate* (or *failure rate*, or *death rate*) function $\lambda(t)$ is the probability per unit time that the component will fail just after time t, given that it has survived up to time t

$$P(T \in dt \,|\, T > t) = \lambda(t)\, dt$$

For relations between λ and the density, survival function, etc., of T, see the table "Random Lifetimes" on page 297.

Expectation from the survival function: For a non-negative random variable T,

$$E(T) = \int_0^{\infty} G(t)dt$$

where $G(t) = P(T > t)$ is the survival function of T.

One-to-one change of variable for densities

Let X be a random variable with density $f_X(x)$ in the range (a, b).

Let $Y = g(X)$ where g is either strictly increasing or strictly decreasing on (a, b). The range of Y is then an interval with endpoints $g(a)$ and $g(b)$. And the density of Y on this interval is

$$f_Y(y) = f_X(x) \Big/ \left|\frac{dy}{dx}\right| \qquad \text{at} \quad x = g^{-1}(y)$$

where dy/dx is the derivative of $y = g(x)$, and g^{-1} is the inverse function of g.

Linear change of variable for densities:

$$f_{aX+b}(y) = \frac{1}{|a|} f_X\left(\frac{y-b}{a}\right)$$

Change of variable principle: If X has the same distribution as Y, then $g(X)$ has the same distribution as $g(Y)$, for any function g.

Cumulative distribution function of X: $F(x) = P(X \le x)$

If the distribution has a *density* $f(x)$, then

$$F(x) = \int_{-\infty}^{x} f(y)\, dy$$

and the density function at x is the derivative of the c.d.f. at x

$$f(x) = \frac{dF(x)}{dx} = F'(x)$$

provided $F'(x)$ is continuous at x.

Percentiles

The kth percentile point of a distribution is the value x such that $F(x) = k/100$, written $x = F^{-1}(k/100)$, where F^{-1} is the *inverse c.d.f.*

Transformation by the inverse c.d.f.

If U has uniform $(0,1)$ distribution, then $F^{-1}(U)$ has c.d.f. F.

Order statistics

If $X_1, \ldots, X_n$ are independent with common density f and c.d.f. F, then the kth *order statistic* $X_{(k)}$, that is, the kth smallest value among the $X_1, \ldots, X_n$, has density

$$f_{X_{(k)}}(x) = nf(x)\binom{n-1}{k-1}(F(x))^{k-1}(1-F(x))^{n-k}$$

If the X_i have uniform $(0,1)$ distribution, then $X_{(k)}$ has beta $(k, n-k+1)$ distribution.

Review Exercises

1. Suppose atoms of a given kind have an exponentially distributed lifetime with rate λ. Let X_t be the number of atoms still present at time $t \geq 0$, starting from $X_0 = n$. Find formulae in terms of n, t, and λ for a) $E(X_t)$; b) $Var(X_t)$.

2. Find the constant c which makes the function $f(x) = c(x+x^2)$ for $0 < x < 1$ the density of a probability distribution on $(0,1)$. Find the corresponding c.d.f. $F(x)$. Sketch the graphs of $f(x)$ and $F(x)$. Find the expectation μ and standard deviation σ of a random variable X with this distribution. Mark the points μ, $\mu + \sigma$ on your graphs.

3. Let Y_1, Y_2, and Y_3 be three points chosen independently and uniformly from $(0,1)$, and let X be the rightmost (largest) point. Find the c.d.f., density function, and expectation of X.

4. Let X be a random variable with density $f(x) = 0.5e^{-|x|}$ $(-\infty < x < \infty)$. Find:

   a) $P(X < 1)$; b) $E(X)$ and $SD(X)$; c) the c.d.f. of X^2.

5. An ambulance station, 30 miles from one end of a 100-mile road, services accidents along the whole road. Suppose accidents occur with uniform distribution along the road, and the ambulance can travel at 60 miles an hour. Let T minutes be the response time (between when accident occurs and when ambulance arrives).

   a) Find $P(T > 30)$.

   b) Find $P(T > t)$ as a function of t. Sketch its graph.

   c) Calculate the density function of T.

   d) Calculate the mean and standard deviation of T.

   e) What would be a better place for the station? Explain.

6. Electrical components of a particular type have exponentially distributed lifetimes with mean 48 hours. In one application the component is replaced by a new one if it fails before 48 hours, and in case it survives 48 hours it is replaced by a new one anyway. Let T represent the potential lifetime of a component in continuous use, and U the time of such a component in use with the above replacement policy. Sketch the graphs of:

   a) the c.d.f. of T; b) the c.d.f. of U. Is U discrete, continuous, or neither?

   c) Find $E(U)$. [*Hint*: Express U as a function of T.]

   d) Does the replacement policy serve any good purpose? Explain.

7. **Two-sided exponential distribution.** Suppose X with range $(-\infty, \infty)$ has density $f(x) = \alpha e^{-\beta|x|}$ where α and β are positive constants.

   a) Express α in terms of β. b) Find $E(X)$ and $Var(X)$ in terms of β.

   c) Find $P(|X| > y)$ in terms of y and β. d) Find $P(X \leq x)$ in terms of x and β.

8. **The principle of ignoring constants.** In calculating the density of a random variable X, a quick method is to ignore constant factors as you go along, to end up with an answer of the form $P(X \in dx)/dx = f(x)$ with $f(x) = c\,h(x)$ for a known function $h(x)$ and mystery constant c. The point is that provided your calculation has been consistent with the basic rules of probability, the density of X must integrate to 1, so

$$\int c\,h(x)\,dx = \int f(x)\,dx = 1$$

a) Use this identity to evaluate c in terms of $\int h(x)\,dx$.

b) You can often recognize at the end of a calculation that $h(x) = c_1 f_1(x)$ for some named density $f_1(x)$ (e.g. one of the densities displayed in the table on page 477 and some constant c_1. Deduce that then $c = 1/c_1$ and $f(x) = f_1(x)$.

Use this method to evaluate the constant factor c that makes $ch(x)$ a probability density for each of the following functions $h(x)$, assumed to be zero except for the indicated range of x, and find $E(X)$ and $Var(X)$ in each case from the table on page 477.

c) $e^{-\frac{1}{2}x^2}$ $(-\infty < x < \infty)$ d) x $(0 < x < 1)$

e) 1 $(0 < x < 10)$ f) e^{-5x} $(x > 0)$

9. Use the method of Exercise 8 to evaluate the constant factor c that makes $f(x) = ch(x)$ a probability density for each of the following functions $h(x)$, assumed to be zero except for the indicated range of x, where a and b are positive parameters. Also find $E(X)$ and $Var(X)$ in each case:

a) $e^{-(x-a)^2}$ $(-\infty < x < \infty)$; b) $e^{-(x-a)^2/b^2}$ $(-\infty < x < \infty)$;

c) $e^{-ax}x^5 (x > 0)$; d) $e^{-a|x|}$ $(-\infty < x < \infty)$;

e) $x^7(1-x)^9$ $(0 < x < 1)$; f) $x^7(b-x)^9$ $(0 < x < b)$.

10. Evaluate the following integrals:

a) $\displaystyle\int_0^\infty e^{-x^2}dx$; b) $\displaystyle\int_0^1 e^{-x^2}dx$; c) $\displaystyle\int_0^\infty x\,e^{-x^2}dx$; d) $\displaystyle\int_0^\infty x^2\,e^{-x^2}dx$.

11. Evaluate the following integrals:

a) $\displaystyle\int_0^\infty z^3 e^{-z^2}dz$; b) $\displaystyle\int_0^\infty x^7 e^{-2x}dx$; c) $\displaystyle\int_0^{100} x^2(100-x)^2dx$.

12. A Geiger counter is recording background radiation at an average rate of 2 hits per minute; the hits may be modeled as a Poisson process. Let T be the time (in minutes) of the third hit after the machine is switched on. Find $P(1 < T < 3)$.

13. Local calls are coming into a telephone exchange according to a Poisson process with rate λ_{loc} calls per minute. Independently of this, long-distance calls are coming in at a rate of λ_{dis} calls per minute. Write down expressions for probabilities of the following events:

a) exactly 5 local calls and 3 long-distance calls come in a given minute;

b) exactly 50 calls (counting both local and long distance) come in a given three-minute period;

c) starting from a fixed time, the first ten calls to arrive are local.

14. Particles arrive at a Geiger counter according to a Poisson process with rate 3 per minute.

a) Find the chance that less than 4 particles arrive in the time interval 0 to 2 minutes.

b) Let T_n minutes denote the arrival time of the nth particle. Find

$$P(T_1 < 1,\ T_2 - T_1 < 1,\ T_3 - T_2 < 1)$$

c) Find the conditional distribution of the number of arrivals in 0 to 2 minutes, given that there were 10 arrivals in 0 to 4 minutes. Recognize this as a named distribution, and state the parameters.

15. Two Geiger counters record arrivals of radioactive particles. Particles arrive at Counter I according to a Poisson process, at an average rate of 3 per minute. Independently, particles arrive at Counter II at an average rate of 4 per minute, also according to a Poisson process. In a particular one-minute period, the counters recorded at total of 8 arrivals. Given this, what is the chance that each counter recorded four arrivals?

16. Cars arrive at a toll booth according to a Poisson process at a rate of 3 arrivals per minute.

a) What is the probability that the third car arrives within three minutes of the first car?

b) Of the cars arriving at the booth, it is known that over the long run 60% are Japanese imports. What is the probability that in a given ten-minute interval, 15 cars arrive at the booth, and 10 of these are Japanese imports? State your assumptions clearly.

17. Show that T has exponential distribution with rate λ if and only if

$$P(T \leq t) = 1 - e^{-\lambda t} \quad \text{for all} \quad 0 \leq t < \infty$$

18. Bus lines A, B, and C service a particular stop. Suppose the lines come as independent Poisson processes with rates λ_A, λ_B, and λ_C buses per hour respectively. Find expressions for the following probabilities:

a) exactly one A bus, two B buses, and one C bus come to the stop in a given hour;

b) a total of 7 buses come to the stop in a given two hour time period;

c) starting from a fixed time, the first A bus arrives after t hours.

19. A piece of rock contains 10^{20} atoms of a particular substance, each with a half-life of one century. How many centuries must pass before:

a) most likely about 100 atoms remain;

b) there is about a 50% chance that at least one atom remains.

20. **Hazard rates (refers to Section 4.3).** Suppose a component with constant failure rate λ is backed up by a second similar component. When the first component burns out the second is installed, and is thereafter subject to failure at the same rate λ, independently of when it was installed and how long it has been in use. Let T be the total time to failure of both components. Find for T:

a) the density function; b) the survival function; c) the hazard rate function.

d) Suppose $\lambda = 1$ per hour. Given $T \geq 2$ hours, what is the approximate probability of failure in the next minute?

21. Suppose R_1 and R_2 are two independent random variables with the same density function $f(x) = x \exp\left(-\frac{1}{2}x^2\right)$ for $x \geq 0$. Find

(a) the density of $Y = \min\{R_1, R_2\}$; b) the density of Y^2; c) $E(Y^2)$.

22. Let X be a random variable that has a uniform distribution on the interval $(0, a)$.

a) Find the c.d.f. of $Y = \min(X, a/2)$.

b) Is the distribution of Y continuous? Explain. c) Find $E(Y)$.

23. An earthquake of magnitude M releases energy X such that $M = \log X$. For earthquakes of magnitude greater that 3, suppose that $M - 3$ has an exponential distribution with mean 2.

a) Find $E(M)$ and $Var(M)$ for an earthquake of magnitude greater than 3.

b) For an earthquake as in part a), find the density of X.

c) Consider two earthquakes, both of magnitude greater than 3. What is the probability that the magnitude of the smaller earthquake is greater than 4? Assume that the magnitudes of the two earthquakes are independent of each other.

24. Suppose stop lights at an intersection alternately show green for one minute, red for one minute (ignore amber). Suppose a car arrives at the lights at a time distributed uniformly at random relative to this cycle. Let X be the delay of the car at the lights, neglecting any delay due to traffic congestion.

a) Find a formula for the c.d.f. of X, and sketch its graph.

b) Is X discrete, continuous, or neither? c) Find $E(X)$ and $Var(X)$.

d) Suppose that the car encounters a succession of ten such stop lights. Make an independence assumption and use the normal approximation to estimate the probability that the car will be delayed more than four minutes by the lights.

25. Suppose the random variable U is distributed uniformly on the interval $(0, 1)$. Find:

a) the density of the random variable $Y = \min\{U, 1 - U\}$ (indicate where the density is positive);

b) the density of $2Y$; c) $E(Y)$ and $Var(Y)$.

26. Suppose that the weight W_t of a tumor after time t is modeled by the formula $W_t = Xe^{tY}$ where X and Y are independent random variables, X distributed according to a gamma distribution with mean 2 and variance 1, and Y distributed uniformly on 1 to 1.5. Find formulae for: a) $E(W_t)$; b) $SD(W_t)$.

27. Suppose $U_1, U_2, \ldots$ are independent uniform $(0, 1)$ variables, and let N be the first $n \geq 2$ such that $U_n > U_{n-1}$. Show that for $0 \leq u \leq 1$:

a) $P(U_1 \leq u \text{ and } N = n) = \dfrac{u^{n-1}}{(n-1)!} - \dfrac{u^n}{n!} \quad n \geq 2;$

b) $P(U_1 \leq u \text{ and } N \text{ is even}) = 1 - e^{-u}$.

c) $E(N) = e$.

28. A point is chosen uniformly at random from the circumference of a circle of diameter 1. Let X be the length of the chord joining the random point to an arbitrary fixed point on the circumference. Find: a) the c.d.f. of X; b) $E(X)$; c) $Var(X)$.

29. A gambling game works as follows. A random variable X is produced; you win \$1 if $X > 0$ and you lose \$1 if $X < 0$. Suppose first that X has a normal $(0, 1)$ distribution. Then the game is clearly "fair". Now suppose the casino gives you the following option. You can make X have a normal $(b, 1)$ distribution, but to do so you have to pay \$$cb$ which is not returned to you even if you win. Here $c > 0$ is set by the casino, but you can choose any $b > 0$.

a) For what values of c is it advantageous for you to use this option?

b) For these values of c, what value of b should you choose?

30. A manufacturing process produces ball bearings with diameters which are independent and normally distributed with mean 0.250 inches and SD 0.001 inches. In a high-precision application, 16 bearings are arranged in a ring. The specifications are that:

(i) each bearing must be between 0.249 and 0.251 inches in diameter;

(ii) the sum of the diameters of the 16 bearings must be between 3.995 and 4.005 inches.

a) What is the expected number of bearings which must be produced by the process to obtain 16 satisfying specification (i)?

b) Given 16 bearings obtained like this, what is the chance that they meet specification (ii)?

[*Hint* for b): Write $x^2\phi(x) = x[x\phi(x)]$ and use integration by parts to show that

$$\int_{-z}^{z} x^2\phi(x)dx = 2\Phi(z) - 1 - 2z\phi(z). \text{]}$$

31. The skew-normal pseudo-density. Referring to the end of Section 3.3, let

$$\Phi_\theta(z) = \Phi(z) - \frac{\theta}{6}(z^2 - 1)\phi(z)$$

This is the substitute for the normal c.d.f. $\Phi(z)$ which for $\frac{\theta}{6} \neq 0$ typically gives a better approximation than $\Phi(z)$ to the c.d.f. of a random variable with mean zero, variance 1 and third moment θ.

a) Let $\phi_\theta(z) = \frac{d}{dz}\Phi_\theta(z)$. Show $\phi_\theta(z) = [1 - \frac{\theta}{6}(3z - z^3)]\phi(z)$.

b) Show that for every θ

$$\int_{-\infty}^{\infty} \phi_\theta(z)dz = 1; \quad \int_{-\infty}^{\infty} z\phi_\theta(z)dz = 0; \quad \int_{-\infty}^{\infty} z^2\phi_\theta(z)dz = 1; \quad \int_{-\infty}^{\infty} z^3\phi_\theta(z)dz = \theta$$

[So $\phi_\theta(z)$ is very like the probability density of a distribution with mean zero, variance 1 and third moment θ. This explains the choice $\theta = \text{Skewness}(X) = E(X_*^3)$ in the skew-normal approximation to the distribution of a standardized variable $X_* = (X - \mu)/\sigma$.]

c) Show that ϕ_θ is negative for large negative z if $\theta > 0$, and negative for large positive z if $\theta < 0$. So for $\theta \neq 0$, $\phi_\theta(z)$ is in fact *not* a probability density. It may be called instead a *pseudo-density*.

d) Find a probability in the Poisson(9) distribution whose normal approximation with continuity and skewness corrections is a negative number.

e) Explain carefully why, despite c) and d) the functions $\phi_{1/3}(z)$ and $\Phi_{1/3}(z)$ provide practically useful approximations to the Poisson(9) and other distributions which are roughly normal in shape but slightly skewed.

5
Continuous Joint Distributions

The *joint distribution* of a pair of random variables X and Y is the probability distribution over the plane defined by

$$P(B) = P((X, Y) \in B)$$

for subsets B of the plane. So $P(B)$ is the probability that the random pair (X, Y) falls in the set B. Joint distributions for discrete random variables were considered in Section 3.1. This chapter shows how these ideas for discrete random variables are extended to two or more continuously distributed random variables with sums replaced by integrals.

Section 5.1 concerns the simplest kind of continuous joint distribution, a *uniform* distribution defined by relative areas. Section 5.2 introduces the concept of a *joint density function.* Joint probabilities are then defined by volumes under a density surface. The important special case of independent normal variables is studied in Section 5.3. Then Section 5.4 deals with a general technique for finding the distribution of a function of two variables.

5.1 Uniform Distributions

The uniform distribution on an interval was discussed in Section 4.1. The idea extends to higher dimensions with relative lengths replaced by relative areas or relative volumes. For example, a random point (X, Y) in the plane has uniform distribution on D, where D is a region of the plane with finite area, if:

(i) (X, Y) is certain to lie in D;

(ii) the chance that (X, Y) falls in a subregion C of D is proportional to the area of C

$$P((X, Y) \in C) = \frac{\text{area } (C)}{\text{area } (D)} \qquad \text{for} \quad C \subset D$$

Here is an important observation:

Independent Uniform Variables

If X and Y are independent random variables, each uniformly distributed on an interval, then (X, Y) is uniformly distributed on a rectangle.

To see why, suppose X and Y are independent and uniformly distributed on, say, $(0, a)$ and $(0, b)$, respectively. For intervals A and B the event $(X \in A, Y \in B)$ is the event that (X, Y) falls in the rectangle $A \times B$, as shown in the following Venn diagram:

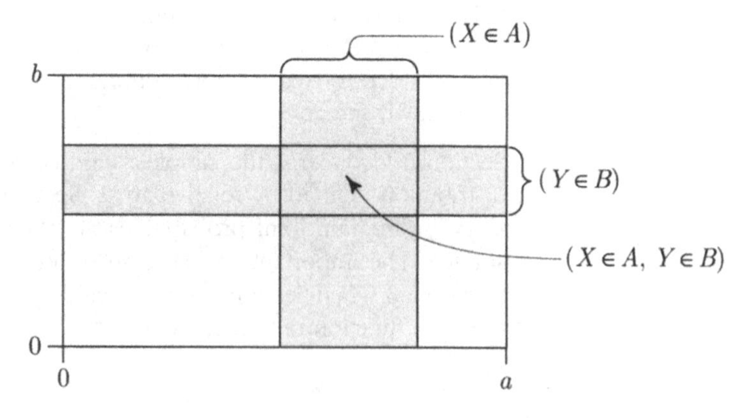

So for any rectangle $A \times B$

$$P((X,Y) \in A \times B) = P(X \in A, Y \in B)$$
$$= P(X \in A)P(Y \in B) \quad \text{by independence of } X \text{ and } Y$$
$$= \frac{\text{length }(A)}{a} \cdot \frac{\text{length }(B)}{b} \quad \text{by assumed uniform distributions of } X \text{ and } Y$$
$$= \frac{\text{area }(A \times B)}{ab}$$

Thus the probability that $(X,Y) \in C$ is the relative area of C in $(0,a) \times (0,b)$ for every rectangle C. The same must then be true for finite unions of rectangles, by the addition rule of probability and for area, hence also for any set C whose area can be defined by approximating with unions of rectangles. *Conclusion*: (X,Y) has uniform distribution on the rectangle $(0,a) \times (0,b)$.

The above observation allows probabilities involving two independent uniform variables X and Y to be found geometrically in terms of areas. The key step is correct identification of areas in the plane corresponding to events in question. Skill at doing this is essential for all further work in this chapter.

Example 1. Probabilities for two independent uniform random variables.

Suppose X and Y are independent uniform $(0,1)$ random variables.

Problem 1. Find $P(X^2 + Y^2 \leq 1)$.

Solution. Proceed by 3 steps as in the diagram below:

- Draw a unit square with coordinates X, Y.
- Notice that $X^2 + Y^2 = 1$ gives the equation of a circle of radius 1.
- Recognize $(X^2 + Y^2 \leq 1)$ as the region inside both the square and circle.
- Use the formula for the area of a circle to get $P(X^2 + Y^2 \leq 1) = \frac{\pi}{4}$.

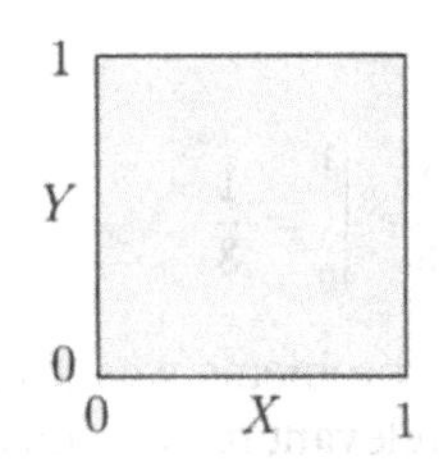

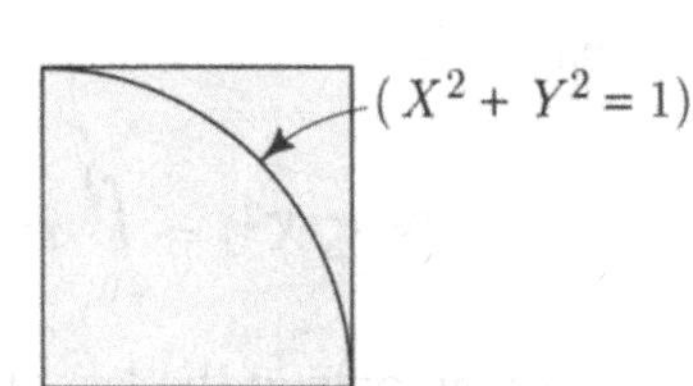

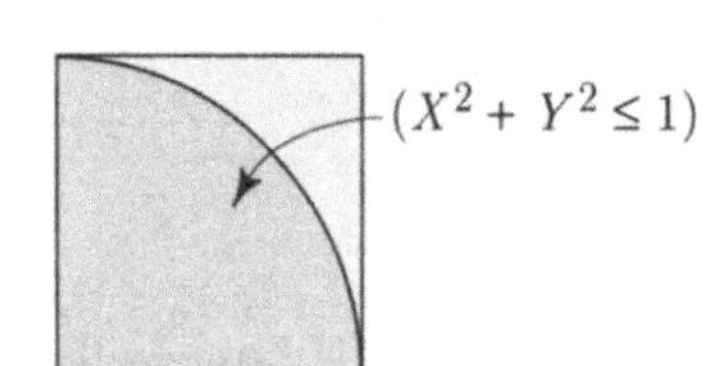

Problem 2. Find the conditional probability $P(X^2 + Y^2 \leq 1 | X + Y \geq 1)$.

Solution. After first identifying $X^2 + Y^2 \leq 1$ as above, next:

- Recognize $(X + Y = 1)$ as the line through the points $(0,1)$ and $(1,0)$.

- Deduce that $(X+Y \geq 1)$ is the shaded region above this line.

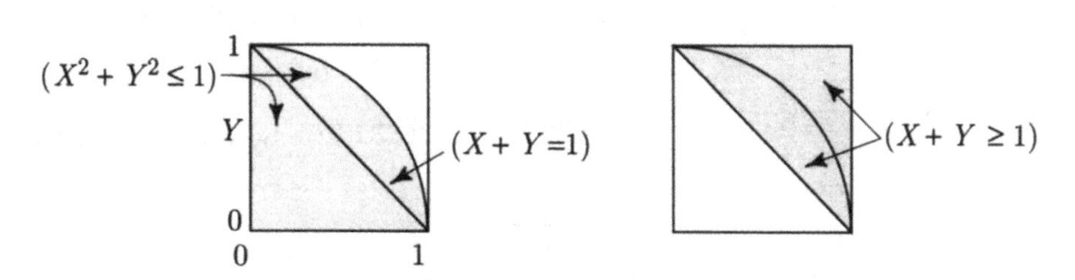

- Now compute the required relative area:

$$P(X^2+Y^2 \leq 1 | X+Y \geq 1) = \frac{P(X^2+Y^2 \leq 1, X+Y \geq 1)}{P(X+Y \geq 1)}$$
$$= \frac{\pi/4 - 1/2}{1/2} = \frac{\pi}{2} - 1$$

Problem 3. Find $P(Y \leq X^2)$.

Solution.

- Graph $Y = X^2$.
- Recognize $(Y \leq X^2)$ as the region under this graph.
- Compute the area of this region by calculus.

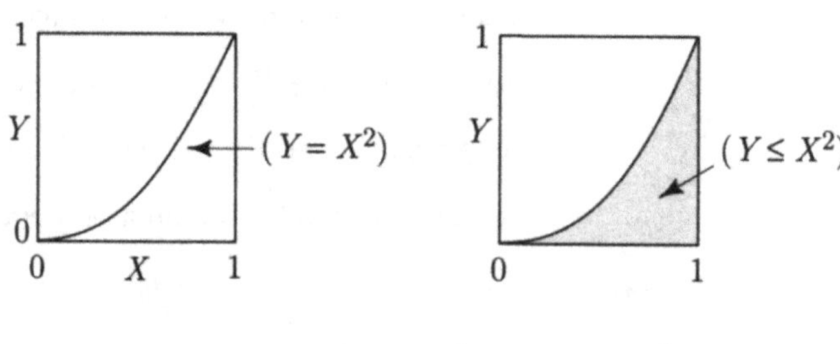

$$P(Y \leq X^2) = \int_0^1 x^2 dx = \frac{1}{3}x^3 \Big|_0^1 = \frac{1}{3}$$

Discussion. Note well how only in the last of these problems was it necessary to resort to calculus to find the area. *Always* sketch the relevant regions first, then look out for familiar shapes, rectangles, triangles, and circles. If all else fails, use calculus.

Example 2. **More probabilities for two independent uniform variables.**
Let X and Y be independent random variables, each uniformly distributed on $(0,1)$. Calculate the following probabilities:

a)
$$P(|X-Y|\le 0.5) = \text{indicated area}$$
$$= 1 - \frac{1}{4} = 0.75$$

b)
$$P\left(\left|\frac{X}{Y}-1\right| \le 0.5\right) = P\left(\frac{2}{3}X \le Y \le 2X\right)$$
$$= \text{indicated area}$$
$$= 1 - \frac{1}{2}\left(\frac{1}{2}+\frac{2}{3}\right) = \frac{5}{12}$$

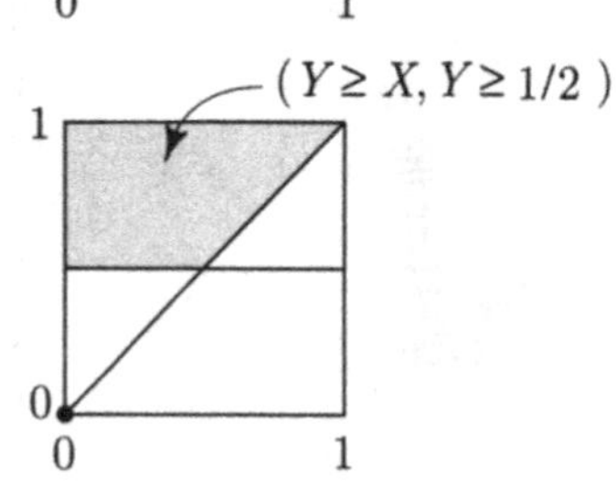

c)
$$P\left(Y \ge X \mid Y \ge \frac{1}{2}\right) = \text{indicated area}/\frac{1}{2}$$
$$= \left(\frac{1}{2}-\frac{1}{8}\right)/\frac{1}{2} = \frac{3}{4}$$

Example 3. **Probability of meeting.**

Problem. Two people try to meet at a certain place between 5:00 P.M. and 5:30 P.M. Suppose that each person arrives at a time distributed uniformly at random in this time interval, independent of the other, and waits for the other at most 5 minutes. What is the probability that they meet?

Solution. Let X and Y be the arrival times measured as fractions of the 30 minute interval, starting from 5:00 P.M. Then X and Y are independent uniform $(0, 1)$ random variables. The people meet if and only if $|X - Y| \le 1/6$.

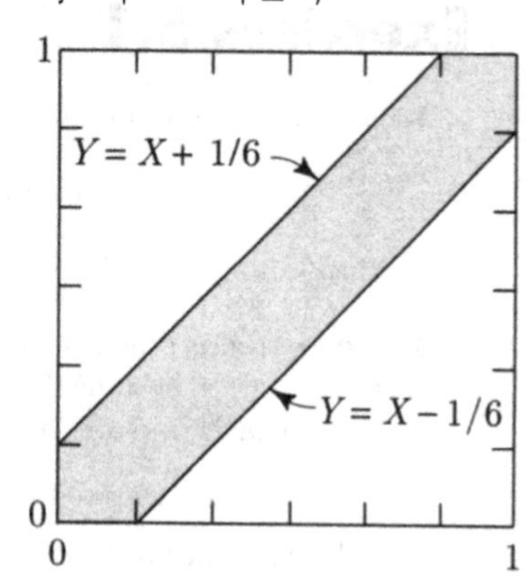

Desired probability = indicated area = $1 - (\frac{5}{6})^2 = \frac{11}{36}$

Uniform Distribution over a Volume

This is the extension of the idea of relative lengths in one dimension and relative areas in two dimensions to relative volumes in three and higher dimensions. If $U_1, \ldots, U_n$ are n independent random variables, with U_i uniformly distributed on an interval (a_i, b_i), then the same argument given earlier for the case $n = 2$ shows that the joint distribution of $(U_1, \ldots, U_n)$ is the uniform distribution defined by relative volumes within the n-dimesional box

$$(a_1, b_1) \times (a_2, b_2) \times \cdots \times (a_n, b_n)$$

whose n-dimensional volume is the product $(b_1 - a_1)(b_2 - a_2) \cdots (b_n - a_n)$ of the lengths of its sides.

To illustrate, a random point in the *unit cube* $(0,1) \times (0,1) \times (0,1)$, with approximately independent coordinates, is obtained by three successive calls of a pseudo-random number generator, say $(\text{RND}_1, \text{RND}_2, \text{RND}_3)$. For any subvolume B of the unit cube bounded by a reasonably smooth surface (e.g., the portion of a box, pyramid, or sphere that lies inside the unit cube) the long-run frequency of times that $(\text{RND}_1, \text{RND}_2, \text{RND}_3)$ is in B will be approximately the volume of B, that is $P(B)$ for the uniform distribution on the unit cube. For example, the long-run frequency of triples $(\text{RND}_1, \text{RND}_2, \text{RND}_3)$ with

$$(\text{RND}_1 - \tfrac{1}{2})^2 + (\text{RND}_2 - \tfrac{1}{2})^2 + (\text{RND}_3 - \tfrac{1}{2})^2 < 1/4$$

is approximately the volume of the subset of the unit cube

$$\{(x, y, z) : 0<x<1,\ 0<y<1,\ 0<z<1,\ (x - \tfrac{1}{2})^2 + (y - \tfrac{1}{2})^2 + (z - \tfrac{1}{2})^2 < 1/4\}$$

This is the volume of a sphere of radius $\frac{1}{2}$ centered at $(\frac{1}{2}, \frac{1}{2}, \frac{1}{2})$, which is $\frac{4}{3}\pi\left(\frac{1}{2}\right)^3 = \frac{\pi}{6}$.

Exercises 5.1

1. Let (X, Y) have uniform distribution on the set

$$\{(x, y) : 0 < x < 2 \text{ and } 0 < y < 4 \text{ and } x < y\}.$$

Find: a) $P(X < 1)$; b) $P(Y < X^2)$.

2. A metal rod is l inches long. Measurements on the length of this rod are equal to l plus random error. Assume that the errors are uniformly distributed over the range -0.1 inch to $+0.1$ inch, and are independent of each other.

a) Find the chance that a measurement is less than $1/100$ of an inch away from l.

b) Find the chance that two measurements are less than $1/100$ of an inch away from each other.

3. Suppose X and Y are independent and uniformly distributed on the unit interval $(0, 1)$. Find:

$$P(Y \geq \frac{1}{2} | Y \geq 1 - 2X).$$

4. Let X and Y be independent random variables each uniformly distributed on $(0, 1)$. Find:

a) $P(|X - Y| \leq 0.25)$; b) $P(|X/Y - 1| \leq 0.25)$; c) $P(Y \geq X | Y \geq 0.25)$.

5. A very large group of students takes a test. Each of them is told his or her percentile rank among all students taking the test.

a) If a student is picked at random from all students taking the test, what is the probability that the student's percentile rank is over 90%?

b) If two students are picked independently at random, what is the probability that their percentile ranks differ by more than 10%?

6. A group of 10 people agree to meet for lunch at a cafe between 12 noon and 12:15 P.M. Assume that each person arrives at the cafe at a time uniformly distributed between noon and 12:15 P.M., and that the arrival times are independent of each other.

a) Jack and Jill are two members of the group. Find the probability that Jack arrives at least two minutes before Jill.

b) Find the probability of the event that the first of the 10 persons to arrive does so by 12:05 P.M., and the last person arrives after 12:10 P.M.

7. Let X and Y be two independent uniform $(0, 1)$ random variables. Let M be the smaller of X and Y. Let $0 < x < 1$.

a) Represent the event $(M \geq x)$ as the region in the plane, and find $P(M \geq x)$ as the area of this region.

b) Use your result in a) to find the c.d.f. and density of M. Sketch the graph of these functions.

8. Let $U_{(1)}, \ldots, U_{(n)}$ be the values of n independent uniform $(0, 1)$ random variables arranged in increasing order. Let $0 \leq x < y \leq 1$.

a) Find and justify a simple formula for $P(U_{(1)} > x \text{ and } U_{(n)} < y)$.

b) Find a formula for $P(U_{(1)} \leq x \text{ and } U_{(n)} < y)$.

9. A triangle problem. Suppose a straight stick is broken in three at two points chosen independently at random along its length. What is the chance that the three sticks so formed can be made into the sides of a triangle?

5.2 Densities

The concept of a *joint probability density function* $f(x, y)$ for a pair of random variables X and Y is a natural extension of the idea of a one-dimensional probability density function studied in Chapter 4. The function $f(x, y)$ gives the density of probability per unit area for values of (X, Y) near the point (x, y).

FIGURE 1. A joint density surface. Here a particular joint density function given by the formula $f(x, y) = 5!\,x(y - x)(1 - y) \quad (0 < x < y < 1)$, is viewed as the height of a surface over the unit square $0 \le x \le 1, 0 \le y \le 1$. As explained later in Example 3, two random variables X and Y with this joint density are the second and fourth smallest of five independent uniform $(0, 1)$ variables. But for now the source and special form of this density are not important. Just view it as a typical joint density surface.

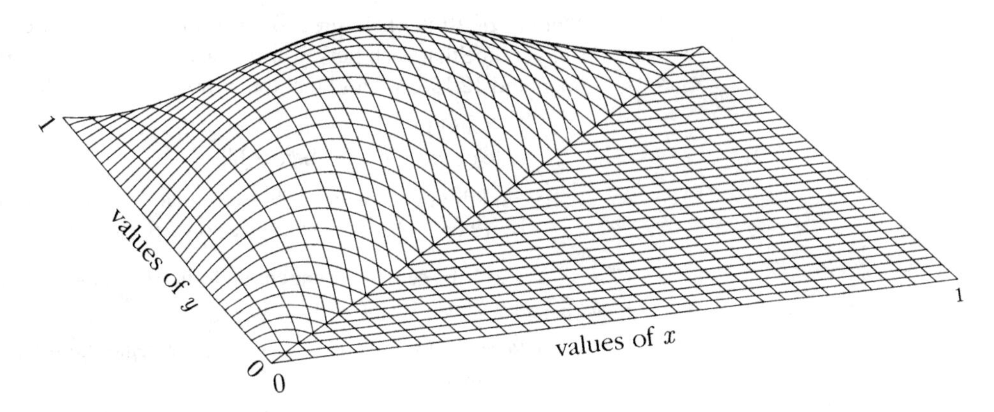

Examples in the previous section show how any event determined by two random variables X and Y, like the event $(X > 0.25$ and $Y > 0.5)$, corresponds to a region of the plane. Now instead of a uniform distribution defined by relative areas, the probability of region B is defined by the volume under the density surface over B. This volume is an integral

$$P((X, Y) \in B) = \iint_B f(x, y)\,dx\,dy$$

This is the analog of the familiar area under the curve interpretation for probabilities obtained from densities on a line. Examples to follow show how such integrals can be computed by repeated integration, change of variables, or symmetry arguments. Uniform distribution over a region is now just the special case when $f(x, y)$ is constant over the region and zero elsewhere. As a general rule, formulae involving joint densities are analogous to corresponding formulae for discrete joint distributions described in Section 3.1. See pages 348 and 349 for a summary.

FIGURE 2. **Volume representing a probability.** The probability $P(X > 0.25$ and $Y > 0.5)$, for random variables X and Y with the joint density of Figure 1. The set B in this case is $\{(x, y) : x > 0.25$ and $y > 0.5\}$. You can see the volume is about half the total volume under the surface. The exact value, found later in Example 3, is $27/64$.

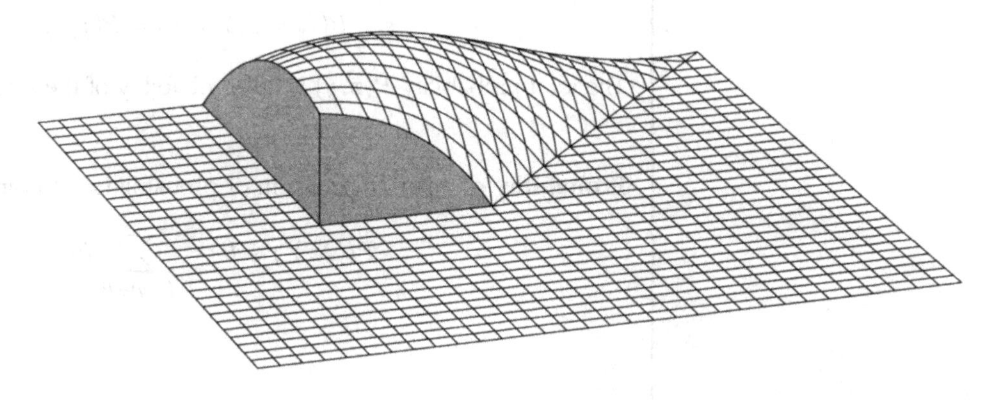

Informally, if (X, Y) has joint density $f(x, y)$, then there is the *infinitesimal probability formula*

$$P(X \in dx, Y \in dy) = f(x, y)dx\,dy$$

This means that the probability that the pair (X, Y) falls in an infinitesimal rectangle of width dx and height dy near the point (x, y) is the probability density at (x, y) multiplied by the area $dx\,dy$ of the rectangle.

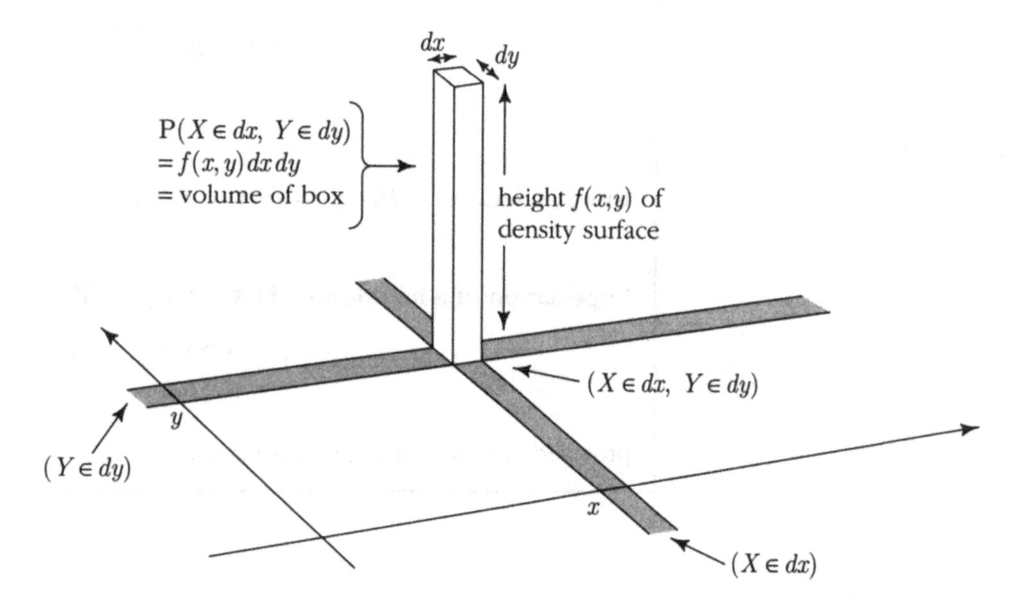

Discrete Joint Distribution

Probability of a point:

$$P(X = x, Y = y) = P(x, y)$$

The joint probability $P(x, y)$ is the probability of the single point (x, y).

Probability of a set B: The sum of probabilities of points in B

$$P((X, Y) \in B) = \sum_{(x,y)\in B} P(x, y)$$

Constraints: Non-negative with total sum 1

$$P(x, y) \geq 0 \quad \text{and} \quad \sum_{\text{all } x} \sum_{\text{all } y} P(x, y) = 1$$

Marginals:

$$P(X = x) = \sum_{\text{all } y} P(x, y)$$

$$P(Y = y) = \sum_{\text{all } x} P(x, y)$$

Independence: $P(x, y) = P(X = x)P(Y = y)$ (for all x and y)

Expectation of a function g of (X,Y), e.g., XY,

$$E(g(X, Y)) = \sum_{\text{all } x} \sum_{\text{all } y} g(x, y)P(x, y)$$

provided the sum converges absolutely.

Joint Distribution Defined by a Density

Infinitesimal probability:

$$P(X \in dx, Y \in dy) = f(x,y)\,dx\,dy$$

The joint density $f(x,y)$ is the probability per unit area for values near (x,y).

Probability of a set B: The volume under the density surface over B

$$P((X,Y) \in B) = \iint_B f(x,y)\,dx\,dy$$

Constraints: Non-negative with total integral 1

$$f(x,y) \geq 0 \qquad \text{and} \qquad \int_{-\infty}^{\infty}\int_{-\infty}^{\infty} f(x,y)\,dx\,dy = 1$$

Marginals:

$$f_X(x) = \int_{-\infty}^{\infty} f(x,y)dy$$

$$f_Y(y) = \int_{-\infty}^{\infty} f(x,y)dx$$

Independence: $f(x,y) = f_X(x)f_Y(y)$ (for all x and y)

Expectation of a function g of (X,Y), e.g., XY

$$E(g(X,Y)) = \iint g(x,y)f(x,y)dx\,dy$$

provided the integral converges absolutely.

The infinitesimal probability formula

$$P(X \in dx, Y \in dy) = f(x,y)dx\,dy$$

is really shorthand for a limiting statement about the ratio of probability per unit area for small areas, which, strictly speaking, holds only at points (x, y) such that the joint density is continuous at (x, y). But the infinitesimal formula conveys the right intuitive idea, and can be manipulated to obtain useful formulae which turn out to be valid even without assuming that the joint density is continuous.

Marginal densities. If (X, Y) has a joint density $f(x, y)$ in the plane, then each of the random variables X and Y has a density on the line. These are called the *marginal densities*. As shown in the preceding display, the marginal densities can be calculated from the joint density by integral analogs of the discrete formulae for marginal probabilities as row and column sums in a joint distribution table. Probabilities of discrete points are replaced by densities, and sums by integrals.

Independence. In general, random variables X and Y are called independent if

(1) $P(X \in A, Y \in B) = P(X \in A)P(Y \in B)$ for all choices of sets A and B.

Joint Density for Independent Variables

Random variables X and Y with joint density $f(x, y)$ are independent if and only if the joint density is the product of the two marginal densities:

(2) $$f(x,y) = f_X(x)f_Y(y) \qquad \text{(for all } x \text{ and } y\text{)}$$

Intuitively (2) follows from (1) by taking A to be a small interval $(x, x + dx)$ near x, B a small interval $(y, y + dy)$ near y, to obtain

(3) $$P(X \in dx, Y \in dy) = P(X \in dx)P(Y \in dy)$$

$$\text{so} \quad f(x,y)\,dx\,dy = f_X(x)\,dx\,f_Y(y)\,dy$$

Cancelling the differentials dx and dy leaves the product formula for densities. Conversely, (1) is obtained from (2) by integration.

Example 1. Uniform on a triangle.

Suppose (X, Y) is uniformly distributed over the region $\{(x, y) : 0 < x < y < 1\}$.

Problem 1. Find the joint density of (X, Y).

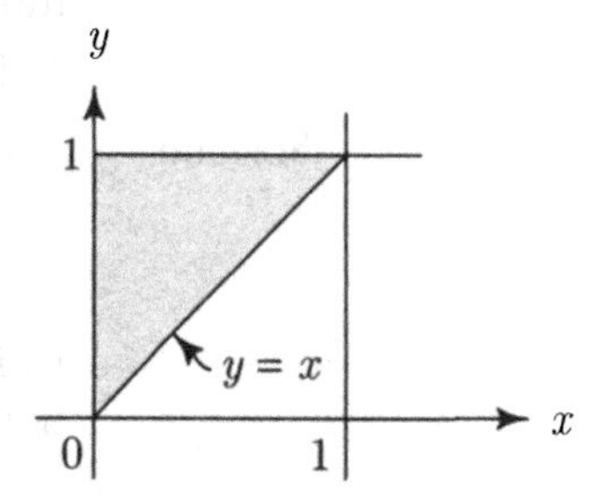

Solution. By the assumption, $f(x, y) = c$ for $0 < x < y < 1$ and 0 elsewhere. Because the triangle has area $\frac{1}{2}$, $c = 2$.

Problem 2. Find the marginal densities $f_X(x)$ and $f_Y(y)$.

Solution.

$$\begin{aligned} f_X(x) &= \int_{-\infty}^{\infty} f(x, y)\, dy \\ &= \int_{y=x}^{y=1} 2dy \qquad \text{since } f(x, y) = 2 \text{ for } 0 < x < y < 1, \quad 0 \text{ elsewhere} \\ &= 2(1 - x) \qquad \text{for } 0 < x < 1 \quad \text{and } 0 \text{ elsewhere.} \end{aligned}$$

$$\begin{aligned} f_Y(y) &= \int_{-\infty}^{\infty} f(x, y)\, dx \\ &= \int_{x=0}^{x=y} 2dx \qquad \text{since } f(x, y) = 2 \text{ for } 0 < x < y < 1 \quad 0 \text{ elsewhere} \\ &= 2y \qquad \text{for } 0 < y < 1 \quad \text{and } 0 \text{ elsewhere.} \end{aligned}$$

Problem 3. Are X and Y independent?

Solution. No, since $f(x, y) \neq f_X(x) f_Y(y)$.

Problem 4. Find $E(X)$ and $E(Y)$.

Solution.

$$E(X) = \int_{-\infty}^{\infty} x f_X(x)\, dx = \int_0^1 2x(1 - x)\, dx = \frac{1}{3}$$

$$E(Y) = \int_{-\infty}^{\infty} y f_Y(y)\, dy = \int_0^1 2y^2\, dy = \frac{2}{3}$$

Problem 5. Find $E(XY)$.

Solution.

$$E(XY) = \iint_{R^2} xy f(x, y)\, dx\, dy = 2 \int_{y=0}^{1} dy \int_{x=0}^{y} xy\, dx = 2 \int_{y=0}^{1} \frac{y^3}{2}\, dy = \frac{1}{4}$$

Remark. You can show that the joint distribution of X and Y considered here is that of $X = \min(U, V)$, $Y = \max(U, V)$, where U and V are independent uniform $(0, 1)$ variables. Example 3 gives a more difficult derivation of this kind.

Example 2. Independent exponential variables.

Problem. Let X and Y be independent and exponentially distributed random variables with parameters λ and μ, respectively. Calculate $P(X < Y)$.

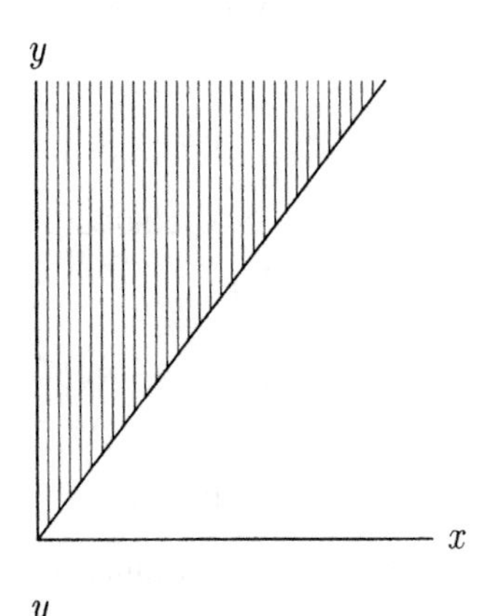

Solution. The joint density is

$$f(x, y) = (\lambda e^{-\lambda x})(\mu e^{-\mu y}) = \lambda\mu e^{-\lambda x - \mu y}$$

by independence. And $P(X < Y)$ is found by integration of this joint density over the set $\{(x, y) : x < y\}$:

$$\begin{aligned} P(X < Y) &= \iint_{x<y} \lambda\mu\, e^{-\lambda x - \mu y} dx\, dy \\ &= \int_{x=0}^{\infty} dx \int_{y=x}^{\infty} \lambda\mu\, e^{-\lambda x - \mu y} dy \\ &= \int_{x=0}^{\infty} \lambda e^{-\lambda x - \mu x} dx \\ &= \frac{\lambda}{\lambda + \mu} \end{aligned}$$

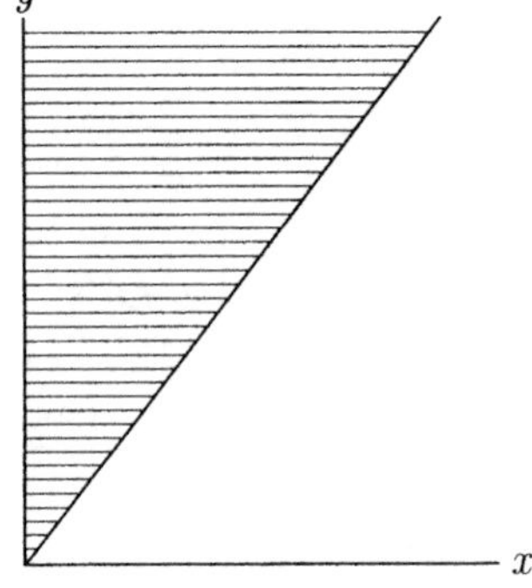

Remark. Done in the other order, the integral is

$$\int_{y=0}^{\infty} dy \int_{x=0}^{y} \lambda\mu\, e^{-\lambda x - \mu y} dx$$

which simplifies to the same answer. As a general rule, provided the integrand is positive, as always when finding probabilities, double integrals done in either order produce the same result.

Example 3. Joint distribution of order statistics.

Suppose $U_{(1)} < U_{(2)} < \cdots < U_{(5)}$ are the order statistics of 5 independent uniform $(0, 1)$ variables $U_1, \ldots, U_5$, so $U_{(i)}$ is the ith smallest of $U_1, \ldots, U_5$, as, for example, in the following diagram:

Problem 1. Find the joint density of $U_{(2)}$ and $U_{(4)}$.

Solution. This is very like the calculation of the density of $U_{(i)}$ done in Section 4.6. The following diagram shows one way of getting $U_{(2)}$ in dx and $U_{(4)}$ in dy for $0 < x < y < 1$:

$0 \quad U_2 \quad x\ \overset{dx}{U_4} \quad U_3 \quad y\ \overset{dy}{U_1} \quad U_5$

$$\begin{aligned}
&P(U_{(2)} \in dx, U_{(4)} \in dy) \\
&= P(\text{ one } U_i \text{ in } (0,x), \text{ one in } dx, \text{ one in } (x,y), \text{ one in } dy, \text{ one in } (y,1)\ ) \\
&= 5!\, P(U_2 \in (0,x), U_4 \in dx, U_3 \in (x,y), U_1 \in dy, U_5 \in (y,1)) \\
&= 5!\, x\, dx (y-x)\, dy (1-y)
\end{aligned}$$

Here the 5! is the number of different ways of deciding which variables fall in which intervals. The conclusion is that the joint density of $U_{(2)}$ and $U_{(4)}$ is

$$P(U_{(2)} \in dx, U_{(4)} \in dy)/dx\, dy = \begin{cases} 5!\, x(y-x)(1-y) & \text{for } 0 < x < y < 1 \\ 0 & \text{elsewhere} \end{cases}$$

This is the density surface shown in Figure 1 on page 346.

Problem 2. Find $P(U_{(2)} > 1/4 \text{ and } U_{(4)} > 1/2)$.

Solution. The volume representing this probability is shown in Figure 2 on page 347. This is the volume under the density surface over the area shaded in the diagram at right. This area is the intersection of:

(i) the region representing the event; and

(ii) the region where the density is strictly positive.

This determines the ranges of integration. The required probability is thus

$$\begin{aligned}
&5! \int_{y=1/2}^{1} \int_{x=1/4}^{y} x(y-x)(1-y)\, dx\, dy \\
&= 5! \int_{y=1/2}^{1} (1-y)\, dy \left[\frac{1}{2}x^2 y - \frac{1}{3}x^3 \right] \Bigg|_{1/4}^{y} \\
&= 5! \int_{y=1/2}^{1} (1-y)\, dy \left[\frac{y^3}{6} - \frac{y}{2^5} - \frac{1}{3 \times 2^6} \right] = \frac{27}{64}
\end{aligned}$$

by straightforward integration of the polynomial.

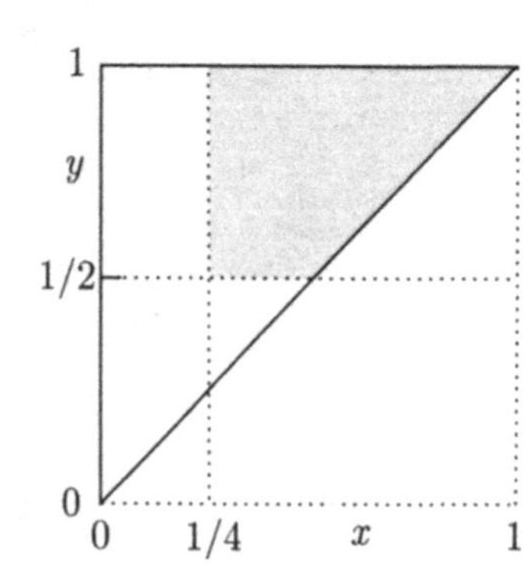

Exercises 5.2

1. Suppose that (X,Y) is uniformly distributed over the region $\{(x,y): 0 < |y| < x < 1\}$. Find:

a) the joint density of (X,Y); b) the marginal densities $f_X(x)$ and $f_Y(y)$.

c) Are X and Y independent? d) Find $E(X)$ and $E(Y)$.

2. Repeat Exercise 1 for (X,Y) with uniform distribution over $\{(x,y): 0 < |x|+|y| < 1\}$.

3. A random point (X,Y) in the unit square has joint density $f(x,y) = c(x^2 + 4xy)$ for $0 < x < 1$ and $0 < y < 1$, for some constant c.

a) Evaluate c. b) Find $P(X \le a)$, $0 < a < 1$. c) Find $P(Y \le b)$, $0 < b < 1$.

4. For random variables X and Y with joint density function

$$f(x,y) = 6e^{-2x-3y} \qquad (x,y > 0)$$

and $f(x,y) = 0$ otherwise, find:

a) $P(X \le x, Y \le y)$; b) $f_X(x)$; c) $f_Y(y)$.

d) Are X and Y independent? Give a reason for your answer.

5. Let X be exponentially distributed with rate λ, independent of Y, which is exponentially distributed with rate μ. Find $P(X \ge 3Y)$.

6. Let X and Y have joint density

$$f(x,y) = \begin{cases} 90(y-x)^8 & 0 < x < y < 1 \\ 0 & \text{otherwise} \end{cases}$$

a) Find $P(Y > 2X)$. b) Find the marginal density of X.

c) Fill in the blanks (explain briefly):
The joint density f above is the joint density of the ________ and ________ of ten independent uniform (0, 1) random variables.

7. Two points are picked independently and uniformly at random from the region inside a circle. Let R_1 and R_2 be the distances of these points from the center of the circle. Find $P(R_2 \le R_1/2)$.

8. Random variables X and Y have joint density

$$f_{X,Y}(x,y) = \begin{cases} c(y^2 - x^2)e^{-y} & -y \le x \le y, \quad y > 0 \\ 0 & \text{otherwise} \end{cases}$$

Here c is a constant.

a) Show that Y has a gamma density, and hence deduce that $c = 1/8$.

b) Find the density of $4Y^3$.

c) Explain why $E(|X|)$ is at most 4.

9. Minimum and maximum of two independent exponentials. Let $X = \min(S,T)$ and $Y = \max(S,T)$ for independent exponential(λ) variables S and T. Let $Z = Y - X$.

a) Find the joint density of X and Y. Are X and Y independent?

b) Find the joint density of X and Z. Are X and Z independent?

c) Identify the marginal distributions of X and Z.

10. Minimum and maximum of n independent exponentials. Let $X_1, X_2, \ldots, X_n$ be independent, each with exponential (λ) distribution. Let $V = \min(X_1, X_2, \ldots, X_n)$ and $W = \max(X_1, X_2, \ldots, X_n)$. Find the joint density of V and W.

11. Suppose X and Y are independent random variables such that X has uniform $(0, 1)$ distribution, Y has exponential distribution with mean 1. Calculate:

a) $E(X + Y)$; b) $E(XY)$; c) $E[(X - Y)^2]$; d) $E(X^2 e^{2Y})$.

12. Let T_1 and T_5 be the times of the first and fifth arrivals in a Poisson process with rate λ, as in Section 4.2. Find the joint density of T_1 and T_5.

13. Uniform spacings. Let $X = \min(U, V)$ and $Y = \max(U, V)$ for independent uniform$(0, 1)$ variables U and V. Find the distributions of

a) X; b) $1 - Y$; c) $Y - X$.

14. Let U_1, U_2, U_3, U_4, U_5 be independent, each with uniform distribution on $(0, 1)$. Let R be the distance between the minimum and the maximum of the U_i's. Find

a) $E(R)$;

b) the joint density of the minimum and maximum of the U_i's;

c) $P(R > 0.5)$

15. C.d.f.'s in two dimensions. The *cumulative joint distribution function* of random variables X and Y is the function of x and y defined by $F(x, y) = P(X \leq x, Y \leq y)$.

a) Find a formula in terms of $F(x, y)$ for $P(a < X \leq b,\ c < Y \leq d)$.

b) For X and Y with joint density $f(x, y)$, express $F(x, y)$ in terms of f.

c) For X and Y with joint density $f(x, y)$, express $f(x, y)$ in terms of F.

These are analogs of formulae of Section 4.5 for cumulative distribution functions in one dimension. They are not used much, as there are few joint distributions for which there is an explicit formula for $F(x, y)$. But here are two examples.

d) Find $F(x, y)$ in terms of the marginal c.d.f's for independent X and Y.

e) Find $F(x, y)$ for X the minimum and Y the maximum of n independent uniform $(0, 1)$ variables, and $0 < x < y < 1$. Deduce the joint density of X and Y.

16. Suppose X_1, X_2, X_3 are independent exponential random variables with parameters $\lambda_1, \lambda_2, \lambda_3$ respectively. Evaluate $P(X_1 < X_2 < X_3)$.

17. Let (X, Y) be picked uniformly from the unit disc $R^2 \leq 1$, where $R^2 = X^2 + Y^2$. Find:

a) the joint density of R and X;

b) repeat a) for a point (X, Y, Z) picked at random from inside the unit sphere $R^2 \leq 1$, where now $R^2 = X^2 + Y^2 + Z^2$.

18. Suppose X_1, X_2 are independent random variables with the same density function.

a) Evaluate $P(X_1 < X_2)$.

b) Continuing, suppose X_1, X_2, X_3 are independent random variables with the same density function. Evaluate $P(X_{i_1} < X_{i_2} < X_{i_3})$ where (i_1, i_2, i_3) is a given permutation of $(1, 2, 3)$.

19. Let Lat be the latitude, Lon the longitude of the point of impact of the next meteorite that strikes the Earth's surface. Measure Lat in degrees from $-90°$ (South Pole) to $+90°$ (North Pole), and measure Lon similarly from $-180°$ to $+180°$. Assuming the point of impact is uniformly distributed over the Earth's surface, find

a) the density of Lon; b) the density of Lat;

c) the joint density of Lat and Lon. d) Are Lat and Lon independent?

20. Let X and Y be independent and uniform $(0, 1)$ and let $R = \sqrt{X^2 + Y^2}$. Show that:

a) $$f_R(r) = \begin{cases} \frac{\pi}{2} r & 0 \le r \le 1 \\ 2r \left[\frac{\pi}{4} - \arccos(1/r) \right] & 1 \le r \le \sqrt{2} \end{cases}$$

b) $$F_R(r) = \begin{cases} \frac{1}{4}\pi r^2 & 0 \le r \le 1 \\ \sqrt{r^2 - 1} + \left[\frac{\pi}{4} - \arccos(1/r) \right] r^2 & 1 \le r \le \sqrt{2} \end{cases}$$

c) Show without explicitly calculating $E(R)$ that

$$\sqrt{\frac{1}{2}} < E(R) < \sqrt{\frac{2}{3}}$$

d) (*Hard.*) Show that $E(R) \approx 0.765$.

21. Suppose two points are picked at random from the unit square. Let D be the distance between them. The main point of this problem is to find $E(D)$. This is hard to do exactly by calculus. But some information about $E(D)$ can be obtained as follows.

a) It is intuitively clear that $E(D)$ must be greater than $E(D_{\text{center}})$, where D_{center} is the distance from one point picked at random to the center of the square, and less than $E(D_{\text{corner}})$, the expected distance of one point from a particular corner of the square. Assuming this to be the case, find the values of these bounds on $E(D)$ using the results of Exercise 20.

b) Compute $E(D^2)$ exactly.

c) Deduce from b) a better upper bound for $E(D)$.

d) Computer simulation of 10,000 pairs of points gave mean distance 0.5197, and mean square distance 0.3310. Use these results to find an approximate 95% confidence interval for the unknown value of $E(D)$.

5.3 Independent Normal Variables

The most important properties of the normal distribution involve two or more independent normal variables. Suppose first that X and Y are independent, each with standard normal density function

(a) $$\phi(z) = ce^{-\frac{1}{2}z^2} \quad \text{where the formula} \quad c = \frac{1}{\sqrt{2\pi}}$$

taken for granted up to now, will be verified in this section. The joint density of X and Y is given by

(b) $$f(x, y) = \phi(x)\phi(y) = c^2 e^{-\frac{1}{2}(x^2+y^2)}$$

The key property of this joint density is that it is a function of $r^2 = x^2 + y^2$, where r is the radial distance from the origin of the point (x, y). This makes the graph of this joint density a round bell-shaped surface over the (x, y) plane, with cross sections proportional to the standard normal curve.

FIGURE 1. Perspective plot of the joint density of X and Y.

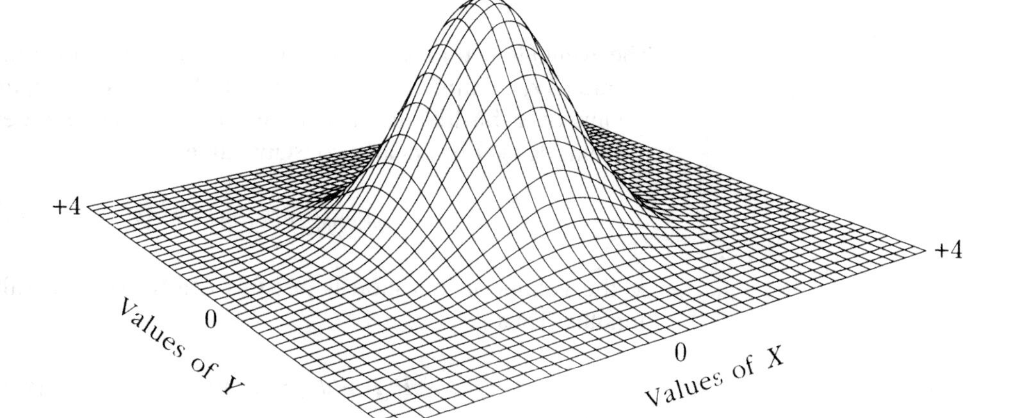

The rotational symmetry of this bivariate distribution obtained from two independent normal variables is a very special property. It can be shown that this property distinguishes the normal distribution from all other probability distributions on the line. And this rotational symmetry is the key to understanding several important properties of the normal distribution, now considered in turn.

Evaluation of the Constant of Integration

The value of the constant c in the normal density (a) is found as a byproduct of calculating the distribution of the random variable

$$R = \sqrt{X^2 + Y^2}$$

which is the distance from the origin of a random point (X, Y) with joint density $\phi(x)\phi(y)$.

FIGURE 2. Geometry of X, Y, and R.

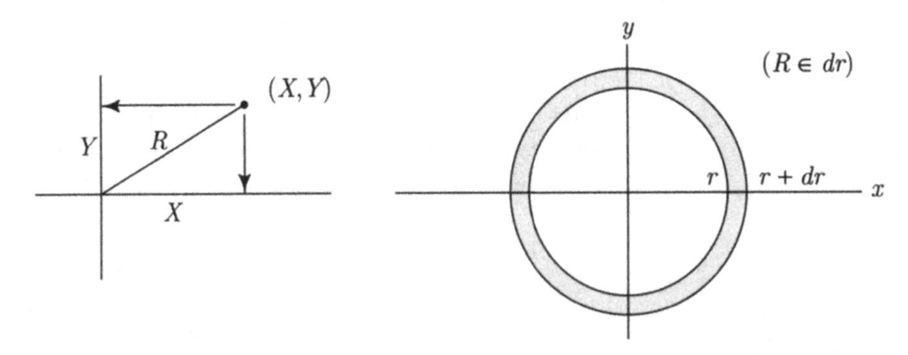

The event $(R \in dr)$ corresponds to (X, Y) falling in an annulus of infinitesimal width dr, radius r, circumference $2\pi r$, and area $2\pi r\, dr$, as in Figure 2. And $P(R \in dr)$ is the volume over this infinitesimal annulus beneath the joint density. But on the annulus the joint density has nearly constant value

$$\phi(x)\phi(y) = c^2 e^{-\frac{1}{2}(x^2+y^2)} = c^2 e^{-\frac{1}{2}r^2}$$

so the volume in question is just this nearly constant value times the area of the annulus. Thus

$$P(R \in dr) = 2\pi\, r\, dr\, c^2 e^{-\frac{1}{2}r^2} \qquad (r > 0)$$

This shows that R has probability density function

$$f_R(r) = 2\pi r c^2 e^{-\frac{1}{2}r^2}$$

The integral of this density from 0 to ∞ must be 1:

$$1 = \int_0^\infty 2\pi r c^2 e^{-\frac{1}{2}r^2} dr = -2\pi c^2 e^{-\frac{1}{2}r^2}\Big|_0^\infty = 2\pi c^2$$

This makes

$$2\pi c^2 = 1 \qquad \text{and} \qquad c = 1/\sqrt{2\pi}$$

So the constant of integration in the normal density involves π, due to the fact that the joint density of two independent standard normal variables is constant on circles centered at the origin.

The distribution of R appearing here, with density function

(c1) $$f_R(r) = r\, e^{-\frac{1}{2}r^2} \qquad (r > 0)$$

and c.d.f.

(c2) $$F_R(r) = \int_0^r s e^{-\frac{1}{2}s^2} ds = 1 - e^{-\frac{1}{2}r^2} \qquad (r > 0)$$

is called the *Rayleigh distribution.*

FIGURE 3. Density of the Rayleigh distribution of R.

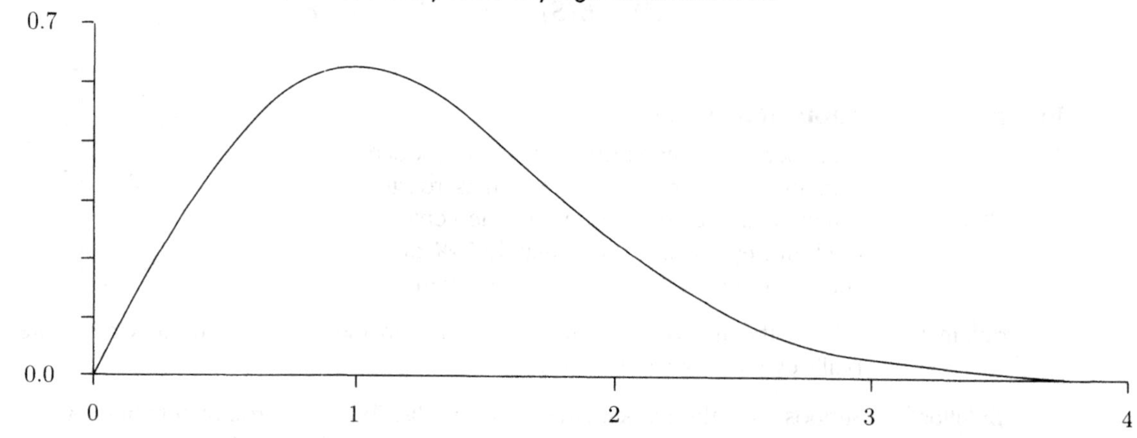

Calculating the Variance of the Standard Normal Distribution

Since $E(X) = 0$ by symmetry, the variance of a standard normal random variable X is

$$\sigma^2 = E(X^2) = \frac{1}{\sqrt{2\pi}} \int_{-\infty}^{\infty} x^2 e^{-\frac{1}{2}x^2} dx$$

This integral can be reduced by an integration by parts to the integral of the standard normal density (exercise). But two independent standard normal variables X and Y

can also be used to show that $\sigma^2 = 1$. This, too, involves the radial random variable R. Because $R^2 = X^2 + Y^2$,

$$E(R^2) = E(X^2) + E(Y^2) = 2E(X^2)$$

using the fact that X and Y have the same distribution. So

$$\sigma^2 = E(X^2) = \frac{1}{2}E(R^2)$$

But $S = R^2$ has density given by the change of variable formula

$$\begin{aligned} f_S(s) &= f_R(r) \Big/ \frac{ds}{dr} \qquad (s = r^2 > 0) \\ &= re^{-\frac{1}{2}r^2}/2r \qquad (s = r^2 > 0) \qquad \text{by (c1)} \\ &= \frac{1}{2}e^{-\frac{1}{2}s} \qquad (s > 0) \end{aligned}$$

Since this is the exponential density with parameter $\lambda = 1/2$,

$$E(R^2) = E(S) = 1/\lambda = 2 \qquad \text{so} \quad \sigma^2 = 1$$

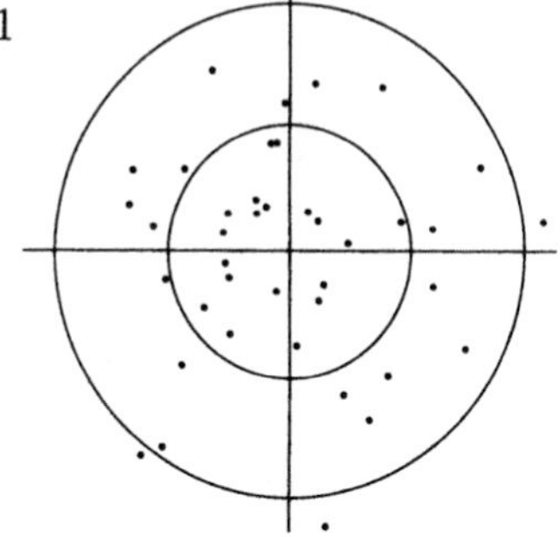

Example 1. Shots at a target.

An expert marksman firing at a target produces a random scatter of shots which is roughly symmetrically distributed about the center of the bull's eye, with approximately 50% of the shots in the bull's eye, as in the diagram.

Problem 1. What is the approximate fraction of shots inside a circle with the same center as the bull's eye, but twice the radius?

Solution. Suppose that the marksman's shots are distributed approximately like (X, Y), where X and Y are independent normal random variables with mean 0 and variance σ^2. This would give such a symmetric distribution. By measuring distances in standard units, that is, relative to σ, we may as well assume $\sigma = 1$. Then the formulae obtained above for the distribution of $R = \sqrt{X^2 + Y^2}$ apply directly. Let r denote the radius of the bull's eye, measured in standard units. Using the normal approximation, the probability of each shot hitting the bull's eye would be

$$F_R(r) = 1 - e^{-\frac{1}{2}r^2}$$

from formula (c2) on page 359. Estimating this probability as 50% from the empirical data gives

$$e^{-\frac{1}{2}r^2} = 1/2 \qquad \text{so} \qquad r = \sqrt{2\log(2)} = 1.177\ldots \text{standard units}$$

Similarly, the fraction of shots inside a circle of twice the radius of the bull's eye should be approximately

$$F_R(2r) = 1 - e^{-\frac{1}{2}(2r)^2} = 1 - (e^{-\frac{1}{2}r^2})^4 = 1 - (1/2)^4 = \frac{15}{16} = 0.9375$$

Problem 2. What is the approximate average distance of the marksman's shots from the center of the bull's eye?

Solution. Using the law of large numbers, this average should be approximately

$$\begin{aligned}
E(R) &= \int_0^\infty r f_R(r) dr = \int_0^\infty r^2 e^{-\frac{1}{2}r^2} dr \qquad \text{by (c1) on page 359} \\
&= \frac{1}{2}\int_{-\infty}^\infty x^2 e^{-\frac{1}{2}x^2} dx \qquad \text{by symmetry} \\
&= \frac{\sqrt{2\pi}}{2}\int_{-\infty}^\infty x^2 \phi(x) dx \qquad \text{by definition of } \phi(x) \\
&= \sqrt{\frac{\pi}{2}} \qquad \text{because standard normal variance is 1} \\
&\approx 1.253 \text{ standard units} \\
&\approx 1.253/1.177 = 1.065 \text{ times the bull's eye radius } r
\end{aligned}$$

Linear Combinations and Rotations

Linear combinations of independent normal variables are always normally distributed. This important fact is another consequence of the rotational symmetry of the joint distribution of independent standard normal random variables X and Y. To see why, let X_θ be the first coordinate of (X, Y) relative to new coordinate axes set up at angle θ relative to the original X and Y axes, as in Figure 4.

As the diagram shows,

$$X_\theta = X \cos\theta + Y \sin\theta$$

But due to the rotational symmetry of the joint distribution, it is clear without calculation that the probability distribution of X_θ must be the same as that of X, namely, standard normal, no matter what the angle θ of rotation. For example, the event $x \le X_\theta \le x + \Delta x$ corresponding to (X, Y), falling in the area shaded in the left diagram of Figure 5, must have the same probability as the event $x \le X \le x + \Delta x$ corresponding to (X, Y), falling in the area shaded in the right diagram, because the shape of the bivariate normal density is the same over the two shaded regions. So,

$$P(x \le X_\theta \le x + \Delta x) = P(x \le X \le x + \Delta x)$$

FIGURE 4. Projection X_θ onto axis at angle θ to X-axis: $X_\theta = X \cos\theta + Y \sin\theta$.

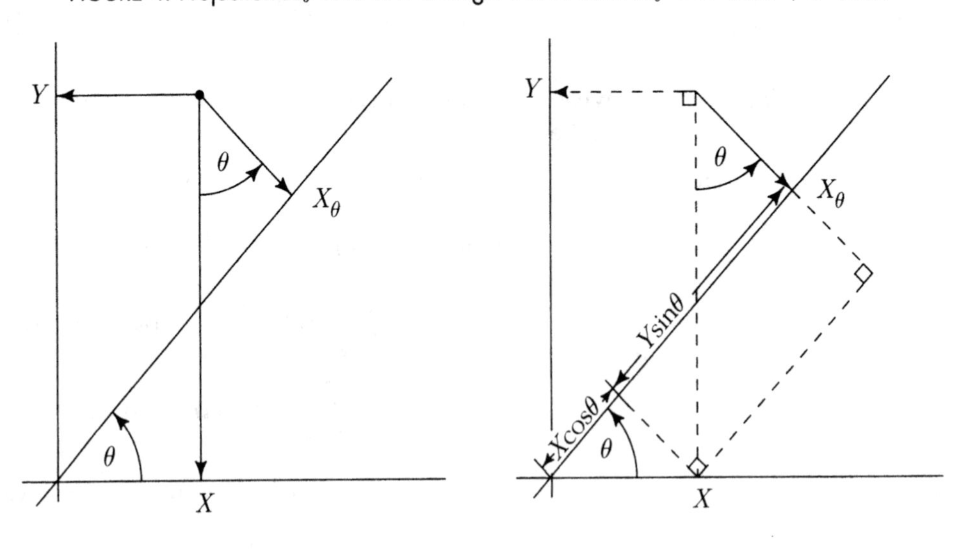

FIGURE 5. Events $(x \le X_\theta \le x + \Delta x)$ and $(x \le X \le x + \Delta x)$. Rotational symmetry of the joint density implies these two events have the same probability.

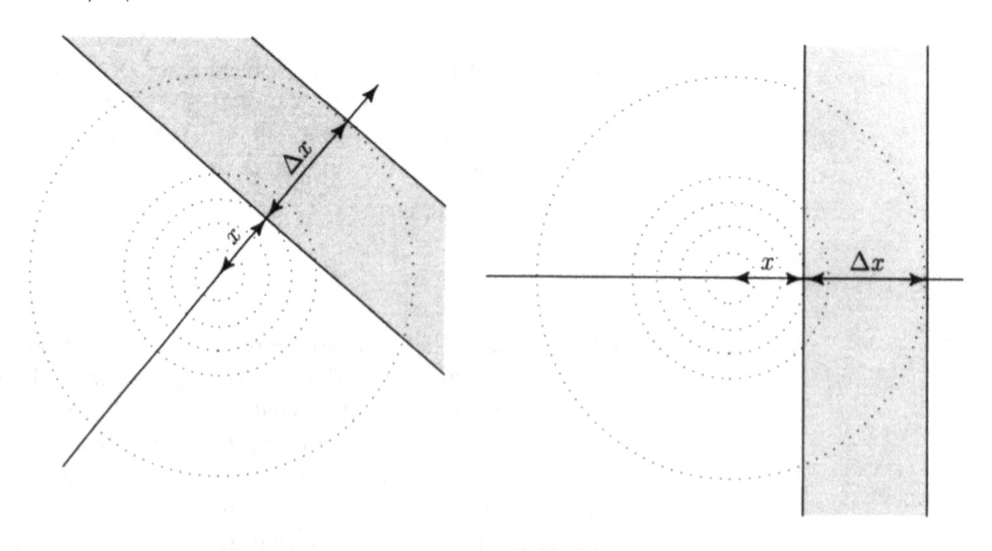

for every x and Δx. This shows that X_θ has normal $(0,1)$ distribution, for every θ. Since $\cos\theta$ and $\sin\theta$ may be arbitrary numbers α and β, subject only to the constraint that $\alpha^2+\beta^2=1$, the rotational symmetry of the joint distribution of two independent normal variables X and Y implies that:

(d) *If X and Y are two independent normal $(0,1)$ random variables, then $\alpha X+\beta Y$ has normal $(0,1)$ distribution for all α and β with $\alpha^2+\beta^2=1$.*

In particular, taking $\alpha=\beta=1/\sqrt{2}$, corresponding to rotation by 45°:

(e) *If X and Y are independent normal $(0,1)$ random variables, then $(X+Y)/\sqrt{2}$ has normal $(0,1)$ distribution.*

If Z has normal $(0,1)$ distribution, then σZ has normal $(0,\sigma^2)$ distribution. Taking $\sigma=\sqrt{2}$, (e) implies:

(f) *If X and Y are independent normal $(0,1)$ random variables, then $X+Y$ has normal $(0,2)$ distribution.*

This argument extends to give the following general conclusion, which includes (d), (e), and (f), as special cases.

Sums of Independent Normal Variables

If X and Y are independent with normal (λ,σ^2) and normal (μ,τ^2) distributions, then $X+Y$ has normal $(\lambda+\mu,\sigma^2+\tau^2)$ distribution.

Proof. Recall that X has normal (λ,σ^2) distribution if and only if $(X-\lambda)/\sigma$ has normal $(0,1)$ distribution. Transform all the variables to standard units by letting

$$U=(X-\lambda)/\sigma \quad\text{and}\quad V=(Y-\mu)/\tau \quad\text{and}\quad W=\frac{X+Y-(\lambda+\mu)}{\sqrt{\sigma^2+\tau^2}}$$

Then U and V are independent normal $(0,1)$ random variables. By algebra,

$$W=\alpha U+\beta V \quad\text{where}\quad \alpha^2=\frac{\sigma^2}{\sigma^2+\tau^2} \quad\text{and}\quad \beta^2=\frac{\tau^2}{\sigma^2+\tau^2} \quad\text{so}\quad \alpha^2+\beta^2=1$$

Apply (d) above with (U,V) instead of (X,Y) to deduce that W has normal $(0,1)$ distribution. So $X+Y=(\lambda+\mu)+\sqrt{\sigma^2+\tau^2}W$ has normal $(\lambda+\mu,\sigma^2+\tau^2)$ distribution.
□

Several Independent Normal Variables

The result that the sum of two independent normal variables is normal extends to sums and linear combinations of several independent normal random variables, by repeated applications of the result for two variables. For example, if $X_1, \ldots, X_n$ are independent and normal $(0, 1)$, then $X_1 + \cdots + X_n$ has normal $(0, n)$ distribution, with standard deviation $\sqrt{n}$.

Example 2. Linear combinations of normals.

For $\sigma = 1, 2, 3$ suppose X_σ has normal $(0, \sigma^2)$ distribution, and these three random variables are independent.

Problem 1. Find $P(X_1 + X_2 + X_3 < 4)$.

Solution. Let $S = X_1 + X_2 + X_3$. Then S has normal $(0, 1^2 + 2^2 + 3^2)$ distribution, and if $Z = S/\sqrt{14}$ is S standardized, the problem is just to find

$$P(S < 4) = P(Z < 4/\sqrt{14}) = \Phi(4/\sqrt{14}) \approx 0.857$$

Problem 2. Find $P(4X_1 - 10 < X_2 + X_3)$.

Solution. Rearranging the statement of the inequality shows this is the same as

$$P(4X_1 - X_2 - X_3 < 10) = P(L < 10) \qquad \text{where } L = 4X_1 - X_2 - X_3$$

Since the linear combination L has normal distribution with mean 0 and variance $4^2 \times 1^2 + (-1)^2 \times 2^2 + (-1)^2 \times 3^2 = 29$, the probability is

$$P(L < 10) = \Phi(10/\sqrt{29}) \approx 0.968$$

The Chi-Square Distribution

By the same calculation as in two dimensions, the joint density of n independent normal variables at every point on the sphere of radius r in n-dimensional space is $(1/\sqrt{2\pi})^n \exp\left(-\frac{1}{2}r^2\right)$. This joint density is symmetric with respect to arbitrary rotations of the coordinates in n-dimensional space, or *spherically symmetric*. So a cloud of points (or a galaxy of stars), in ordinary 3-dimensional space, with approximately independent normally distributed coordinates with common variance, appears spherical when viewed at a distance, from any perspective. For independent standard normal Z_i let

$$R_n = \sqrt{Z_1^2 + \cdots + Z_n^2}$$

denote the distance of $(Z_1, \ldots, Z_n)$ from the origin in n-dimensional space. The n-dimensional volume of a thin spherical shell of thickness dr at radius r is $c_n r^{n-1}\, dr$

where c_n is the $(n-1)$-dimensional volume of the "surface" of a sphere of radius 1 in n dimensions. (For $n = 3$, $c_3 = 4\pi$, by the formula $4\pi r^2$ for the surface area of a sphere of radius r in 3 dimensions.) The same argument used in two dimensions shows that

$$P(R_n \in dr) = c_n\, r^{n-1} \left(1/\sqrt{2\pi}\right)^n e^{-\frac{1}{2}r^2} dr \qquad (r > 0) \tag{1}$$

A change of variable allows the constant c_n to be evaluated by recognizing that the density of $R_n^2 = Z_1^2 + \cdots + Z_n^2$ is the gamma $(n/2, 1/2)$ density introduced in Section 4.2:

$$f_{R_n^2}(t) = (2^{n/2}\Gamma(n/2))^{-1} t^{(n/2)-1} e^{-t/2} \qquad (t > 0) \tag{2}$$

Exercise 15 and Chapter 5 Review Exercise 26 give formulae for c_n and $\Gamma(n/2)$.

Statisticians call this gamma $(n/2, 1/2)$ distribution of R_n^2 the *chi-square* distribution with n *degrees of freedom*. The chi-square distribution provides a useful test of *goodness of fit*, that is, how well data from an empirical distribution of n observations conform to the model of random sampling from a particular theoretical distribution. If there are only two categories, say success and failure, the model of independent trials with probability p of success is tested using the normal approximation to the binomial distribution. But for data in several categories the problem is how to combine the tests for different categories in a reasonable way. This problem was solved as follows by the statistician Karl Pearson (1857–1936). For a finite number of categories m, let N_i denote the number of results in category i. Under the hypothesis that the N_i are counting results of independent trials with probability p_i for category i on each trial, it turns out that no matter what the probabilities p_i, for large enough n the so-called *chi-square statistic*

$$\sum_{i=1}^{m} \frac{(N_i - n\,p_i)^2}{n\,p_i}$$

that is the sum over categories of (observed − expected)^2/expected, has distribution that is approximately chi-square with $m-1$ degrees of freedom. In statistical jargon, a value of the statistic higher than the 95th percentile point on the chi-square distribution with $m-1$ degrees of freedom would "reject the hypothesis at the 5% level". Unusually small values of the chi-square statistic are sometimes taken as evidence to suggest that an observer fudged the data to suit the hypothesis. The exact joint distribution of the N_i under the hypothesis of randomness is multinomial with parameters n and $p_1, \ldots, p_m$. The above result can be derived from a multivariate form of the normal approximation to the binomial. The joint distribution of $N_1, \ldots, N_m$ is essentially $m-1$ dimensional due to the constraint $N_1 + \cdots + N_m = n$. This is why the relevant chi-square distribution has $m-1$ degrees of freedom.

For tables of the chi-square distribution, and similar chi-square tests of other hypotheses such as independence, consult a statistics book. The mean, standard deviation and skewness of the chi-square distribution of R_n^2 with n degrees of freedom are easily calculated (Exercise 15):

$$E(R_n^2) = n, \qquad SD(R_n^2) = \sqrt{2n} \qquad \text{and} \qquad \text{Skewness}(R_n^2) = 4/\sqrt{2n}$$

For large n the chi-square distribution is approximately normal, by the central limit theorem. Because the skewness is quite large even for moderate values of n, the normal approximation with skewness correction gives the better approximation

$$P(R_n^2 \leq x) \approx \Phi(z) - \frac{\sqrt{2}}{3\sqrt{n}}(z^2 - 1)\phi(z) \quad \text{where} \quad z = (x - n)/\sqrt{2n} \text{ and } x > 0$$

TABLE 1. **Distribution of radial distance in three dimensions.** The probability that a point with independent standard normal coordinates in three dimensions lies inside a sphere of radius r, that is, $P(R_3 \leq r) = P(R_3^3 \leq r^2)$, was obtained by numerical integration of the density. These probabilities are shown along with their approximations obtained using the skew–normal approximation to the chi-square (3) distribution of R_3^2. The approximations are surprisingly good considering the small value of n.

radius r	1	2	3	4
probability $P(R_3 \leq r)$	0.199	0.739	0.971	0.999
skew-normal approximation	0.233	0.741	0.966	1.000

Exercises 5.3

1. Continuing Example 1, calculate the following, where all distances are measured in standard units:

a) the probability of a shot falling inside a circle of radius $1/2$;

b) the probability of a shot falling in the region of the positive quadrant between radii 1 and 2;

c) the approximate average absolute distance of the shots from the horizontal line through the center of the bull's eye;

d) the probability that a shot hit within distance r of the vertical axis through the center (r = radius of bull's eye in standard units);

e) the probability of hitting a square touching the outside of the bull's eye;

f) the probability of hitting a square touching the inside of the bull's eye;

g) the probability of hitting a rectangle of sides r and $2r$ positioned as shown relative to the bull's eye.

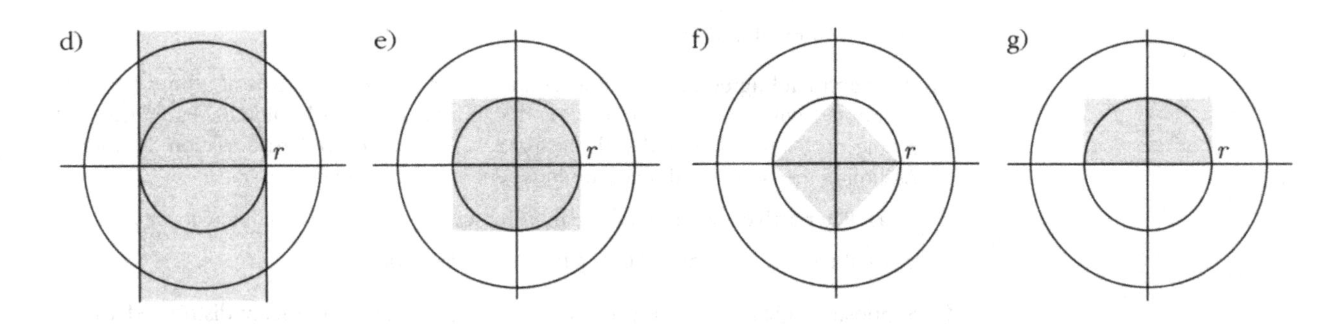

2. Let X and Y be independent random variables, with $E(X) = 1$, $E(Y) = 2$, $Var(X) = 3$, and $Var(Y) = 4$.

a) Find $E(10X^2 + 8Y^2 - XY + 8X + 5Y - 1)$.

b) Assuming all variables are normally distributed, find $P(2X > 3Y - 5)$.

3. W, X, Y and Z are independent standard normal random variables. Find (no integrations are necessary!)

a) $P(W + X > Y + Z + 1)$; b) $P(4X + 3Y < Z + W)$;

c) $E(4X + 3Y - 2Z^2 - W^2 + 8)$; d) $SD(3Z - 2X + Y + 15)$.

4. Suppose the true weight of a standard weight is 10 grams. It is weighed twice independently. Suppose that the first measurement is a normal random variable X with $E(X) = 10$ g and $SD(X) = 0.2$ g, and that the second measurement is a normal random variable Y with $E(Y) = 10$ g and $SD(Y) = 0.2$ g.

a) Compute the probability that the second measurement is closer to 10 g than the first measurement.

b) Compute the probability that the second measurement is smaller than the first, but not by more than 0.2 g.

5. Let X and Y be independent and normally distributed, X with mean 0 and variance 1, Y with mean 1. Suppose $P(X > Y) = 1/3$. Find the standard deviation of Y.

6. Let X and Y be independent standard normal variables. Find:

a) $P(3X + 2Y > 5)$; b) $P(\min(X, Y) < 1)$;

c) $P(|\min(X, Y)| < 1)$; d) $P(\min(X, Y) > \max(X, Y) - 1)$.

7. Suppose the AC Transit bus is scheduled to arrive at my corner at 8:10 A.M., but its actual arrival time is a normal random variable with mean 8:10 A.M., and standard deviation 40 seconds. Suppose I try to arrive at the corner at 8:09, but my arrival time is actually normally distributed with mean 8:09 A.M., and standard deviation 30 seconds.

a) What percentage of the time do I arrive at the corner before the bus is scheduled to arrive?

b) What percentage of the time do I arrive at the corner before the bus does?

c) If I arrive at the stop at 8:09 A.M. and the bus still hasn't come by 8:12 A.M., what is the probability that I have already missed it?

(State your assumptions carefully.)

8. Peter and Paul agree to meet at a restaurant at noon. Peter arrives at a time normally distributed with mean 12:00 noon, and standard deviation 5 minutes. Paul arrives at a time normally distributed with mean 12:02 P.M., and standard deviation 3 minutes. Assuming the two arrival times are independent, find the chance that

a) Peter arrives before Paul; b) both men arrive within 3 minutes of noon;

c) the two men arrive within 3 minutes of each other.

9. Suppose heights in a large population are approximately normally distributed with a mean of 5 feet 10 inches and an SD of 2 inches. Suppose a group of 100 people is picked at random from this population.

a) What is the probability that the tallest person in this group is over 6 feet 4 inches tall?

b) What is the probability that the average height of people in the group is over 5 feet 10.5 inches?

c) Suppose instead that the distribution of heights in the population was not normal, but some other distribution with the given mean and SD. To which of the problems a) and b) would the answer still be approximately the same? Explain carefully.

10. In a large corporation, people over age thirty have an annual income whose distribution can be approximated by a normal distribution with mean $60,000 and standard deviation $10,000. The incomes of those under age thirty are also approximately normal, but with mean $40,000 and standard deviation $10,000.

a) Two people are selected at random from those over age thirty. What is the chance that the average of their two incomes is over $65,000?

b) One person is selected at random from those over thirty, and independently, one person is selected at random from those under thirty. What is the chance that the younger's income exceeds the older's?

c) What is the chance that the smaller of the two incomes in b) exceeds $50,000?

11. Einstein's model for Brownian motion. Suppose that the X coordinate of a particle performing Brownian motion has normal distribution with mean 0 and variance σ^2 at time 1. Let X_t be the X displacement after time t. Assume the displacement over any time interval has a normal distribution with parameters depending only on the length of the interval, and that displacements over disjoint time intervals are independent.

a) Find the distribution of X_t.

b) Let (X_t, Y_t) represent the position at time t of a particle moving in two dimensions. Assume that X_t and Y_t are independent Brownian motions starting at 0 at time $t = 0$. Find the distribution of $R_t = \sqrt{X_t^2 + Y_t^2}$, and give the mean and standard deviation in terms of σ and t.

c) Suppose a particle performing Brownian motion (X_t, Y_t) as in b) has an X coordinate after one second which has mean 0 and standard deviation one millimeter (mm). Calculate the probability that the particle is more than 2 mm from the point $(0, 0)$ after one second.

12. Suppose two shots are fired at a target. Assume each shot hits with independent normally distributed coordinates, with the same means and equal unit variances.

a) Find the mean of the distance between the points where the two shots strike.

b) Find the variance of the same random variable.

13. Independence of radial and angular parts. Let X and Y be independent normal $(0, \sigma^2)$ random variables. Let (R, Θ) be (X, Y) in polar coordinates, so $X = R\cos\Theta$, $Y = R\sin\Theta$.

a) Show that R and Θ are independent, and that Θ has uniform $(0, 2\pi)$ distribution.

b) Let R and Θ now be arbitrary random variables such that R/σ has the Rayleigh distribution (c1), Θ has uniform $(0, 2\pi)$ distribution, and R and Θ are independent. Explain why the random variables $X = R\cos\Theta$ and $Y = R\sin\Theta$ must be independent normal $(0, \sigma^2)$.

c) Find functions h and k such that if U and V are independent uniform $(0, 1)$ random variables, then $X = \sigma h(U)\cos[k(V)]$ and $Y = \sigma h(U)\sin[k(V)]$ are independent normal $(0, \sigma^2)$. [This gives a means of simulating normal random variables using a computer random number generator. Try generating a random scatter of independent bivariate normally distributed pairs if you have random numbers available. It should look like the scatter in Example 1.]

14. Let X and Y be independent standard normal variables. Suppose they are transformed into polar coordinates, $X = R\cos\Theta$ and $Y = R\sin\Theta$ with $0 < \Theta < 2\pi$ and $0 < R < \infty$, as in Exercise 13.

a) Derive the distribution of $2\Theta \ \mathit{mod}\ 2\pi$. [The quantity $x \ \mathit{mod}\ a$ denotes the remainder when x is divided by a.]

b) Derive the joint distribution of $R\cos 2\Theta$ and $R\sin 2\Theta$.

c) Show that both

$$\frac{2XY}{\sqrt{X^2+Y^2}} \quad \text{and} \quad \frac{X^2-Y^2}{\sqrt{X^2+Y^2}}$$

have the standard normal distribution. Are they independent?

15. Chi-square distributions. These are the special case of half-integer gamma distributions which come from sums of squares of independent standard normal variables. Show:

a) If Z has standard normal distribution, then Z^2 has gamma $(1/2, 1/2)$ distribution, and $\Gamma(1/2) = \sqrt{\pi}$.

b) If n is an odd integer, then $\quad \Gamma(n/2) = \dfrac{\sqrt{\pi}(n-1)!}{2^{n-1}(\frac{n-1}{2})!}$

c) If X has normal $(0, \sigma^2)$ distribution, then X^2 has gamma $(1/2, 1/2\sigma^2)$ distribution.

d) If $Z_1, \ldots, Z_n$ are independent standard normal random variables, then ${Z_1}^2 + \cdots + {Z_n}^2$ has gamma $(n/2, 1/2)$ distribution, also known as the *chi-square distribution with n degrees of freedom*, or chi-square (n) distribution.

e) If $Y_1, \ldots, Y_n$ are independent chi-square random variables with $k_1, \ldots, k_n$ degrees of freedom, respectively, then $Y_1 + \cdots + Y_n$ has chi-square $(k_1 + \cdots + k_n)$ distribution.

f) The mean, variance and skewness of the chi-square (n) distribution are as stated on page 366.

16. Poisson formula for the chi-square $(2m)$ c.d.f. For $m = 1, 2, \ldots$ let R_{2m}^2 have chi-square $(2m)$ distribution. Use the connection between the gamma distribution and the Poisson process to find formulae in terms of appropriate Poisson probabilities for:

a) the c.d.f. of R_{2m}^2; b) the c.d.f. of R_{2m}.

c) Check that your formulae agree with the formulae in the text for $m = 1$. Now make a table of $P(R_4 \le r)$ for $r = 1, \ldots 5$.

17. Skew-normal approximation to the chi-square distribution. Let R_n^2 have chi-square (n) distribution.

a) Find the approximation to $P(R_4 \le r)$ for $r = 1, \ldots 5$ obtained from the skew-normal approximation to the distribution of R_4^2. Compare to the exact results found in Exercise 16.

b) Find both the plain normal approximation and the skew-normal approximation to $P(R_{10}^2 \le 9.34) = 0.500$. Which approximation is better?

18. Suppose a large number n identical molecules are distributed independently at random in a box with sides of 1 centimeter. Let X, Y, Z be the coordinates in centimeters of the center of mass of the n molecules at a particular instant, relative to the center of the box. Thus,

$$X = (X_1 + \cdots + X_n)/n$$

and so on, where (X_i, Y_i, Z_i) are the coordinates of the ith molecule in centimeters. Let $R = \sqrt{X^2 + Y^2 + Z^2}$ be the distance of the center of mass of the n molecules from the center of the box. Given that for the chi-square distribution with 3 degrees of freedom the 95th percentile is at 7.82, find approximately the value of r such that R is 95% sure to be smaller than r.

5.4 Operations (Optional)

Many applications require calculation of the distribution of some random variable Z which is a function of X and Y, where X and Y are random variables with some joint density $f(x, y)$. Here the function of X and Y might be, for example, $X + Y$, XY, X/Y, $\max(X, Y)$, $\min(X, Y)$, or $\sqrt{X^2 + Y^2}$. This kind of calculation has been done in special cases in previous sections. For example, maxima and minima in Section 4.5, sums and $\sqrt{X^2 + Y^2}$ for normal variables in Section 5.3. This section gives a general technique for computing such distributions by integration.

Calculating the whole distribution of a function of X and Y can sometimes be tedious. So keep in mind that for some purposes it may be enough to calculate an expectation. The expectation of a function of X and Y can always be expressed as an integral with respect to the density of (X, Y). For example, for the product XY,

$$\begin{aligned} E(XY) &= \iint xy f(x, y) dx\, dy \\ &= E(X)E(Y) \qquad \text{if } X \text{ and } Y \text{ are independent} \end{aligned}$$

despite the fact that there are very few examples where the whole distribution of a product of independent random variables can be found explicitly.

One method of finding the distribution of $Z = g(X, Y)$ is to find the c.d.f. $P(Z \leq z)$ by integration of $f(x, y)$ over the region in the (x, y) plane where $g(x, y) \leq z$. Provided this integral can be evaluated fairly explicitly, the density of Z can then be found by differentiation of the c.d.f. Usually a quicker method of finding the distribution of Z is to anticipate that Z will have a density function f_Z, and to find this density $f_Z(z) = P(Z \in dz)/dz$ by integrating the joint density of X and Y over the subset $(Z \in dz)$ in the (X, Y) plane. This technique gives integral formulae for the density for the sum $X + Y$, for other linear combinations like $X - Y$, and for the product XY, and ratio X/Y. The formulae for sums and ratios will now be worked out in detail. Results for other operations are similar and left as exercises.

Distribution of Sums

A good deal has already been said on this topic. Recall the addition rule for expectation

$$E(X + Y) = E(X) + E(Y) \qquad \text{whatever the joint distribution of } X \text{ and } Y$$

the addition rule for variances in the case of independence, and the central limit theorem governing the asymptotic distribution for the sum of a large number of independent and identically distributed terms. Also, the exact distribution of sums has been computed in special cases by a variety of methods. The following table reviews some important examples:

Distribution of terms	Distribution of sum	See Section
n independent Bernoulli (p)	binomial (n, p)	2.1
independent Poisson (μ_i)	Poisson $(\Sigma\mu_i)$	3.5
independent normal (μ_i, σ_i^2)	normal $(\Sigma\mu_i, \Sigma\sigma_i^2)$	5.3
r independent geometric (p)	negative binomial (r, p)	3.4
r independent exponential (λ)	gamma (r, λ)	4.2

In the discrete case the distribution of the sum of random variables is determined by the formula

$$P(X + Y = z) = \sum_{\text{all } x} P(X = x, Y = z - x)$$

found in Section 4.1. The following display gives the corresponding formula for densities:

Density of X + Y

If (X, Y) has density $f(x, y)$ in the plane, then $X + Y$ has density on the line

$$f_{X+Y}(z) = \int_{-\infty}^{\infty} f(x, z - x)dx$$

Density Convolution Formula

If X and Y are independent, then

$$f_{X+Y}(z) = \int_{-\infty}^{\infty} f_X(x)f_Y(z - x)dx$$

Note: If the random variables X and Y are non-negative, then the lower limit of integration in the convolution formula can be changed from $-\infty$ to 0, since $f_X(x) = 0$ for all $x < 0$, and the upper limit can be changed from ∞ to z, since $f_Y(z - x) = 0$ for $x > z$.

The convolution formula is the special case of the formula for the density of $X + Y$ when $f(x, y) = f_X(x)f_Y(y)$ by independence. This operation on probability density functions f_X and f_Y is called *convolution.* It leads to a new density, the density of the sum of random variables X and Y, assumed independent.

To avoid confusion about limits of integration in particular examples, sketch the subset of the plane where the joint density is strictly positive, and the line of integration corresponding to $X + Y = z$, as in examples below.

Derivation of the density of $X + Y$. Let $Z = X + Y$. The event $(Z \in dz)$ is shaded in the following diagram:

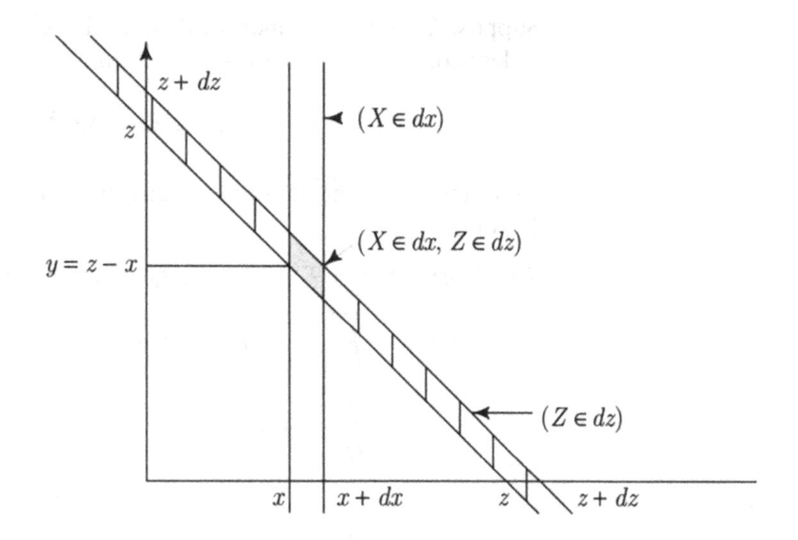

The event $(Z \in dz)$ can be broken up into vertical slices according to the values of X, as suggested by the vertical shading in the diagram. The heavily shaded parallelogram contained in the event $(Z \in dz)$ near the point $(x, z - x)$, represents the intersection of the events $(X \in dx)$ and $(Z \in dz)$, and has area $dx\,dz$. The probability density near this little parallelogram is $f(x, z - x)$, so

(a1) $$P(X \in dx, Z \in dz) = f(x, z - x)dx\,dz$$

This formula gives the joint density of X and Z. The marginal density of $Z = X + Y$ is therefore obtained by integrating out the x-variable

(a2) $$P(Z \in dz) = \left[\int_{-\infty}^{\infty} f(x, z - x)dx\right] dz$$

This gives the boxed formula for the density of $Z = X + Y$. Intuitively, you can think of (a2) as obtained by summing over infinitesimal parallelograms as in (a1).□

Example 1. Sums of independent exponential variables.

In Section 4.2 a Poisson process argument was used to show that the distribution of the sum of r independent exponential (λ) random variables is gamma (r, λ): If $f_{r,\lambda}(t)$ denotes the density of such a sum, then

$$f_{r,\lambda}(t) = \frac{1}{(r-1)!}\lambda^r t^{r-1} e^{-\lambda t} \qquad (t \geq 0)$$

This fact can also be derived using the convolution formula. Here is the calculation for $r = 2$.

Suppose T and U are independent, each exponentially distributed with rate λ. By independence, the joint density of T and U at (t, u) is

$$f(t,u) = f_T(t)f_U(u) = \lambda e^{-\lambda t}\lambda e^{-\lambda u} = \lambda^2 e^{-\lambda(t+u)} \qquad (t, u \geq 0)$$

Note how this joint density is a function of $t + u$. You can see the effect of this in Figure 1.

The density of $S = T + U$ at s is given by the convolution formula

$$\begin{aligned}
f_S(s) &= \int_{-\infty}^{\infty} f_T(t)f_U(s-t)dt \\
&= \int_0^s f_T(t)f_U(s-t)dt \qquad \text{since } f_T(t) = 0 \text{ if } t < 0 \\
&\qquad\qquad\qquad\qquad\qquad \text{and} \quad f_U(s-t) = 0 \text{ if } t > s \\
&= \int_0^s \lambda e^{-\lambda t}\lambda e^{-\lambda(s-t)}dt \\
&= \int_0^s \lambda^2 e^{-\lambda s}dt \\
&= \lambda^2 s e^{-\lambda s} \qquad (s \geq 0)
\end{aligned}$$

See Figure 1. For small s the factor of s makes the density grow linearly near zero. For large s the exponential factor $e^{-\lambda s}$ brings the density down to zero very rapidly.

Another way to derive this density is to argue infinitesimally: Let $s \geq 0$. The probability of $(S \in ds)$ is the integral of the joint density over the infinite strip $((t,u) : s \leq t + u \leq s + ds)$. We need only integrate over the (approximately) rectangular segment $((t,u) : s \leq t + u \leq s + ds,\ t \geq 0, u \geq 0)$, where the joint density is nonzero. This segment has length $\sqrt{2}s$ and width $ds/\sqrt{2}$, and the joint density has nearly constant value $\lambda^2 e^{-\lambda(t+u)} = \lambda^2 e^{-\lambda s}$ for points (t, u) in this segment; so the desired probability is

$$P(S \in ds) = \sqrt{2}s \cdot ds/\sqrt{2} \cdot \lambda^2 e^{-\lambda s} = \lambda^2 s e^{-\lambda s} ds \qquad (s \geq 0)$$

The fact that the sum of r independent exponential (λ) variables has gamma (r, λ) distribution can be derived from the convolution formula by mathematical induction on r.

FIGURE 1. **Distribution of the sum of two independent exponential variables.** Here is a random scatter of points suggesting the joint density of independent exponential variables T and U, along with graphs of the densities of T, U, and $S = T + U$.

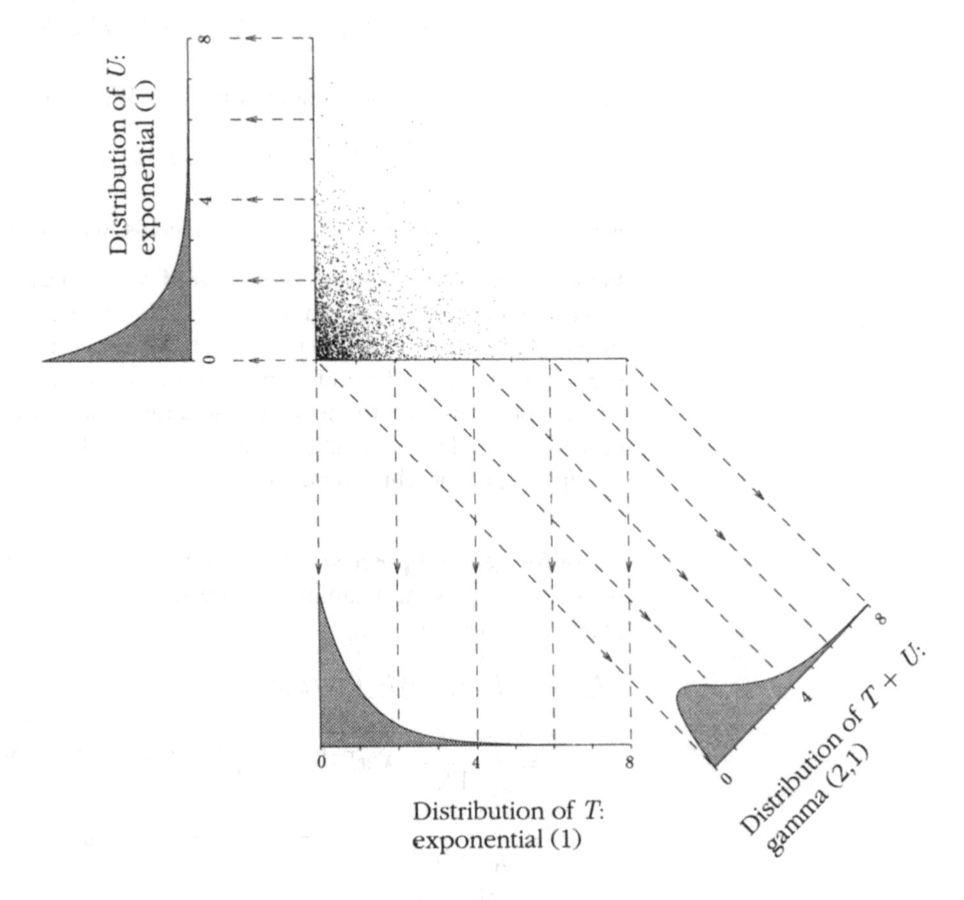

Example 2. Sums of independent gamma variables.

Recall from Section 4.2 that the gamma (r, λ) distribution is defined for every real $r > 0$ by the density

$$f_{r,\lambda}(t) = \begin{cases} [\Gamma(r)]^{-1}\lambda^r t^{r-1} e^{-\lambda t} & t > 0 \\ 0 & t \leq 0 \end{cases} \qquad \text{where} \quad \Gamma(r) = \int_0^\infty t^{r-1} e^{-t} dt$$

If T_r and T_s are independent random variables with gamma (r, λ) and gamma (s, λ) distributions, respectively, then $T_r + T_s$ has gamma $(r + s, \lambda)$ distribution.

Proof for positive integers *r* and *s*. This case follows from the representation of a gamma variable as the sum of independent exponential variables. To see how, note first that the density of an independent sum $T_r + T_s$ is determined by the densities of T_r and T_s, by the convolution formula. So it is enough to derive the result for any convenient pair of independent random variables with gamma (r, λ) and gamma (s, λ) distributions. But the conclusion is obvious if we consider

$$T_r = W_1 + \cdots + W_r \quad \text{and} \quad T'_s = W'_1 + \cdots + W'_s$$

defined by $r + s$ independent exponentials $W_1 \ldots, W_r, W'_1, \ldots, W'_s$. Because then

$$T_r + T'_s = W_1 + \cdots + W_r + W'_1 + \cdots + W'_s$$

is the sum of $r + s$ independent exponentials, with gamma $(r + s, \lambda)$ distribution. □

Proof for positive half-integers r and s. The case $r = n/2$ and $s = m/2$ for positive integers n and m can be derived almost the same way, using the result found in Section 5.3 that the gamma $(n/2, 1/2)$ distribution is the chi-square distribution of the sum of squares of n independent standard normal variables. Adding the sum of squares of n variables to the sum of squares of m variables gives the sum of squares of $n+m$ variables. Changing the rate parameter $1/2$ to a general λ is just a matter of multiplying of the chi-square variables by $1/(2\lambda)$. (See Exercises 5.3.15 and 4.4.2). □

Proof for general positive r and s. For $r > 0$, $s > 0$, let T_r and T_s be independent, with gamma (r, λ) and gamma (s, λ) distributions, and let $Z = T_r + T_s$. Then by the convolution formula

$$\begin{aligned}
f_Z(z) &= \int_0^z f_{T_r}(x) f_{T_s}(z-x)dx \\
&= \int_0^z \frac{1}{\Gamma(r)} \cdot \lambda^r x^{r-1} e^{-\lambda x} \cdot \frac{1}{\Gamma(s)} \lambda^s (z-x)^{s-1} e^{-\lambda(z-x)} dx \\
&= \int_0^z \frac{1}{\Gamma(r)\Gamma(s)} \lambda^{r+s} x^{r-1} (z-x)^{s-1} e^{-\lambda z} dx \\
&= \int_0^1 \frac{1}{\Gamma(r)\Gamma(s)} \lambda^{r+s} (zu)^{r-1} (z - zu)^{s-1} e^{-\lambda z} z\, du \qquad (x = zu, dx = z\, du) \\
&= \frac{1}{\Gamma(r+s)} \lambda^{r+s} z^{r+s-1} e^{-\lambda z} \int_0^1 \frac{\Gamma(r+s)}{\Gamma(r)\Gamma(s)} u^{r-1} (1-u)^{s-1} du \\
&= f_{r+s,\lambda}(z) \int_0^1 \frac{\Gamma(r+s)}{\Gamma(r)\Gamma(s)} u^{r-1} (1-u)^{s-1} du
\end{aligned}$$

where $f_{r+s,\lambda}(z)$ is the gamma $(r+s, \lambda)$ density. The integral on the right is a constant which does not depend on z. Since both $f_Z(z)$ and $f_{r+s,\lambda}(z)$ are probability densities on $(0, \infty)$, integrating both sides with respect to z from 0 to ∞ gives

$$1 = 1 \times \int_0^1 \frac{\Gamma(r+s)}{\Gamma(r)\Gamma(s)} u^{r-1} (1-u)^{s-1} du$$

So the integral must equal 1. Therefore Z has the gamma $(r+s,\lambda)$ density. □

The last line of the previous argument evaluates an important integral:

The Beta Integral

$$B(r,s)=\int_0^1 u^{r-1}(1-u)^{s-1}du=\frac{\Gamma(r)\Gamma(s)}{\Gamma(r+s)} \qquad (r>0, s>0)$$

This evaluation of $B(r,s)$ in terms of the gamma function agrees with the evaluation in Section 4.6 for integer r and s because $\Gamma(r)=(r-1)!$ for positive integers r.

Example 3. Sums of independent uniform variables.

Two terms. Suppose X and Y are independent, each with uniform $(0,1)$ distribution. To find the density of $X+Y$ it is simpler to work directly with a diagram than to use the convolution formula. Here (X,Y) has uniform distribution on the unit square. See Figure 2 on page 380.

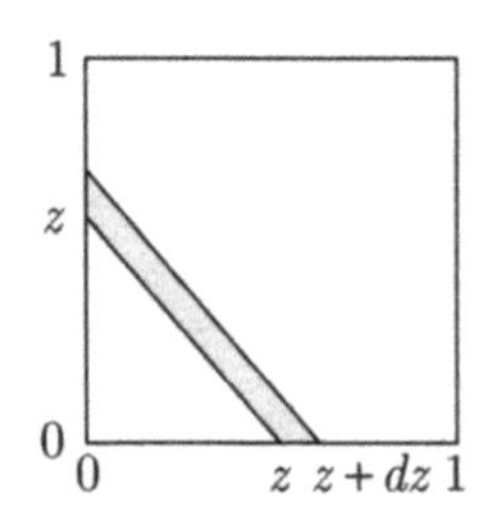

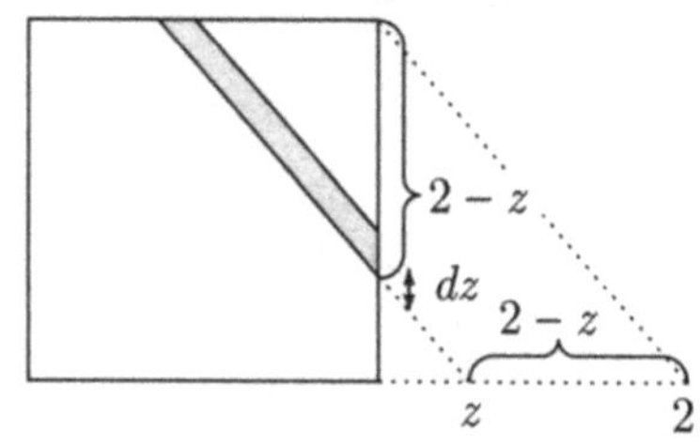

For $0<z<1$, the event $(X+Y\in dz)$ is represented as in the diagram by a shape of area $z\,dz+\frac{1}{2}(dz)^2$, by splitting the area into a parallelogram with altitude z perpendicular to sides of length dz, plus half a square of side dz. Ignoring the $(dz)^2$ as negligible in comparison to dz, gives simply

$$P(Z\in dz)=z\,dz$$

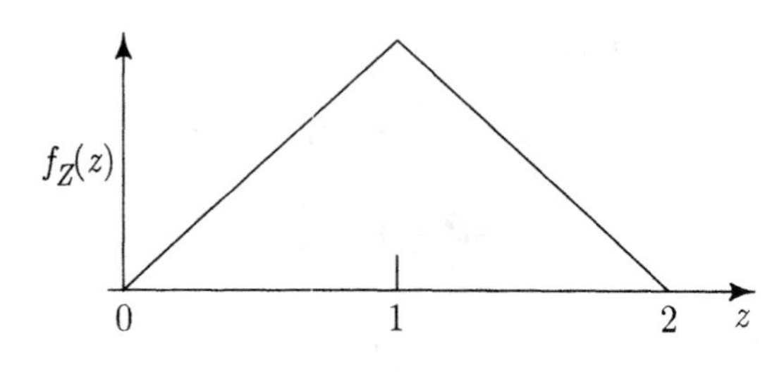

since the total area is 1. Similarly, for $1\le z<2$,

$$P(Z\in dz)=(2-z)dz$$

Thus $Z=X+Y$ has a tent-shaped density,

$$f_Z(z)=\begin{cases} z & 0<z<1\\ 2-z & 1\le z<2\\ 0 & \text{otherwise}\end{cases}$$

Three terms. Consider now $T=X+Y+W$ where X, Y, and W are independent uniform $(0,1)$. The joint distribution of (X,Y,W) is now uniform on a unit cube, and the density of T is proportional to the areas of slices through the cube perpendicular to an axis passing through the long diagonal. As you can convince yourself by handling a real cube, there are now several cases depending on which faces of the cube cut the slicing plane. This 3-dimensional geometry is tricky, but it reduces to two simpler two-dimensional problems.

To compute the density of $T = X + Y + W$ where X, Y, and W are independent uniform $(0,1)$, write $T = X+Y+W = Z+W$, say, where the density of $Z = X+Y$ was found before. The convolution formula gives the density of $T = Z + W$ as an integral

$$\begin{aligned} f_T(t) &= \int_{-\infty}^{\infty} f_Z(z) f_W(t-z)dz \\ &= \int_{t-1}^{t} f_Z(z)dz \qquad \text{since} \quad f_W(t-z) = 1 \text{ for } t-1 < z < t, \quad 0 \text{ else} \\ &= P(t-1 < Z < t) \qquad \text{by definition of } f_Z \end{aligned}$$

So the probability density of T at t turns out in this case to be a probability defined in terms of Z and t. This probability is represented by the shaded areas under the density $f_Z(z)$ in the diagrams that follow. There are 3 cases to consider.

Case 1. $0 < t < 1$. Then $t - 1 < 0$, so

$$f_T(t) = P(t-1 < Z < t) = \frac{1}{2}t^2$$

by the formula for area of a triangle.

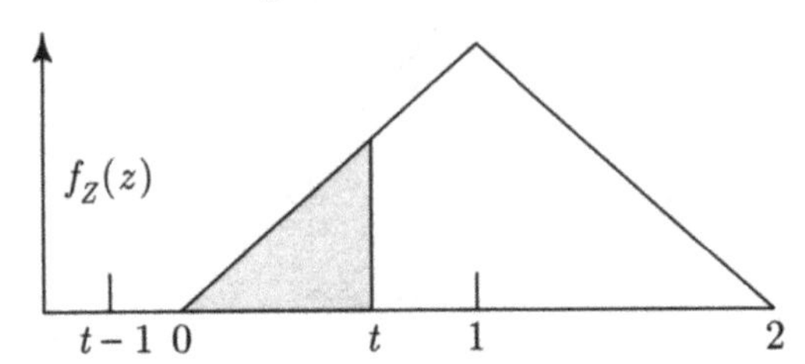

Case 2. $1 < t < 2$. Then $0 < t - 1 < 1$. The relevant area is a unit square less two triangles, hence

$$\begin{aligned} f_T(t) &= P(t-1 < Z < t) \\ &= 1 - \frac{1}{2}(2-t)^2 - \frac{1}{2}(t-1)^2 \\ &= -t^2 + 3t - \frac{3}{2} \end{aligned}$$

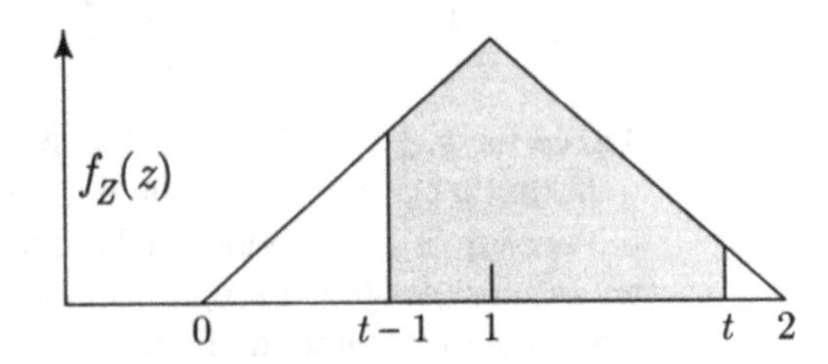

Case 3. $2 < t < 3$. Then $1 < t - 1 < 2$. The relevant area is now another triangle

$$f_T(t) = P(t - 1 < Z < t) = \frac{1}{2}(3 - t)^2$$

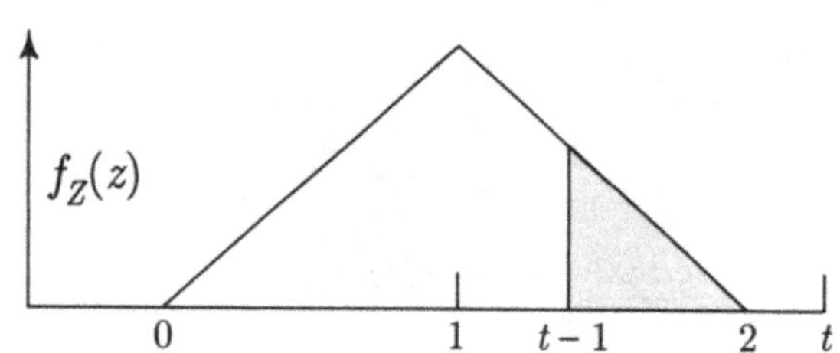

To summarize, the density of the sum $T = X + Y + W$ of three independent uniform $(0, 1)$ variables is $f_T(t)$, as defined above by quadratic functions of t, on each of the intervals $(0, 1)$, $(1, 2)$, and $(2, 3)$, and zero elsewhere. See Figures 2 and 3. Note the symmetric bell shape of the density of T.

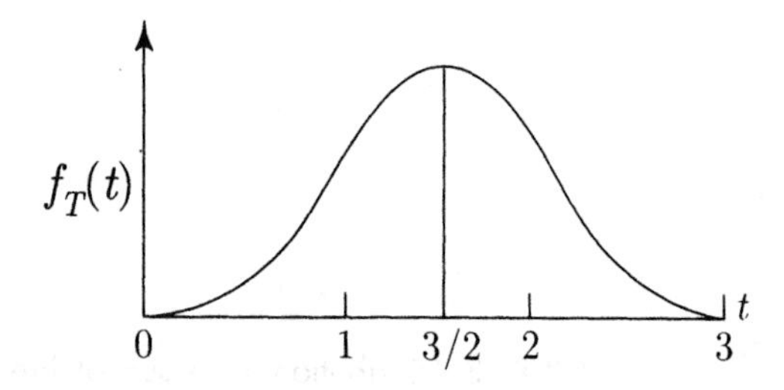

Illustration of Example 3 by numerical calculations. Let $T = X + Y + W$ where X, Y, and W are independent with uniform $(0, 1)$ distribution. Let us find:

a) $P(T < 3/2) = 1/2$ by symmetry of the density of T about $3/2$,

b) $P(1/2 < T < 3/2) = P(T < 3/2) - P(T \leq 1/2)$

$= 1/2 - \int_0^{1/2} \frac{t^2}{2} dt = \frac{23}{48}$ by a) and Case 1 on page 378

c) $P(T > 5/2) = P(T \leq 1/2) = 1/48$ by integral evaluated in b);

d) $E(T) = 3/2$ by symmetry;

e) $SD(T) = \sqrt{3} SD(W)$ where W is uniform $(0, 1)$, by the square root law

$= \sqrt{3} \cdot 1/\sqrt{12}$ by calculation done in Section 4.1

$= 1/2$

FIGURE 2. Distribution of the sum of two independent uniform $(0,1)$ variables X and Y. The joint density of (X,Y) is suggested by a scatter, along with graphs of the densities of X, Y, and $X+Y$.

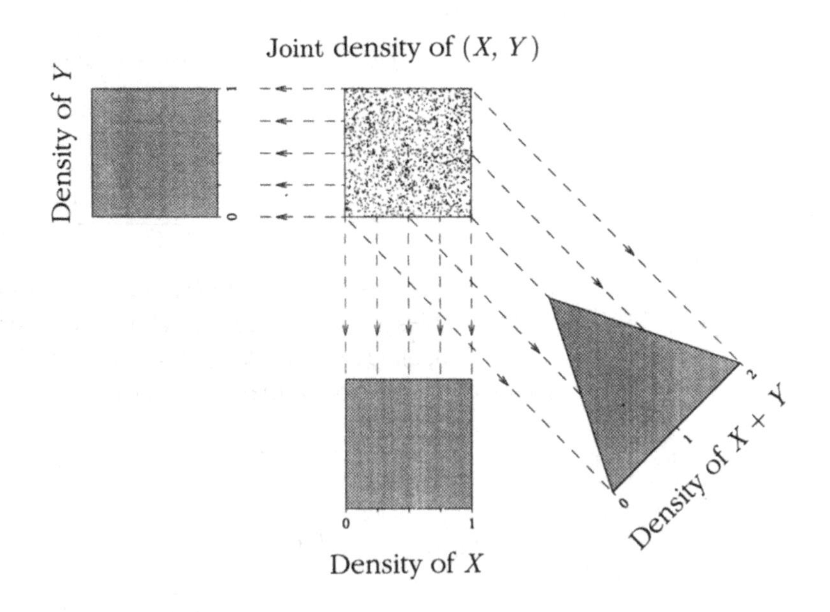

FIGURE 3. Distribution of the sum of three independent uniform $(0,1)$ variables X, Y, and W. The joint density of $(X+Y,W)$ is suggested by a scatter, along with graphs of the densities of $X+Y$, W, and $X+Y+W$.

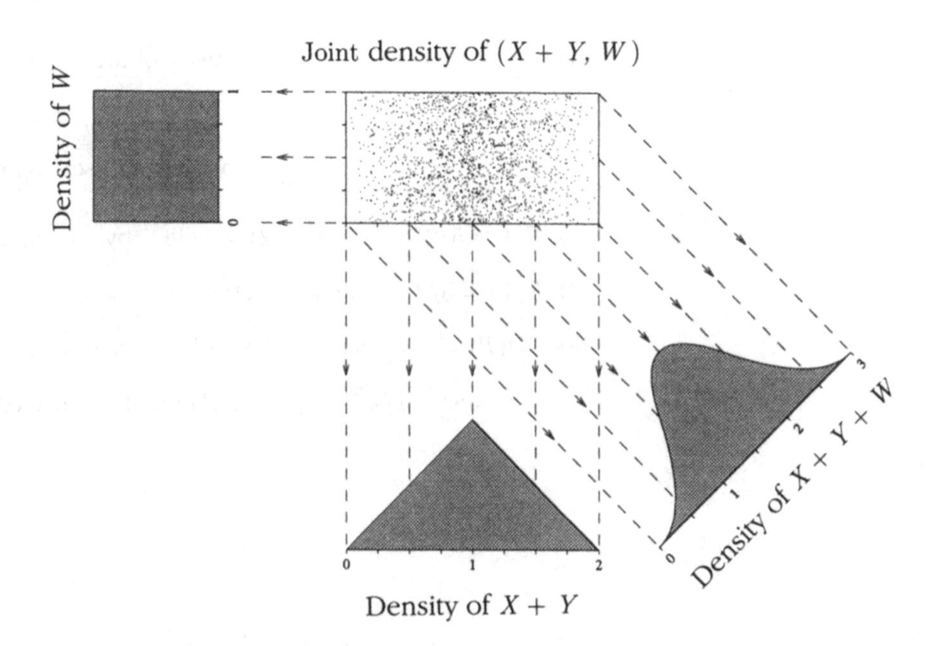

FIGURE 4. **Density of the sum of n independent uniform $(0, 1)$ variables.** The graphs are all centered at the mean with a constant horizontal distance on the page representing one standard unit in each graph. This shows how rapidly the shape of the distribution becomes normal as n increases.

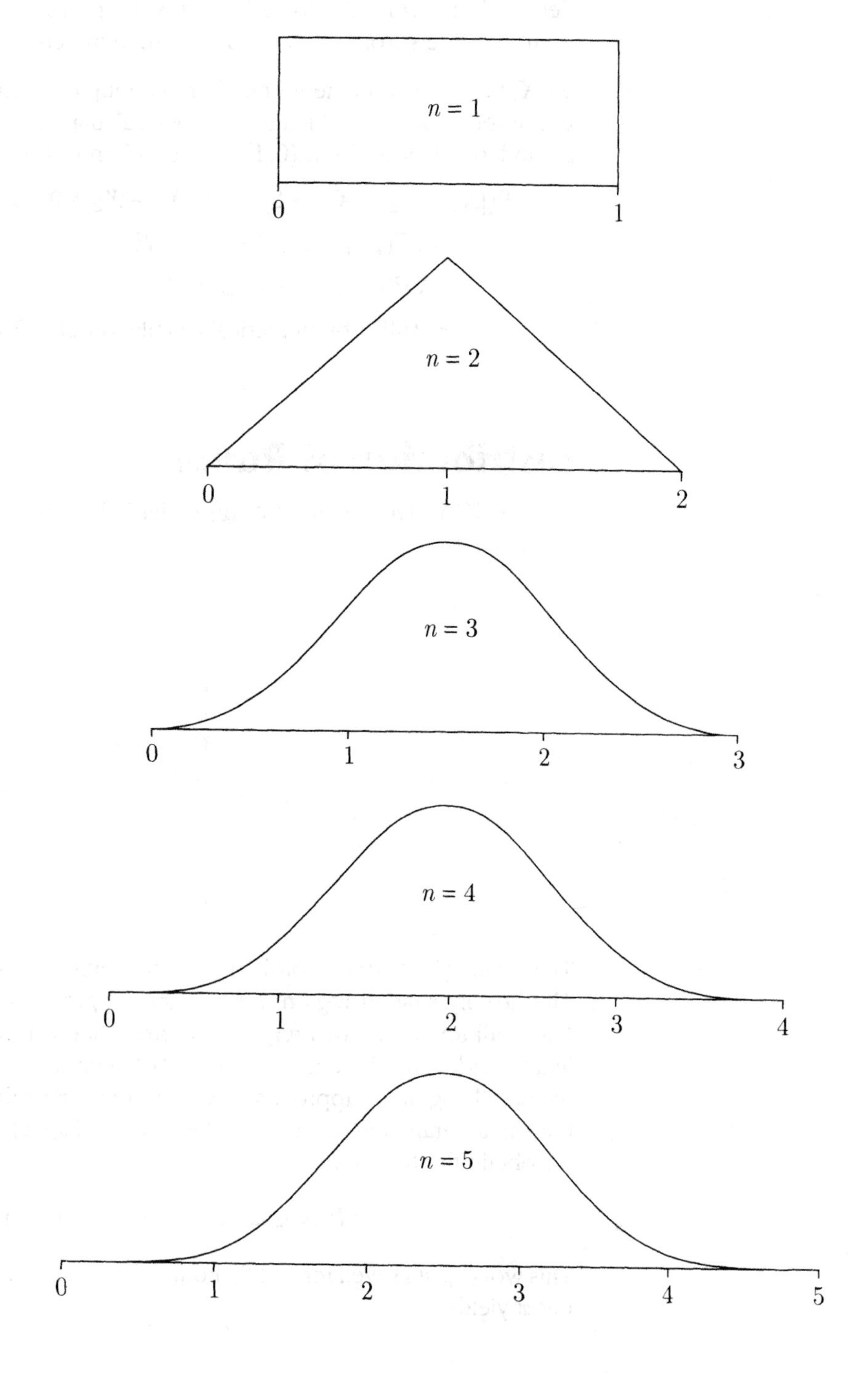

Example 4. Roundoff errors.

Problem. Suppose three numbers are computed, each with a roundoff error known to be smaller than 10^{-6} in absolute value. If the roundoff errors are assumed independent and uniformly distributed, what is the probability that the sum of the rounded numbers differs from the true sum of the numbers by more than 2×10^{-6}?

Solution. Let X_i be the error in the ith number in multiples of 10^{-6}, so the X_i are independent uniform $(-1, 1)$. To reduce to previous calculations, let $U_i = (X_i + 1)/2$, so the U_i are independent uniform $(0, 1)$. The problem is to find

$$\begin{aligned} P(|X_1 + X_2 + X_3| > 2) &= 2P(X_1 + X_2 + X_3 > 2) \quad \text{by symmetry} \\ &= 2P(2U_1 - 1 + 2U_2 - 1 + 2U_3 - 1 > 2) \\ &= 2P(U_1 + U_2 + U_3 > 5/2) \\ &= 2/48 \quad \text{by numerical calculation c) of Example 3 above.} \end{aligned}$$

Distribution of Ratios

Let $Z = Y/X$. The event $(Z \in dz)$ is shaded in the following diagram, for $z > 0$.

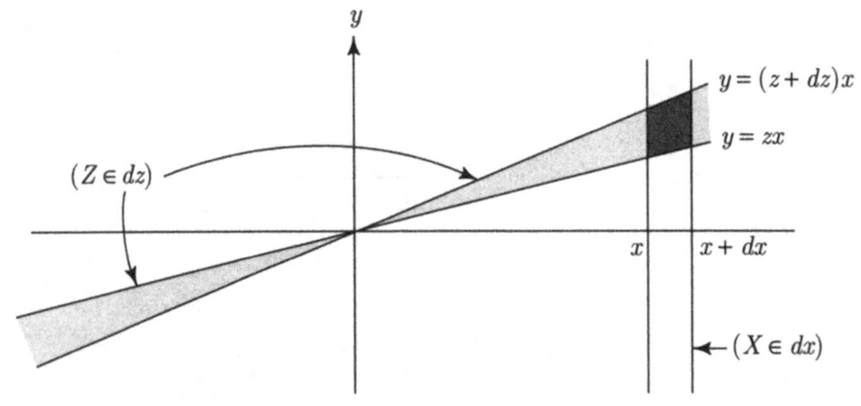

The event $(Z \in dz)$ is broken up into vertical slices according to values of X. The heavily shaded region, near (x, xz), represents the event $(X \in dx, Z \in dz)$. For small dx and dz this region is approximately a parallelogram. The left side has length $|x|dz$, and there is distance dx between the two vertical sides, so the area of the parallelogram is approximately $|x|dz\,dx = |x|dx\,dz$. The probability density over the small parallelogram can be taken to be $f(x, xz)$, so as dx and dz tend to zero we obtain the formula

$$P(X \in dx, Y/X \in dz) = f(x, xz)|x|dx\,dz$$

This works just as well for $z < 0$, though the picture looks a little different. Integrating out x yields

(f) $$f_{Y/X}(z) = \int_{-\infty}^{\infty} |x| f(x, xz)dx$$

As a special case, if X and Y are independent positive random variables, (f) reduces to $f_{Y/X}(z) = 0$ for $z \leq 0$, and

(g) $$f_{Y/X}(z) = \int_{0}^{\infty} x f_X(x) f_Y(xz)dx \qquad (0 < z < \infty)$$

Example 5. **Ratio of independent normal variables.**

Suppose that X and Y are independent and normally distributed with mean 0 and variance σ^2.

Problem. Find the distribution of X/Y.

Solution. We may assume $\sigma = 1$, since

$$\frac{X}{Y} = \frac{X/\sigma}{Y/\sigma} \quad \text{and both} \quad X/\sigma \quad \text{and} \quad Y/\sigma \qquad \text{are standard normal.}$$

By symmetry between X and Y and (f) above

$$\begin{aligned} f_{X/Y}(z) = f_{Y/X}(z) &= \int_{-\infty}^{\infty} |x| f_{X,Y}(x, xz)dx \\ &= \int_{-\infty}^{\infty} |x| \cdot \frac{1}{2\pi} e^{-\frac{x^2 + x^2 z^2}{2}} dx \\ &= \int_{0}^{\infty} \frac{1}{\pi} x e^{-\frac{x^2(z^2+1)}{2}} dx \\ &= \frac{1}{\pi} \cdot \frac{(-1)}{z^2+1} \cdot e^{-\frac{x^2(z^2+1)}{2}} \Big|_0^{\infty} \\ &= \frac{1}{\pi(z^2+1)} \end{aligned}$$

That is, X/Y has Cauchy distribution (see Exercise 4.4.6).

Remark. This calculation illustates the general method, but is a bit heavy handed. In fact the distribution of X/Y is Cauchy whenever the joint distribution of X and Y is symmetric under rotations. See Exercise 14 below.

Exercises 5.4

1. Let X_1 be uniform $(0,1)$ independent of X_2, that is, uniform $(0,2)$. Find:

a) $P(X_1 + X_2 \leq 2)$; b) the density of $X_1 + X_2$; c) the c.d.f. of $X_1 + X_2$.

2. Let S_n be the sum of n independent uniform $(0,1)$ random variables. Find

a) $P(S_2 \leq 1.5)$; b) $P(S_3 \leq 1.5)$; c) $P(S_3 \leq 1.1)$;

d) $P(1.0 \leq S_3 \leq 1.001)$ approximately.

3. A computer job must pass through two queues before it is processed. Suppose the waiting time in the first queue is exponential with rate α, and the waiting time in the second queue is exponential with rate β, independent of the first.

a) Find the density of the total time the job spends waiting in the two queues. Sketch the density in case $\alpha = 1$ and $\beta = 2$.

b) Find the expected total waiting time in terms of α and β.

c) Find the SD of the total waiting time in terms of α and β.

4. A system consists of two components. Suppose each component is subject to failure at constant rate λ, independently of the other, up to when the first component fails. After that moment the remaining component is subject to additional load and to failure at constant rate 2λ.

a) Find the distribution of the time until both components have failed.

b) What are the mean and variance of this distribution?

c) Find the 90th percentile of this distribution.

5. Let X be the number on a die roll, between 1 and 6. Let Y be a random number which is uniformly distributed on $[0,1]$, independent of X. Let $Z = 10X + 10Y$.

a) What is the distribution of Z? Explain.

b) Find $P(29 \leq Z \leq 58)$.

6. Suppose $X_1, X_2, \ldots, X_n$ are independent and X_i has gamma (r_i, λ) distribution. What is the distribution of $X_1 + X_2 + \cdots + X_n$? Explain.

7. Let X and Y have joint density $f(x,y)$. Find formulae for the densities of each of the following random variables: a) XY; b) $X - Y$; c) $X + 2Y$.

8. Let X and Y be independent exponential variables with rates α and β. Find the c.d.f. of X/Y.

9. Find the density of $X = UV$ for independent uniform $(0,1)$ variables U and V.

10. Find the density of $Y = U/V$ for independent uniform $(0,1)$ variables U and V.

11. Find the distribution of $\min(U,V)/\max(U,V)$ for independent uniform$(0,1)$ variables U and V.

12. Let U, V be independent random variables, each uniform on $(0,1)$.

(a) Find the density of $X = -\log\{U(1-V)\}$. b) Compute $E(X)$ and $Var(X)$.

13. Find the density of $Z = X - Y$ for independent exponential (λ) variables X and Y.

14. Let X and Y have a joint distribution which is symmetric under rotations (e.g., uniform on a circle around 0, or uniform on a disc centered at 0). By changing to polar coordinates, show that

a) the distribution of X/Y is Cauchy (see Exercise 4.4.6);

b) the distribution of $X^2/(X^2 + Y^2)$ is arcsine (see Exercise 4.4.8).

15. Let $Z = \min(X, Y)/\max(X, Y)$ for independent exponential (λ) variables X and Y.

a) Explain with little calculation why the distribution of Z does not depend on λ

b) Let $0 < z < 1$. Identify the set $(Z \leq z)$ as a subset of the (x, y) plane, and calculate $P(Z \leq z)$ by integration of the joint density over this subset.

c) Find the density of Z at z for $0 < z < 1$.

16. Consider the c.d.f. of T with gamma (r, λ) distribution, $F(r, \lambda, t) = P(T \leq t)$. Section 4.2 gives a formula for $F(r, \lambda, t)$ for integer r, but for r not an integer there is no simple formula for $F(r, \lambda, t)$.

a) Show that for fixed r and t, $F(r, \lambda, t)$ is an increasing function of λ. [*Hint*: Rescale to the gamma $(r, 1)$ distribution.]

b) Show that for fixed λ and t, $F(r, \lambda, t)$ is a decreasing function of r. [*Hint*: Use sums of independent gamma variables.]

17. Take a unit cube in three dimensions. Cut the cube by a plane perpendicular to the line from its corners $(0, 0, 0)$ and $(1, 1, 1)$, that cuts this line at the point $(t/3, t/3, t/3)$.

a) What is the cross-sectional area of this slice through the cube?

b) Check your answer by describing geometrically the shape of the cross section in the case when $t \leq 1$ and $t = 3/2$.

18. Let f_n be the density function and F_n the c.d.f. of the sum S_n of n independent uniform $(0, 1)$ random variables.

a) Show that $f_n(x) = F_{n-1}(x) - F_{n-1}(x - 1)$.

b) Show that on each of the n intervals $(i - 1, i)$ for $i = 1$ to n, f_n is equal to a polynomial of degree $n - 1$, and F_n is equal to a polynomial of degree n.

c) Find $f_n(x)$ and $F_n(x)$ for $0 \leq x \leq 1$.

d) Find $f_n(x)$ and $F_n(x)$ for $n - 1 \leq x \leq n$.

Find also: e) $P(0 \leq S_4 \leq 1)$; f) $P(1 \leq S_4 \leq 2)$; g) $P(1.5 \leq S_4 \leq 2)$.

19. Let X and Y be independent variables with gamma (r, λ) and gamma (s, λ) distribution, respectively. Show that $X/(X + Y)$ has beta (r, s) distribution, independently of $X + Y$.

Continuous Joint Distributions: Summary

Differential Formula for Joint Density

$$P(X \in dx, Y \in dy) = f(x,y)dx\,dy$$

The density $f(x,y)$ is the probability per unit area for values near (x,y). See pages 348 and 349 of Section 5.2 for properties of joint densities, and comparison with joint distributions in the discrete case.

Central Limit Theorem

Let $X_1, X_2, \ldots$ be a sequence of independent, identically distributed random variables, each with mean μ and variance σ^2. Let $S_n = X_1 + \cdots + X_n$. Provided $\sigma^2 < \infty$, the limit distribution, as $n \to \infty$, of the standardized sum $Z_n = [S_n - n\mu]/(\sqrt{n}\sigma)$ is the standard normal distribution.

Formula for Density of $X + Y$

If (X, Y) has density $f(x,y)$ in the plane, then $X + Y$ has density on the line

$$f_{X+Y}(z) = \int_{-\infty}^{\infty} f(x, z - x)dx.$$

Convolution Formula

If X and Y are independent, then

$$f_{X+Y}(z) = \int_{-\infty}^{\infty} f_X(x) f_Y(z - x)dx.$$

Exact distribution of various functions of particular variables. See distribution summaries.

The Rayleigh Distribution

If X and Y are independent standard normal variables, then $R = \sqrt{X^2 + Y^2}$ has the Rayleigh distribution, with density

$$f_R(r) = re^{-\frac{1}{2}r^2}, \qquad r > 0,$$

and distribution function

$$F_R(r) = 1 - e^{-\frac{1}{2}r^2}, \qquad r > 0.$$

The variable R represents the distance from the origin of the random point (X, Y).

Review Exercises

1. For X and Y independent and uniform $(0,1)$, find $P(Y \geq 1/2 \mid Y \geq X^2)$.

2. For X and Y independent and both uniform $(-1,1)$, find

a) $P(|X+Y| \leq 1)$; b) $E(|X+Y|)$.

3. Coin in a can. A coin of diameter 1 inch is tossed in the air and caught in an empty soup can of bottom radius 3 inches. The coin lies flat on the bottom.

a) What is the chance that the coin covers the center point of the bottom of the can?

Suppose that instead of the soup can, the coin is dropped into a box whose bottom is a square with sides of length 5 inches.

b) What is the chance that the coin covers the center point of the bottom of the box?

c) Consider one of the main diagonals of the bottom of the box. What is the probability that part of the coin crosses that diagonal line?

State any assumptions you make.

4. Let X and Y be independent with uniform $(-1,1)$ distribution. Find

a) $P(X^2+Y^2) \leq r^2$; b) the c.d.f. of $R^2 = X^2 + Y^2$; c) the density of R^2.

5. A point is chosen uniformly at random from a unit square. Let D be the distance of the point from the midpoint of one side of the square. Find a) $P(D \geq \frac{1}{2})$; b) $E(D^2)$.

6. For a particular kind of call, the phone company charges $1 for the first minute or any portion thereof, and one cent per second for time after the first minute. Calculate the approximate value of the long-run average charge per call assuming the distribution of call duration is:

a) exponential with mean 1 minute;

b) exponential with mean 2 minutes;

c) gamma with shape parameter 2 and mean 1 minute.

7. Suppose that $X_1, X_2, \ldots, X_{100}$ are independent random variables, with normal $(\mu, 1)$ distribution, representing 100 measurements whose average $\bar{X} = (X_1 + \cdots + X_{100})/100$ should be close to the number μ. Calculate the probability that $|\bar{X} - \mu| \geq 0.25$.

8. Suppose that $X_1, \ldots, X_{100}$ are independent random variables with common distribution with mean μ and variance 1, but not necessarily normally distributed. Repeat Exercise 7 with these assumptions. Explain why the answer will be approximately the same.

9. Let X be the number of heads in two fair coin tosses. Suppose U has uniform distribution on $(0,1)$, independently of X.

a) Find the density of $X + U$ and sketch its graph.

b) Find an alternate distribution for U such that for any integer-valued random variable X independent of U, the graph of the density of $X + U$ is simply the usual histogram of the distribution of X.

10. Let X, Y be independent exponential random variables with parameters λ and μ.

a) Find the density function for $Z = \min(X, Y)$.

b) Calculate $P(X \geq Y)$.

c) Calculate $P(\frac{1}{2} < X/Y < 2)$, in the case $\lambda = \mu$. [*Hint:* Use the result of b).]

11. Let U and V be two independent uniform $(0, 1)$ random variables. Let $X = U/V$.

a) For $0 < x < 1$, calculate $P(X > x)$.

b) Find the c.d.f. F of the random variable X. Sketch the graph of F.

c) Find the density function f of X. Sketch the graph of f.

12. A marksman fires at the center of a target; he hits a random point (X, Y) (measured relative to the center of the target) such that X and Y are independent normal $(0, a^2)$ random variables. A second marksman fires, and hits at (X', Y') where X' and Y' are independent with normal $(0, b^2)$ distributions. What is the chance that the second marksman hits closer to the center of the target than the first marksman?

13. Suppose (X, Y) is uniformly distributed according to relative arc length on the circumference of the circle $\{(x, y) : x^2 + y^2 = 1\}$. Find the c.d.f. of

a) X; b) Y; c) $X + Y$.

14. Suppose U_1, U_2, U_3 are independent and uniform $(0, 1)$. Find: a) $P(U_1 < U_2 < U_3)$;

b) $E(U_1 U_2 U_3)$; c) $Var(U_1 U_2 U_3)$; d) $P(U_1 U_2 > U_3)$; e) $P(\max(U_1, U_2) > U_3)$.

15. Repeat Exercise 14 for Z_i instead of U_i, where the Z_i are independent normal $(0, 1)$ random variables. Find also:

f) $P(Z_1^2 + Z_2^2 > 1)$ g) $P(Z_1 + Z_2 + Z_3 < 2)$;

h) $P(Z_1/Z_2 < 1)$; i) $P(3Z_1 - 2Z_2 < 4Z_3 + 1)$.

16. A point is picked randomly in space. Its three coordinates X, Y and Z are independent standard normal variables. Let $R = \sqrt{X^2 + Y^2 + Z^2}$ be the distance of the point from the origin. Find

a) the density of R^2; b) the density of R; c) $E(R)$; d) $Var(R)$.

17. Let $X_1, X_2, \ldots,$ be independent normally distributed random variables having mean 0 and variance 1. Use the normal approximation to find:

a) $P(X_1^2 + X_2^2 + \cdots + X_{100}^2 \geq 80)$;

b) a number c such that $P(100 - c \leq X_1^2 + \cdots + X_{100}^2 \leq 100 + c) \approx 0.95$.

18. For X and Y independent normal $(0, 1)$ variables, show that for $r > 0$

$$P(aX + bY \leq r\sqrt{a^2 + b^2} \text{ for all } a, b \geq 0) = \Phi(r) - \frac{1}{4}e^{-\frac{1}{2}r^2}$$

19. Independent Poisson processes. Suppose particles of d different kinds, labeled $k = 1, 2, \ldots, d$, arrive at a counter according to independent Poisson processes at rates λ_k. Let W_k be arrival time of the first particle of kind k. Let K_1 be the kind of the first particle to arrive, K_2 the kind of the second particle to arrive, and so on. So the K_n are discrete random variables with values in the set $\{1, \ldots, d\}$.

a) Express the event $(K_1 = k)$ in terms of the random variables $W_1, \ldots, W_d$.

b) Use this expression to find $p_k = P(K_1 = k), 1 \leq k \leq d$ in terms of $\lambda_1, \ldots, \lambda_d$.

c) Explain informally why $K_1, K_2, \ldots$ are independent with identical distribution.

d) Assuming the result of c), derive the formula for p_k in another way after filling in the blanks in the following statements e), f) and g): After a very long time T,

e) the number of arrivals of type k should be about ________.

f) the number of all arrivals of all types should be about ________.

g) the fraction of all arrivals that are of type k should be about ________.

20. Minimum of independent exponential variables. Let T_1 and T_2 be two independent exponential variables, with rates λ_1 and λ_2. Think of T_i as the lifetime of component i, $i = 1, 2$. Let $T_{\min}$ represent the lifetime of a system which fails whenever the first of the two components fails, so $T_{\min} = \min(T_1, T_2)$. Let $X_{\min}$ designate which component failed first, so $X_{\min}$ has value 1 if $T_1 < T_2$ and value 2 if $T_2 < T_1$. Show:

a) that the distribution of $T_{\min}$ is exponential $(\lambda_1 + \lambda_2)$;

b) that the distribution of $X_{\min}$ is given by the formula $P(X_{\min} = i) = \dfrac{\lambda_i}{\lambda_1 + \lambda_2}$ for $i = 1, 2$;

c) that the random variables $T_{\min}$ and $X_{\min}$ are independent;

d) how these results generalize simply to describe the minimum of n independent exponential random variables with rates $\lambda_1, \ldots, \lambda_n$.

21. Closest point. Consider a Poisson random scatter of points in a plane with mean intensity λ per unit area. Let R be the distance from 0 to the closest point of the scatter.

a) Find formulae for the c.d.f. and density of R, and sketch their graphs.

b) Show that $\sqrt{2\lambda\pi}R$ has the Rayleigh distribution described in Section 5.3

c) Use b) to find formulae for the mean and SD of R from results of Section 5.3.

d) Find the mode and the median of the distribution of R.

22. In Maxwell's model of a gas, molecules of mass m are assumed to have velocity components, V_x, V_y, V_z that are independent, with a joint distribution that is invariant under rotation of the three-dimensional coordinate system. Maxwell showed that V_x, V_y, V_z must have normal $(0, \sigma^2)$ distribution for some σ. Taking this result for granted:

a) find a formula for the density of the kinetic energy

$$K = \frac{1}{2}mV_x^2 + \frac{1}{2}mV_y^2 + \frac{1}{2}mV_z^2$$

b) find the mean and mode of the energy distribution.

23. Let Y be the minimum of three independent random variables with uniform distribution on $(0, 1)$, and let Z be their maximum. Find:

a) $P(Z \leq \frac{2}{3} | Y \geq \frac{1}{3})$; b) $P(Z \leq \frac{2}{3} | Y \leq \frac{1}{3})$.

24. A coin of diameter d is tossed at random on a grid of squares of side s. Making appropriate assumptions, to be stated clearly, calculate:

a) the probability that the coin lands inside some square (i.e., not touching any line);

b) the probability that the coin lands heads inside some square.

Suppose now that the coin is tossed four times. Let X be the number of times it lands inside a square, Y the number of heads. Assume $d = s/2$. Calculate:

c) $P(X = Y)$; d) $P(X < Y)$; e) $P(X > Y)$.

25. Joint distribution of order statistics. Let $V_1 < V_2 < \cdots < V_n$ be the order statistics of n independent uniform $(0,1)$ variables. (Refer to Section 4.6.) Let $1 \leq k < m \leq n$.

a) Find the joint density of V_k and V_m.

Now show that each of the following variables has a beta distribution, and identify the parameters: b) $V_m - V_k$; c) V_k/V_m;

26. Averages of order statistics. Let $V_1, \ldots, V_n$ be the order statistics of n independent uniform $(0,1)$ variables. Let

$A_{\text{all}} = (V_1 + \cdots + V_n)/n$

$A_{1n} = (V_1 + V_n)/2$

$A_{\text{mid}} = V_{(n+1)/2}$ the middle value, where you can assume n is odd.

a) Show that for sufficiently large n, each of these three variables is most likely very close to $1/2$.

b) For all large enough values of n, one of these variables can be expected to be very much closer to $1/2$ than either of the two others. Which one, and why?

c) Confirm your answer to b) for $n = 100$ by finding for each of the A's a good approximation to the probability that it is between 0.49 and 0.51.

27. A box contains n balls numbered $1, \ldots, n$. Balls are drawn at random until the first draw that produces a ball obtained on some previous draw. Let D_n be the random number of draws required. So the possible values of D_n are $2, \ldots, n+1$.

a) Check that for $0 < x < \infty$,

$$\lim_{n\to\infty} P(D_n/\sqrt{n} > x) = e^{-x^2/2}$$

That is to say, the limit distribution of $D_n/\sqrt{n}$ is the Rayleigh distribution.

b) Assuming a switch in the order of the limit and integration can be justified (it can, but do not worry about that), deduce that

$$\lim_{n\to\infty} E(D_n/\sqrt{n}) = \sqrt{\pi/2}$$

c) There seems to be no simpler expression for $E(D_n)$ than a sum of n or $n+1$ terms. But the terms can be arranged in some interesting ways. Show by writing $E(D_n)$ as the sum of the tail probabilities $P(D_n > k)$ in reverse order that

$$E(D_n) = P(X_n \leq n)\, n!\, n^{-n} e^n$$

where X_n is a Poisson random variable with mean n.

d) Deduce the limit of $P(X_n \leq n)$ as $n \to \infty$ from the central limit theorem, then combine b) and c) to give a derivation of Stirling's formula

$$n! \sim \sqrt{2\pi n}\left(\frac{n}{e}\right)^n$$

e) Derive the following formula, which is surprisingly simple in view of c)

$$E(D_n^2 - D_n) = 2n$$

f) Transform the identity e) as in the calculation c) to derive the formula

$$E(|X_n - n|) = \frac{2n^{n+1}}{e^n n!}$$

and give yet another derivation of Stirling's formula, much as in d) above, this time using the central limit theorem instead of a).

28. Volumes in higher dimensions. Use the derivation of the chi-square distribution to derive part a), then use a) for the remaining parts:

a) Find the constant c_n such that the $(n-1)$-dimensional volume of the "surface" of a sphere of radius r in n-dimensional space is $c_n r^{n-1}$.

b) Find d_n so the n-dimensional volume inside a sphere of radius r is $d_n r^n$.

c) An n-dimensional sphere of radius r is packed inside an n-dimensional cube with sides of length $2r$. What proportion p_n of the volume of the cube is inside the sphere?

d) Use Stirling's formula $\Gamma(s) \sim \sqrt{2\pi} s^{s-1/2} e^{-s}$ as $s \to \infty$ to find a simple approximation for p_n for large n. What is the limit of p_n as $n \to \infty$?

e) Interpret p_n probabilistically in terms of n independent uniform $(-1, 1)$ variables.

29. A needle is tossed at random on a grid of equally spaced parallel lines. Assume the needle is so much longer than the spacing between the lines that the possibility of the needle not crossing any line can be neglected. Let X be the distance between the center of the needle and the closest point at which the needle crosses one of the lines. Find:

a) the distribution function of X;

b) the density function of X.

30. Random walk inside squares. Draw a square centered at $(0,0)$ with sides of length 2 parallel to the axes, so the corners are at $(\pm 1, \pm 1)$. Let (X_1, Y_1) be picked uniformly at random from the area inside this square. Given (X_1, Y_1), draw a square centered at (X_1, Y_1), with sides of length 2 parallel to the axes, so the corners are at $(X_1 \pm 1, X_2 \pm 1)$. Let (X_2, Y_2), be picked uniformly at random from the area inside this square, and so on: Given $(X_1, Y_1), \ldots, (X_n, Y_n)$ let (X_{n+1}, Y_{n+1}) be picked uniformly at random from the area inside the square with corners at $(X_n \pm 1, Y_n \pm 1)$. For $n = 300$, use a normal approximation to find the following probabilities:

a) $P(|X_n| > 10)$; b) $P(|Y_n| > 10)$.

c) The probability that (X_n, Y_n) lies outside the square with corners at $(\pm 10, \pm 10)$.

d) The probability that (X_n, Y_n) lies outside the circle of radius 10 centered at $(0,0)$.

31. Random walk inside circles. Fix $r > 0$. Draw a circle centered at $(0,0)$ with radius r. Let (X_1, Y_1) be picked uniformly at random from the area inside this circle. Given (X_1, Y_1), draw a circle with radius r centered at (X_1, Y_1). Let (X_2, Y_2), be picked uniformly at random from the area inside this circle, and so on. Given $(X_1, Y_1), \ldots, (X_n, Y_n)$, let (X_{n+1}, Y_{n+1}) be picked uniformly at random from the area inside the circle around (X_n, Y_n) with radius r.

a) Find r so that for large n the distribution of X_n in this problem is nearly the same as in Exercise 30 for a square of side 2 instead of the circle of radius r. [*Hint*: Find $E[X_1^2]$ by considering $E[Y_1^2]$ as well.]

b) Are X_n and Y_n independent?

c) The point (X_n, Y_n) is projected onto the line rotated an angle θ from the X-axis at $X_n \cos\theta + Y_n \sin\theta$ measured from the origin along this line. Use the normal approximation for sums of independent random variables to show that with r as in part a), for every $\theta \in [0, 2\pi]$ and for large n, the distribution of $X_n \cos\theta + Y_n \sin\theta$ is nearly the same for both the circle of radius r and the square of side 2.

d) It is known that a joint distribution of (X, Y) in the plane is determined by the distributions of all the projections $X \cos\theta + Y \sin\theta$ as θ ranges over $[0, 2\pi]$. In particular if $X \cos\theta + Y \sin\theta$ has standard normal distribution for every θ then X and Y are independent standard normal variables. An approximate version of this result is also true: if $X \cos\theta + Y \sin\theta$ has approximately the standard normal distribution for every θ, then X and Y are approximately independent standard normal variables. Apply this result and part c) to approximate the probability that for r as in part a), and $n = 300$, the point (X_n, Y_n) defined using circles of radius r lies outside the circle of radius 10 centered at the origin.

32. Random walk on circles. Repeat Exercise 31 for the motion defined by picking points at random according to the uniform distribution on the *perimeter* of the circle of radius r, so each new point is at distance r from the previous one, in a random direction.

33. Mixture of discrete and continuous. Repeat Exercise 31 for the motion defined by repeatedly picking points at random according to the uniform distribution (proportional to length) on the perimeter of a square centered at the current point with sides of length $2r$. Note that the distribution of X_n in this case is neither discrete nor continuous but a mixture of the two kinds. The second moment of X_1 is defined by adding the discrete and continuous parts. It can be shown that the usual method of calculating the second moment of X_n is still valid, and that the normal approximation is still correct in the limit of large n. Following parts a) to d) as in Exercise 31,

e) Calculate and plot the graph of the distribution function of X_1.

f) Calculate and plot the graph of the distribution function of X_2.

g) What is the total probability in the discrete part of the distribution of X_n?

34. Ratios of sums of squares.

a) Use the result of Exercise 5.4.19 to show that if X, Y and Z are independent normal $(0, 1)$ random variables, then $X^2/(X^2 + Y^2 + Z^2)$ has beta $(1/2, 1)$ distribution, independent of $X^2 + Y^2 + Z^2$.

b) Suppose that (U, V, W) has uniform distribution on the surface of the unit sphere in three dimensions. Deduce from a) that $U^2/(U^2 + V^2 + W^2)$ has beta $(1/2, 1)$ distribution.

c) What is the distribution of $U^2/(U^2 + V^2 + W^2)$ if (U, V, W) has uniform distribution over the volume inside the unit sphere in three dimensions?

d) Suppose that $U_1, U_2, \ldots, U_n$ are independent uniform $(-1, 1)$ variables. For $1 \leq k \leq n$, let $S_k = U_1^2 + \cdots + U_k^2$. Find the the conditional distribution of S_k/S_n given that $S_n \leq 1$.

6 Dependence

This chapter treats features of a joint distribution which give insight into the nature of dependence between random variables. Sections 6.1 and 6.2 concern conditional distributions and expectations in the discrete case. Then parallel formulae for the density case are developed in Section 6.3. Covariance and correlation are introduced in Section 6.4. All these ideas are combined in Section 6.5 in a study of the bivariate normal distribution.

6.1 Conditional Distributions: Discrete Case

This section translates into the language of random variables the conditioning ideas of Section 1.4. The dependence between two variables X and Y can be understood in terms of the marginal distribution of X and the conditional distribution of Y given $X = x$, which may be a different distribution for each possible value x of X. Given this information, the distribution of Y is found by the rule of average conditional probabilities, and the conditional distribution of X given $Y = y$ is found by Bayes' rule.

Example 1. Number of successes in a random number of trials.

Suppose a fair die is rolled. Then as many fair coins are tossed as there are spots showing on the die.

Problem 1. Find the distribution of the number of heads showing among the coins.

Solution. Let Y denote the number of heads showing among the coins. The problem is to calculate the probabilities

$$P(Y = y) = P(y \text{ heads}) \qquad (y = 0, 1, 2, \ldots, 6)$$

Let X represent the number showing on the die. If $X = x$, that is to say the die rolls x, then x coins are tossed, so the chance of y heads given the die rolls x is given by the binomial formula for the probability of y successes in x trials with probability $1/2$ of success on each trial:

$$P(y \text{ heads} \,|\, \text{die rolls } x) = P(y \text{ heads in } x \text{ fair coin tosses}) = \binom{x}{y} 2^{-x}$$

where $\binom{x}{y} = 0$ if $x < y$. In random variable notation,

$$P(Y = y \,|\, X = x) = \binom{x}{y} 2^{-x}$$

This formula states that the *conditional distribution of* Y *given* $X = x$ is the binomial distribution with parameters $n = x$ and $p = 1/2$.

FIGURE 1. Conditional distribution of Y given $X = x$ for $x = 1, 2, \ldots, 6$ in Example 1.

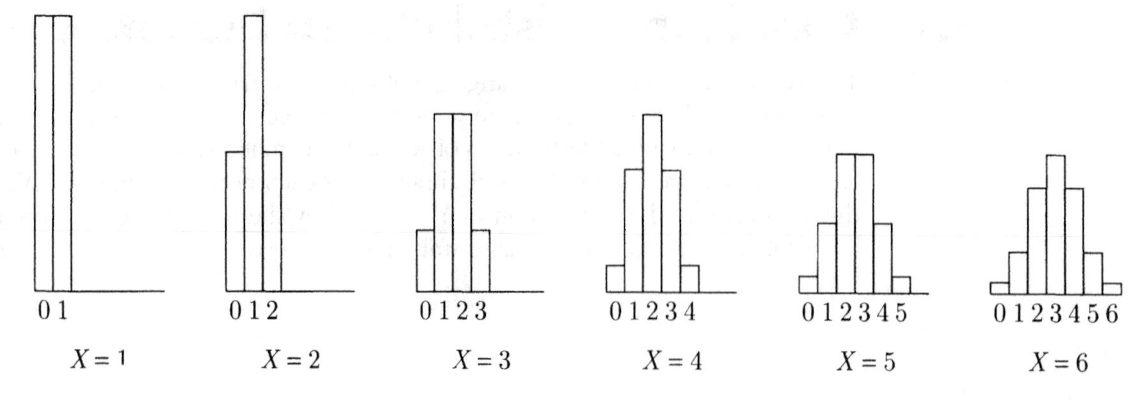

The assumption that the die is fair specifies the unconditional distribution of X:

$$P(X = x) = P(\text{die rolls } x) = 1/6 \qquad (x = 1, 2, \ldots, 6)$$

These ingredients are combined by the rule of average conditional probabilities to give $P(Y = y)$, the unconditional probability of getting y heads:

$$\begin{aligned} P(Y = y) = P(y \text{ heads}) &= \sum_{x=1}^{6} P(\text{die rolls } x \text{ and } y \text{ heads}) \\ &= \sum_{x=1}^{6} P(y \text{ heads} \,|\, \text{die rolls } x) P(\text{die rolls } x) \\ &= \sum_{x=1}^{6} P(Y = y \,|\, X = x) P(X = x) \\ &= \frac{1}{6} \sum_{x=1}^{6} \binom{x}{y} 2^{-x} \qquad\qquad (0 \le y \le 6) \end{aligned}$$

where $\binom{x}{y} = 0$ if $x < y$. For example,

$$P(Y = 0) = \frac{1}{6}\left[\frac{1}{2} + \frac{1}{2^2} + \cdots + \frac{1}{2^6}\right] = \frac{1}{6} \times \frac{63}{64} = \frac{63}{384}$$

$$P(Y = 4) = \frac{1}{6}\left[\binom{4}{4}\frac{1}{2^4} + \binom{5}{4}\frac{1}{2^5} + \binom{6}{4}\frac{1}{2^6}\right] = \frac{29}{384}$$

and so on. Continuing in this way we obtain $P(Y = y)$ for each $y = 0, 1, 2, \ldots, 6$, as shown in Table 1.

TABLE 1. Probability $P(Y = y)$ of getting y heads.

y	0	1	2	3	4	5	6
$P(Y = y)$	$\frac{63}{384}$	$\frac{120}{384}$	$\frac{99}{384}$	$\frac{64}{384}$	$\frac{29}{384}$	$\frac{8}{384}$	$\frac{1}{384}$

Example 1 introduces the important idea of conditional distributions.

Conditional Distribution of Y Given $X = x$

For each possible value x of X, as y varies over all possible values of y, the probabilities $P(Y = y \,|\, X = x)$ form a probability distribution, depending on x, called the *conditional distribution of* Y *given* $X = x$.

The given value x of X can be thought of a as *parameter* in the distribution of Y given $X = x$. In Example 1, the distribution of Y given $X = x$ is the binomial distribution with parameters $n = x$ and $p = 1/2$.

According to the rule of average conditional probabilities, the unconditional distribution of Y, found in Example 1, is the *average* or *mixture* of these conditional

Example 1 introduces the important idea of conditional distributions.

Conditional Distribution of Y Given $X = x$

For each possible value x of X, as y varies over all possible values of y, the probabilities $P(Y = y \mid X = x)$ form a probability distribution, depending on x, called the *conditional distribution of* Y *given* $X = x$.

The given value x of X can be thought of as a *parameter* in the distribution of Y given $X = x$. In Example 1, the distribution of Y given $X = x$ is the binomial distribution with parameters $n = x$ and $p = 1/2$.

According to the rule of average conditional probabilities, the unconditional distribution of Y, found in Example 1, is the *average* or *mixture* of these conditional distributions, with equal weights $1/6$ defined by the uniform distribution of X. This distribution of Y may be called the *overall, marginal,* or *unconditional* distribution of Y, to distinguish it from the conditional distributions used to calculate it. The key step in the calculation of Example 1 was the following:

Rule of Average Conditional Probabilities

$$P(Y = y) = \sum_{x} P(Y = y \mid X = x)P(X = x)$$

This is just a basic rule of probability expressed in random variable notation. The rule holds for every pair of discrete random variables X and Y defined in the same probabilistic setting. The method of finding the distribution of a random variable Y by using this formula is called *conditioning on the value of* X. Note that in the sum for $P(Y = y)$ the term

$$P(Y = y \mid X = x)P(X = x) = P(X = x, Y = y)$$

is the generic entry in the joint probability table for X and Y. See Table 2 for example. You can use the above formula to calculate the distribution of a random variable Y if you can find a random variable X such that you either know or can easily calculate:

(i) the distribution of X;

(ii) the conditional probabilities $P(Y = y \mid X = x)$ for all possible values x of X.

If Y is determined by some two-stage or multistage process, the distribution of Y can often be calculated this way by letting X be the result of the first stage.

Example 1. (Continued.)

As in the previous example, let Y be the number of heads in X fair coin tosses, where X is uniformly distributed on $\{1, \ldots, 6\}$.

Problem 1. Find the conditional distribution of X given $Y = y$ for $y = 0, 1, \ldots 6$.

Solution. The problem now is to find $P(X = x \mid Y = y)$ as x varies, for each possible value y of Y. These conditional probabilities are calculated using Bayes' rule, as in Section 1.5. All that is new here is the random variable notation and terminology. As a start, the division rule for conditional probabilities gives

$$P(X = x \mid Y = y) = \frac{P(X = x, Y = y)}{P(Y = y)}$$

where the joint probabilities

$$P(X = x, Y = y) = P(Y = y \mid X = x) P(X = x)$$

are the individual terms in the sum used previously to calculate $P(Y = y)$. Substituting the values of $P(X = x)$ and $P(Y = y \mid X = x)$, the joint probabilities $P(X = x, Y = y)$ are displayed in Table 2.

TABLE 2. Joint distribution table for (X, Y).

		Possible values x for X						Marginal distn. of Y
		1	2	3	4	5	6	
Possible values y for Y	0	$\frac{1}{6}\frac{1}{2}$	$\frac{1}{6}\frac{1}{4}$	$\frac{1}{6}\frac{1}{8}$	$\frac{1}{6}\frac{1}{16}$	$\frac{1}{6}\frac{1}{32}$	$\frac{1}{6}\frac{1}{64}$	$\frac{63}{384}$
	1	$\frac{1}{6}\frac{1}{2}$	$\frac{1}{6}\frac{2}{4}$	$\frac{1}{6}\frac{3}{8}$	$\frac{1}{6}\frac{4}{16}$	$\frac{1}{6}\frac{5}{32}$	$\frac{1}{6}\frac{6}{64}$	$\frac{120}{384}$
	2	0	$\frac{1}{6}\frac{1}{4}$	$\frac{1}{6}\frac{3}{8}$	$\frac{1}{6}\frac{6}{16}$	$\frac{1}{6}\frac{10}{32}$	$\frac{1}{6}\frac{15}{64}$	$\frac{99}{384}$
	3	0	0	$\frac{1}{6}\frac{1}{8}$	$\frac{1}{6}\frac{4}{16}$	$\frac{1}{6}\frac{10}{32}$	$\frac{1}{6}\frac{20}{64}$	$\frac{64}{384}$
	4	0	0	0	$\frac{1}{6}\frac{1}{16}$	$\frac{1}{6}\frac{5}{32}$	$\frac{1}{6}\frac{15}{64}$	$\frac{29}{384}$
	5	0	0	0	0	$\frac{1}{6}\frac{1}{32}$	$\frac{1}{6}\frac{6}{64}$	$\frac{8}{384}$
	6	0	0	0	0	0	$\frac{1}{6}\frac{1}{64}$	$\frac{1}{384}$
	Marginal distn. of X	$\frac{1}{6}$	$\frac{1}{6}$	$\frac{1}{6}$	$\frac{1}{6}$	$\frac{1}{6}$	$\frac{1}{6}$	1

In column x of the table you see numbers proportional to the binomial $(x, 1/2)$ probabilities forming the conditional distribution of Y given $X = x$. The constant of proportionality is $1/6$, which is the marginal probability of $(X = x)$. Similarly, in row y of the table you see numbers proportional to the conditional distribution of X given $Y = y$. The conditional probabilities themselves are obtained by dividing the numbers in the row y by the constant factor $P(Y = y)$, their sum, which appears in the margin. For example, the conditional distribution of X given $Y = 2$ is displayed in Table 3.

TABLE 3. Conditional distribution of X given $Y = 2$.

x	1	2	3	4	5	6
$P(X = x \mid Y = 2)$	0	$\frac{16}{99}$	$\frac{24}{99}$	$\frac{24}{99}$	$\frac{20}{99}$	$\frac{15}{99}$

So, given two heads, the number of coins tossed is equally likely to be either 3 or 4, and these are the most likely values.

Similar tables of the conditional distributions are easily made for other values y of Y. Here is a graphical display of all seven of these conditional distributions using histograms.

FIGURE 3. Conditional distribution of X given $Y = y$.

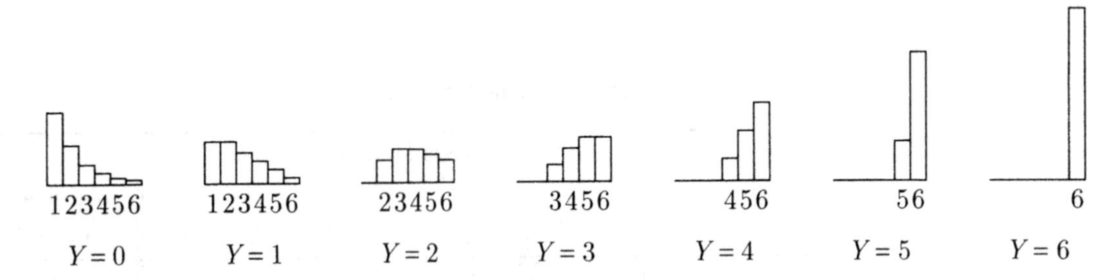

Exercises 6.1

1. Suppose I toss three coins. Some of them land heads and some land tails. Those that land tails I toss again. Let X be the number of heads showing after the first tossing, Y the total number showing after the second tossing, including the X heads appearing on the first tossing. So X and Y are random variables such that $0 \le X \le Y \le 3$ no matter how the coins land. Write out distribution tables and sketch histograms for each of the following distributions:

   a) the distribution of X;

   b) the conditional distribution of Y given $X = x$ for $x = 0, 1, 2, 3$;

c) the joint distribution of X and Y (no histogram in this case);

d) the distribution of Y;

e) the conditional distribution of X given $Y = y$ for $y = 0, 1, 2, 3$.

f) What is the best guess of the value of X given $Y = y$ for $y = 0, 1, 2, 3$? That is, for each y, choose x depending on y to maximize $P(X = x|Y = y)$.

g) Suppose the random experiment generating X and Y is repeated independently over and over again. Each time you observe the value of Y, and then guess the value of X using the rule found in f). Over the long run, what proportion of times will you guess correctly?

2. In a particular town 10% of the families have no children, 20% have one child, 40% have two children, 20% have three children, and 10% have four. Let T represent the total number of children, and G the number of girls, in a family chosen at random from this town. Assuming that children are equally likely to be boys or girls, find the distribution of G. Display your answer in a table and sketch the histogram.

3. Suppose the names of all the children in the town of Exercise 2 are put into a hat, and a name is picked out at random. So now a child is picked at random instead of a family being picked at random. Let U be the total number of children in the family of the child chosen at random.

a) Find the distribution of U. Why is this distribution different from the distribution of T in Exercise 2?

b) What is the probability that the child picked at random comes from a family consisting of two girls and a boy?

c) Is this the same as the probability that a family picked at random consists of two girls and a boy? Calculate and explain.

4. Let $A_1, \ldots, A_{20}$ be independent events each with probability $1/2$. Let X be the number of events among the first 10 which occur and let Y be the number of events among the last 10 which occur. Find the conditional probability that $X = 5$, given that $X+Y = 12$.

5. Let X_1 and X_2 be independent Poisson random variables with parameters λ_1 and λ_2.

a) Show that for every $n \geq 1$, the conditional distribution of X_1, given $X_1+X_2 = n$, is binomial, and find the parameters of this binomial distribution.

b) The number of eggs laid by a certain kind of insect follows a Poisson distribution quite closely. It is known that two such insects have laid their eggs in a particular area. If the total number of eggs in the area is 150, what is the chance that the first insect laid at least 90 eggs? (State your assumptions, and give approximate decimal answer.)

6. Conditioning independent Poisson variables on their sum. Let N_i be independent Poisson variables with parameters λ_i. Think of the N_i as the number of points of a Poissson scatter in disjoint parts of the plane with areas λ_i, where the mean intensity is one point per unit area.

a) What is the conditional joint distribution of $(N_1, \ldots, N_m)$ given $N_1 + \ldots + N_m = n$? [*Hint*: See Exercise 5 for a special case.]

b) Suppose now that N has Poisson(λ) distribution, and given $N = n$ the conditional joint distribution of some m-tuple of random variables $(N_1, \ldots, N_m)$ is exactly what you found in part a). What can you conclude about the unconditional distribution of $(N_1, \ldots, N_m)$?

7. Poissonization of the binomial distribution. Let N have Poisson (λ) distribution. Let X be a random variable with the following property: for every n, the conditional distribution of X given $(N = n)$ is binomial (n, p).

a) Show that the unconditional distribution of X is Poisson, and find its parameter.

It is known that X-rays produce chromosome breakages in cells. The number of such breakages usually follows a Poisson distribution quite closely, where the parameter depends on the time of exposure, etc. For a particular dosage and time of exposure, the number of breakages follows the Poisson (0.4) distribution. Assume that each breakage heals with probability 0.2, independently of the others.

b) Find the chance that after such an X-ray, there are 4 healed breakages.

8. Independence in Poissonization of the binomial distribution. Suppose you roll a random number of dice. If the number of dice follows the Poisson (λ) distribution, show that the number of sixes is independent of the number of nonsixes. [*Hint*: Let N be the number of dice, X the number of sixes, and Y the number of nonsixes. Exercise 7 gives you the marginal distributions of X and Y. To show that the joint distribution of X and Y is the product of the marginals, show

$$P(X = x, Y = y) = P(N = x + y, X = x, Y = y)$$

and then use the multiplication rule.]

9. Conditional independence. Random variables X and Y are called *conditionally independent given* Z if given the value of Z, X, and Y are independent. That is,

$$P(X = x, Y = y \mid Z = z) = P(X = x \mid Z = z) P(Y = y \mid Z = z)$$

for all possible values x, y, and z. Prove that X and Y are conditionally independent given Z if and only if the conditional distribution of Y given $X = x$ and $Z = z$ is a distribution which depends only on z:

$$P(Y = y \mid X = x, Z = z) = P(Y = y \mid Z = z)$$

for all possible values x, y, and z. Give a further equivalent condition in terms of the conditional distribution of X given $Y = y$ and $Z = z$.

10. Conditional independence (continued). Suppose as in Example 5 of Section 3.1 that two sequences of n draws with replacement are made from a box containing an unknown number of red tickets among a total of 10 tickets. Regard the number of red tickets in the box as the value of a random variable R, with probability distribution $P(R = r) = \pi_r$, $r = 0, 1, \ldots, 10$. Let X_1 be the number of red tickets in the first n draws, and X_2 the number in the second n draws. Assuming that X_1 and X_2 are conditionally independent and binomially distributed given $R = r$, find expressions for the following:

a) $P(R = r, X_1 = x_1, X_2 = x_2)$; b) $P(R = r \mid X_1 = x_1)$;

c) $P(X_2 = x_2 \mid R = r, X_1 = x_1)$; d) $P(X_2 = x_2 \mid X_1 = x_1)$.

e) Calculate numerical values for the conditional probabilities in d) assuming that $\pi_r = 1/11$ for $r = 0, 1, \ldots, 10$ and $n = 1$. Are X_1 and X_2 independent?

6.2 Conditional Expectation: Discrete Case

Conditional expectations are simply expectations relative to conditional distributions.

Conditional Expectation Given an Event

The *conditional expectation of a random variable Y given an event A*, denoted by $E(Y \mid A)$, is the expectation of Y under the conditional probability distribution given A:

$$E(Y \mid A) = \sum_{\text{all } y} y P(Y = y \mid A)$$

This is just the definition of $E(Y)$, with probabilities replaced by conditional probabilities given A. Intuitively, $E(Y \mid A)$ is the expected value of Y, given the information that event A has occurred.

Example 1. Conditioning on at most 2 heads on 4 coin tosses.

Let Y be the number of heads in four tosses of a fair coin. Calculate the conditional expectation of Y given 2 or less heads. What is the long-run interpretation of this quantity?

Solution. Here the conditioning event is $A = (Y \leq 2)$. Since Y has the binomial $(4, \frac{1}{2})$ distribution

$$P(Y = y) = \binom{4}{y} \Big/ 2^4 \qquad (y = 0 \text{ to } 4)$$

$$P(Y \leq 2) = (1 + 4 + 6)/16 = 11/16$$

Hence

$$P(Y = y \mid Y \leq 2) = \binom{4}{y} \Big/ 11 \qquad (y = 0, 1, 2)$$

and

$$E(Y \mid Y \leq 2) = \sum_{y=0}^{2} y \binom{4}{y} \Big/ 11 = (1 \cdot 4 + 2 \cdot 6)/11 = 16/11$$

The long-run interpretation is that if you repeatedly toss four fair coins, the long-run average number of heads, averaging only over the trials that produce 0, 1, or 2 heads, will be $16/11$.

Properties of Conditional Expectation

For a fixed conditioning event A, conditional expectation has familiar properties of expectation like linearity. For instance, there is the addition rule

$$E(X+Y\,|\,A) = E(X\,|\,A) + E(Y\,|\,A)$$

and so on. For a fixed random variable Y, as A varies, there is a useful generalization of the rule of average conditional probabilities, a rule of *average conditional expectations*: If $A_1, \ldots, A_n$ is a partition of the whole outcome space, then

$$E(Y) = \sum_{i=1}^{n} E(Y\,|\,A_i)P(A_i)$$

In the special case when Y is an indicator random variable, say $Y = I_B$, the indicator of event B, this reduces to the rule of average conditional probabilities

$$P(B) = \sum_{i=1}^{n} P(B\,|\,A_i)P(A_i)$$

The general case can be derived from this special case by linear operations. It is most convenient for applications to express the general rule as follows, for the partition generated by values of a discrete random variable X:

Rule of Average Conditional Expectations

For any random variable Y with finite expectation and any discrete random variable X,

$$E(Y) = \sum_{\text{all } x} E(Y\,|\,X=x)P(X=x)$$

This formula is also called the *formula for $E(Y)$ by conditioning on X*. This formula gives a useful method of calculating expectations, as shown by the examples below. The next box introduces a useful short notation:

Definition of $E(Y|X)$

The *conditional expectation of Y given X*, denoted $E(Y\,|\,X)$, is the function of X whose value is $E(Y\,|\,X=x)$ if $X=x$.

Here $E(Y \mid X)$ is actually a random variable, since by definition it is a particular function of X, and a function of a random variable defines another random variable. It can be shown that $E(Y \mid X)$ is the best predictor of Y based on X, in the sense of mean-square error. That is to say, $E(Y \mid X)$ is the function $g(X)$ that minimizes the mean square prediction error $E[(Y - g(X))^2]$. See Exercise 17. Because $E(Y \mid X)$ is a random variable, it makes sense to consider its expectation. The result is stated in the next box.

Expectation is the Expectation of the Conditional Expectation

$$E(Y) = E[E(Y \mid X)]$$

This is a condensed form of the rule of average conditional expectations, obtained by application to $g(x) = E(Y \mid X = x)$ of the formula

$$E[g(X)] = \sum_{\text{all } x} g(x)P(X = x)$$

Examples

Example 2. Tossing a random number of coins.

As in Example 1 of the previous section, let Y be the number of heads in X tosses of a fair coin, where X is generated by a fair die roll.

Problem 1. Find the conditional expectation of Y given $X = x$.

Solution. Since the conditional distribution of Y given $X = x$ is binomial with parameters $n = x$ and $p = 1/2$, the conditional expectation of Y given $X = x$ is the mean of the binomial(n, p) distribution, that is np, for $n = x$ and $p = 1/2$:

$$E(Y \mid X = x) = x/2 \qquad (x = 1, 2, \ldots, 6)$$

Problem 2. Find $E(Y)$.

Solution. Since from the previous solution $E(Y \mid X) = X/2$, and $E(X) = 3.5$

$$E(Y) = E[E(Y \mid X)] = E(X/2) = E(X)/2 = (3.5)/2 = 1.75$$

Discussion. Of course, the expectation of Y can also be calculated from the distribution of Y, shown in Table 1 of Section 6.1. But the method of conditioning on X gives the result more quickly. Also, the method of computing $E(Y)$ by conditioning on a suitable random variable X can be applied in problems where it is difficult to obtain a formula for the distribution of Y.

Problem 3. Find $E(X|Y = 2)$

Solution. There is no simple formula for $E(X \mid Y = y)$ as a function of y in this problem. But these conditional expectations can be calculated one by one from the various conditional distributions of X given $Y = y$ for $y = 0$ to 6. Using the conditional distribution of X given $Y = 2$ displayed in Table 3 of Section 6.1 gives

$$E(X|Y = 2) = (2\times16 + 3\times24 + 4\times24 + 5\times20 + 6\times15)/99 = 390/99 \approx 3.94$$

Example 3. Number of girls in a family.

Suppose the number of children in a family is a random variable X with mean μ, and given $X = n$ for $n \geq 1$, each of the n children in the family is a girl with probability p and a boy with probability $1 - p$.

Problem. What is the expected number of girls in a family?

Solution. Intuitively, the answer should be $p\mu$. To show this is correct, let G be the random number of girls in a family. Given $X = n$, G is the sum of n indicators of events with probability p, so

$$E(G|X = n) = np$$

Note that this is correct even for $n = 0$. By conditioning on X,

$$E(G) = \sum_n E(G|X = n)P(X = n) = p\sum_n nP(X = n) = p\mu$$

Remark. In short notation,

$$E(G|X) = pX$$

$$E(G) = E\left[E(G|X)\right] = E(pX) = pE(X)$$

Example 4. Success counts in overlapping series of trials.

Let S_n be the number of successes in n independent trials with probability p of success on each trial.

Problem. Calculate $E(S_m \mid S_n = k)$ for $m \leq n$.

Solution. Since $S_m = X_1 + \cdots + X_m$ where X_j is the indicator of success on the jth trial

$$E(S_m \mid S_n = k) = \sum_{j=1}^{m} E(X_j \mid S_n = k) \quad \text{where}$$

$$\begin{aligned} E(X_j \mid S_n = k) &= P(j\text{th trial is a success} \mid S_n = k) \\ &= \frac{P(j\text{th trial is a success}, S_n = k)}{P(S_n = k)} \\ &= \frac{P(j\text{th trial success}, k-1 \text{ of other } n-1 \text{ trials are successes})}{P(S_n = k)} \\ &= \frac{p\binom{n-1}{k-1}p^{k-1}(1-p)^{n-k}}{\binom{n}{k}p^k(1-p)^{n-k}} \quad \text{using independence and the binomial distribution} \\ &= \frac{k}{n} \quad \text{so} \end{aligned}$$

$$E(S_m \mid S_n = k) = \frac{mk}{n}$$

Discussion. In short notation, the conclusion is that for $1 \le m \le n$

$$E(S_m \mid S_n) = \frac{m}{n} S_n$$

This is a rather intuitive formula. It says that given S_n successes in n trials, the number of successes to be expected in m of the trials is proportional to m. The formula can be derived in other ways. By symmetry, $E(X_j \mid S_n)$ must be the same for all j, and equal to $E(X_1 \mid S_n)$. Since

$$S_n = E(S_n \mid S_n) = \sum_{j=1}^{n} E(X_j \mid S_n) = nE(X_1 \mid S_n)$$

it follows that $E(X_1 \mid S_n) = S_n/n$ and hence

$$E(S_m \mid S_n) = \sum_{j=1}^{m} E(X_j \mid S_n) = mE(X_1 \mid S_n) = \frac{m}{n} S_n$$

This argument shows that formula

$$E(S_m \mid S_n) = \frac{m}{n} S_n \qquad (1 \le m \le n)$$

holds whenever S_n is a sum of n independent and identically distributed variables $X_1, \ldots, X_n$. In fact all that is required is that the variables $X_1, \ldots, X_n$ are *exchangeable*, as defined in Section 3.6. This is an example where a conditional expectation can be calculated using symmetry and linearity, even though there is no nice formula for the conditional distribution.

Treating a conditioned variable as a constant. When computing conditional probabilities or expectations given $X = x$, the random variable X may be treated as if it were the constant x. Intuitively, this is quite obvious: on the restricted outcome space $(X = x)$, the random variable X has only one value, namely, x. To illustrate, if g is a function of two random variables X and Y, the conditional distribution of $g(X,Y)$ given $X = x$, is the same as the conditional distribution of $g(x,Y)$ given $X = x$. And if g has numerical values

$$E\left[g(X,Y)\,|\,X = x\right] = E\left[g(x,Y)\,|\,X = x\right]$$

For instance

$$E[XY\,|\,X = x] = E[xY\,|\,X = x] = xE[Y\,|\,X = x]$$

which reads in short notation

$$E[XY\,|\,X] = XE[Y\,|\,X]$$

Another example is

$$E[aX + bY\,|\,X = x] = E[ax + bY\,|\,X = x] = ax + bE[Y\,|\,X = x]$$

which reads in short notation

$$E[aX + bY\,|\,X] = aX + bE[Y\,|\,X]$$

Example 5. **Conditional expectation of a sum given one of the terms.**

Suppose X and Y are independent.

Problem. Find $E(X + Y\,|\,X = x)$.

Solution.

$$\begin{aligned} E(X + Y\,|\,X = x) &= E(X\,|\,X = x) + E(Y\,|\,X = x) \\ &= x + E(Y) \end{aligned}$$

Here $E(X\,|\,X = x) = x$ because X may be treated as the constant x given $X = x$. And $E(Y\,|\,X = x)$ is the mean of the conditional distribution of Y given $X = x$, and by independence this is just the unconditional distribution of Y with mean $E(Y)$.

Exercises 6.2

1. Let X_1 and X_2 be the numbers on two independent fair-die rolls. Let X be the minimum and Y the maximum of X_1 and X_2. Calculate: a) $E(Y|X = x)$; b) $E(X|Y = y)$.

2. Repeat Exercise 1 above, with X_1 and X_2 independent and uniformly distributed on $\{1, 2, \ldots, n\}$.

3. Repeat Exercise 1 with X_1 and X_2 two draws without replacement from $\{1, 2, \ldots, n\}$.

4. An item is selected randomly from a collection labeled $1, 2, ..., n$. Denote its label by X. Now select an integer Y uniformly at random from $\{1, \ldots, X\}$. Find:

a) $E(Y)$; b) $E(Y^2)$; c) $SD(Y)$; d) $P(X + Y = 2)$.

5. Suppose an event A is independent of a pair of random variables X_1 and X_2, whose c.d.f's are F_1 and F_2. Define a random variable X by:

$$X = \begin{cases} X_1 & \text{if } A \text{ occurs} \\ X_2 & \text{if } A \text{ does not occur} \end{cases}$$

Find and justify formulae for:

a) the c.d.f. $F(x)$ of X, in terms of $F_1(x)$, $F_2(x)$, and $p = P(A)$;

b) $E(X)$ in terms of $E(X_1)$, $E(X_2)$, and p.

c) $Var(X)$ in terms of $E(X_1)$, $E(X_2)$, $Var(X_1)$, $Var(X_2)$ and p.

6. Suppose that N is a Poisson random variable with parameter μ. Suppose that given $N = n$, random variables $X_1, X_2 ...., X_n$ are independent with uniform $(0, 1)$ distribution. So there are a random number of X's.

a) Given $N = n$, what is the probability that all the X's are less than t?

b) What is the (unconditional) probability that all the X's are less than t?

c) Let $S_N = X_1 + \cdots + X_N$ denote the sum of the random number of X's. (If $N = 0$ then $S_N = 0$.) Find $P(S_N = 0)$. Explain.

d) Find $E(S_N)$.

7. Suppose that N is a counting random variable, with values $\{0, 1, \ldots, n\}$, and that given $(N = k)$, for $k \geq 1$, there are defined random variables $X_1, \ldots, X_k$ such that

$$E(X_j | N = k) = \mu \qquad (1 \leq j \leq k)$$

Define a random variable S_N by

$$S_N = \begin{cases} X_1 + X_2 + \cdots + X_k & \text{if } (N = k),\ 1 \leq k \leq n \\ 0 & \text{if } (N = 0) \end{cases}$$

Show that $E(S_N) = \mu E(N)$.

8. Suppose that each individual in a population produces a random number of children, and the distribution of the number of children has mean μ. Starting with one individual, show, using the result of Exercise 7, that the expected number of descendants of that individual in the nth generation is μ^n.

9. Let T_i be the place at which the ith good element appears in a random ordering of $N - k$ bad elements and k good ones. Use the results of Exercise 3.6.13 to calculate:

a) $E(T_1 | T_2 = j)$; b) $E(T_2 | T_1 = j)$;

c) $E(T_h|T_i = j)$ first for $h < i$, then for $h > i$.

10. What is the expected number of black balls among $n \leq b+w+d$ balls drawn at random from a box containing b black balls, w white balls, and d balls drawn at random from another box of b_0 black balls and w_0 white balls? Assume all draws are made without replacement.

11. A deck of cards is cut into two halves of 26 cards each. As it turns out, the top half contains 3 aces and the bottom half just one ace. The top half is shuffled, then cut into two halves of 13 cards each. One of these packs of 13 cards is shuffled into the bottom half of 26 cards, and from this pack of 39 cards, 5 cards are dealt. What is the expected number of aces among these 5 cards?

12. Conditional expectations in Polya's urn scheme. An urn contains 1 black and 2 white balls. One ball is drawn at random and its color noted. The ball is replaced in the urn, together with an additional ball of its color. There are now four balls in the urn. Again, one ball is drawn at random from the urn, then replaced along with an additional ball of its color. The process continues in this way.

a) Let B_n be the number of black balls in the urn just before the nth ball is drawn. (Thus B_1 is 1.) For $n \geq 1$, find $E(B_{n+1}|B_n)$.

b) For $n \geq 1$, find $E(B_n)$. [*Hint*: $E(B_1) = 1$; now use part a) and induction on n.]

c) For $n \geq 1$, what is the expected proportion of black balls in the urn just before the nth ball is drawn?

13. Conditioning on the number of successes in Bernoulli trials. Let $S_n = X_1 + \cdots + X_n$ be the number of successes in of n independent Bernoulli(p) trials $X_1, X_2, \ldots, X_n$.

a) For $1 \leq m \leq n$, show that the conditional distribution of S_m, the number of successes in the first m trials, given $S_n = k$, is identical to the distribution of the number of good elements in a random sample of size m without replacement from a population of k good and $n - k$ bad elements.

b) Use the result of a) to rederive the result of Example 4 that $E(S_m \,|\, S_n = k) = mk/n$.

c) Find $Var(S_m \,|\, S_n = k)$.

14. Sufficiency of the number of successes in Bernoulli trials. Let $S_n = X_1 + \cdots + X_n$ be the number of successes in n independent Bernoulli (p) trials $X_1, X_2, \ldots, X_n$. As a continuation of Exercise 13, show that conditionally given $S_n = k$, the sequence of zeros and ones $X_1, \ldots, X_n$ is distributed like an exhaustive sample without replacement from a population of k ones and $n - k$ zeros. [Note that this conditional distribution does not depend on p. In the language of statistics, when p is an unknown parameter S_n is called a *sufficient statistic* for p. If you want to estimate an unknown p given observed values of $X_1, \ldots, X_n$, and are committed to the assumption of Bernoulli (p) trials, it makes no sense to use any aspect of the data besides S_n in the estimation problem, because given $S_n = k$, the parameter p does not affect the distribution of the data at all. One natural estimate of p given the data is S_n/n, the observed proportion of successes. But other functions of S_n may be considered. See Exercise 6.3.15.]

15. Let Π be a random proportion between 0 and 1, for example, the proportion of black balls in an urn picked at random from some population of urns. Let S be the number of successes in n Bernoulli trials, which given $\Pi = p$ are independent with probability p, for example, the number of black balls in n draws at random, with replacement from the urn picked at random.

a) Find a formula for $E(S)$ in terms of n and $E(\Pi)$.

b) Find a formula for $Var(S)$ in terms of n, $E(\Pi)$, and $Var(\Pi)$.

c) For given n and $E(\Pi) = p$, say, which distribution of Π makes $Var(S)$ as large as possible? Which as small as possible? Prove your answers using your answer to b).

16. Expectation of a product by conditioning. Let X and Y be random variables, and let h be a function of X. Show that

$$E\left[h(X)Y\right] = E\left[h(X)E(Y|X)\right]$$

[*Hint:* Look at $E(h(X)Y|X = x)$.] *Remark:* This identity, for indicator functions $h(x)$, is used in more advanced treatments of probability to define conditional expectations given a continuous random variable X.

17. Prediction by functions. Suppose you want to predict the value of a random variable Y. Instead of just trying to predict the value of Y by a constant, as was done in Section 3.2, suppose that some additional information pertinent to the prediction of Y is available. For instance, you might know the value of some other random variable X, whose joint distribution with Y is assumed known. The problem here is to predict the value of Y by a function of X, call it $g(X)$. Once the value x of X is known, the value $g(x)$ of $g(X)$ can be calculated and used to predict the unknown value of Y.
One measure of the goodness of the predictor $g(X)$ is its *mean square error* (MSE)

$$MSE(g(X)) = E[(Y - g(X))^2]$$

It is a measure of, on average, how far off the prediction is. Show that $g(X) = E(Y|X)$ minimizes the MSE. [*Hint:* Condition on the value of X

$$E[(Y - g(X))^2] = \sum_x E[(Y - g(X))^2|X = x]P(X = x)$$

and minimize each term in the sum separately.]

18. Conditional variance. Define $Var(Y|X)$, the *conditional variance of* Y *given* X, to be the random variable whose value, if $(X = x)$, is the variance of the conditional distribution of Y given $X = x$. So $Var(Y|X)$ is a function of X, namely $h(X)$, where $h(x) = E(Y^2|X = x) - [E(Y|X = x)]^2$. Show that

$$Var(Y) = E[Var(Y|X)] + Var[E(Y|X)]$$

In words, the variance is the expectation of the conditional variance plus the variance of the conditional expectation.

6.3 Conditioning: Density Case

This section treats conditional probabilities given the value of a random variable X with a continuous distribution. In the discrete case, the conditional probability of an event A, given that X has value x, is defined by

$$P(A|X=x) = \frac{P(A, X=x)}{P(X=x)}$$

whenever $P(X = x) > 0$. In the continuous case $P(X = x) = 0$ for every x, so the above formula gives the undefined expression $0/0$. This must be replaced, as in the usual calculus definition of a derivative dy/dx, by the following:

Infinitesimal Conditioning Formula

$$P(A|X=x) = \frac{P(A, X \in dx)}{P(X \in dx)}$$

Intuitively, $P(A \mid X = x)$ should be understood as $P(A \mid X \in dx)$, the chance of A given that X falls in a very small interval near x. It is assumed here that in the limit of small intervals this chance does not depend on what interval is chosen near x. So, like a derivative dy/dx, $P(A|X \in dx)$ is a function of x, hence the notation $P(A|X = x)$. In terms of limits,

$$P(A|X=x) = \lim_{\Delta x \to 0} P(A|X \in \Delta x) = \lim_{\Delta x \to 0} \frac{P(A, X \in \Delta x)}{P(X \in \Delta x)}$$

where Δx stands for an interval of length Δx containing the point x. It is assumed here that the limit exists, except perhaps for a finite number of exceptional points x such as endpoints of an interval defining the range of X, or places where the density of X has a discontinuity. See the book *Probability and Measure* by P. Billingsley for a rigorous treatment of conditioning on a continuously distributed variable.

Most often, the event A of interest is determined by some random variable Y, for instance, $A = (Y > 3)$. If (X, Y) has a joint density $f(x, y)$, then $P(A|X = x)$ can be found by integration of the conditional density of Y given $X = x$, defined as follows:

Conditional Density of Y given $X = x$

For random variables X and Y with joint density $f(x, y)$, for each x such that the marginal density $f_X(x) > 0$, the *conditional density of Y given $X = x$* is the probability density function with dummy variable y defined by

$$f_Y(y \mid X = x) = f(x, y)/f_X(x)$$

Intuitively, the formula for $f_Y(y \mid X = x)$ is justified by the following calculation of the chance of $(Y \in dy)$ given $X = x$:

$$\begin{aligned} P(Y \in dy \mid X = x) &= P(Y \in dy \mid X \in dx) \\ &= \frac{P(X \in dx, Y \in dy)}{P(X \in dx)} \\ &= \frac{f(x, y)\, dx\, dy}{f_X(x)\, dx} \\ &= f_Y(y \mid X = x) dy \end{aligned}$$

The formula $\int f(x, y) dy = f_X(x)$, the marginal density of X, implies that

$$\int f_Y(y \mid X = x) dy = 1$$

So for each fixed x with $f_X(x) > 0$, the formula for $f_Y(y \mid X = x)$ gives a probability density in y. This conditional density given x defines a probability distribution parameterized by x, called the *conditional distribution of Y given $X = x$*. In examples, this will often be a familiar distribution, for example, a uniform or a normal distribution, with parameters depending on x.

The conditional density of Y given $X = x$ can be understood geometrically by taking a vertical slice through the joint density surface at x, and renormalizing the resulting function of y by its total integral, which is $f_X(x)$. Conditional probabilities given $X = x$ of events determined by X and Y can be calculated by integrating with respect to this conditional density. For example

$$P(Y > b \mid X = x) = \int_b^\infty f_Y(y \mid X = x) dy$$

$$P(Y > 2X \mid X = x) = \int_{2x}^\infty f_Y(y \mid X = x) dy$$

Such expressions are obtained formally from their discrete analogs by replacing a sum by an integral, and replacing the probability of an individual point by the value of a density times an infinitesimal length. See the display at the end of this section for details of this analogy.

FIGURE 1. Joint, marginal, and conditional densities.

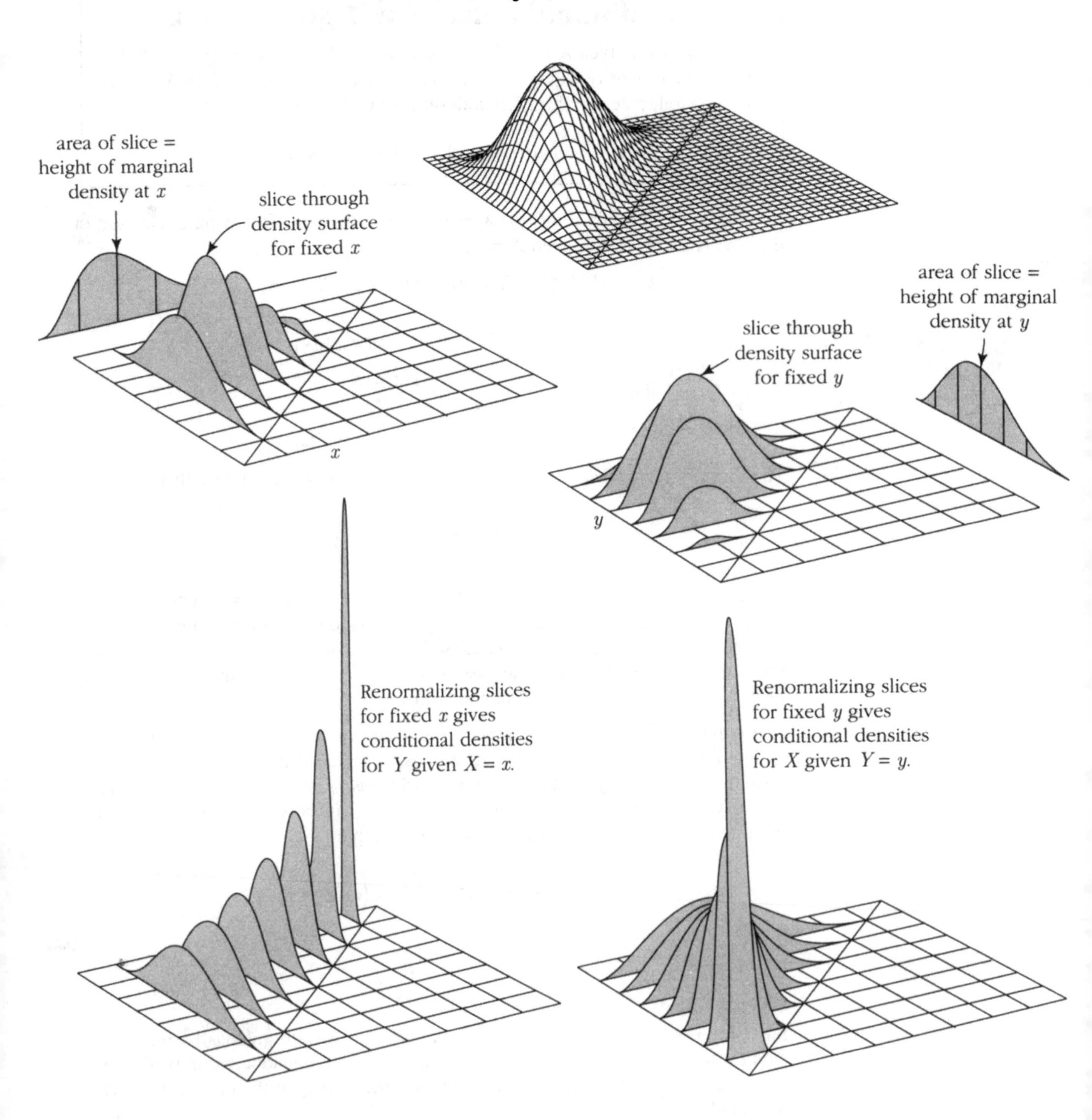

Key to Figure 1

Top: Joint density surface. This is a perspective projection of the surface

$$z = f(x, y)$$

defined by a particular joint density function $f(x, y)$.

Middle left: Slices for some values of X and the marginal density of X. Here are seven slices, or cross sections through the density surface for given values X ranging from $1/8$ to $7/8$. (The last two are so low that they are invisible.) The probability that X falls in a short interval of length Δ near x is the volume of such a slice of thickness Δ, which for small enough Δ is essentially Δ times the area of the slice at x. This area equals

$$\int f(x, y) dy = f_X(x)$$

the height of the *marginal density of X at x*, graphed at back. This marginal density shows how probability is distributed between slices according to the distribution of X. The heights of the vertical segments shown in the graph of the marginal density are proportional to the areas of corresponding slices.

Middle right: Slices for some values Y and the marginal density of Y. Here are perpendicular slices through the density surface for given values of Y. The area of the slice at y equals

$$\int f(x, y)\, dx = f_Y(y),$$

the height of the *marginal density of Y at y*, shown at right.

Bottom left: Conditional density of Y for some given values of X. Rescaling each section of the diagram above by its total area, the marginal density of X at x, gives the *conditional density of Y given $X = x$*, shown here using the same vertical scale as for the marginal densities in the middle diagrams. Given $X = x$, Y is distributed with density proportional to the section of the density surface $f(x, y)$ through x. Dividing by the total area of the section through x gives the conditional density of Y given $X = x$. Note how the shape of the two invisible sections in the middle left diagram can now be seen, due to the normalization of each section by its total area. The marginal density of Y (see middle right) is the average of all the conditional densities of Y given $X = x$ weighted according to the marginal distribution of X (middle left).

Bottom right: Conditional density of X for some given values of Y. These are interpreted just as above, with the roles of X and Y switched.

Example 1. Uniform on a triangle.

Problem. Suppose that a point (X, Y) is chosen uniformly at random from the triangle $\{(x, y) : x \geq 0, y \geq 0, x + y \leq 2\}$. Find $P(Y > 1 | X = x)$.

To illustrate the basic concepts, three slightly different solutions will be presented.

Solution 1. *Informal approach.* Intuitively, it seems obvious that given $X = x$, the random point (X, Y) should be regarded as uniformly distributed on the vertical line segment $\{(x, y) : y \geq 0, x + y \leq 2\}$ with length $2 - x$. This is the conditional distribution of (X, Y) given $X = x$. If x is between 0 and 1, the portion of this segment above $y = 1$ has length $(2 - x) - 1 = 1 - x$. Otherwise, no portion of the segment is above $y = 1$. So the answer is

$$P(Y > 1 | X = x) = \begin{cases} (1 - x)/(2 - x) & 0 \leq x < 1 \\ 0 & \text{otherwise} \end{cases}$$

Solution 2. *Definition of conditional probability.* To see that Solution 1 agrees with the formal definition

$$P(Y > 1 | X = x) = \lim_{\Delta x \to 0} P(Y > 1 | X \in \Delta x)$$

look at the following diagram which shows the events $(Y > 1)$ and $(X \in \Delta x) = (x \leq X \leq x + \Delta x)$:

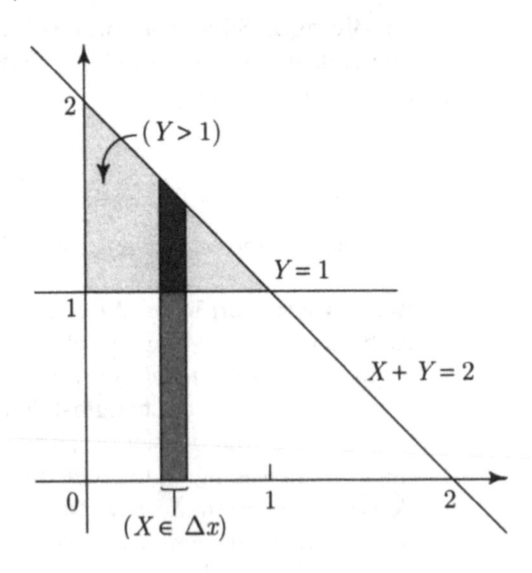

Since the triangle has area 2, the probability of an event is half its area. So, for $0 \leq x < 1, x + \Delta x \leq 1$, there are the exact formulae

$$P(X \in \Delta x) = \frac{1}{2}\Delta x(2 - x - \frac{1}{2}\Delta x)$$

$$P(Y > 1, X \in \Delta x) = \frac{1}{2}\Delta x(1 - x - \frac{1}{2}\Delta x)$$

Therefore, for $0 \leq x < 1$,

$$\begin{aligned} P(Y > 1 \mid X \in \Delta x) &= \frac{P(Y > 1, X \in \Delta x)}{P(X \in \Delta x)} \\ &= \frac{1 - x - \frac{1}{2}\Delta x}{2 - x - \frac{1}{2}\Delta x} \\ &\to \frac{1 - x}{2 - x} \quad \text{as} \quad \Delta x \to 0 \end{aligned}$$

This verifies the formula of Solution 1 for $0 \leq x < 1$. The formula for $x \geq 1$ is obvious because the event $(Y > 1, X \in \Delta x)$ is empty if $x \geq 1$.

Solution 3. *Calculation with densities.* Let us recalculate $P(Y > 1 \mid X = x)$ using the conditional density $f_Y(y \mid X = x)$. The uniform distribution on the triangle makes the joint density

$$f(x, y) = \begin{cases} 1/2 & x \geq 0, y \geq 0, x + y \leq 2 \\ 0 & \text{otherwise} \end{cases}$$

So for $0 \leq x \leq 2$,

$$f_X(x) = \int_0^\infty f(x, y)dy = \int_0^{2-x} \frac{1}{2}dy = \frac{1}{2}(2 - x)$$

and

$$f_Y(y \mid X = x) = \begin{cases} \dfrac{f(x, y)}{f_X(x)} = \dfrac{1}{2 - x} & 0 \leq y \leq 2 - x \\ 0 & \text{otherwise} \end{cases}$$

That is, given $X = x$ for $0 \leq x \leq 2$, Y has uniform $(0, 2 - x)$ distribution, as is to be expected intuitively. So

$$P(Y > 1 \mid X = x) = \begin{cases} \displaystyle\int_1^{2-x} \frac{dy}{2 - x} = \frac{1 - x}{2 - x} & 0 \leq x \leq 1 \\ 0 & \text{otherwise} \end{cases}$$

as before.

Discussion. The point of the first solution is that conditional distributions are often intuitively obvious, and once identified they can be used to find conditional probabilities very quickly. The second solution shows how this kind of calculation is justified by the formal definition. This method is not recommended for routine calculations. The third solution is essentially a more detailed version of the first. While rather pedantic in the present problem, this kind of calculation is essential in more difficult problems where you cannot guess the answer by intuitive reasoning.

Rules for conditional densities. These are analogs of corresponding rules in the discrete case. Note that every concept defined by the distribution of a real-valued random variable Y, in particular, the notions of density function, distribution function, expectation, variance, moments, and so on, can be considered for conditional distributions, just as well as for unconditional ones. There is just an extra parameter, x, the given value of X.

When the density of X is known, and a conditional density for Y given $X = x$ is specified for each x in the range of X, the joint density of X and Y is calculated by the following rearrangement of the formula $f_Y(y \,|\, X = x) = f(x, y)/f_X(x)$.

Multiplication Rule for Densities

$$f(x, y) = f_X(x) f_Y(y \,|\, X = x)$$

Example 2. **Gamma and uniform.**

Suppose X has gamma $(2, \lambda)$ distribution, and that given $X = x$, Y has uniform $(0, x)$ distribution.

Problem 1. Find the joint density of X and Y.

Solution. By the definition of the gamma distribution

$$f_X(x) = \begin{cases} \lambda^2 x e^{-\lambda x} & x > 0 \\ 0 & x \leq 0 \end{cases}$$

and from the uniform $(0, x)$ distribution of Y given $X = x$

$$f_Y(y \,|\, X = x) = \begin{cases} 1/x & 0 < y < x \\ 0 & \text{otherwise} \end{cases}$$

So by the multiplication rule for densities

$$f(x, y) = f_X(x) f_Y(y \,|\, X = x) = \begin{cases} \lambda^2 e^{-\lambda x} & 0 < y < x \\ 0 & \text{otherwise} \end{cases}$$

Problem 2. Find the marginal density of Y.

Solution. Integrating out x in the joint density gives the marginal density of Y: for $y > 0$

$$f_Y(y) = \int_0^\infty f(x, y)\, dx = \int_y^\infty \lambda^2 e^{-\lambda x}\, dx = \lambda e^{-\lambda y}$$

The density is of course 0 for $y \leq 0$. That is to say, Y has exponential (λ) distribution.

Problem 3. Show that X and Y have the same joint distribution as T_2 and T_1, where T_1 is the first arrival time and T_2 is the second arrival time in a Poisson arrival process with rate λ.

Solution. That X has the same distribution as T_2, and that Y has the same distribution as T_1, follows from the above calculation and the result of Section 4.2 that the ith arrival time in a Poisson process with rate λ has gamma(i, λ) distribution. That the *joint* distribution of X and Y is the same as the joint distribution of T_2 and T_1 requires a little more calculation, because a joint distribution is not determined by its marginals. The simplest way to verify this is to observe that for $0 < y < x$

$$P(T_1 \in dy, T_2 \in dx)$$

is the probability of no arrivals in the time interval $[0, y]$ of length y, one arrival in time dy, no arrivals in the time interval $[y + dy, x]$ of length $x - y - dy \approx x - y$, and finally one arrival in dx. By independence and Poisson distribution of counts in disjoint intervals, and neglecting a term of order $(dy)^2$, this event has probability

$$e^{-\lambda y} \lambda \, dy \, e^{-\lambda(x-y)} \lambda \, dx = \lambda^2 e^{-\lambda x} dy \, dx$$

Dividing the last expression by $dy \, dx$ shows that the joint density of (T_2, T_1) at (x, y) with $0 < y < x$ is identical to the joint density found in Problem 1. Since obviously $P(T_1 < T_2) = 1$, the joint density of (T_2, T_1) can be taken to be zero except if $0 < y < x$. Thus (T_2, T_1) has the same joint density function as (X, Y), hence the same joint distribution.

Problem 4. For T_1 and T_2 the first two arrival times in a Poisson process with rate λ, find the conditional distribution of T_1 given $T_2 = x$.

Solution. Since according to the solution of the previous problem, T_2 and T_1 have the same joint density as X and Y, found in Problem 1, the conditional distribution of T_1 given $T_2 = x$ is identical to the conditional distribution of Y given $X = x$, which was given at the start, that is to say, uniform on $(0, x)$.

Averaging Conditional Probabilities

For a random variable X with density f_X, the rule of average conditional probabilities becomes the following:

Integral Conditioning Formula

$$P(A) = \int P(A \mid X = x) f_X(x) \, dx$$

The integral breaks up the probability of A according to the values of X:

$$P(A \mid X = x) f_X(x)\, dx = P(A \mid X \in dx) P(X \in dx) = P(A, X \in dx)$$

Just as in the discrete case, $P(A \mid X = x)$ is often specified in advance by the formulation of a problem. Then $P(A)$ can be calculated by the integral conditioning formula, assuming also that the distribution of X is known. Bayes' rule then gives the conditional density of X given that A has occurred:

$$P(X \in dx \mid A) = \frac{P(X \in dx) P(A \mid X = x)}{P(A)} = \frac{f_X(x) P(A \mid X = x)}{P(A)}\, dx$$

The following example shows how the integral conditioning formula arises naturally by taking limits of discrete problems. In this example, as is often the case, the limits defined by integrals are much easier to work with than the discrete sums. The example makes precise the idea of independent trials with probabilty p of success in a setting where it makes clear sense to think of p as picked at random from some distribution before the trials are performed. In the first problem p is picked from a discrete uniform distribution on $N+1$ evenly spaced points in $[0,1]$. Passing to the limit as $N \to \infty$ leads to p that is uniformly distributed on $[0,1]$. Bayesian statisticians view this as a model for independent trials with unknown probability of success.

Example 3. Discrete uniform–binomial.

Suppose there are $N+1$ boxes labeled by $b = 0, 1, 2, \ldots, N$. Box b contains b black and $N - b$ white balls. A box is picked uniformly at random, and then n balls are drawn at random with replacement from whatever box is picked (the same box for each of the n draws). Let S_n denote the total number of black balls that appear among the n balls drawn.

Problem 1. Find the distribution of S_n.

Solution. Let Π denote the proportion of black balls in the box picked. Let G_N denote the grid of $N+1$ possible values p of Π:

$$G_N = \{0, \frac{1}{N}, \frac{2}{N}, \ldots, \frac{N-1}{N}, 1\}$$

For each $p \in G_N$ the binomial formula for n independent trials with probability p of success on each trial gives

$$P(S_n = k \mid \Pi = p) = \binom{n}{k} p^k (1-p)^{n-k}$$

Averaging with respect to the uniform distribution of Π over the $N+1$ values in G_N, and substituting $p = b/N$, gives the unconditional distribution of S_n:

$$P(S_n = k) = \sum_{p \in G_N} \binom{n}{k} p^k (1-p)^{n-k} \frac{1}{N+1} \tag{1}$$

$$= \binom{n}{k} \frac{1}{(N+1)N^n} \sum_{b=0}^{N} b^k (N-b)^{n-k}$$

It is hard to simplify this expression further. But the expression is easily evaluated for small values of n and N. To illustrate, for $N = n = 2$ the result is shown in the next table. The limiting behavior for large N is the subject of the next problem.

Distribution of S_2 for $N = 2$

k	0	1	2
$P(S_2 = k)$	$\frac{5}{12}$	$\frac{2}{12}$	$\frac{5}{12}$

Problem 2. For a fixed value of n, find the limiting distribution of S_n, the number of black balls that appear in n draws, as the number of boxes N tends to ∞.

Solution. Expression (1) for $P(S_n = k)$ is $\binom{n}{k}$ times a discrete approximation to the beta integral

$$B(k+1, n-k+1) = \int_0^1 p^k (1-p)^{n-k} dp$$

The approximation in (1) is obtained by taking the average value of the function $p^k(1-p)^{n-k}$ at $N+1$ evenly spaced points p, beween 0 and 1. In the limit as $N \to \infty$, the discrete average converges to the continuous integral. Using the expression for the beta integral in terms of the gamma function, and $\Gamma(m+1) = m!$ for integers m, gives

$$B(k+1, n-k+1) = \frac{\Gamma(k+1)\Gamma(n-k+1)}{\Gamma(k+1+n-k+1)} = \binom{n}{k}^{-1} \frac{1}{n+1} \tag{2}$$

The conclusion is that as $N \to \infty$

$$P(S_n = k) \to \binom{n}{k} \binom{n}{k}^{-1} \frac{1}{n+1} = \frac{1}{n+1}$$

for every $0 \le k \le n$. That is, the limiting distribution of S_n as $N \to \infty$ is uniform on $\{0, 1, \ldots, n\}$.

Example 4. Continuous uniform–binomial.

Suppose that Π is picked uniformly at random from $(0,1)$. Given that $\Pi = p$, let S_n be the number of successes in n independent trials with probability p of success on each trial.

Problem 1. Find the distribution of S_n.

Solution. By the limiting result obtained in the previous example as $N \to \infty$, the answer must be uniform on $\{0, 1, \ldots n\}$. This can be derived directly in the continuous model using the integral conditioning formula. Since the density of Π is $f_\Pi(p) = 1$ for $0 < p < 1$, and 0 otherwise,

$$P(S_n = k) = \int P(S_n = k \,|\, \Pi = p) f_\Pi(p) dp \tag{3}$$

$$= \int_0^1 \binom{n}{k} p^k (1-p)^{n-k} dp$$

$$= \frac{1}{n+1}$$

by evaluation of the beta integral as in the previous problem.

Discussion. Note the close parallel between the expression (3) for $P(S_n = k)$ obtained by the integral conditioning formula for Π with uniform distribution on $(0,1)$, and the corresponding expression (1) for $P(S_n = k)$ in the previous example for Π with uniform distribution on the set of $N+1$ values in G_N. All that happens is that the sum is replaced by an integral, and $1/(N+1)$, which is both the probability of each point in G_N and the difference between adjacent points in G_N, is replaced by the calculus differential dp representing the probability that the uniform variable falls in an infinitesimal length dp near p.

Problem 2. Find the conditional distribution of Π given that $S_n = k$.

Solution. Using Bayes' rule, for $0 < p < 1$,

$$P(\Pi \in dp \,|\, S_n = k) = \frac{P(\Pi \in dp) P(S_n = k \,|\, \Pi = p)}{P(S_n = k)}$$

$$= (n+1) \binom{n}{k} p^k (1-p)^{n-k} dp$$

This is the density at p of the beta distribution with parameters $k+1$ and $n-k+1$, times dp. Conclusion: the conditional distribution of Π given $S_n = k$ is beta$(k+1,\ n-k+1)$.

Problem 3. In the above setup, given that n trials have produced k successes, what is the probability that the next trial is a success?

Solution. Given $\Pi = p$ and $S_n = k$, the next trial is a success with probability p, by the assumption of independent trials with constant probability p of success given $\Pi = p$. Given just $S_n = k$, the value of Π is unknown. Rather, Π is a random variable with beta $(k+1,\ n-k+1)$ distribution. By the integral conditioning formula, the required probability is the conditional expectation of Π given $S_n = k$, which is $(k+1)/(n+2)$, by the formula $a/(a+b)$ for the mean of the beta (a, b) distribution. In detail:

$$P(\text{next trial a success} \mid S_n = k)$$

$$= \int_0^1 P(\text{next trial a success} \mid S_n = k, \Pi = p) f_\Pi(p \mid S_n = k) dp$$

$$= \int_0^1 p f_\Pi(p \mid S_n = k) dp = E(\Pi \mid S_n = k) = \frac{k+1}{n+2}$$

Discussion. In particular, for $k = n$, given n successes in a row, the chance of one more success is $(n+1)/(n+2)$. This formula, for the probability of one more success given a run of n successes in independent trials with unknown success probability assumed uniformly distributed on $(0, 1)$, is known as *Laplace's law of succession.* Laplace illustrated his formula by calculating the probability that the sun will rise tomorrow, given that it has risen daily for 5000 years, or $n = 1,826,213$ days. But this kind of application is of doubtful value. Both the assumption of independent trials with unknown p and the uniform prior distribution of p make little sense in this context.

Example 5. Simulation of uniform–binomial.

Suppose you have available a random number generator which you are willing to believe generates independent uniform $(0, 1)$ variables $U_0, U_1, \ldots$.

Problem 1. How could you simulate a pair of values from the joint distribution of Π and S_n considered above, with Π uniform on $(0, 1)$, and S_n binomial(n, p) given $\Pi = p$?

Solution. Set

$$\Pi = U_0, \text{ and } S_n = \sum_{i=1}^{n} I(U_i < U_0)$$

where $I(U_i < U_0)$ is an indicator variable that is 1 if $(U_i < U_0)$ and 0 otherwise. If $\Pi = p$, then $S_n = \sum_{i=1}^{n} I(U_i < p)$ is the sum of n independent indicator variables, each of which is 1 with probability p and 0 with probability $1-p$, exactly as required.

Problem 2. Use this construction to calculate $P(S_n = k)$ without integration.

Solution. By construction of S_n from $U_0, U_1, \ldots, U_n$

$(S_n = 0)$ if and only if U_0 is the smallest of the $U_0, U_1, \ldots, U_n$

$(S_n = 1)$ if and only if U_0 is the second smallest of the $U_0, U_1, \ldots, U_n$

$\ldots \qquad \ldots \qquad \ldots$

$(S_n = n)$ if and only if U_0 is the largest of the $U_0, U_1, \ldots, U_n$

Since all $(n+1)!$ possible orderings of the $U_0, U_1, \ldots, U_n$ are equally likely, each of these events has the same probability $1/(n+1)$.

Remark. This calculation is closely related to the distribution of order statistics treated in Section 4.6. For $j = 1, \ldots, n+1$, let $U_{(j)}$ denote the jth smallest of the $n+1$ variables $U_0, \ldots, U_n$. Then the event $S_n = j - 1$, that there are exactly $j - 1$ values U_i less than U_0, is identical to the event $U_{(j)} = U_0$, that the jth smallest of the U_i equals U_0. The solution of Problem 2 in Example 4 now translates into the following: the conditional distribution of U_0, or of $U_{(j)}$, given that $U_{(j)} = U_0$, is beta $(j, n-j+2)$. By symmetry, the same is true for U_k instead of U_0 for any $1 \le k \le n$. Consequently, the distribution of $U_{(j)}$, the jth smallest of $n+1$ independent uniform $(0,1)$ variables, is beta $(j, n-j+2)$, independently of K, where K is the random index k such that $U_k = U_{(j)}$. This agrees with the result of Section 4.6, with the present $n+1$ and j instead of n and k in that section.

Independence

In the continuous case, just as in the discrete case, it can be shown that each of the following conditions is equivalent to independence of random variables X and Y:

- the conditional distribution of Y given $X = x$ does not depend on x;
- the conditional distribution of X given $Y = y$ does not depend on y.

By integration with respect to the distribution of X, the common conditional distribution of Y given $X = x$ then equals the unconditional distribution of Y. That is to say, for all subsets B in the range of Y

$$P(Y \in B \mid X = x) = P(Y \in B)$$

Similarly for all subsets A in the range of X

$$P(X \in A \mid Y = y) = P(X \in A)$$

These are variations of the basic definition of independence of X and Y, which is

$$P(X \in A, Y \in B) = P(X \in A)P(Y \in B)$$

for all subsets A and B in the ranges of X and Y respectively. When X and Y have densities, X and Y are independent if and only if $f_Y(y \mid X = x) = f_Y(y)$ for all x and y, and again if and only if $f_X(x \mid Y = y) = f_X(x)$ for all x and y. So the general multiplication rule for densities reduces in this case to the formula

$$f(x, y) = f_X(x) f_Y(y)$$

for independent variables X and Y. This formula was applied in Section 5.2.

Conditional Expectations

The conditional expectation of Y given $X = x$, denoted $E(Y \mid X = x)$, is defined as the expectation of Y relative to the conditional distribution of Y given $X = x$. More generally, for a function g, assuming that Y has a conditional density $f_Y(y \mid X = x)$,

$$E[g(Y) \mid X = x] = \int g(y) f_Y(y \mid X = x) dy$$

Taking $g(y) = y$ gives $E(Y \mid X = x)$. And integrating the conditional expectation with respect to the distribution of X gives the unconditional expectation

$$E[g(Y)] = \int E[g(Y) \mid X = x] f_X(x)\, dx$$

These formulae are extensions to general functions g of the basic conditional probability formulae, which are the special cases when g is an indicator. As a general rule, all the basic properties of conditional expectations, considered in the discrete case in Section 6.2, remain valid in the density case.

Example 6. Uniform distribution on a triangle.

Problem. Suppose, as in Example 1, that (X, Y) is chosen uniformly at random from the triangle $\{(x, y) : x \geq 0, y \geq 0, x + y \leq 2\}$. Find $E(Y \mid X)$ and $E(X \mid Y)$.

Solution. As argued before, given $X = x$, for $0 < x < 2$, Y has uniform distribution on $(0, 2 - x)$. Since the mean of this conditional distribution is $(2 - x)/2$,

$$E(Y \mid X = x) = (2 - x)/2$$

In short notation

$$E(Y \mid X) = (2 - X)/2$$

Similarly, because joint density of X and Y is symmetric in x and y,

$$E(X \mid Y) = (2 - Y)/2$$

Conditioning Formulae: Discrete Case

Multiplication rule: The joint probability is the product of the marginal and the conditional

$$P(X = x, Y = y) = P(X = x)P(Y = y \,|\, X = x)$$

Division rule: The conditional probability of $Y = y$ given $X = x$ is

$$P(Y = y \,|\, X = x) = \frac{P(X = x, Y = y)}{P(X = x)}$$

Bayes' rule:

$$P(X = x \,|\, Y = y) = \frac{P(Y = y \,|\, X = x)P(X = x)}{P(Y = y)}$$

Conditional distribution of Y given $X = x$: Sum the conditional probabilities

$$P(Y \in B \,|\, X = x) = \sum_{y \in B} P(Y = y \,|\, X = x)$$

Conditional expectation of $g(Y)$ given $X = x$: Sum g against the conditional probabilities

$$E\,(g(Y) \,|\, X = x) = \sum_{\text{all } y} g(y)P(Y = y \,|\, X = x)$$

Average conditional probability:

$$P(B) = \sum_{\text{all } x} P(B \,|\, X = x)P(X = x)$$

$$P(Y = y) = \sum_{\text{all } x} P(Y = y \,|\, X = x)P(X = x)$$

Average conditional expectation:

$$E(Y) = \sum_{\text{all } x} E(Y \,|\, X = x)P(X = x)$$

Conditioning Formulae: Density Case

Multiplication rule: The joint density is the product of the marginal and the conditional

$$f(x, y) = f_X(x) f_Y(y \mid X = x)$$

Division rule: The conditional density of Y at y given $X = x$ is

$$f_Y(y \mid X = x) = \frac{f(x, y)}{f_X(x)}$$

Bayes' rule:

$$f_X(x \mid Y = y) = \frac{f_Y(y \mid X = x) f_X(x)}{f_Y(y)}$$

Conditional distribution of Y given $X = x$: Integrate the conditional density

$$P(Y \in B \mid X = x) = \int_B f_Y(y \mid X = x) dy$$

Conditional expectation of $g(Y)$ given $X = x$: Integrate g against the conditional density:

$$E\left(g(Y) \mid X = x\right) = \int g(y) f_Y(y \mid X = x) dy$$

Average conditional probability:

$$P(B) = \int P(B \mid X = x) f_X(x)\, dx$$

$$f_Y(y) = \int f_Y(y \mid X = x) f_X(x)\, dx$$

Average conditional expectation:

$$E(Y) = \int E(Y \mid X = x) f_X(x)\, dx$$

Exercises 6.3

1. Suppose X has uniform $(0,1)$ distribution and $P(A|X=x)=x^2$. What is $P(A)$?

2. Let X and Y have the following joint density:

$$f(x,y)=\begin{cases} 2x+2y-4xy & \text{for } 0\le x\le 1 \text{ and } 0\le y\le 1 \\ 0 & \text{otherwise} \end{cases}$$

a) Find the marginal densities of X and Y.

b) Find $f_Y(y\,|\,X=\frac{1}{4})$. c) Find $E(Y\,|\,X=\frac{1}{4})$.

3. Let (X,Y) be as in Example 1. Find a formula for $P(Y\le y|X=x)$.

4. Suppose X, Y are random variables with joint density

$$f(x,y)=\begin{cases} \lambda^3 x e^{-\lambda y} & \text{for } 0<x<y \\ 0 & \text{otherwise} \end{cases}$$

a) Find the density of Y. What is $E(Y)$? b) Compute $E(X|Y=1)$.

5. Suppose (X,Y) has uniform distribution on the triangle shown in the diagram. For x between -1 and 1, find:

a) $P(Y\ge\frac{1}{2}|X=x)$;

b) $P(Y<\frac{1}{2}|X=x)$;

c) $E(Y|X=x)$;

d) $Var(Y|X=x)$.

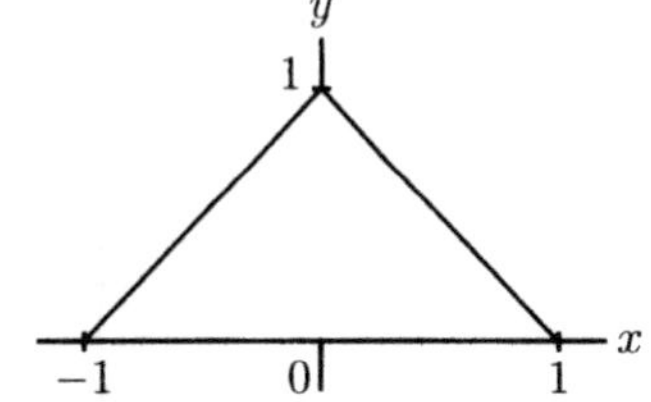

6. Suppose X, Y are random variables with joint density

$$f_{X,Y}(x,y)=\frac{1}{2\pi\sqrt{x(y-x)}}\,e^{-y/2} \qquad (0<x<y)$$

a) Find the distribution of Y. [*Hint*: For integration use the substitution $x=y\,s$.]

b) Compute $E(X|Y=1)$.

7. Suppose that Y and Z are random variables with the following joint density:

$$f(y,z)=\begin{cases} k(z-y) & \text{for } 0\le y\le z\le 1 \\ 0 & \text{otherwise} \end{cases}$$

for some constant k. Find:

a) the marginal distribution of Y; b) $P\left(Z<\frac{2}{3}|Y=\frac{1}{2}\right)$.

8. The random variable X has a uniform distribution on $(0,1)$. Given that $X=x$, the random variable Y is binomial with parameters $n=5$ and $p=x$.

a) Find $E(Y)$ and $E(Y^2)$. b) Find $P(Y=y \text{ and } x<X<x+dx)$.

c) Find the density of X given $Y=y$. Do you recognize it? If yes, as what?

9. Let A and B be events and let Y be a random variable uniformly distributed on $(0,1)$. Suppose that, conditional on $Y = p$, A and B are independent, each with probability p. Find:

a) the conditional probability of A given that B occurs;

b) the conditional density of Y given that A occurs and B does not.

10. Conditioning a Poisson process on the number of arrivals in a fixed time. Let T_1 and T_5 be the time of the first and fifth arrivals in a Poisson process with rate λ, as in Section 4.2.

a) Find the conditional density of T_1 given that there are 10 arrivals in the time interval $(0,1)$.

b) Find the conditional density of T_5 given that there are 10 arrivals in the time interval $(0,1)$.

c) Recognize the answers to a) and b) as named densities, and find the parameters.

11. Suppose X has uniform distribution on $(-1,1)$ and, given $X = x$, Y is uniformly distributed on $(-\sqrt{1-x^2}, \sqrt{1-x^2})$. Is (X,Y) then uniformly distributed over the unit disk $\{(x,y) : x^2 + y^2 < 1\}$? Explain carefully.

12. Suppose there are ten atoms, each of which decays by emission of an α-particle after an exponentially distributed lifetime with rate 1, independently of the others. Let T_1 be the time of the first α-particle emission, T_2 the time of the second. Find:

a) the distribution of T_1;

b) the conditional distribution of T_2 given T_1;

c) the distribution of T_2.

13. Let X and Y be independent random variables, X with uniform distribution on $(0,3)$, Y with Poisson (λ) distribution. Find:

a) a formula in terms of λ for $P(X < Y)$;

b) the conditional density of X given $X < Y$, and sketch its graph in the cases $\lambda = 1, 2, 3$;

c) $E(X|X < Y)$.

14. Bayesian sufficiency. Let $S_n = X_1 + \cdots + X_n$ be the number of successes in a sequence of n independent Bernoulli (p) trials $X_1, X_2, \ldots, X_n$ with unknown success probability p. Regard p as the value of a random variable Π whose prior distribution has some density $f(p)$ on $(0,1)$. Show that the conditional (posterior) distribution of Π given $X_1 = x_1, \ldots, X_n = x_n$, for any particular sequence of zeros and ones $x_1, \ldots, x_n$ with $x_1 + \cdots + x_n = k$, depends only the observed number of successes k in the n trials, and not on the order in which the k successes and $n - k$ failures appear. Deduce that this conditional distribution is identical to the posterior distribution of Π given $S_n = k$. [This is another expression of the fact that S_n is a sufficient statistic for p. See Exercise 6.2.14.]

15. Beta-binomial. As in Exercise 14 let $S_n = X_1 + \cdots + X_n$ be the number of successes in a sequence of n independent Bernoulli (p) trials $X_1, X_2, \ldots, X_n$, with unknown success probability p, regarded as the value of a random variable Π.

a) Suppose the prior distribution of Π is beta (r, s) for some $r > 0$ and $s > 0$. Show that the posterior distribution of Π given $S_n = k$ is beta $(r + k, s + n - k)$. [*Hint for quick solution*: It is enough to show that the posterior density is *proportional* to the beta $(r + k, s + n - k)$ density. See Chapter 4 Review Exercise 8.]

b) Using the fact that the total integral of the beta $(r+k, s+n-k)$ density is 1, find a formula for the unconditional probability $P(S_n = k)$.

c) Check your result in part b) agrees with the distribution of S_n found in Example 4 in the case $r = s = 1$.

d) For general r and s find the posterior mean $E(\Pi \mid S_n = k)$ and the posterior variance $Var(\Pi | S_n = k)$.

e) Suppose n is very large and the observed proportion of successes $\hat{p} = k/n$ is not very close to either 0 or 1. Show that no matter what r and s, provided n is large enough, $E(\Pi | S_n = k) \approx \hat{p}$ and $Var(\Pi | S_n = k) \approx \hat{p}(1 - \hat{p})/n$.

[It can be shown that the posterior distribution of Π given $S_n = k$ is approximately normal under the assumptions in e). So
for large enough n, the conditional distribution of the unknown value of p, given the observed proportion of successes $\hat{p}$ in n trials, is approximately normal with mean $\hat{p}$ and standard deviation $\sqrt{\hat{p}(1-\hat{p})}/\sqrt{n}$,
regardless of the prior parameters r and s. The same conclusion holds for any strictly positive and continuous prior density $f(p)$ instead of a beta prior. In the long run, any reasonable prior opinion is overwhelmed by the data. The italicized assertion should be compared to the following paraphrase of the normal approximation to the binomial distribution:
for large enough n, the distribution of proportion of successes $\hat{p}$ in n trials, given the probability p of success on each trial, is approximately normal with mean p and standard deviation $\sqrt{p(1-p)}/\sqrt{n}$.
While the assertions are very similar, and both true, it is not a trivial matter to pass from one to the other. There is a big conceptual difference between, on the one hand, the distribution of $\hat{p}$ for a fixed and known value of p, which has a clear frequency interpretation in terms of repeated blocks of n trials with the same p, and on the other hand, the posterior distribution of p given $\hat{p}$, which while intuitive from a subjective standpoint, is almost impossible to interpret in terms of long-run frequencies. Long-run frequency of what? The problem is that for large n, in any model of repeated blocks of n trials, the exact value of $\hat{p}$ observed in the first block will typically not be observed even once again until after a very large number of blocks have been examined. The number of blocks required to find the first repeat is of order $\sqrt{n}$ if the same p is used in each block, and order n if p is randomized for each block using the prior distribution: this is because the probability of the most likely values of $\hat{p}$ is of order $1/\sqrt{n}$ in the first case, by the normal approximation to the binomial, and order $1/n$ in the second case, as typified when the prior is uniform on $(0, 1)$ and the distribution of $\hat{p}$ is uniform on the $n + 1$ possible multiples of $1/n$. Either way, it is hard to make a convincing frequency interpretation of the conditional distribution of p given an exact observed value of $\hat{p}$.]

16. Negative binomial distribution for number of accidents. Consider a large population of individuals subject to accidents at various rates. Suppose the empirical distribution of accident rates over the whole population is well approximated by the gamma (r, α) distribution for some $r > 0$ and $\alpha > 0$. Suppose that given an individual has accident rate λ per day, the number of accidents that individual has in t days has Poisson (λt) distribution. Let Λ be the accident rate and N be the number of accidents in t days for an individual picked at random from this population. So Λ has gamma (r, α) distribution, and given $\Lambda = \lambda$, N has Poisson (λt) distribution.

a) Show by integration that

$$P(N = k) = \frac{\Gamma(r+k)}{\Gamma(r)k!} p^r q^k \quad (k = 0, 1, 2 \ldots) \text{ where } p = \alpha/(t+\alpha), \quad q = t/(t+\alpha)$$

b) Evaluate $\Gamma(r+k)/\Gamma(r)$ as a product of k factors. Deduce that if r is a positive integer, the distribution of N is the same as the distribution of the number of failures before the rth success in Bernoulli (p) trials, as found in Section 3.4.

[In general, the distribution of N defined in a) is called the *negative binomial*(r, p) distribution, now defined for arbitrary $r > 0$ and $0 < p < 1$. The terminology is explained by the following relation between this distribution and the binomial expansion for the negative power $-r$.]

c) Show, either by conditioning on Λ, or from a) and b), that N has generating function

$$E(z^N) = p^r(1 - zq)^{-r} \qquad (|z| < 1)$$

d) Find $E(N)$ and $E(N^2)$ in terms of r and p by conditioning on Λ. Deduce a formula for $Var(N)$. Check for integer r that your results agree with those obtained in Section 3.4.

e) Derive $E(N)$ and $Var(N)$ another way by differentiating the generating function. (Refer to Exercise 3.4.22.)

f) Show that for each integer $k \geq 0$, the conditional density of Λ given $N = k$ is a gamma density, and find its parameters.

17. Sums of independent negative binomial variables. Consider, as in Exercise 16, a large population of individuals subject to accidents at various rates. Suppose now that an individual picked at random from the population is subject to one kind of accident at rate Λ_1 per day, and another kind of accident at rate Λ_2 per day, where Λ_1 and Λ_2 are independent gamma variables with parameters (r_1, α) and (r_2, α) for some $\alpha > 0$. Assume that given $\Lambda_1 = \lambda_1$ and $\Lambda_2 = \lambda_2$ the two types of accidents occur according to independent Poisson processes with rates λ_1 and λ_2. Let N_1 and N_2 be the numbers of accidents of these two kinds the individual has in t days.

a) Describe the joint distribution of N_1 and N_2.

b) What is the distribution of $N_1 + N_2$? [*Hint:* No calculation required. Use results about sums of independent random variables with gamma or Poisson distributions.] Check your conclusion is consistent with the mean and variance formulae of Exercise 16.

c) Suppose $X_i, 1 \leq i \leq k$ are k independent random variables, and that X_i has negative binomial (r_i, p) distribution for some $r_i > 0$, $0 < p < 1$. What is the distribution of $X_1 + \cdots + X_n$? Explain carefully how your conclusion follows from parts a) and b).

d) Derive the result of c) another way using generating functions [see Chapter 3 Review Exercise 34].

6.4 Covariance and Correlation

Covariance is a quantity which appears in calculation of the variance of a sum of possibly dependent random variables. This quantity is useful in variance calculations, but like variance is hard to interpret intuitively. Correlation is a standardized covariance which is easier to interpret. It provides a measure of the degree of linear dependence between two variables. In Section 3.3, the formula

$$Var(X+Y) = Var(X) + Var(Y) \qquad \text{if } X \text{ and } Y \text{ are independent}$$

was derived from the more general formula

$$Var(X+Y) = Var(X) + Var(Y) + 2E\left[(X-\mu_X)(Y-\mu_Y)\right]$$

where $\mu_X = E(X)$ and $\mu_Y = E(Y)$. For independent random variables, the last term vanishes. In general, for two random variables X and Y with finite second moments, there is the following:

Definition of Covariance

The *covariance of* X *and* Y, denoted $Cov(X,Y)$, is the number

$$Cov(X,Y) = E\left[(X-\mu_X)(Y-\mu_Y)\right]$$

where $\quad \mu_X = E(X), \quad \mu_Y = E(Y)$

Alternative Formula

$$Cov(X,Y) = E(XY) - E(X)E(Y)$$

Variance of a Sum

$$Var(X+Y) = Var(X) + Var(Y) + 2Cov(X,Y)$$

Proof of alternative formula for covariance. Expand

$$(X-\mu_X)(Y-\mu_Y) = XY - \mu_X Y - X\mu_Y + \mu_X\mu_Y$$

and take expectations.□

Variance. Notice that $Cov(X,X) = Var(X)$, so these formulae for covariance are extensions of old formulae for variance.

Independence. If X and Y are independent then $Cov(X,Y) = 0$.

Warning. $Cov(X,Y) = 0$ *does not* imply X and Y are independent. See Exercises.

Indicators

Let $X = I_A$ be the indicator of event A, and $Y = I_B$ the indicator of another event B. These could be events in any outcome space, where there is given a probability distribution P. In this case

$$XY = I_A I_B = I_{AB}$$

is the indicator of the intersection of the events A and B. Thus

$$E(I_A) = P(A); \qquad E(I_B) = P(B); \qquad E(I_A I_B) = P(AB)$$

$$Cov(I_A, I_B) = P(AB) - P(A)P(B)$$

This covariance is

positive	iff	$P(AB) > P(A)P(B)$, when A and B are called *positively dependent*;
zero	iff	$P(AB) = P(A)P(B)$, when A and B are *independent;*
negative	iff	$P(AB) < P(A)P(B)$, when A and B are called *negatively dependent.*

In the case of positive dependence, learning that B has occurred increases the chance of A:

$$P(A|B) > P(A) \quad \text{and vice versa} \quad P(B|A) > P(B)$$

For negative dependence, learning that B has occurred decreases the chance of A:

$$P(A|B) < P(A) \quad \text{and vice versa} \quad P(B|A) < P(B)$$

These formulations of positive and negative dependence are easily seen to be equivalent to those in the box, by using the formula for $P(A|B)$, and rearranging inequalities. The most extreme case of positive dependence is if A is a subset of B, with $0 < P(A) \leq P(B) < 1$. Then, given that A occurs, B is certain to occur. In this case, given that B occurs, A is more likely to occur than before

$$P(A|B) = P(AB)/P(B) = P(A)/P(B) > P(A)$$

The most extreme case of negative dependence is if A and B are mutually exclusive events B with $P(A) > 0$ and $P(B) > 0$. Then, given that A occurs, B cannot occur, and vice versa.

Example 1. Draws with and without replacement.

Consider two draws at random from a box of b black balls and w white balls, where $b > 0$, $w > 0$. Let Black_i and White_i denote the events of getting a black or a white ball on the ith draw, $i = 1, 2$. Then you can check that the dependence between pairs of these events from different draws is affected by whether the sampling is done with or without replacement, as shown in the following table.

Dependence Between Events on Different Draws

Pairs of events	Sampling with replacement	Sampling without replacement
$\text{Black}_1, \text{Black}_2$	independent	$-$ dependent
$\text{Black}_1, \text{White}_2$	independent	$+$ dependent
$\text{White}_1, \text{White}_2$	independent	$-$ dependent
$\text{White}_1, \text{Black}_2$	independent	$+$ dependent

The Sign of the Covariance

As a general rule, the sign of $Cov(X, Y)$ is *positive* if above-average values of X tend to be associated with above-average values of Y, and below-average values of X with below-average values of Y. The random variable $(X - \mu_X)(Y - \mu_Y)$ is then most likely positive, with a positive expectation.

The sign of $Cov(X, Y)$ is *negative* if above-average values of X tend to be associated with below-average values of Y, and vice versa. Then $(X - \mu_X)(Y - \mu_Y)$ is most likely negative, with a negative expectation.

$Cov(X, Y)$ is *zero* only in special cases when there is no such association between the variables X and Y. Then $(X - \mu_X)(Y - \mu_Y)$ has positive values balanced by negative values, and expected value zero.

While the sign of the covariance can be interpreted as above, its magnitude is hard to interpret. It is easier to interpret the *correlation of X and Y*, denoted here by $Corr(X, Y)$, which is defined as follows:

Definition of Correlation

$$Corr(X, Y) = \frac{Cov(X, Y)}{SD(X)SD(Y)}$$

Assume now that neither X nor Y is a constant, so $SD(X)SD(Y) > 0$. The sign of $Cov(X, Y)$ is then the same as the sign of $Corr(X, Y)$.

Conditions for *X* and *Y* to be Uncorrelated

The following three conditions are equivalent:

$$Corr(X,Y) = 0$$

$$Cov(X,Y) = 0$$

$$E(XY) = E(X)E(Y)$$

in which case X and Y are called *uncorrelated.* Independent variables are uncorrelated, but uncorrelated variables are not necessarily independent.

Let X^* and Y^* now denote X and Y rescaled to standard units. So

$$X^* = (X - \mu_X)/SD(X) \quad \text{and} \quad Y^* = (Y - \mu_Y)/SD(Y)$$

Then

$$E(X^*) = E(Y^*) = 0 \quad \text{and} \quad SD(X^*) = SD(Y^*) = 1$$

by the scaling properties of E and SD. And you can check that

$$Corr(X,Y) = Cov(X^*,Y^*) = E(X^*Y^*)$$

So correlation is a kind of standardized covariance that is unaffected by changes of origin or units of measurement. See Exercises.

Correlations are between -1 and +1

$$-1 \le Corr(X,Y) \le 1$$

no matter what the joint distribution of X and Y.

Proof. Since $E(X^{*2}) = E(Y^{*2}) = 1$

$$0 \le E(X^* - Y^*)^2 = 1 + 1 - 2E(X^*Y^*)$$
$$0 \le E(X^* + Y^*)^2 = 1 + 1 + 2E(X^*Y^*)$$

Thus $-1 \le E(X^*Y^*) \le 1$, and $Corr(X,Y) = E(X^*Y^*)$ by the preceding discussion. □

Correlations of ± 1. The proof that correlations are between ± 1 shows $Corr(X,Y) = +1$ if and only if $E(X^* - Y^*)^2 = 0$, that is, if and only if $X^* = Y^*$ with probability one. This means there are constants a and b with $a > 0$ such that

$$Y = aX + b$$

with probability 1. That is to say, a correlation of $+1$ indicates a deterministic linear relationship between X and Y with positive slope. Similarly, a correlation of -1 indicates a deterministic linear relationship between X and Y with negative slope. Correlations between -1 and $+1$ indicate intermediate degrees of linear association between the two variables.

Example 2. Empirical correlations.

Like expectation and variance, covariance and correlation are generalizations to random variables of corresponding notions for empirical variables. Suppose $(x_1, y_1), \ldots, (x_n, y_n)$ is a list of n pairs of numbers, and (X, Y) is one of these pairs picked uniformly at random. Then the joint distribution of (X, Y) puts probability $1/n$ at each of the pairs, as suggested by the scatter diagram:

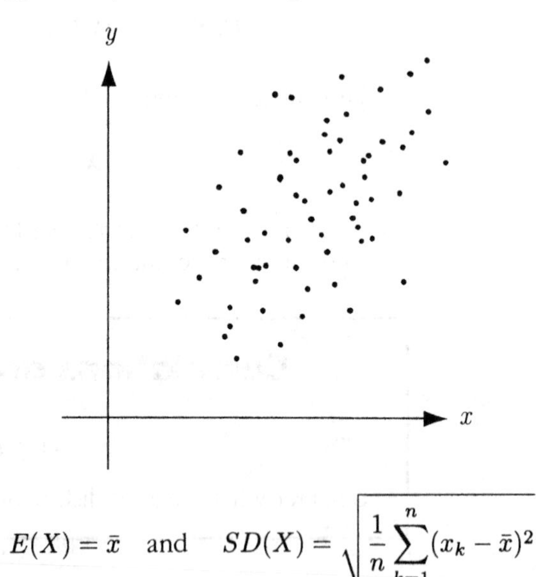

$$E(X) = \bar{x} \quad \text{and} \quad SD(X) = \sqrt{\frac{1}{n}\sum_{k=1}^{n}(x_k - \bar{x})^2}$$

and similarly for Y instead of X. Also

$$E(XY) = \frac{1}{n}\sum_{k=1}^{n} x_k y_k \quad \text{so}$$

$$Cov(X,Y) = E(XY) - E(X)E(Y) \quad \text{and} \quad Corr(X,Y) = \frac{Cov(X,Y)}{SD(X)SD(Y)}$$

can be computed from the list of number pairs. If the list of number pairs is a list of empirical measurements, or a sample of some kind, these may be called empirical or sample quantities. These quantities are all defined in terms of averages, which may be expected to converge to theoretical expectations as the sample size n increases, under conditions of random sampling. For example, the empirical correlation of n observed values of independent random variables $(X_1, Y_1), (X_2, Y_2), \ldots, (X_n, Y_n)$, all with the same joint distribution, will most probably be close to the theoretical correlation of X_1 and Y_1, provided n is sufficiently large. Thus a correlation in a theoretical model is often estimated by an empirically observed correlation based on a random sample. In particular, the empirical correlation of two variables over a large population can be estimated this way by the procedure of random sampling.

Example 3. **Correlation and distribution of the sum.**

This example shows in a simple case how the distribution of the sum of random variables X and Y is affected by their correlation. Suppose a gambler can bet on the value of a number U chosen uniformly at random from the numbers $1, 2, \ldots, 8$. The gambler can choose any set A of four numbers, such as $A = \{1, 2, 3, 4\}$, and place an even-money bet of \$1 on A. So the gambler wins \$1 if $U \in A$, and loses \$1 if $U \in A^c$. Let $\$X$ denote the gambler's net gain from this contract. Then, X has value $+1$ if $U \in A$, -1 if $U \in A^c$. In terms of indicators,

$$X = 2I_A - 1$$

Clearly $E(X) = 0$. The bet is fair no matter what set A the gambler chooses, because $P(A) = P(A^c) = 1/2$ for every set of four numbers A.

Suppose now that in addition to placing a bet on A, the gambler is also free to place at the same time a similar bet on a second set of four numbers B, for example $B = \{1, 3, 5, 7\}$. Let

$$Y = 2I_B - 1$$

denote the net gain to the gambler from this second bet. Then the gambler's overall gain from the placement of the two bets is the sum

$$S = X + Y$$

Notice that the distribution of X and the distribution of Y are the same, uniform on $\{-1, 1\}$, regardless of the gambler's choice of sets A and B. But the distribution of S is affected by the degree of dependence between X and Y, which is governed in turn by the amount of overlap between A and B. Clearly, $E(S)$ is zero no matter what the choice of A and B. But $SD(S)$ is affected by the gambler's choice of A and B. This standard deviation gives an indication of the likely size of the fluctuation in the gambler's fortune due to the combined bet.

Problem. Find how the standard deviation of S is determined by the choice of A and B.

Solution. Use the addition rule for variance

$$\begin{aligned} Var(S) &= Var(X) + Var(Y) + 2Cov(X,Y) \\ &= 2 + 2Corr(X,Y) \end{aligned}$$

because $SD(X) = SD(Y) = 1$, so $Corr(X,Y) = Cov(X,Y)$ in this case. Because $X = 2I_A - 1$, $Y = 2I_B - 1$, and the correlation coefficient is unchanged by linear transformations,

$$Corr(X,Y) = Corr(I_A, I_B) = \frac{Cov(I_A, I_B)}{SD(I_A)SD(I_B)} = \left(P(AB) - \frac{1}{4}\right) \bigg/ \frac{1}{4} = 4P(AB) - 1$$

This used $P(A) = P(B) = 1/2$, which makes $SD(I_A) = SD(I_B) = 1/2$. Using the earlier expression for $Var(S)$ this gives

$$SD(S) = \sqrt{8P(AB)} = \sqrt{\#(AB)}$$

where $\#(AB)$ is the number of points in the intersection of A and B, so $P(AB) = \#(AB)/8$.

Discussion. The formula shows that the larger the overlap between A and B, the larger will be the likely size of the fluctuation in the gambler's fortune as a result of betting on both A and B. This is intuitively clear if you think about the following special cases:

Case $\#(AB) = 0$, $Corr(X,Y) = -1$, $SD(S) = 0$. This means $B = A^c$. Then $Y = -X$, because whatever is gained on one bet is lost on the other. So $S = X + Y = 0$. This is a strategy of extreme hedging, with zero result.

Case $\#(AB) = 1$, $Corr(X,Y) = -1/2$, $SD(S) = 1$. Intuitively, this is still hedging. The two bets tend to cancel each other.

Case $\#(AB) = 2$, $Corr(X,Y) = 0$, $SD(S) = \sqrt{2}$. In this case A and B are independent. Therefore, so too are the indicator random variables I_A and I_B, and the random variables $X = 2I_A - 1, Y = 2I_B - 1$ representing the net gains from the two bets. So the net effect of betting on both A and B in one game is the same as the effect of betting on A in one game, then betting on A again in a second game, independent of the first. The distribution of S in this case is the familiar binomial $(2, 1/2)$ distribution, but centered at 0 and rescaled by a factor of 2, because

$$S = X + Y = 2(I_A + I_B) - 2$$

where $I_A + I_B$ is the number of successes in two independent trials with probability $1/2$ of success on each trial, with binomial $(2, 1/2)$ distribution. The appearance of $\sqrt{2}$ as the standard deviation in this case illustrates the square root law for the standard deviation of the sum of $n = 2$ independent variables.

Case $\#(AB) = 3$, $Corr(X, Y) = 1/2$, $SD(S) = \sqrt{3}$. This is a bolder strategy.

Case $\#(AB) = 4$, $Corr(X, Y) = 1$, $SD(S) = 2$. Now $A = B$. All the gambler's eggs are in one basket. This is the boldest strategy for the gambler, effectively doubling the stake on A from \$1 to \$2.

Example 4. Red and black.

Let N_R be the number of reds that appear, N_B the number of blacks, in n spins of a roulette wheel that has proportion r of its numbers red, proportion b black, and the rest of its numbers green. (So $r + b < 1$. For a Nevada roulette wheel, as described at the end of Section 1.1, $r = b = 18/38$.)

Problem. Find $Corr(N_R, N_B)$.

Solution. Notice first, without calculation, that the answer ought to be negative for the usual case with $r+b \approx 1$. If $r+b = 1$ (no green numbers on the wheel) then $N_B = n - N_R$ which makes $Corr(N_R, N_B) = -1$. For $r + b \approx 1$ this relation is still approximately correct, so you should expect a correlation close to -1. Since N_R is a binomial (n, r) random variable,

$$E(N_R) = nr \text{ and } SD(N_R) = \sqrt{nr(1-r)}$$

and similarly for N_B, with b instead of r. Since

$$Cov(N_R, N_B) = E(N_R N_B) - E(N_R)E(N_B)$$

to calculate

$$Corr(N_R, N_B) = \frac{Cov(N_R, N_B)}{SD(N_R)SD(N_B)}$$

the only missing ingredient is $E(N_R N_B)$. You might try to calculate this from the joint distribution of N_R and N_B, but you will find this a frightful task. It is difficult to calculate even the variance of N_R directly from its binomial distribution, and the covariance with N_B is worse. The way around this difficulty is to use the connection between $Cov(N_R, N_B)$ and the variance of $N_R + N_B$

$$Var(N_R + N_B) = Var(N_R) + Var(N_B) + 2\,Cov(N_R, N_B)$$

The point is that $N_R + N_B$ is just the number of spins which are either red or black, which is a binomial $(n, r + b)$ random variable, with variance $n(r + b)(1 - r - b)$. Rearrange the equation and substitute all the variances to get

$$Cov(N_R, N_B) = \frac{1}{2}n\left[(r + b)(1 - r - b) - r(1 - r) - b(1 - b)\right] = -nrb,$$

hence,

$$Corr(N_R, N_B) = \frac{-nrb}{\sqrt{nr(1-r)}\sqrt{nb(1-b)}} = -\sqrt{\frac{rb}{(1-r)(1-b)}}$$

Discussion. In particular, for a Nevada roulette wheel,

$$r/(1-r) = b/(1-b) = 18/20 = 0.9 \qquad \text{so}$$

$$Corr(N_R, N_B) = -0.9$$

Note the interesting fact that the correlation does not depend at all on the number of spins n, only on the proportions of red and black. Also, the correlation is always negative, no matter what the proportions r and b.

Example 5. Correlations in the multinomial distribution.

Suppose the joint distribution of $(N_1, \ldots, N_m)$ is multinomial with parameters n and $(p_1, \ldots, p_m)$.

Problem. Find $Corr(N_i, N_j)$.

Solution. Call results in category i red, results in category j black, and results in all other categories green. Then the joint distribution of N_i and N_j is is the same as the joint distribution of N_R and N_B in the previous problem, for $r = p_i$, $b = p_j$. Since the correlation between two variables is determined by their joint distribution (by definition of correlation and the change of variable principle) this choice of r and b makes $Corr(N_i, N_j) = Corr(N_R, N_B)$. That is to say, from the solution of the previous problem,

$$Corr(N_i, N_j) = -\sqrt{\frac{p_i p_j}{(1-p_i)(1-p_j)}}$$

Correlation and Conditioning

An important connection between the ideas of correlation and conditioning is brought out by the following example.

Example 6. Sharkey's Casino.

At Sharkey's Casino the roulette wheels spin an average of one thousand times a day. Every day, Sharkey records the total numbers of red and black spins for the day on a computer. One day he notices that over the years he has been keeping data, the correlation between the number of reds and number of blacks has come out around

+0.8, rather than around −0.9 as predicted by the above calculation. Sharkey is very concerned that his roulette wheels are not obeying the laws of chance, and that someone might take advantage of it.

Problem. Should Sharkey get new roulette wheels?

Solution. Despite the fact that no matter what the number of spins n, the correlation between numbers of reds and blacks is −0.9, this does not imply that the same is true for a random number of spins, say N, the number of spins in a day picked at random at Sharkey's. While the expected value of N may be estimated as 1000 based on the long-run average of 1000 spins a day, it is reasonable to expect some spread in the distribution of N due to fluctuations in the number of customers and the rate of play. Since to a first approximation $N_B \approx \frac{18}{38}N$, $N_R \approx \frac{18}{38}N$, both N_B and N_R are positively correlated with N. If there is enough spread in the distribution of N, this will make for a positive correlation between N_B and N_R. So Sharkey need not be concerned, provided his data give a standard deviation of N consistent with a correlation of +0.8 between N_R and N_B.

To find the precise relation between $SD(N)$ and $Corr(N_R, N_B)$, for N_R and N_B, now numbers of reds and blacks in a random number N of spins, use the formula

$$Cov(N_R, N_B) = E(N_R N_B) - E(N_R)E(N_B)$$

where each expectation can be computed by conditioning on N. First, if N is treated as a constant, then by previous calculations,

$$E(N_R) = Nr \qquad E(N_B) = Nb$$

$$E(N_R N_B) = E(N_R)E(N_B) + Cov(N_R, N_B) = N^2 rb - Nrb$$

For random N, these are *conditional* expectations given N. But since expectations are expectations of conditional expectations, this gives

$$\begin{aligned}
E(N_R) &= E(N)r, \qquad E(N_B) = E(N)b \\
E(N_R N_B) &= E(N^2)rb - E(N)rb, \qquad \text{hence} \\
Cov(N_R, N_B) &= E(N_R N_B) - E(N_R)E(N_B) \\
&= rb\left[E(N^2) - E(N) - [E(N)]^2\right] \\
&= rb\left[Var(N) - E(N)\right]
\end{aligned}$$

In particular, $Cov(N_R, N_B)$ will be positive provided $Var(N) > E(N)$. Thus for $E(N) = 1000$, if $SD(N) > \sqrt{1000} \approx 32$, there will be a positive correlation between N_R and N_B. The same method of calculation gives

$$Var(N_B) = b^2 Var(N) + b(1-b)E(N)$$

For $b = r$ this gives

$$\begin{aligned} Corr(N_R, N_B) &= \frac{b^2[Var(N) - E(N)]}{b^2 Var(N) + b(1-b)E(N)} \\ &= \frac{9\,Var(N) - 9000}{9\,Var(N) + 10,000} \quad \text{for} \quad b = \frac{18}{38}, \quad E(N) = 1,000. \end{aligned}$$

If $Var(N) = 0$ this simplifies to -0.9 as before. But as $Var(N)$ increases the correlation increases, and approaches 1 for large values of $Var(N)$. Set $Corr(N_R, N_B) = \rho$ and solve for $SD(N) = \sqrt{Var(N)}$ to get

$$\begin{aligned} SD(N) &= \sqrt{\frac{9000 + 10,000\rho}{9(1-\rho)}} \\ &= \sqrt{\frac{17,000}{9 \times 0.2}} \qquad \text{for } \rho = 0.8 \\ &\approx 100 \end{aligned}$$

So a correlation of 0.8 between N_R and N_B is consistent with a standard deviation of about 100 for the number of spins per day. Provided that is the case, Sharkey need not be concerned.

Discussion. The example makes the important point that two variables, like N_R and N_B, may be positively correlated due to association with some third variable, like N, even if there is zero or negative correlation between the two variables for a fixed value of N. Here is another example. For children of a fixed age, the correlation between height and reading ability would most likely come out around zero. But if you looked at children of ages from 5 to 10, there would be a high positive correlation between height and reading ability, because both variables are closely associated with age. For data variables, looking at distributions or relationships between some variables for a fixed value of another variable, N say, is called *controlling for N*. In a probability model the corresponding thing is *conditioning on N*. Whether or not you condition or control on one variable typically has major effects on relationships between other variables.

The calculations in the example show in general that for two mutually exclusive outcomes in independent trials, like red and black at roulette, the counts of results of the two kinds that occur in any fixed number of trials will be negatively correlated. If the number of trials N is random, the two counts will be positively or negatively correlated according to whether $Var(N) > E(N)$ or $Var(N) < E(N)$. In the case where $Var(N) = E(N)$, the two counts will be uncorrelated. In particular this is the case if N has a Poisson distribution. Then the two counts are actually independent. See Exercise 6.1.8.

Variance of a Sum of *n* Variables

The general formula involving covariance for the variance of a sum of two random variables has the following extension to n variables. The formula shows that the simple addition rule for the variance of a sum of independent random variables works just as well for uncorrelated ones, but in general there are $\binom{n}{2}$ covariance terms to be considered as well.

Variance of a Sum of *n* Variables

$$Var\left(\sum_k X_k\right) = \sum_k Var(X_k) + 2\sum_{j<k} Cov(X_j X_k)$$

where $\sum_k$ denotes a sum of n terms from $k = 1$ to n, and $\sum_{j<k}$ denotes a sum of $\binom{n}{2}$ terms indexed by j and k with $1 \le j < k \le n$.

Proof: The variance of the sum is by definition the expectation of

$$\left[\sum_k X_k - E(\sum_k X_k)\right]^2 = \left[\sum_k X_k - \sum_k \mu_k\right]^2 \qquad \text{where } \mu_k = E(X_k)$$

$$= \left[\sum_k (X_k - \mu_k)\right]^2$$

$$= \sum_k (X_k - \mu_k)^2 + 2\sum_{j<k}(X_j - \mu_j)(X_k - \mu_k)$$

by the algebraic identity

$$\left(\sum a_k\right)^2 = \sum_k a_k^2 + 2\sum_{j<k} a_j a_k$$

applied to $a_k = X_k - \mu_k$. Now use the linearity of expectation and the definition of $Cov(X_j, X_k)$. In the sum over all $j < k$, there are exactly $\binom{n}{2}$ terms, one for each way of choosing two indices $j < k$ from the set $\{1, 2, \ldots, n\}$. □

Example 7. Variance of sample averages.

Let $x(1), x(2), \ldots, x(N)$ be a list of N numbers. Think of $x(k)$ as representing the height of the kth individual in a population of size N. Let

$$\bar{x} = \frac{1}{n}\sum_{k=1}^{n} x(k) \quad \text{and} \quad \sigma^2 = \frac{1}{n}\sum_{k=1}^{n}[x(k) - \bar{x}]^2$$

So $\bar{x}$ is the *population mean*, and σ^2 is the *population variance*. Let $X_1, X_2, \ldots, X_n$ be the heights obtained in a random sample of size n from this population. More formally, for $i = 1, 2, \ldots, n$, the ith height in the sample is $X_i = x(K_i)$, where $K_1, K_2, \ldots, K_n$ is a random sample of size n from the index set $\{1, 2, \ldots, N\}$. This random sample might be taken either with replacement or without replacement. Either way, each random index K_i has uniform distribution over $\{1, 2, \ldots, N\}$, by symmetry. So each X_i is distributed according to the distribution of the list of heights in the total population, with

$$E(X_i) = \bar{x} \quad \text{and} \quad SD(X_i) = \sigma \qquad (i = 1, 2, \ldots, n)$$

Let

$$\bar{X}_n = (X_1 + X_2 + \ldots + X_n)/n$$

be the *sample average*. This is the average height of individuals in the sample of size n. Note that this is a random variable: repeating the sampling procedure will typically produce a different sample average. Whereas $\bar{x}$, the population average, is a constant. Since $E(X_i) = \bar{x}$ for $i = 1, 2, \ldots, n$, the rules of expectation imply that also

$$E(\bar{X}_n) = \bar{x}$$

still no matter whether the sampling is done with or without replacement. In the case with replacement, the random variables X_i are independent, all with standard deviation σ, so

$$SD(\bar{X}_n) = \sigma/\sqrt{n} \qquad \text{(with replacement)}$$

by the square root law of Section 3.3. So the average height in a random sample of size n is most likely only a few multiples of $\sigma/\sqrt{n}$ away from the population average $\bar{x}$. If σ can be bounded or estimated, this gives an indication of the quality of the sample average $\bar{X}_n$ as an estimator of the unknown population average $\bar{x}$.

Intuitively, for sampling without replacement, $\bar{X}_n$ should provide a better estimate of $\bar{x}$ than for sampling with replacement. In this case, the random variables $X_1, \ldots, X_n$ turn out to be negatively correlated, which affects the formula for $SD(\bar{X}_n)$. The problem is how to correct for the dependence.

Problem. Calculate $SD(\bar{X}_n)$ for sampling without replacement.

Solution. Let $S_n = X_1 + \cdots + X_n$, so $\bar{X}_n = S_n/n$. Then

$$\begin{aligned} Var(S_n) &= \sum_j Var(X_j) + 2\sum_{j<k} Cov(X_j, X_k) \\ &= n\sigma^2 + n(n-1)\,Cov(X_1, X_2), \end{aligned}$$

because $Cov(X_j, X_k) = Cov(X_1, X_2)$ by the symmetry of sampling without replacement discussed in Section 3.6: (X_j, X_k) is for every $j < k$ a simple random sample of size 2, with the same distribution as (X_1, X_2). This formula for $Var(S_n)$ holds for every sample size n with $1 \leq n \leq N$. But for $n = N$

$$S_N = x_1 + x_2 + \cdots + x_N$$

is constant, because in a complete sample of the population each element appears exactly once, so the sum defining S_N is just the sum on the right done in a random order. Thus $Var(S_N) = 0$. Comparison with the previous formula for $Var(S_n)$, in the case where $n = N$, shows

$$Cov(X_1, X_2) = -\sigma^2/(N-1)$$

hence

$$Var(S_n) = n\sigma^2 \left[1 - \frac{n-1}{N-1}\right]$$

and

$$SD(\bar{X}_n) = SD(S_n)/n = \frac{\sigma}{\sqrt{n}}\sqrt{\frac{N-n}{N-1}}$$

Discussion. This shows that the standard deviation for the average in sampling without replacement is the corresponding standard deviation for sampling with replacement, reduced by the *correction factor* $\sqrt{\frac{N-n}{N-1}}$. The same is true for the sum as well as the average, by scaling.

The same correction factor appears in the formula for the variance of the hypergeometric distribution, calculated in Section 3.6. Though covariances are not used in that calculation, it is still a special case of the current example, with $x_j = 0$ or 1 for every j.

It is remarkable that the same correction factor works no matter what the distribution of the empirical variable x. The correction factor takes care of the slight negative correlation between terms, which also does not depend on the distribution of x:

$$Corr(X_j, X_k) = \frac{Cov(X_j, X_k)}{SD(X_j)SD(X_k)} = -1/(N-1)$$

The correlation is negative because observation of a large value of X_j removes a large value from the population, and tends to make large values of X_k less likely. Similarly, small values of X_j tend to make small values of X_k less likely. This means there is a greater tendency for the deviations $X_j - E(X_j)$ to cancel each other out for sampling without replacement than for sampling with replacement, when these deviations are independent. This reduces the likely size of the deviation for the sum

$$S_n - E(S_n) = \sum_{j=1}^{n} (X_j - E(X_j))$$

Ultimately, for $n = N$, the deviation of S_N is zero, which was the key to calculating the correction factor.

Bilinearity of Covariance

The following formulae for covariances of linear combinations of variables are easily derived from the definition. These formulae can often be used to simplify covariance calculations.

$$Cov(X, Y + Z) = Cov(X, Y) + Cov(X, Z)$$

$$Cov(W + X, Y) = Cov(W, Y) + Cov(X, Y)$$

For constants a and b

$$Cov(aX, Y) = a\,Cov(X, Y) \quad \text{and} \quad Cov(X, bY) = b\,Cov(X, Y)$$

and so on for linear combinations of several variables. For example

$$Cov(aW + bX, cY + dZ) = a\,c\,Cov(W, Y) + a\,d\,Cov(W, Z) + b\,c\,Cov(X, Y) + b\,d\,Cov(X, Z)$$

To summarize:

Covariance is Bilinear

$$Cov\left(\sum_i a_i X_i\,,\ \sum_j b_j Y_j\right) = \sum_i \sum_j a_i\, b_j\, Cov(X_i, Y_j)$$

Here the a_i and b_j are arbitrary constants. If there are n terms in the sum over i and m terms in the sum over j there are nm terms in the double sum on the right side. Taking $n = m$, $a_i = b_i = 1$ and $X_i = Y_i$ for $1 \le i \le n$, this formula reduces to the formula for the variance of $\sum_i X_i$.

Exercises 6.4

1. Suppose A, B are two events such that $P(A) = 0.3$, $P(B) = 0.4$, and $P(A \cup B) = 0.5$.

a) Find $P(A|B)$. b) Are A and B independent, positively or negatively dependent?

c) Find $P(A^c B)$. d) Let $X = I_A$, $Y = I_B$. Find $Corr(X, Y)$.

2. Use the formula $P(A) = P(A|B)P(B) + P(A|B^c)P(B^c)$ to prove:

a) if $P(A|B) = P(A|B^c)$ then A and B are independent;

b) if $P(A|B) > P(A|B^c)$ then A and B are positively dependent;

c) if $P(A|B) < P(A|B^c)$ then A and B are negatively dependent.

Now prove the converses of a), b), and c).

3. Suppose that the failures of two components are positively dependent. If the first component fails, does that make it more or less likely that the second component works? What if the first component works?

4. Let (X, Y) have uniform distribution on the four points $(-1, 0), (0, 1), (0, -1), (1, 0)$. Show that X and Y are uncorrelated but not independent.

5. Let X have uniform distribution on $\{-1, 0, 1\}$ and let $Y = X^2$. Are X and Y uncorrelated? Are X and Y independent? Explain carefully.

6. Let X_1 and X_2 be the numbers on two independent fair die rolls, $X = X_1 - X_2$ and $Y = X_1 + X_2$. Show that X and Y are uncorrelated, but not independent.

7. Let X_2 and X_3 be indicators of independent events with probabilities $1/2$ and $1/3$, respectively.

a) Display the joint distribution table of $X_2 + X_3$ and $X_2 - X_3$.

b) Calculate $E(X_2 - X_3)^3$.

c) Are X_2 and X_3 uncorrelated? Prove your answer.

8. You have N boxes labeled Box1, Box2, ..., BoxN, and you have k balls. You drop the balls at random into the boxes, independently of each other. For each ball the probability that it will land in a particular box is the same for all boxes, namely $1/N$. Let X_1 be the number of balls in Box1 and X_N be the number of balls in BoxN. Calculate $Corr(X_1, X_N)$.

9. Suppose n cards numbered $1, 2, \ldots, n$ are shuffled and k of the cards are dealt. Let S_k be the sum of the numbers on the k cards dealt. Find formulae in terms of n and k for:

a) the mean of S_k; b) the variance of S_k.

10. **Overlapping counts.** A fair coin is tossed 300 times. Let H_{100} be the number of heads in the first 100 tosses, and H_{300} the total number of heads in the 300 tosses. Find $Corr(H_{100}, H_{300})$.

11. Let T_1 and T_3 be the times of the first and third arrivals in a Poisson process with rate λ. Find $Corr(T_1, T_3)$.

12. Suppose α, β, γ denote the proportions of Democrats (D), Republicans (R) and Others (O) in a large population of voters. (So $0 \leq \alpha, \beta, \gamma \leq 1$ and $\alpha+\beta+\gamma = 1$.) An individual is selected at random from the population. Write $X = 1, Y = 0, Z = 0$ if that individual is D, write $X = 0, Y = 1, Z = 0$ if the individual is R and write $X = 0, Y = 0, Z = 1$ if the individual is O. Find:

a) $E(X)$, $E(Y)$; b) $Var(X)$, $Var(Y)$; c) $Cov(X, Y)$.

Suppose next that n individuals are selected independently and randomly with replacement from the population. The total number of D's may be written, $D_n = X_1 + \ldots + X_n$. Similarly let $R_n = Y_1 + \ldots + Y_n$. and let $O_n = Z_1 + \ldots + Z_n$. Let $D_n - R_n$ denote the excess of D's over R's selected. Find d) $E(D_n - R_n)$; e) $Var(D_n - R_n)$.

13. Let A and B be two possible results of a trial, not necessarily mutually exclusive. Let N_A be the number of times A occurs in n independent trials, N_B the number of times B occurs in the same n trials. True or false and explain: If N_A and N_B are uncorrelated, then they are independent.

14. Show that for any two random variables X and Y

$$|SD(X) - SD(Y)| \leq SD(X + Y) \leq |SD(X) + SD(Y)|$$

15. **Covariance is bilinear.** Show from the definition of covariance that:

a) $Cov(X, Y + Z) = Cov(X, Y) + Cov(X, Z)$

b) $Cov(W + X, Y) = Cov(W, Y) + Cov(X, Y)$

c) $Cov(\sum_i X_i, \sum_j Y_j) = \sum_i \sum_j Cov(X_i, Y_j)$

d) Use c) to rederive the formula for $Cov(N_R, N_B)$ in Example 6.

16. **Invariance of the correlation coefficient under linear transformations.** Show that for arbitrary random variables X and Y, and constants a, b, c, d with $a \neq 0$, $c \neq 0$,

$$Corr(aX + b, cY + d) = \begin{cases} Corr(X, Y) & \text{if } a \text{ and } c \text{ have the same sign} \\ -Corr(X, Y) & \text{if } a \text{ and } c \text{ have opposite signs.} \end{cases}$$

Thus the correlation coefficients are affected only by the sign of a linear change of variable. They are therefore unaffected by shifts of origin or changes of units.

17. Show that for indicator random variables I_A and I_B of events A and B

$$Corr(I_A, I_B) = Corr(I_{A^c}, I_{B^c}) = -Corr(I_A, I_{B^c}) = -Corr(I_{A^c}, I_B)$$

Deduce that if A and B are positively dependent, then so are A^c and B^c, but A and B^c are negatively dependent, as are A^c and B.

18. Random variables $X_1, \ldots, X_n$ are *exchangeable* if their joint distribution is the same, no matter what order they are presented (see Section 3.6). Show that if $X_1, \ldots, X_n$ are exchangeable, then

$$Var(\sum_{k=1}^{n} X_k) = n\,Var(X_1) + n(n-1)\,Cov(X_1, X_2)$$

19. A box contains 5 nickels, 10 dimes, and 25 quarters. Suppose 20 draws are made at random without replacement from this box. Let X be the total sum obtained in these 20 draws. Calculate: a) $E(X)$; b) $SD(X)$;

c) $P(X \leq \$3)$ using the normal approximation.

d) Can you imagine why these calculations might give results inconsistent with long-run repetitions of the sampling experiment? For each of a) and c), say whether your reasoning would suggest higher or lower long-run averages.

20. Correlation and conditioning. A random variable X assumes values x_1 and x_2 with probabilities p_1 and p_2, where $p_1 + p_2 = 1$. Given $X = x_i$, random variable Y has mean equal to μ_i and SD equal to σ_i. Find formulae in terms of x_i, p_i, μ_i, and σ_i, $i = 1, 2$, for the following quantities:

a) $E(X)$; b) $E(Y)$; c) $SD(X)$; d) $SD(Y)$; e) $Cov(X, Y)$; f) $Corr(X, Y)$.

Indicate how these formulae could be generalized to the case of X with n possible values $x_1, \ldots, x_n$.

21. A box contains 5 red balls and 8 blue ones. A random sample of size 3 is drawn *without* replacement. Let X be the number of red balls and let Y be the number of blue balls selected. Compute: a) $E(X)$; b) $E(Y)$; c) $Var(X)$; d) $Cov(X, Y)$.

22. Suppose there were m married couples, but that d of these $2m$ people have died. Regard the d deaths as striking the $2m$ people at random. Let X be the number of surviving couples. Find:

a) $E(X)$; b) $Var(X)$.

23. Linear prediction and the correlation coefficient. For random variables X and Y, the *linear prediction problem* for predicting Y based on knowledge of X is the problem of finding a linear function of X, $\beta X + \gamma$, which minimizes the *mean square* of the prediction error

$$MSE = E[Y - (\beta X + \gamma)]^2$$

(Compare with Exercise 6.2.17 where the predictor of Y could be an arbitrary function of X.) This exercise derives the basic formulae for the best linear predictor according to this criterion.

a) Expand out the MSE using algebra, and regard it as a quadratic function of γ and β with coefficients involving the numbers $E(X)$, $E(Y)$, $E(XY)$, etc.

b) Differentiate this function with respect to γ to show that for fixed β, the unique γ which minimizes the MSE is $\hat{\gamma}(\beta) = E(Y) - \beta E(X)$. What is the resulting minimal MSE called when $\beta = 0$?

c) Consider now the MSE as a function of β, with $\gamma = \hat{\gamma}(\beta)$ the best γ for the given β. Differentiate this function with respect to β, and show that it is minimized at $\hat{\beta} = Cov(X, Y)/Var(X)$ where it is assumed that $Var(X) > 0$.

d) Deduce that the unique pair (β, γ) which minimizes the MSE is $(\hat{\beta}, \hat{\gamma}(\hat{\beta}))$.

e) Let $\hat{Y} = \hat{\beta} X + \hat{\gamma}$ now denote this best linear predictor. Show that

$$E(\hat{Y}) = E(Y); \quad Var(\hat{Y}) = \hat{\beta}^2 Var(X); \quad E[\hat{Y}(Y - \hat{Y})] = 0$$

f) Deduce that the variance of Y can be decomposed into the sum of the variance of the best predictor $\hat{Y}$ and the minimum MSE according to the formula

$$Var(Y) = Var(\hat{Y}) + E[(Y - \hat{Y})^2]$$

with $Var(\hat{Y}) = \rho^2 Var(Y)$ and $E[(Y - \hat{Y})^2] = (1 - \rho^2) Var(Y)$ where $\rho = Corr(X, Y)$.

g) It is customary to express the slope $\hat{\beta}$ of the best linear predictor $\hat{Y} = \hat{\beta}X + \hat{\gamma}$ in terms of ρ. Show that $\hat{\beta} = \rho SD(Y)/SD(X)$ and that the intercept $\hat{\gamma}$ is then uniquely determined by the requirement that the line $y = \hat{\beta}x + \hat{\gamma}$ passes through the point $(E(X), E(Y))$.

h) Let $Y^* = (Y - E(Y))/SD(Y)$, $X^* = (X - E(X))/SD(X)$. Show that the best linear predictor of Y^* based on X^* is just ρX^*. So the correlation coefficient ρ is simply the slope of the best linear predictor when the variables are expressed in standard units.

6.5 Bivariate Normal

The radially symmetric bivariate normal distribution corresponding to independent normal variables was considered in Section 5.3. This section uses the tools of previous sections to analyze correlated normal variables by making a linear transformation to the simpler case of independent variables.

FIGURE 1. Bivariate normal scatter.
The diagram shows points picked at random from a bivariate distribution, in which the coordinates X and Y each have the same normal distribution, but are not independent. The two variables are *positively correlated*, which makes the cloud elliptical, sloping upward to the right and downwards to the left.

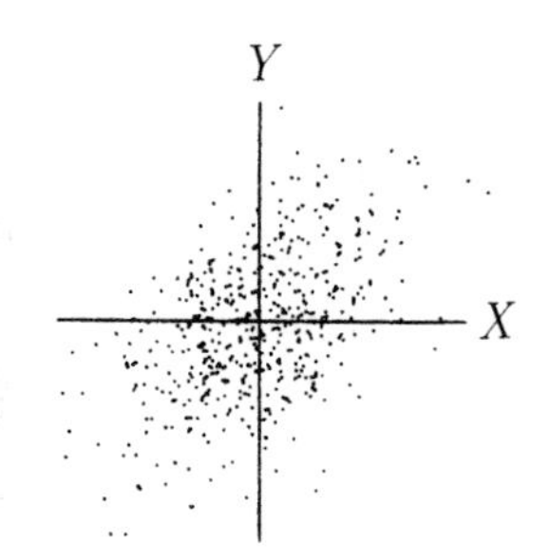

Clouds of data like this are very common in statistical analysis. They were first examined by the British scientist Francis Galton (1822–1911), who studied relations between variables like a father's height and his son's height. To display visually how two variables are related, a ***scatter diagram*** like Figure 1 may be used. In such a diagram, data pairs are represented by plotting a point at the coordinates of each pair. The hereditary connection between a father's height and his son's height makes the variables positively correlated—taller fathers tend to have taller sons, taller sons tend to have taller fathers. But the relation is not a rigid one, since the son's height is not a deterministic function of his father's height. The dependence between the two variables is more interesting and subtle. When variables are measured in their standard units, this dependence shows up in a scatter diagram as a tendency to form an elliptical cloud along a diagonal. The cloud has a major axis along the line $Y = X$ at 45° to the axes in the case of positive correlation, and a major axis along the perpendicular line $Y = -X$ in the case of negative correlation.

FIGURE 2. Bivariate normal scatters for various correlations ρ.

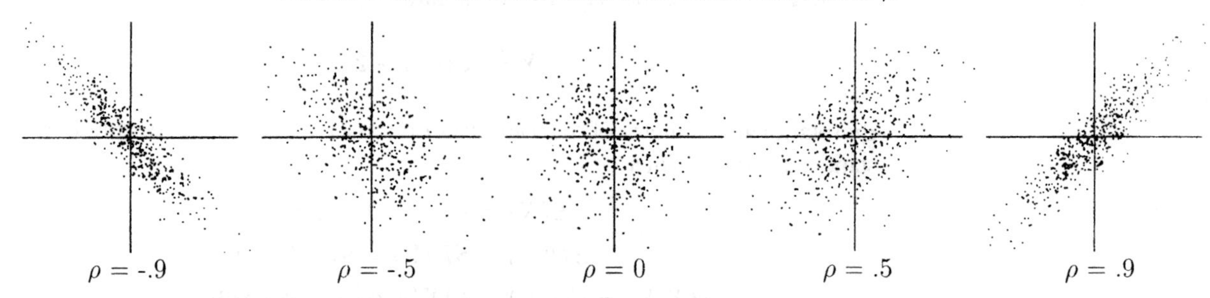

The object now is to describe this kind of dependence between variables by representing correlated normal variables as linear functions of independent ones. This is a powerful technique which is the basis for much statistical analysis of two or more variables. A basic ingredient is the correlation coefficient, denoted here by ρ, often also by r:

$$\rho = Corr(X,Y) = E(X^*Y^*)$$

where X^* is X in standard units, and Y^* is Y in standard units. This correlation ρ is a theoretical quantity, defined by expected values or integrals with respect to a bivariate distribution. In practice, such correlations are usually estimated by the corresponding empirical correlation obtained from data, with the empirical distribution of a data list $(x_1, y_1), \ldots, (x_n, y_n)$ instead of the theoretical distribution, and averages instead of expectations.

Constructing Correlated Normal Variables

To get a pair of correlated standard normal variables X and Y, start with a pair of independent standard normal variables, say X and Z. Let Y be the projection of (X, Z) onto an axis at an angle θ to the X-axis, as in the left-hand diagram:

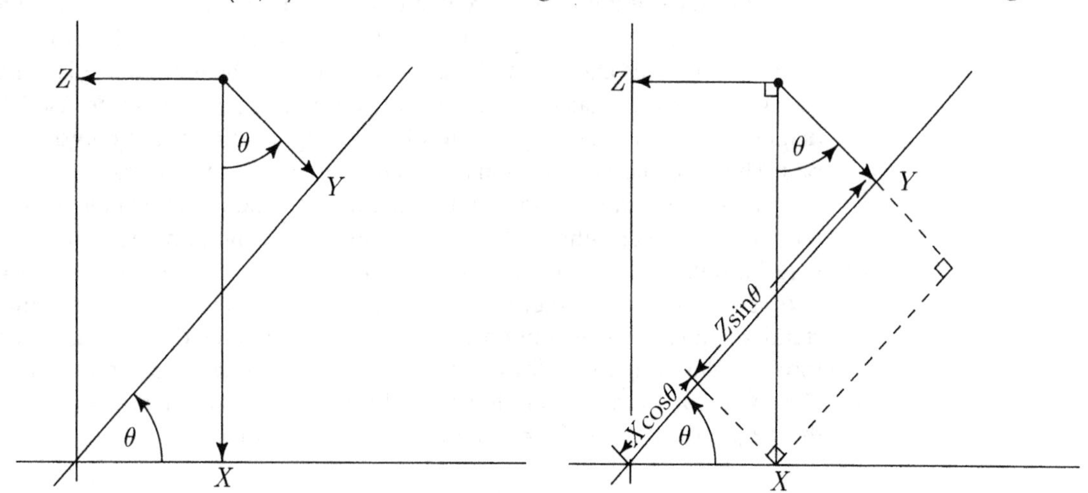

By the geometry of the right-hand diagram

$$Y = X\cos\theta + Z\sin\theta$$

By rotational symmetry of the joint distribution of X and Z, the distribution of Y is standard normal. Thus

$$\begin{aligned}
E(X) = E(Y) &= E(Z) = 0 \\
SD(X) = SD(Y) &= SD(Z) = 1 \\
\rho(X,Y) = E(XY) &= E[X(X\cos\theta + Z\sin\theta)] \\
&= E(X^2)\cos\theta + E(XZ)\sin\theta \\
&= \cos\theta
\end{aligned}$$

since $E(X^2) = 1$, and $E(XZ) = E(X)E(Z) = 0$ by independence of X and Z. To summarize, X and Y are standard normal variables with correlation $\rho = \cos\theta$. Note the special cases

$\theta = 0$	when $\rho = 1$	$Y = X$
$\theta = \pi/2$	when $\rho = 0$	$Y = Z$ is independent of X
$\theta = \pi$	when $\rho = -1$	$Y = -X$

For each ρ between -1 and 1, there is an angle $\theta = \arccos\rho$, which makes X and Y have correlation ρ. Then $\cos\theta = \rho$, $\sin\theta = \sqrt{1-\rho^2}$, and

$$Y = \rho X + \sqrt{1-\rho^2}Z$$

where X and Z are independent normal (0, 1). The joint distribution of X and Y so defined is the ***standard bivariate normal distribution with correlation*** ρ.

Standard Bivariate Normal Distribution

X and Y have standard bivariate normal distribution with correlation ρ if and only if

$$Y = \rho X + \sqrt{1-\rho^2}Z$$

where X and Z are independent standard normal variables.

Marginals. Both X and Y have standard normal distribution.

Conditionals. Given $X = x$, Y has normal $(\rho x, 1-\rho^2)$ distribution. Given $Y = y$, X has normal $(\rho y, 1-\rho^2)$ distribution.

Joint density. The joint density of X and Y is

$$f(x,y) = \frac{1}{2\pi\sqrt{1-\rho^2}} \exp\left\{-\frac{1}{2(1-\rho^2)}(x^2 - 2\rho xy + y^2)\right\}$$

Independence. For X and Y with standard bivariate normal distribution, X and Y are independent if and only if $\rho = 0$.

The next two pages display the geomerty of linear transformation from (X, Z) to (X, Y). Following these pages is a discussion of the results presented in the above box.

FIGURE 3. Geometry of the bivariate normal distribution. Properties of the standard bivariate normal distribution with correlation ρ may be understood in terms of the simplest case $\rho = 0$ by the geometry of the linear transformation $(X, Z) \mapsto (X, Y)$, displayed here for $\theta = 60°$, so

$$\rho = \cos\theta = \frac{1}{2}, \qquad \sqrt{1-\rho^2} = \sin\theta = \frac{\sqrt{3}}{2} \quad \text{and} \quad Y = \frac{1}{2}X + \frac{\sqrt{3}}{2}Z.$$

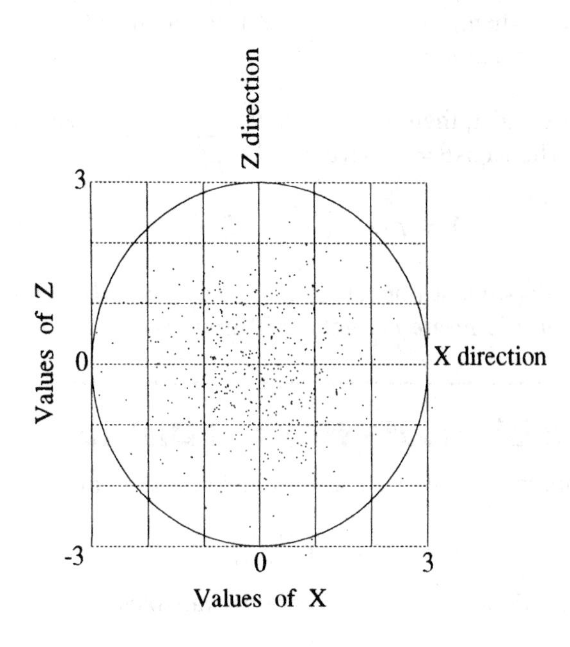

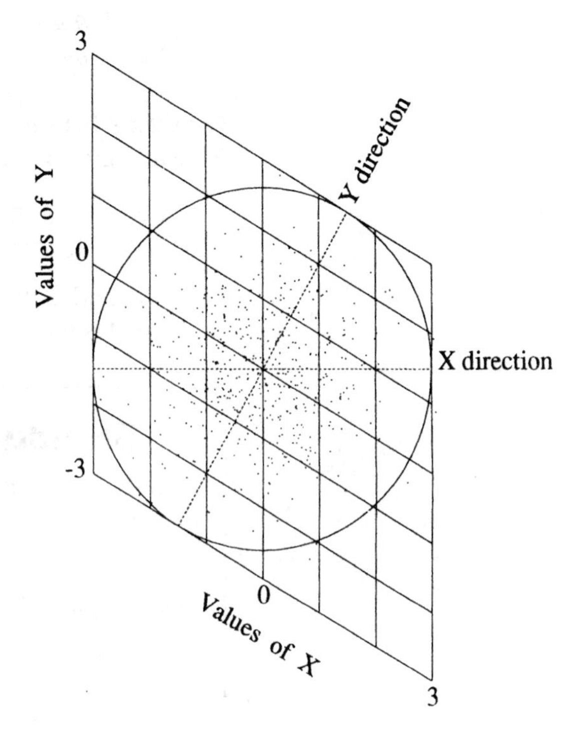

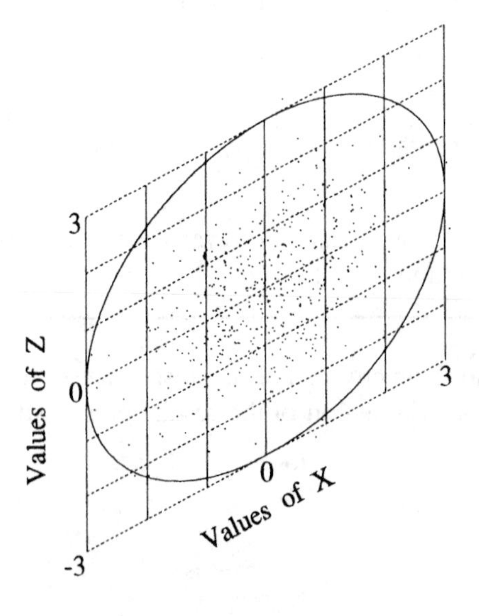

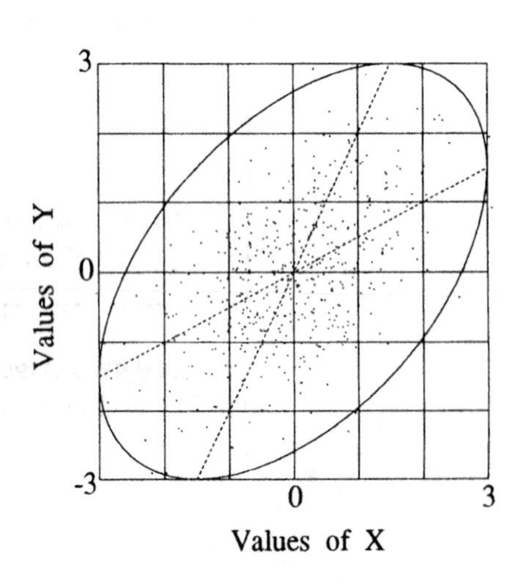

Key to Figure 3.

Top left panel. This shows a computer-generated scatter of 500 points picked at random according to the joint distribution of X and Z, plotted in the usual way with rectangular X and Z coordinates. This is a roughly circular cloud, due to the rotational symmetry of the distribution of two independent standard normals. The circle is the contour of constant density for (X, Z), of radius 3 standard units, containing 98.9% of the probability. The vertical lines represent the events $X = 0, \pm 1, \pm 2, \pm 3$. The dashed horizontal lines represent $Z = 0, \pm 1, \pm 2, \pm 3$.

Top right panel. This is the same scatter in the (X, Z) plane, but with the diagonal lines $Y = 0, \pm 1, \pm 2, \pm 3$. The Y direction is the dotted line at angle $\theta = 60°$ to the horizontal X direction. The diagonals $Y = constant$ are at angle θ to the vertical lines $X = constant$.

Bottom right panel. This is the image of the top right panel after shearing and shrinking to represent X and Y by new rectangular axes. Each point in the top scatter is transformed into one in the bottom scatter. Thus the cloud becomes a random scatter of 500 points picked at random according to the bivariate normal distribution of X and Y, with correlation $\rho = \cos\theta$. Think of the lines in the top right panel as a lattice of rigid rods attached by pins. Keep the vertical axis $X = 0$ fixed, and shear the lattice so the diagonals become horizontal. This makes a lattice of squares of side $1/\sin\theta$. Now shrink everything by a factor of $\sin\theta$ to get the bottom-right panel.

The shearing which turns the diamonds into squares turns the circle into an ellipse, with major axis on the 45-degree line through the new origin. This is an ellipse of constant density for (X, Y). The images of the dotted lines in the old X and Y directions are the dotted lines $Y = \rho X$ and $X = \rho Y$. These are the *regression lines* discussed further in the next paragraph.

Bottom left panel. This is the image of the top left panel by the same transformation from (X, Z) to (X, Y). The ellipse and the cloud of points are the same as in the bottom right panel. But now the lines representing $X = 0, \pm 1, \pm 2, \pm 3$ are shown, along with those representing $Z = 0, \pm 1, \pm 2, \pm 3$. The line $Z = 0$ plays a particularly important role. This is the *regression line*. The equation of this line $Z = 0$ in the (X, Y) plane is

$$Y = \rho X$$

where ρ is the correlation. Geometrically, this is the line of midpoints of vertical sections of the ellipse. Statistically, it is the best predictor of Y based on X.

The properties of the standard bivariate normal distribution stated in the box on page 451 all follow from the basic representation

$$Y = \rho X + \sqrt{1-\rho^2} Z \tag{1}$$

in terms of independent standard normal X and Z.

Conditionals. The formula for the distribution of Y given $X = x$ is immediate from (1). Conditioning on X does not affect the distribution of Z. And given $X = x$ you can treat X in (1) as the constant x, so Y is then just a linear transformation of the standard normal variable Z with coefficients involving ρ and x. This gives the conditional distribution of Y given $X = x$. The distribution of X given $Y = y$ follows by symmetry, or from (1′) below.

Symmetry. The standard bivariate normal distribution of (X, Y) is symmetric with respect to switching X and Y. This can be seen from the formula for the joint density, which is a symmetric function of x and y, or from the geometric description of X and Y. This symmetry is obscured in formula (1) however. You should check as an exercise that (1) has a dual

$$X = \rho Y + \sqrt{1-\rho^2}\, Z' \tag{1′}$$

where Z' is a linear combination of X and Z that is independent of Y.

Joint density. The derivation of this is an exercise: Write out the formulae for the marginal and conditional densities, multiply, and simplify. There is no point remembering this formula. Rather, take the following:

Advice. Do not attempt to compute bivariate normal probabilities or expectations by integrating against the joint density. It is always simpler to rewrite the problem in terms of independent variables X and Z, using (1). This technique is used in all the examples below.

Bivariate Normal Distribution

Random variables U and V have *bivariate normal distribution* with parameters $\mu_U, \mu_V, \sigma_U^2, \sigma_V^2$, and ρ if and only if the standardized variables

$$X = (U - \mu_U)/\sigma_U \qquad Y = (V - \mu_V)/\sigma_V$$

have standard bivariate normal distribution with correlation ρ. Then

$$\rho = Corr(X, Y) = Corr(U, V)$$

and U and V are independent if and only if $\rho = 0$.

Examples

The point of the following examples is to show how any problem involving random variables U and V with a bivariate normal distribution can be solved by a simple three-step procedure:

- **Step 1.** Express U and V in terms of the standardized variables X and Y.
- **Step 2.** Write $Y = \rho X + \sqrt{1-\rho^2} Z$ to reduce the problem to one involving two independent standard normal variables X and Z.
- **Step 3.** Solve the reduced problem involving X and Z by exploiting independence or rotational symmetry.

Example 1. Fathers and sons.

Galton's student Karl Pearson carried out a study on the resemblances between parents and children. He measured the heights of 1078 fathers and sons, and found that the sons averaged one inch taller than the fathers:

Fathers:	mean height: 5′9″	SD: 2″
Sons:	mean height: 5′10″	SD: 2″
	correlation: 0.5	

Problem 1. Predict the height of the son of a father who is 6′2″ tall.

Solution. Assume that the data are approximately bivariate normal in distribution. Then the parameters can be estimated by the corresponding empirical measurements.

Let X be the father's height in standard units, and Y be the son's height in standard units. The assumption of a bivariate normal distribution makes

$$Y = \rho X + \sqrt{1-\rho^2} Z$$

where Z is standard normal independent of X. The natural prediction for Y given $X = x$ is

$$E(Y|X=x) = \rho x$$

Here the given value of X is

$$x = 6'2'' \text{ converted to standard units}$$
$$= (6'2'' - 5'9'')/2'' = 2.5 \text{ standard units}$$

So the predicted value of Y is

$$E(Y|X=x) = 0.5 \times 2.5 = 1.25 \text{ standard units,}$$

That is,

$$\text{predicted son's height} = 5'10'' + 2''Y$$
$$= 5'10'' + 2'' \times 1.25 = 6'\,0.5''$$

Discussion. Though the father is exceptionally tall (height $6'2''$), the son is not predicted to be $6'2''$, but only $6'0.5''$ tall. Galton called this phenomenon ***regression to the mean.***

Problem 2. What is the chance that your prediction is off by more than 1 inch?

Solution. Since 1 inch is 0.5 times the SD of sons' heights, and we are given $X = 2.5$, the problem in standard units is to find

$$P(|Y - \rho X| > 0.5 \,|\, X = 2.5).$$

But since $Y - \rho X = \sqrt{1-\rho^2}Z$ is independent of X with normal $(0, 1-\rho^2)$ distribution, where

$$\sqrt{1-\rho^2} = \sqrt{0.75} \approx 0.87,$$

this is the same as

$$\begin{aligned}
P(|Y - \rho X| > 0.5) &= P(\sqrt{1-\rho^2}|Z| > 0.5) \\
&= P(|Z| > 0.5/\sqrt{1-\rho^2}) \\
&= P(|Z| > 0.5/0.87) \\
&= 2[1 - \Phi(0.5/0.87)] \approx 2[1 - \Phi(0.57)] \approx 0.57
\end{aligned}$$

So with about 57% chance, the prediction will be off by more than an inch.

Problem 3. Estimate the height of a father whose son is $6'0.5''$ tall.

Solution. From above, $6'0.5''$ is the mean height of sons of $6'2''$ fathers. So you might guess that $6'2''$ was the mean height of fathers of $6'0.5''$ sons. But this is wrong, because a given father's height corresponds to a vertical slice through the scatter, whereas a given son's height corresponds to a horizontal slice, which is something quite different. See diagrams. The roles of X and Y must simply be switched in the calculation of Problem 1. The son's height of $6'0.5''$ is 1.25 in standard units. So

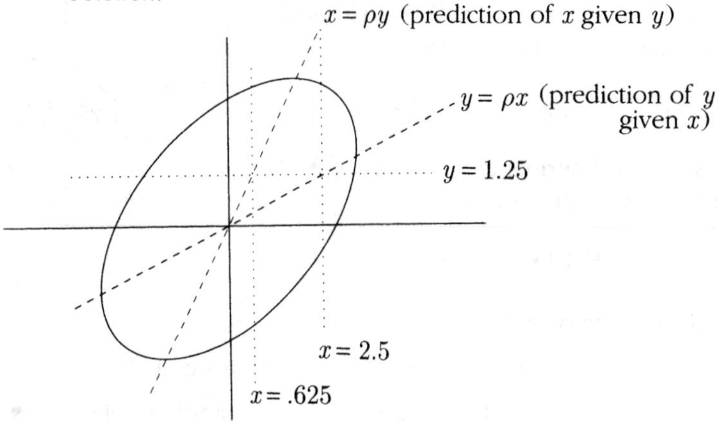

$$\begin{aligned}
\text{estimated father's height} &= 0.5 \times 1.25 \quad \text{in standard units} \\
&= 0.625 \text{ in standard units} \\
&= 5'9'' + 0.625 \times 2'' = 5'10.25''
\end{aligned}$$

Example 2. The probability that both variables are above average.

Problem 1. For the data in Example 1, what fraction of father–son pairs have both father and son of above average height?

Solution. Expressed in terms of the standardized variables X and Y, the problem is to find $P(X \geq 0, Y \geq 0)$. In principle, the answer can be computed as a double integral

$$\iint_{\text{positive quadrant}} f(x, y)\, dx\, dy$$

where $f(x, y)$ is the standard bivariate normal density with $\rho = 0.5$. But, as usual, it is easier to first express X and Y in terms of independent standard normal variables X and Z:

$$Y = \rho X + \sqrt{1 - \rho^2} Z$$

Now the problem is to find

$$P(X \geq 0, Y \geq 0) = P(X \geq 0, \rho X + \sqrt{1 - \rho^2} Z \geq 0)$$
$$= P\left(X \geq 0, Z \geq \frac{-\rho}{\sqrt{1 - \rho^2}} X\right)$$

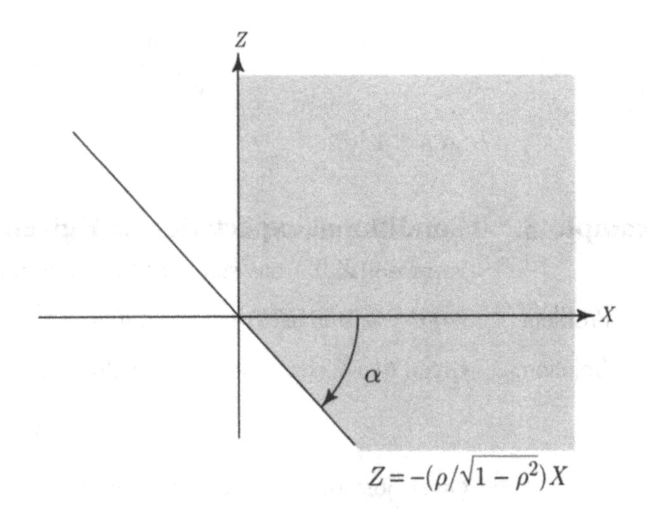

The diagram shows the (X, Z) plane, with the line $Z = -\rho/\sqrt{1 - \rho^2} X$. The shaded region corresponds to the event above. The slope of the line is $-\rho/\sqrt{1 - \rho^2}$.

So for α as in the diagram, considered a negative angle,

$$\tan\alpha = \frac{-\rho}{\sqrt{1-\rho^2}}$$
$$= \frac{-0.5}{\sqrt{0.75}} = -1/\sqrt{3}$$

So $\alpha = -30°$. Thus the angle at the corner of the shaded region is $-\alpha + 90° = 120°$. By rotational symmetry, the chance that (X, Z) lies in the shaded region is the ratio of angles $120°/360° = 1/3$. So

$$P(X \geq 0, Y \geq 0) = 1/3$$

In other words, about one-third of the father–son pairs had both father and son above average height.

Problem 2. Suppose you have data on two variables with a bivariate normal distribution, and 3/8 of the data is above average in both variables. Estimate ρ.

Solution. Transform to standard units and use the same linear change of variable as in the solution of the previous problem. Now

$$\frac{3}{8} = \frac{135°}{360°}$$

so the angle of the corner at the origin is 135°. Thus α in the diagram is $-45°$, and by the previous solution

$$\frac{-\rho}{\sqrt{1-\rho^2}} = \tan\alpha = \tan(-45°) = -1$$

So $\rho = 1/\sqrt{2}$.

Example 3. Conditional expectation of Y given X in an interval.

Suppose (X, Y) has standard bivariate normal density with correlation ρ.

Problem. For $a < b$, find $E(Y \mid a < X < b)$.

Solution. Given that X has a particular value $x \in (a, b)$, the expected value of Y is

$$E(Y \mid X = x) = \rho x.$$

Given just $(a < X < b)$ the precise value of X is unknown. But by the rule of average conditional expectations, $E(Y \mid a < X < b)$ can be found by integration of the conditional expectation $E(Y \mid X = x) = \rho x$ with respect to the conditional density of X given $a < X < b$. This gives

$$E(Y \mid a < X < b) = \int_a^b \rho x f_X(x \mid a < X < b)\, dx$$

where for $a < x < b$

$$f_X(x \mid a < X < b)\,dx = P(X \in dx \mid a < X < b) = \frac{P(X \in dx, a < X < b)}{P(a < X < b)} = \frac{P(X \in dx)}{P(a < X < b)} = \frac{\frac{1}{\sqrt{2\pi}}e^{-\frac{1}{2}x^2}}{\Phi(b) - \Phi(a)}\,dx$$

Substituting this expression gives

$$E(Y \mid a < X < b) = \int_a^b \rho x \frac{\frac{1}{\sqrt{2\pi}}e^{-\frac{1}{2}x^2}}{\Phi(b) - \Phi(a)}\,dx = \frac{\frac{\rho}{\sqrt{2\pi}}\left[e^{-\frac{1}{2}a^2} - e^{-\frac{1}{2}b^2}\right]}{\Phi(b) - \Phi(a)}$$

Example 4. Midterm and final.

Midterm and final scores in a large class have an approximately bivariate normal distribution, with parameters

midterm scores:	mean: 65	SD: 18
final scores:	mean: 60	SD: 20
	correlation: 0.75	

Problem. Estimate the average final score of students who were above average on the midterm.

Solution. Let X and Y denote the midterm and final scores in standard units. The event "midterm score above average" is the same as the event $X > 0$. Take $a = 0$ and $b = \infty$ in the previous example to get

$$E(Y|X > 0) = \frac{\rho}{\sqrt{2\pi}}\left[\frac{1-0}{0.5}\right] = \frac{0.75 \times 2}{\sqrt{2\pi}} \approx 0.6$$

So the average final score of those who scored above average on the midterm is 0.6, in standard units. Thus the required score is

$$60 + 20 \times 0.6 = 72$$

Linear Combinations of Several Independent normal variables

The standard bivariate normal distribution was defined as the joint distribution of a particular pair of linear combinations of independent standard normal variables X and Z, namely, X and $\rho X + \sqrt{1-\rho^2}Z$. While this representation seems at first artificial, the examples show how it is the basis of all calculations involving the more general bivariate normal distribution, which is obtained by allowing arbitrary means and variances, but insisting that the two standardized variables are standard bivariate normal.

The rotational symmetry of the joint distribution of two independent standard normal variables Z_1 and Z_2 implies that the joint distribution of any two linear combinations of Z_1 and Z_2, say

$$V = a_1 Z_1 + a_2 Z_2 \qquad \text{and} \qquad W = a_1 Z_1 + a_2 Z_2$$

is bivariate normal. By reducing to this case by scaling, the same conclusion is obtained for any two independent normal variables Z_1 and Z_2 (not necessarily standard). It can be shown that this extends to linear combinations of any number of independent normal variables Z_i:

Two Linear Combinations of Independent Normal Variables

Let

$$V = \sum_i a_i Z_i \qquad \text{and} \qquad W = \sum_i b_i Z_i$$

be two linear combinations of independent normal (μ_i, σ_i^2) variables Z_i. Then the joint distribution of V and W is bivariate normal.

Granted this, the parameters of the bivariate normal distribution of V and W are easily computed:

$$\mu_V = \sum_i a_i \mu_i \quad \text{and} \quad \mu_W = \sum_i b_i \mu_i$$

$$\sigma_V^2 = \sum_i a_i^2 \sigma_i^2 \quad \text{and} \quad \sigma_W^2 = \sum_i b_i^2 \sigma_i^2$$

$$Cov(V, W) = \sum_i a_i b_i \sigma_i^2$$

$$\rho = Cov(V, W)/\sigma_V \sigma_W$$

Thus the bivariate normal distribution adequately describes the dependence between any two linear combinations of independent normal variables. In particular, this discussion implies the following result:

Independence of Linear Combinations

Two linear combinations $V = \sum_i a_i Z_i$ and $W = \sum_i b_i Z_i$ of independent normal(μ_i, σ_i^2) variables Z_i are independent if and only if they are uncorrelated, that is, if and only if $\sum_i a_i b_i {\sigma_i}^2 = 0$.

Just as the bivariate normal distribution is the joint distribution of two linear combinations of independent normal variables, the *multivariate normal distribution* is the joint distribution of several linear combinations of independent normal variables. It can be shown that several linear combinations of independent normal variables are mutually independent if and only if the covariance between every pair of them is zero. This is a special and important property of normally distributed random variables. It makes covariance and correlation perfectly suited to the analysis of linear combinations of such variables. Keep in mind however, that in general uncorrelated random variables are not necessarily independent.

Exercises 6.5

1. Here is a summary of Pre-SAT and SAT scores of a large group of students.

PSAT scores:	average: 1200	SD: 100
SAT scores:	average: 1300	SD: 90
	correlation: 0.6	

Assume the data are approximately bivariate normal in distribution.

a) Of the students who scored 1000 on the PSAT, about what percentage scored above average on the SAT?

b) Of the students who scored below average on the PSAT, about what percentage scored above average on the SAT?

c) About what percentage of students got at least 50 points more on the SAT than on the PSAT?

2. Data from a large population indicate that the heights of mothers and daughters in this population follow the bivariate normal distribution with correlation 0.5. Both variables have mean 5 feet 4 inches, and standard deviation 2 inches. Among the daughters of above average height, what percent were shorter than their mothers?

3. Heights and weights of a large group of people follow a bivariate normal distribution, with correlation 0.75. Of the people in the 90th percentile of weights, about what percentage are above the 90th percentile of heights?

4. Suppose X and Y are standard normal variables. Find an expression for $P(X+2Y \leq 3)$ in terms of the standard normal distribution function Φ,

a) in case X and Y are independent;

b) in case X and Y have bivariate normal distribution with correlation $1/2$.

5. Let X and Y have bivariate normal distribution with parameters μ_X, μ_Y, ${\sigma_X}^2$, ${\sigma_Y}^2$, and ρ. Let $P(X > \mu_X, Y > \mu_Y) = q$. Find:

a) a formula for q in terms of ρ; b) a formula for ρ in terms of q.

6. Let X and Y be independent standard normal variables.

a) For a constant k, find $P(X > kY)$.

b) If $U = \sqrt{3}X + Y$, and $V = X - \sqrt{3}Y$, find $P(U > kV)$.

c) Find $P(U^2 + V^2 < 1)$.

d) Find the conditional distribution of X given $V = v$.

7. Let X and Y have bivariate normal distribution with parameters μ_X, μ_Y, ${\sigma_X}^2$, ${\sigma_Y}^2$, and ρ.

a) Show that X and Y are independent if and only if they are uncorrelated.

b) Find $E(Y|X = x)$. c) Find $Var(Y|X = x)$.

d) Show that for constants a, b, and c, $aX + bY + c$ has a normal distribution. Find its mean and variance in terms of the parameters of X and Y.

e) Show that if $\mu_X = \mu_Y = 0$, then $X\cos\theta + Y\sin\theta$ and $-X\sin\theta + Y\cos\theta$ are independent normal variables, where

$$\theta = \frac{1}{2}\cot^{-1}\left[\frac{{\sigma_X}^2 - {\sigma_Y}^2}{2\rho\sigma_X\sigma_Y}\right]$$

Explain the geometric significance of θ in terms of the axes of an ellipse of constant density for (X, Y).

8. Let X_1 and X_2 be two independent standard normal random variables. Define two new random variables as follows: $Y_1 = X_1 + X_2$ and $Y_2 = \alpha X_1 + 2X_2$. You are not given the constant α but it is known that $Cov\{Y_1, Y_2\} = 0$. Find

a) the density of Y_2; b) $Cov\{X_2, Y_2\}$.

9. Suppose that W has normal (μ, σ^2) distribution. Given that $W = w$, suppose that Z has normal $(a\,w + b, \tau^2)$ distribution.

a) Show the joint distribution of W and Z is bivariate normal, and find its parameters.

b) What is the distribution of Z?

c) What is the conditional distribution of W given $Z = z$?

10. Show that if V and W have a bivariate normal distribution then

a) every linear combination $aV + bW$ has a normal distribution;

b) every pair of linear combinations $(aV + bW,\ cV + dW)$ has a bivariate normal distribution.

c) Find the parameters of the distributions obtained in a) and b) in terms of the parameters of the joint distribution of V and W.

11. Show that for standard bivariate normal variables X and Y with correlation ρ,

$$E(\max(X,Y)) = \sqrt{\frac{1-\rho}{\pi}}$$

12. Suppose that the magnitude of a signal received from a satellite is

$$S = a + bV + W$$

where V is a voltage which the satellite is measuring, a and b are constants, and W is a noise term. Suppose V and W are independent and normally distributed with means 0 and variances ${\sigma_V}^2$ and ${\sigma_W}^2$.

a) Find $Corr(S, V)$.

b) Given that $S = s$, what is the distribution of V?

c) What is the best estimate of V given $S = s$?

d) If this estimate is used repeatedly for different values of S coming from a sequence of independent values of V and W with the given normal distributions, what is the long-run average absolute value of the error of estimation?

13. Find a formula in terms of ρ for the ratio of the lengths of the axes of an ellipse of constant density in the standard bivariate normal distribution with correlation ρ. (Let the ratio be the length of the axis at $+45°$ over the length of the axis at $-45°$.)

Check your answer by measurement with a ruler in Figure 3 in the case where $\rho = 1/2$. [*Hint*: Let $\rho = \cos\theta$ and reason from Figure 3 that an ellipse of constant density is the image in the (X, Y) plane of the unit circle in the (X, Z) plane. Now consider the images of the points $(\cos\theta/2,\ \sin\theta/2)$ and $(\cos(\theta/2 + \pi/2),\ \sin(\theta/2 + \pi/2))$ in the (X, Y) plane which end up on the $\pm 45°$ lines in the (X, Z) plane, and use trigonometric identities.]

Dependence: Summary

Conditional Distributions: Let X be a discrete random variable. The conditional probability of an event A given $X = x$ is

$$P(A|X = x) = \frac{P(A, X = x)}{P(X = x)}$$

by the division rule of Section 1.4.

For continuously distributed X, there is instead the *infinitesimal conditioning formula*

$$P(A|X = x) = \frac{P(A, X \in dx)}{P(X \in dx)}$$

Understand $P(A|X = x)$ as the chance of A given that X falls in a very small interval near x.

If X and Y are discrete random variables, the conditional probability of $Y = y$ given $X = x$ is

$$P(Y = y|X = x) = \frac{P(X = x, Y = y)}{P(X = x)}$$

If X and Y are continuous random variables with joint density $f_{X,Y}$, the conditional density of Y at y given $X = x$ is $f_Y(y|X = x)$ where

$$f_Y(y|X = x)dy = P(Y \in dy|X \in dx) = \frac{f_{X,Y}(x, y)dx\, dy}{f_X(x)dx} = \frac{f_{X,Y}(x, y)}{f_X(x)}dy$$

Once you have conditioned on $X = x$, you can treat the random variable X as the constant x. Conditional distributions given $X = x$ behave exactly like ordinary distributions, with the constant x as a parameter.

Conditional expectation: The *conditional expectation of Y given $X = x$*, denoted $E(Y|X = x)$, is defined as the expectation of Y relative to the conditional distribution of Y given $X = x$.

The *conditional expectation of Y given X*, denoted $E(Y|X)$, is a random variable, whose value is $E(Y|X = x)$ if $(X = x)$. Thus the random variable $E(Y|X)$ is a function of the random variable X, namely, $f(X)$, where $f(x) = E(Y|X = x)$ for every x.

Expectation is the expectation of conditional expectation: $E(Y) = E[E(Y|X)]$.

See boxes on pages 424 and 425 for important properties of conditional distributions and expectations, and a comparison of the discrete and continuous cases.

Independence: Random variables X and Y are independent if and only if for all subsets B in the range of Y, and all x

$$P(Y \in B|X = x) = P(Y \in B)$$

That is, the conditional distribution of Y given $X = x$ does not depend on x.

Equivalently, X and Y are independent if the conditional distribution of X given $Y = y$ does not depend on y.

Covariance and correlation: $Cov(X,Y) = E\Big[[X-E(X)][Y-E(Y)]\Big] = E(XY)-E(X)E(Y)$

$Var(X+Y) = Var(X) + Var(Y) + 2Cov(X,Y)$

$Corr(X,Y) = \dfrac{Cov(X,Y)}{SD(X)SD(Y)} \in [-1,1]$

X and Y independent $\Longrightarrow Corr(X,Y) = 0$ *but not conversely.*

X and Y uncorrelated $\Longleftrightarrow Corr(X,Y) = 0$
$\Longleftrightarrow Cov(X,Y) = 0 \Longleftrightarrow E(XY) = E(X)E(Y).$

Bivariate normal: X and Y have standard bivariate normal distribution with correlation ρ if and only if

$$Y = \rho X + \sqrt{1-\rho^2}Z,$$

where X and Z are independent standard normal variables.

Marginals. Both X and Y have standard normal distribution.

Conditionals.
Given $X = x$, Y has normal $(\rho x, 1-\rho^2)$ distribution.
Given $Y = y$, X has normal $(\rho y, 1-\rho^2)$ distribution.

X and Y have bivariate normal distribution with parameters $\mu_X, \mu_Y, {\sigma_X}^2, {\sigma_Y}^2$, and ρ if and only if the standardized variables $X^* = (X - \mu_X)/\sigma_X$ and $Y^* = (Y - \mu_Y)/\sigma_Y$ have standard bivariate normal distribution with correlation ρ. Conditional distributions in this case are derived from the standardized case by a linear change of variable. All probabilities and expectations for bivariate normal variables are found by a linear change of variable to independent standard normal variables.

Independence. X and Y with bivariate normal distribution are independent if and only if they are uncorrelated.

Review Exercises

1. Let X and Y be independent random variables. Suppose X has Poisson distribution with parameter λ_1, and Y has Poisson distribution with parameter λ_2.

a) Given that $X + Y = 100$, what are the possible values of X?

b) For each possible value k, find $P(X = k \mid X + Y = 100)$.

c) Take $\lambda_1 = 1$ and $\lambda_2 = 99$. Given $X + Y = 100$, estimate the chance that X is 4 or 5 or 6.

2. Let N denote the number of children in a randomly picked family. Suppose N has geometric distribution:

$$P(N = n) = (1/3)(2/3)^{n-1} \qquad (n = 1, 2, 3, \ldots)$$

and suppose each child is equally likely to be male or female. Let X be the number of male children and Y the number of female children, in a randomly picked family:

a) Find the joint distribution of (X, Y).

b) Given $Y = 0$, what is the most likely value of X?

c) What is the conditional expectation of X given $Y = 0$?

3. A list of $2n$ numbers has mean μ and variance σ^2. Suppose that n numbers are picked at random from the list. Let A_n be the average of these n numbers, B_n the average of the other numbers. Find: a) $E(A_n - B_n)$; b) $SD(A_n - B_n)$.

4. Suppose X and Y have joint density function

$$f(x, y) = \begin{cases} c/x^3 & x > y > 1 \\ 0 & \text{otherwise} \end{cases}$$

where c is a constant.

a) Find c. b) Find the marginal density of X.

c) What is the conditional distribution of Y given $X = x$?

5. Suppose X and Y are random variables with joint density in the plane $f(x, y) = ce^{-(x^2+xy+y^2)}$ where c is a constant. a) Find c. b) Find $Corr(X, Y)$.

6. Let X and Y be independent exponential random variables each with mean 1. Find
a) the joint density of $X + Y$ and $X - Y$;
b) $Corr(X + Y, X - Y)$.

7. Suppose that a point (X, Y) is chosen according to the uniform distribution on the triangle with vertices $(0, 0), (0, 1), (1, 0)$. Calculate:

a) the mean and variance of X;

b) the conditional mean and variance of X given that $Y = 1/3$;

c) the mean and variance of $\max(X, Y)$;

d) the mean and variance of $\min(X, Y)$.

8. Let Y have exponential distribution with mean 0.5. Let X be such that, conditional on $Y = y$, X has exponential distribution with mean y. Find:

(a) the joint density of (X, Y); b) $E(X)$; c) $Corr(X, Y)$.

9. Let X, Y, and Z be independent uniform $(0, 1)$ variables. Find $P[(X/Y) > (Y/Z)]$.

10. Let T_A, T_B, and T_C be the failure times of components A, B, and C. Assume these are independent exponential random variables with rates α, β, and γ, respectively.

a) What is the distribution of the time until the first failure?

b) What is the probability that the first component to fail is component C?

c) Given that the first component to fail is component C, what is the distribution of the time between the first and second failures?

d) Write a formula for the (unconditional) c.d.f. of the time between the first and second failures.

11. Insurance claims arrive at an insurance company according to a Poisson process with rate λ. The amount of each claim has exponential distribution with rate μ, independently of times and amounts of all other claims. Let X_t denote the accumulated total of claims between time 0 and time t. Find simple formulae for

a) $E(X_t)$; $E(X_t^2)$; c) $SD(X_t)$; d) $Corr(X_s, X_t)$ for $s \leq t$.

12. An elevator has an occupancy warning of no more than 26 people and of total weight no more than 4000 pounds. For the population of users, suppose weights are approximately normal with mean 150 pounds and standard deviation 30 pounds.

a) What is the probability that the total weight of a random sample of 26 people from the population exceeds 4000 pounds?

b) Suppose next that the people are carrying things and that the weight of these for an individual of weight X pounds, is approximately normal with mean $0.05X$ pounds and standard deviation 2 pounds. What is the probability that the total weight in the elevator now exceeds 4000 pounds?

c) The dimensions of the floor of the elevator are 54 inches by 92 inches. Suppose the amount of floor space needed by users is normally distributed with mean μ square inches and standard deviation 0.1μ. Find μ such that the probability 20 people can be accommodated is 0.99.

13. a) Let X and Y be two random variables with finite and nonzero variances. Show that $X - Y$ and $X + Y$ are uncorrelated if and only if $Var(X) = Var(Y)$.

b) Let X and Y have standard bivariate normal distribution with correlation 0.6. Find $P(X - Y < 1, X + Y > 2)$.

14. Heights. A population consists of 50% men and 50% women. The empirical distribution of heights over the population yields the following statistics:

	Average	Standard deviation
Men's heights	67 inches	3 inches
Women's heights	63 inches	3 inches

a) What is the average height over the whole population?

b) What is the standard deviation of heights over the whole population?

c) Suppose that men's heights are approximately normally distributed, and that women's heights are as well. Calculate the approximate proportion of individuals in the whole population with heights between 63 and 67 inches.

d) Repeat c), assuming instead that heights are normally distributed over the whole population. Explain why the answers are slightly different.

e) Suppose that a man and a woman are picked at random from this population. Making assumptions as in c), what is the probability that the man is taller than the woman? [*Hint:* No integration required!]

15. Sums of normals in the positive quadrant. Let X and Y be two independent standard normal variables.

a) Calculate $P(X \geq 0, Y \geq 0, X + Y \leq 1)$.

b) Find the conditional density of $X + Y$ given $X \geq 0$ and $Y \geq 0$, and sketch its graph.

c) Find, approximately, the median and the mode of this distribution.

16. Rainfall. Suppose that the distribution of annual rainfall in a particular place, measured in inches, is approximately gamma with shape parameter $r = 3$. If the mean annual rainfall is 20 inches, find approximations to the following:

a) the probability of more than 35 inches of rain in any particular year;

b) the probability that in ten consecutive years, it never rains more than 35 inches, assuming different years are independent;

c) still assuming independence of different years, the probability that the record rainfall over the last 20 years is exceeded in at least one of the next ten years, assuming the record rainfall over the last 20 years, R_{20} say, is known;

d) same as c), but assuming the value of R_{20} is unknown.

17. Symmetry under rotations.

a) Suppose the joint distribution of X and Y is symmetric under rotations. Are X and Y necessarily independent? Are they necessarily uncorrelated? Explain by arguments or examples.

b) Suppose (X, Y) is a point picked at random from the unit circle $X^2 + Y^2 = 1$. Calculate $E(X^2)$, $E(Y^2)$, and $E(XY)$.

c) Suppose U is uniformly distributed on $(0, 1)$, $X = \cos 2\pi U$, $Y = \sin 2\pi U$. Are X and Y uncorrelated? Are X and Y independent? Explain carefully the connection betweeen b) and c).

18. Maxima and minima of normal variables.
Calculate the expected values of $\max(X, Y)$ and $\min(X, Y)$:

a) if X and Y are independent standard normal variables;

b) if X and Y are independent normal (μ, σ^2);

c) if X and Y are standard bivariate normal with correlation ρ.

19. Suppose you sample with replacement n times from a population of n elements.

a) What fraction of the n elements should you expect to see in the sample?

b) For example, what fraction of all $\binom{52}{5}$ poker hands should you expect to see in $\binom{52}{5}$ independent deals?

c) Compute the variance of the fraction in a), and show that it is less than $1/4n$.

d) Evaluate for the example in b), and estimate the chance that your prediction in b) is off by more than 1%.

20. Craps. Find the expectated total number of times Y the pair of dice must be rolled in a craps game (see Exercise 3.4.8) by conditioning on the result of the first roll.

21. I toss a coin which lands heads with probability p. Let W_H be the number of tosses till I get a head, W_{HH} the number of tosses till I get two heads in a row, and W_{HHH} the number of tosses till I get three heads in a row. Find:

a) $E(W_H)$; b) $E(W_{HH})$ [*Hint*: condition on whether the first toss was heads or tails]; c) $E(W_{HHH})$ [*Hint*: condition on W_T].

d) Generalize to find the expected number of tosses to obtain m heads in a row.

22. Long runs of heads. In the play *Rosencrantz and Guildenstern are dead* by Tom Stoppard, the results of 101 apparently fair coin tosses are recorded: 100 heads in a row, followed by a tail. Suppose a fair coin is tossed independently once every second. About how many years do you expect it would take before 100 heads in a row came up? How long for it to be 99% sure that such a run will have appeared?

23. Suppose an insect lays a Poisson (λ) number of eggs. Suppose each egg hatches with probability p and dies with probability q, independently of each other egg. Show that the number of eggs that hatch and the number of eggs that die are independent Poisson random variables, and find their parameters.

24. I roll a random number of dice. If the number of dice rolled has the Poisson (12) distribution, find (and justify your answers)

a) the expectation of the total number of spots showing;

b) the standard deviation of the total number of spots showing.

25. Suppose the number of accidents in an interval of time has Poisson (λ) distribution. Suppose that in each accident there are k persons injured with probability p_k, independently of all other accidents. Let N_k be the number of accidents in which k persons are injured.

a) What is the joint distribution of N_1 and N_2?

b) Let M be the total number of persons injured. Find formulae for $E(M)$ and $SD(M)$ in terms of $p_1, p_2, \ldots$ and λ.

26. Distinguishing points in a Poisson scatter. In practical situations, if two points in a scatter are closer than some distance δ, it may not be possible to distinguish them. Suppose that this is the case, and that there is a Poisson scatter over the unit square, with intensity λ. Show that the probability of the event D, that all points in the scatter can be distinguished, is at least $1 - \frac{\pi}{2}\lambda^2\delta^2$.

[*Hint.* Show that $P(D|N = 2) \geq 1 - \pi\delta^2$ and $P(D|N = 3) \geq (1 - \pi\delta^2)(1 - 2\pi\delta^2)$ and so on. Use the inequality

$$(1 - \alpha)(1 - \beta) \geq 1 - (\alpha + \beta) \qquad (\alpha > 0,\ \beta > 0)$$

repeatedly, to obtain

$$P(D|N) \geq 1 - \frac{1}{2}N(N - 1)\pi\delta^2]$$

27. Inhomogeneous Poisson scatter. Let Q be a probability distribution over a set S, $\lambda > 0$. Consider a random scatter of points over the set S, where a Poisson (λ) number N of points are distributed independently at random according to Q. More formally, for B a subset of S, let $N(B) = 0$ if $N = 0$, and

$$N(B) = \sum_{i=1}^{n} I(X_i \in B) \qquad \text{if} \quad (N = n), \quad n = 1, 2, \ldots$$

where $X_1, \ldots, X_n$ are conditionally independent with common distribution Q given $(N = n)$, and N has Poisson (λ) distribution. Prove that
for disjoint $B_1, \ldots, B_j$, *the* $N(B_1), \ldots, N(B_j)$ *are mutually independent Poisson random variables with parameters* $\lambda(B_1), \ldots, \lambda(B_j)$ *where* $\lambda(B) = \lambda Q(B)$.
[*Hint:* Start by considering the case of B_1 and B_2 with $B_1 + B_2 = S$, and calculate $P(N(B_1) = n_1, N(B_2) = n_2)$ by conditioning on $N = n_1 + n_2$. Argue that, in general, it suffices to consider a partition $B_1, \ldots, B_j$ of S, and proceed similarly. The multinomial coefficients $n!/(n_1!n_2! \cdots n_j!)$ should appear.]
Note. Such a collection of random variables $N(B)$ is called a *Poisson process with intensity measure* $\lambda(B)$ *on* S. For S the unit square and $\lambda(B) = \lambda \times Area(B)$ this is a construction of the Poisson scatter over the unit square considered in Section 3.5. Such a scatter is called *homogeneous*. If $Q(B)$ is not the uniform distribution, the scatter is called *inhomogeneous*. Note that if $Q\{s\} > 0$ for a point $s \in S$, there may be more than one "hit" counted at s. In particular, if Q is a discrete measure with probabilities $q_1, \ldots, q_n$ at points $s_1, \ldots, s_n$, then $N(s_1), \ldots, N(s_n)$ are independent Poisson random variables with parameters $\lambda q_1, \ldots, \lambda q_n$.
Illustration. Suppose you roll a Poisson (λ) number N of dice. Then the number of times each of the six faces appears is an independent Poisson $(\lambda/6)$ random variable. And the number of odd faces and the number of even faces are two independent Poisson $(\lambda/2)$ random variables. But if you throw a fixed number n of dice these numbers are dependent, because they must add up to n.

28. You and I both toss a fair coin N times. You get X heads and I get Y heads.

a) If $P(X = Y)$ is approximately 10%, then approximately how large must N be?

b) The normal approximation says $P(\left|X - \frac{1}{2}N\right| \leq \frac{1}{2}\sqrt{N}) \approx 68\%$.
Given $X = Y$, is the conditional probability that $\left|X - \frac{1}{2}N\right| \leq \frac{1}{2}\sqrt{N}$ still about 68%, somewhat larger than 68%, or somewhat smaller than 68%? Explain which, *without* doing detailed calculations.

29. Variance of discrete order statistics. Let T_i be the place at which the ith good element appears in a random ordering of k good and $N - k$ bad elements. From Exercise 3.6.13, the mean of T_i is $E(T_i) = i(N + 1)/(k + 1)$. Calculate $SD(T_i)$ by the following steps.

a) Let $\alpha(k, N) = E(T_1(T_1 - 1))$, $1 \le k \le N$. Show by conditioning on whether the first element is good or bad that

$$\alpha(k, N) = (N - k)\left[\frac{2}{k+1} + \frac{\alpha(k, N-1)}{N}\right]$$

b) Deduce that

$$\alpha(k, N) = \frac{2(N+1)(N-k)}{(k+1)(k+2)}$$

c) Deduce that

$$Var(T_1) = \frac{(N+1)(N-k)k}{(k+1)^2(k+2)}$$

d) Check the case $k = 1$ by calculating $Var(T_1)$ directly from the distribution of T_1.

e) Let $W_i = T_{i+1} - T_i$, $i = 1, \ldots, k+1$, where $T_0 = 0$ and $T_{k+1} = N + 1$. Use the exchangeability of $W_1, \ldots, W_{k+1}$ to show that for each $i = 1, \ldots, k+1$

$$Var(T_i) = i\,Var(T_1) + i(i-1)\,Cov(W_1, W_2)$$

Deduce that

$$Cov(W_1, W_2) = -Var(T_1)/k$$

and hence that

$$Var(T_i) = \frac{i(k+1-i)(N+1)(N-k)}{(k+1)^2(k+2)}$$

f) Give an intuitive explanation of why $SD(T_i) = SD(T_{k+1-i})$.

g) Suppose that $T_1, \ldots, T_4$ are the places at which the aces appear in a well-shuffled deck of 52 cards. Find numerical values of $E(T_i)$ and $SD(T_i)$ for $i = 1, \ldots, 4$.

30. Let $V_1, \ldots, V_n$ be the order statistics of n independent uniform $(0, 1)$ variables. Let
$A_{\text{all}} = (V_1 + \cdots + V_n)/n$, average of all the order statistics,
$A_{\text{ext}} = (V_1 + V_n)/2$, average of the extremes,
$A_{\text{mid}} = V_{(n+1)/2}$, the middle value, where you can assume n is odd.

a) Show that for large n, each of the A's is most likely very close to $1/2$.

b) For large n, one of the A's can be expected to be very much closer to $1/2$ than the two others. Which one, and why?

c) For $n = 101$ find for each of the A's a good approximation to the probability that it is between .49 and .51.

31. From discrete to continuous spacings. Let $U_{(1)} < U_{(2)} < \ldots < U_{(n)}$ be the order statistics of n independent uniform $(0, 1)$ variables $U_1, \ldots, U_n$. Let $V_1 = U_{(1)}$, $V_i = U_{(i)} - U_{(i-1)}$ for $1 \le i \le n$, and let $V_{n+1} = 1 - U_{(n)}$. Imagine the unit interval is cut into subintervals at each of the n random points U_i for $1 \le i \le n$. Then $V_1, V_2, \ldots, V_{n+1}$ are the lengths of the $n+1$ subintervals so obtained, in order from left to right. This model for cutting an interval at random is of interest in genetics. The V_i could represent the relative lengths of strands obtained by random cutting of a long molecule such as DNA. For a positive integer $N > n$ let $U'_1, \ldots, U'_{N-n}$ denote $N - n$ more uniform $(0, 1)$ variables, independent of each other and of the cut points $U_1, \ldots, U_n$. For $1 \le i \le n+1$ let N_i denote the number of U'_i that fall in the interval $(U_{(i-1)}, U_{(i)})$ of length V_i (where $U_{(0)} = 0$ and $U_{(n+1)} = 1$ to make the definition work for N_1 and N_{n+1}).

a) Show that the joint distribution of $N_1, \ldots, N_{n+1}$ is identical to the joint distribution of the discrete spacings $W_1, \ldots, W_{n+1}$ derived from a random ordering of n aces and $N - n$ nonaces as in Exercise 3.6.13. That is to say, $(N_1, \ldots, N_{n+1})$ has uniform distribution over the set of all $(n+1)$–tuples of non-negative integers $(n_1, \ldots, n_{n+1})$ with $n_1 + \cdots + n_{n+1} = N - n$. In particular, $N_1, \ldots, N_{n+1}$ are exchangeable.

b) Conditionally given the continuous spacings $(V_1, \ldots, V_{n+1})$, the sequence $(N_1, \ldots, N_{n+1})$ is distributed like the number of results in each of $n+1$ categories in a sequence of $N-n$ independent trials with probability V_i of a result in category i on each trial. Explain why this is so. Deduce that for large N, N_i/N is almost equal to V_i for each i with overwhelmingly high probability. It follows that in the limit as $N \to \infty$ for fixed n, as discussed at the end of Exercise 3.6.12, the joint distribution of the normalized discrete spacings $(N_1/N, N_2/N, \ldots, N_{n+1}/N)$ converges to the joint distribution of the continuous spacings $V_1, V_2, \ldots, V_{n+1}$.

(Keep in mind that the distribution of N_i depends on N, so N_i/N does not just tend to zero: the sum over i of the N_i/N is identically equal to 1.) Since the N_i/N are exchangeable for every N, it follows that the V_i are exchangeable, something that is not obvious in the continuous model.

32. Joint distribution of continuous spacings. Continuing with the same notation as in Exercise 31,

a) Show that for $v_i \geq 0$ with $v_1 + \cdots + v_{n+1} = v \leq 1$

$$\lim_{N\to\infty} P(N_i/N \geq v_i \text{ for every } 1 \leq i \leq n+1) = (1-v)^n$$

by explicit evaluation of the limit, using Exercise 3.6.15 and the fact that $(N)_k \sim N^k$ as $N \to \infty$ for every $k = 1, 2, \ldots$. This yields the corresponding probabilty for the continuous model: for $v_i \geq 0$ with $v_1 + \cdots + v_{n+1} = v \leq 1$

$$P(V_i \geq v_i \text{ for every } 1 \leq i \leq n+1) = (1-v)^n$$

b) Show that the V_i have identical distribution with

$$P(V_i \geq v) = (1-v)^n \qquad (0 \leq v \leq 1)$$

c) Deduce that V_i has beta $(1, n)$ distribution.

33. Maximum and minimum spacings. Continuing with the notation of the preceding exercises, let $V_{\min} = \min_i V_i$ where the min is over $1 \leq i \leq n+1$.

a) Show that $V_{\min}$ has the same distribution as $V_1/(n+1)$. Deduce the mean and variance of $V_{\min}$ from the mean and variance of the beta $(1, n)$ distribution.

b) Let $V_{\max} = \max_i V_i$. Parallel to the discrete formula of Exercise 3.6.16, show that for $0 \leq v \leq 1$

$$P(V_{\max} \geq v) = \sum_{i=1}^{n+1} (-1)^{i-1} \binom{n+1}{i} (1 - iv)_+^n$$

where $(1 - iv)_+^n$ equals $(1 - iv)^n$ if $iv \leq 1$, and equals 0 otherwise.

c) Deduce by integration of this tail probability from 0 to 1 that

$$E(V_{\max}) = \sum_{i=1}^{n+1} (-1)^{i-1} \binom{n+1}{i} \frac{1}{i(n+1)}$$

It is intuitively clear, and can be verified analytically, that as the number of cuts $n \to \infty$, $V_{\max} \to 0$, which forces the distribution of $V_{\max}$ to pile up around zero. But the rate of convergence is rather slow.

d) Find the numerical values of $E(V_{\min})$ and $E(V_{\max})$ for $n = 1, \ldots, 10$.

34. Dirichlet distribution. A sequence of random variables $(Q_1, \ldots, Q_m)$ has *Dirichlet distribution* with parameters $(r_1, \ldots, r_m)$ if $Q_i \geq 0$, $Q_1 + \cdots + Q_m = 1$, and

$$\frac{P(Q_i \in dq_i, 1 \leq i \leq m-1)}{dq_1 dq_2 \cdots dq_{m-1}} = \frac{\Gamma(r_1 + \cdots + r_m)}{\Gamma(r_1) \cdots \Gamma(r_m)} \prod_{i=1}^{m} q_i^{r_i - 1} \qquad (q_i \geq 0,\ q_1 + \cdots + q_m = 1)$$

For $m = 2$, (Q_1, Q_2) has Dirichlet distribution with parameters r and s if and only if $Q_2 = 1 - Q_1$ for Q_1 with beta (r, s) distribution. So the Dirichlet distribution is a multivariate extension of the beta distribution. There is a multivariate version of the result of Exercise 5.4.19: If $Y_i, 1 \leq i \leq m$ are independent with gamma (r_i, λ) distributions, $\sum = \sum_i Y_i$ and $Q_i = Y_i / \sum$, then $(Q_1, \ldots, Q_m)$ has Dirichlet distribution with parameters $(r_1, \ldots, r_m)$, independently of $\sum$, which has gamma (r, λ) distribution for $r = r_1 + \cdots + r_m$. Assuming this result, deduce the following properties of this Dirichlet distribution of $(Q_1, \ldots, Q_m)$:

a) The marginal distribution of Q_i is beta $(r_i, r - r_i)$.

b) For $i \neq j$ the distribution of $Q_i + Q_j$ is beta $(r_i + r_j, r - r_i - r_j)$. Similarly for any finite sum of at most $m - 1$ different Q_i.

c) The joint distribution of the continuous spacings $V_1, \cdots, V_{n+1}$ derived from n independent uniform$(0, 1)$ random variables as in Exercises 31 and 32 is Dirichlet with parameters $r_i = 1$ for $1 \leq i \leq m = n + 1$.

35. Dirichlet–multinomial. Suppose that $X_1, X_2, \ldots$ is a sequence of independent trials with m possible values $\{1, \ldots, m\}$, with probability q_i for value i on each trial. The parameters $(q_1, \ldots, q_m)$ are unknown, and regarded as the values of random variables $(Q_1, \ldots, Q_m)$. Suppose the prior distribution of $(Q_1, \ldots, Q_m)$ is Dirichlet with parameters $(r_1, \ldots, r_m)$, as in Exercise 34. After n trials, let N_i be the number of results i, that is the number of times that $X_j = i$ for $1 \leq j \leq n$. So the conditional distribution of $(N_1, \ldots, N_m)$ given $(Q_1, \ldots, Q_m)$ is multinomial with parameters n and $(Q_1, \ldots, Q_m)$.

a) Show that the posterior distribution of $(Q_1, \ldots, Q_m)$ given the results $(N_1, \ldots, N_m)$ of n trials is Dirichlet with parameters $(r_1 + N_1, \ldots, r_m + N_m)$.

b) Find a formula for the unconditional probability $P(N_i = n_i \text{ for } 1 \leq i \leq m)$ for any sequence of m non-negative integers n_i with $n_1 + \cdots + n_m = n$.
[*Hint:* Use the fact that the total integral of the Dirichlet joint density with parameters $(r_1 + n_1, \ldots, r_m + n_m)$ is 1].

c) Deduce in particular that if $r_i = 1$ for $1 \leq i \leq m$ then the unconditional distribution of $(N_1, \ldots, N_m)$ is uniform over its range of possible values.

d) Explain the result of part c) without integration by reference to Exercise 31.

Distribution Summaries

Discrete

name and range	$P(k) = P(X = k)$ for $k \in$ range	mean	variance
uniform on $\{a, a+1, \ldots, b\}$	$\frac{1}{b-a+1}$	$\frac{a+b}{2}$	$\frac{(b-a+1)^2-1}{12}$
Bernoulli (p) on $\{0, 1\}$	$P(1) = p$; $P(0) = 1-p$	p	$p(1-p)$
binomial (n, p) on $\{0, 1, \ldots, n\}$	$\binom{n}{k} p^k (1-p)^{n-k}$	np	$np(1-p)$
Poisson (μ) on $\{0, 1, 2, \ldots\}$	$\frac{e^{-\mu}\mu^k}{k!}$	μ	μ
hypergeometric (n, N, G) on $\{0, \ldots, n\}$	$\frac{\binom{G}{k}\binom{N-G}{n-k}}{\binom{N}{n}}$	$\frac{nG}{N}$	$n\left(\frac{G}{N}\right)\left(\frac{N-G}{N}\right)\left(\frac{N-n}{N-1}\right)$
geometric (p) on $\{1, 2, 3 \ldots\}$	$(1-p)^{k-1}p$	$\frac{1}{p}$	$\frac{1-p}{p^2}$
geometric (p) on $\{0, 1, 2 \ldots\}$	$(1-p)^k p$	$\frac{1-p}{p}$	$\frac{1-p}{p^2}$
negative binomial (r, p) on $\{0, 1, 2, \ldots\}$	$\binom{k+r-1}{r-1} p^r (1-p)^k$	$\frac{r(1-p)}{p}$	$\frac{r(1-p)}{p^2}$

Continuous

† undefined.

name	range	density $f(x)$ for $x \in$ range	c.d.f. $F(x)$ for $x \in$ range	Mean	Variance
uniform (a, b)	(a, b)	$\frac{1}{b-a}$	$\frac{x-a}{b-a}$	$\frac{a+b}{2}$	$\frac{(b-a)^2}{12}$
normal $(0, 1)$	$(-\infty, \infty)$	$\frac{1}{\sqrt{2\pi}} e^{-\frac{1}{2}x^2}$	$\Phi(x)$	0	1
normal (μ, σ^2)	$(-\infty, \infty)$	$\frac{1}{\sqrt{2\pi}\sigma} e^{-\frac{1}{2}(x-\mu)^2/\sigma^2}$	$\Phi\left(\frac{x-\mu}{\sigma}\right)$	μ	σ^2
exponential (λ) = gamma $(1, \lambda)$	$(0, \infty)$	$\lambda e^{-\lambda x}$	$1 - e^{-\lambda x}$	$1/\lambda$	$1/\lambda^2$
gamma (r, λ)	$(0, \infty)$	$\Gamma(r)^{-1} \lambda^r x^{r-1} e^{-\lambda x}$	$1 - e^{-\lambda x} \sum_{k=0}^{r-1} \frac{(\lambda x)^k}{k!}$ for integer r	r/λ	r/λ^2
chi-square (n) =gamma $\left(\frac{n}{2}, \frac{1}{2}\right)$	$(0, \infty)$	$\Gamma(\frac{n}{2})^{-1} (\frac{1}{2})^{\frac{n}{2}} x^{\frac{n}{2}-1} e^{-\frac{x}{2}}$	as above for $\lambda = \frac{1}{2}$, $r = \frac{n}{2}$ if n is even	n	$2n$
Rayleigh	$(0, \infty)$	$x e^{-\frac{1}{2}x^2}$	$1 - e^{-\frac{1}{2}x^2}$	$\sqrt{\frac{\pi}{2}}$	$\frac{4-\pi}{2}$
beta (r, s)	$(0, 1)$	$\frac{\Gamma(r+s)}{\Gamma(r)\Gamma(s)} x^{r-1}(1-x)^{s-1}$	see Exercise 4.6.5 for integer r and s	$\frac{r}{r+s}$	$\frac{rs}{(r+s)^2(r+s+1)}$
Cauchy	$(-\infty, \infty)$	$\frac{1}{\pi(1+x^2)}$	$\frac{1}{2} + \frac{1}{\pi}\arctan(x)$	†	†
arcsine =beta $(1/2, 1/2)$	$(0, 1)$	$\frac{1}{\pi\sqrt{x(1-x)}}$	$\frac{2}{\pi}\arcsin(\sqrt{x})$	$\frac{1}{2}$	$\frac{1}{8}$

Beta

Parameters: $r > 0$ and $s > 0$

Range: $x \in [0, 1]$

Density function:

$$P(X_{r,s} \in dx)/dx = \frac{1}{B(r,s)} x^{r-1}(1-x)^{s-1} \quad (0 \le x \le 1)$$

where

$$B(r,s) = \int_0^1 x^{r-1}(1-x)^{s-1}dx = \frac{\Gamma(r)\Gamma(s)}{\Gamma(r+s)}$$

is the *beta function*, and $\Gamma(r)$ is the *gamma function* (see gamma distributions).

Cumulative distribution function: (Exercises in Section 4.6.) No simple general formula for r or s not an integer. See tables of the incomplete beta function. For integers r and s

$$P(X_{r,s} \le x) = \sum_{i=r}^{r+s-1} \binom{r+s-1}{i} x^i (1-x)^{r+s-1-i} \qquad (0 \le x \le 1)$$

Mean and standard deviation: (4.6)

$$E(X_{r,s}) = \frac{r}{r+s} \qquad SD(X_{r,s}) = \frac{\sqrt{rs}}{(r+s)\sqrt{(r+s+1)}}$$

Special cases:

- $r = s = 1$: The uniform $[0, 1]$ distribution.
- $r = s = 1/2$: The arcsine distribution.

Sources and applications:

- Order statistics of uniform variables (4.6).
- Ratios of gamma variables (5.4).
- Bayesian inference for unknown probabilities.

Normal approximation:

- Good for large r and s.

Binomial

Parameters:

n = number of trials $(n = 1, 2, \ldots)$

p = probability of success on each trial $(0 \leq p \leq 1)$

Range: $k \in \{0, 1, \ldots, n\}$

Probability function: (2.1)

$$P(k) = P(S = k) = \binom{n}{k} p^k (1-p)^{n-k} \quad (k = 0, 1, \ldots, n)$$

$$\text{where} \quad S = \begin{pmatrix} \text{number of successes in } n \text{ independent trials with} \\ \text{probability } p \text{ of success on each trial} \end{pmatrix}$$

$$= X_1 + \cdots + X_n \quad \text{where } X_i = \text{indicator of success on trial } i.$$

Mean and standard deviation: (3.2, 3.3)

$$E(S) = \mu = np \qquad SD(S) = \sigma = \sqrt{np(1-p)}$$

Mode: (2.1) $\quad \text{int}(np + p)$

Consecutive odds ratios: (2.1)

$$\frac{P(k)}{P(k-1)} = \frac{(n-k+1)}{k} \cdot \frac{p}{1-p} \qquad \text{(decreasing)}$$

Special case: (1.3)

Binomial $(1, p) \equiv$ Bernoulli (p), distribution of the indicator of an event A with probability $P(A) = p$.

Normal approximation: (2.2, 2.3)

If $\sigma = \sqrt{np(1-p)}$ is sufficiently large

$$P(k) \approx \frac{1}{\sigma}\phi\left(\frac{k-\mu}{\sigma}\right)$$

where $\phi(z)$ is the standard normal density function

$$P(a \text{ to } b) \approx \Phi\left(\frac{b + \frac{1}{2} - \mu}{\sigma}\right) - \Phi\left(\frac{a - \frac{1}{2} - \mu}{\sigma}\right)$$

where Φ is the standard normal cumulative distribution function.

Poisson approximation: (2.4)

If p is close to zero

$$P(k) \approx e^{-\mu}\mu^k/k! \quad \text{where} \quad \mu = np$$

Exponential

Parameter: $\lambda > 0$, the *rate* of an exponential random variable T.

Range: $t \in [0, \infty)$

Density function: (4.2)

$$P(T \in dt)/dt = \lambda e^{-\lambda t} \quad (t \geq 0)$$

Cumulative distribution function: (4.2)

$$P(T \leq t) = 1 - e^{-\lambda t} \quad (t \geq 0)$$

Often T is interpreted as a lifetime.

Survival function:

$$P(T > t) = e^{-\lambda t} \quad (t \geq 0)$$

Mean and Standard Deviation: (4.2)

$$E(T) = 1/\lambda \qquad SD(T) = 1/\lambda$$

Interpretation of λ:

$$\lambda = P(T \in dt \,|\, T > t)/dt$$

is the constant *hazard rate* or chance per unit time of death given survival to time t. See Section 4.3 for a discussion of non-constant hazard rates.

Characterizations:

- Only distribution with constant hazard rate.
- Only distribution with the *memoryless property*

$$P(T > t + s \,|\, T > t) = P(T > s)$$

for all $s, t > 0$

Sources:

- Time until the next arrival in a Poisson process with rate λ.
- Approximation to geometric (p) distribution for small p.
- Approximation to beta $(1, s)$ distribution for large s.
- Spacings and shortest spacings of uniform order statistics.

Gamma

Parameters: $r > 0$ (shape) $\lambda > 0$ (rate or inverse scale)

Range: $t \in [0, \infty)$

Density function: (4.2, 5.4) $P(T_{r,\lambda} \in dt)/dt = \Gamma(r)^{-1}\lambda^r t^{r-1} e^{-\lambda t} \quad (t \geq 0)$
where $\Gamma(r) = \int_0^\infty t^{r-1} e^{-t} dt$ is the *gamma function*. *Note:* $\Gamma(r) = (r-1)!$ for integer r.

Cumulative distribution function: (4.2)
No formula for non-integer r. See tables of the incomplete gamma function. For integer r

$$P(T_{r,\lambda} \leq t) = P(N_{t,\lambda} \geq r) = 1 - \sum_{k=0}^{r-1} e^{-\lambda t} \frac{(\lambda t)^k}{k!}$$

where $N_{t,\lambda}$ denotes the number of points up to time t in a Poisson process with rate λ, and has Poisson (λt) distribution.

Mean and standard deviation: (4.2) $E(T_{r,\lambda}) = r/\lambda \quad SD(T_{r,\lambda}) = \sqrt{r}/\lambda$

Special cases:

- gamma $(1, \lambda)$ is exponential (λ).
- gamma $(n/2, 1/2)$ is chi-square (n), the distribution of the sum of the squares of n independent standard normals.

Sources:

- Sum of r independent exponential (λ) variables.
- Time until the rth arrival in a Poisson process with rate λ.
- Bayesian inference for unknown Poisson rates.
- Approximation to negative binomial (r, p) for small p.
- Approximation to beta (r, s) for large s.

Transformations: (Notation: $X \sim F$ means X is a random variable with distribution F.)

Scaling: $T \sim$ gamma $(r, \lambda) \iff \lambda T \sim$ gamma $(r, 1)$

Sums: For independent $T_i \sim$ gamma (r_i, λ) $\quad \sum_i T_i \sim$ gamma $(\sum_i r_i, \lambda)$

Ratios: For independent $T_{r,\lambda}$ and $T_{s,\lambda}$

$$\frac{T_{r,\lambda}}{T_{r,\lambda} + T_{s,\lambda}} \sim \text{beta}(r, s) \text{ independent of the sum } T_{r,\lambda} + T_{s,\lambda} \sim \text{gamma } (r+s, \lambda)$$

Higher moments: For $s > 0$ $\quad E[(T_{r,\lambda})^s] = \dfrac{\Gamma(r+s)}{\Gamma(r)\lambda^s}$

Normal approximation: If r is sufficiently large, the distribution of the standardized gamma variable $Z_{r,\lambda} = [T_{r,\lambda} - E(T_{r,\lambda})]/SD(T_{r,\lambda})$ is approximately standard normal.

Geometric and Negative Binomial

Geometric

Parameter: $p =$ success probability.

Range: $n \in \{1, 2, \ldots\}$

Definition: Distribution of the waiting time T to first success in independent trials with probability p of success on each trial.

Probability function: (1.6, 3.4)

$$P(n) = P(T = n) = (1-p)^{n-1}p \quad (n = 1, 2, \ldots)$$

Let $F = T - 1$ denote the number of failures before the first success. The distribution of F is the geometric distribution on $\{0, 1, 2, \ldots\}$.

Tail probabilities:

$$P(T > n) = P(\text{first } n \text{ trials are failures}) = (1-p)^n$$

Mean and Standard Deviation: (3.4)

$$E(T) = 1/p \qquad SD(T) = \sqrt{(1-p)}/p$$

Negative Binomial

Parameters: $p =$ success probability, $\quad r =$ number of successes.

Range: $n \in \{0, 1, 2, \ldots\}$

Definition: Distribution of the number of failures F_r before the rth success in Bernoulli trials with probability p of success on each trial.

Probability function: (3.4)

$$P(F_r = n) = P(T_r = n + r) = \binom{n+r-1}{r-1} p^r (1-p)^n \quad (n = 0, 1, \ldots)$$

where T_r is the waiting time to the rth success. The distribution of $T_r = F_r + r$ is the negative binomial distribution on $\{r, r+1, \ldots\}$.

Mean and standard deviation: (3.4)

$$E(F_r) = r(1-p)/p \qquad SD(F_r) = \sqrt{r(1-p)}/p$$

Sum of geometrics: The sum of r independent geometric (p) random variables on $\{0, 1, 2, \ldots\}$ has negative binomial (r, p) distribution.

Hypergeometric

Parameters: n = sample size
N = total population size
G = number of good elements in population

Range: $g \in \{0, 1, \ldots, n\}$

Definition: The *hypergeometric* (n, N, G) *distribution* is the distribution of the number S of good elements in a random sample of size n without replacement from a population of size N with G good elements and $B = N - G$ bad ones.

Probability function: (2.5)

$$P(g) = P(S = g) = \binom{n}{g} \frac{(G)_g (B)_b}{(N)_n} = \frac{\binom{G}{g}\binom{B}{b}}{\binom{N}{n}}$$

is the chance of getting g good elements and b bad elements in the random sample of size n. Here $b = n - g$. The random variable is

$$S = \text{number of good elements in sample} = X_1 + \cdots + X_n$$

where X_i = indicator of the event that the ith element sampled is good. These indicators are dependent due to sampling without replacement. But each indicator has the same Bernoulli(p) distribution, where

$$p = G/N = P(X_i = 1) = P(i\text{th element is good}) \quad \text{for each } i = 1, \ldots, n$$

Compare with the binomial (n, p) distribution of S for sampling with replacement, when the indicators are independent.

Mean and standard deviation: (3.6, 6.4)

$$E(S) = \mu = np \qquad SD(S) = \sigma = \sqrt{np(1-p)} \cdot \sqrt{\frac{N-n}{N-1}}$$

Note: Mean is the same as for sampling with replacement. But the SD is decreased by the *correction factor* of $\sqrt{(N-n)/(N-1)}$.

Normal approximation: As for binomial if σ is large enough, for σ as above with correction factor.

Poisson approximation: As for binomial if $p = G/N$ sufficiently small but both G and N are large.

Binomial approximation: Ignores the distinction between sampling with and without replacement. Works well if $n \ll N$ and both G and B are large.

Conditioned binomial: Let S_n be the number of successes in n independent trials which are part of a larger sequence of N independent trials. Then no matter what the probability of success p, provided it is the same on all trials, the conditional distribution of S_n given $S_N = G$ is hypergeometric (n, N, G).

Normal

Standard Normal

Range: $z \in (-\infty, \infty)$

Standard normal density function: (2.2, 4.1)

$$P(Z \in dz)/dz = \phi(z) = \frac{1}{\sqrt{2\pi}} e^{-\frac{1}{2}z^2} \quad (-\infty < z < \infty)$$

Standard normal cumulative distribution function:

$$P(Z \leq z) = \Phi(z) = \int_{-\infty}^{z} \phi(x)dx$$

No simpler formula—use a normal table (Appendix 5).

$$\Phi(-z) = 1 - \Phi(z) \qquad \text{(by symmetry and rule of complements)}$$

Mean: 0 **Standard deviation:** 1

Other moments:

$$E(Z^m) = 0, \quad m \text{ odd} \qquad \text{(by symmetry)}$$

$$E(Z^{2m}) = \frac{(2m)!}{m!2^m}, \quad m = 0, 1, 2, \ldots$$

$$E(|Z|) = \sqrt{\frac{2}{\pi}}$$

$$E(e^{tZ}) = e^{t^2/2}$$

Sources:

- Approximation to standardized sums of independent random variables (2.2, 3.3, 4.1, 5.4).
- Standardized normal (μ, σ^2) (4.1).
- Approximation to binomial, Poisson, negative binomial, gamma, beta. See summaries of these distributions for conditions under which the approximation is good.

Transformations: (5.3, 5.4) (*Notation*: $X \sim F$ means X is a random variable with distribution F.) Let $Z_1, Z_2, \ldots$ be independent standard normal.

Linear: $(Z_1 + Z_2)/\sqrt{2} \sim$ standard normal.

$$\textstyle\sum_i \alpha_i Z_i \sim \text{standard normal} \quad \text{iff} \quad \sum_i \alpha_i^2 = 1 \qquad \text{(rotational symmetry)}$$

Quadratic: $Z^2 \sim$ gamma $(1/2, 1/2)$

$$Z_1^2 + Z_2^2 + \cdots + Z_n^2 \sim \text{gamma } (n/2, 1/2) \equiv \text{chi-square } (n)$$

Ratios: $Z_1/Z_2 \sim$ Cauchy $(0, 1)$

Normal (μ, σ^2)

$$X = \mu + \sigma Z \quad \text{where} \quad Z \sim \text{normal } (0, 1)$$

Note: All formulae follow from this linear change of variable.

Mean: μ **Standard deviation:** σ

Density function: (4.1)

$$P(X \in dx)/dx = \frac{1}{\sigma}\phi((x - \mu)/\sigma) \quad (-\infty < x < \infty)$$

Cumulative distribution function:

$$P(X \le x) = \Phi\left(\frac{x - \mu}{\sigma}\right) \quad (-\infty < x < \infty)$$

Sources:

- Approximation to distribution of heights, weights, etc., over human and biological populations.
- Measurement errors.
- Random fluctuations.

Sums: (5.4)
If X_i are independent normal (μ_i, σ_i^2), then $\sum_i X_i$ is normal $\left(\sum_i \mu_i, \sum_i \sigma_i^2\right)$.

Bivariate normal (6.5)

X and Y have standard bivariate normal distribution with correlation ρ if and only if

$$Y = \rho X + \sqrt{1 - \rho^2} Z$$

where X and Z are independent standard normal variables.

Poisson

Parameter: $\mu =$ mean number.

Range: $k \in \{0, 1, 2, \ldots\}$

Probability function: (2.4, 3.5)

$$P(k) = P(N_\mu = k) = e^{-\mu}\mu^k/k! \quad (k = 0, 1, \ldots)$$

where N_μ is the number of arrivals in a given time period in a Poisson arrival process, or the number of points in a given area in a Poisson random scatter, when the expected number is μ.

Mean and standard deviation: (3.5)

$$E(N_\mu) = \mu \qquad SD(N_\mu) = \sqrt{\mu}$$

Sources:

- Poisson process.
- Approximation to binomial as $p \to 0$ with $\mu = np$ (2.4).
- From independent exponential variables $W_1, W_2, \ldots$ with rate 1. Let

$$N_\mu = \{\text{first } n \text{ such that } W_1 + \cdots + W_n > \mu\} - 1$$

Transformations: (3.5)

Sums: Let $N_{\mu_1}, N_{\mu_2}, \ldots, N_{\mu_n}$ be independent Poisson. Their sum has Poisson distribution with parameter $\sum_{i=1}^{n} \mu_i$.

Poissonization of the binomial: (3.5)

If $S_N =$ number of successes and $F_N =$ number of failures in N trials, where N has Poisson (μ) distribution, and given N the trials are independent with probability p of success on each trial, then S_N and F_N are independent with

$$S_N \sim \text{Poisson } (\mu p) \qquad F_N \sim \text{Poisson } (\mu q) \qquad p + q = 1$$

Notation: $X \sim F$ means X is a random variable having distribution F. Similarly for multinomial trials. A consequence is the following:

Binomial distribution of Poisson terms given their sum:

If N_α and N_β are independent Poisson variables with mean α and β, then the conditional distribution of N_α given $N_\alpha + N_\beta = n$ is binomial $\left(n, \dfrac{\alpha}{\alpha + \beta}\right)$

Normal approximation: Good for large μ.

Uniform

Uniform distribution on a finite set: (1.1)

This is the distribution of a point picked at random from a finite set Ω, so that all points are equally likely to be picked. For $A \subset \Omega$,

$$P(A) = \frac{\#(A)}{\#(\Omega)}$$

Special cases:

- Bernoulli (1/2) distribution: uniform distribution on $\{0, 1\}$.
- The number on a fair-die roll: uniform distribution on $\{1, 2, 3, 4, 5, 6\}$.
- Uniform distribution on $\{1, 2, \ldots, n\}$: Let X have uniform distribution on the integers 1 to n. Then

$$E(X) = \frac{n+1}{2} \qquad SD(X) = \sqrt{\frac{n^2 - 1}{12}}$$

Uniform distribution on an interval: (4.1)

Parameters: $a < b$, endpoints of the interval.

Range: $x \in (a, b)$ (or (a, b), or $(a, b]$ or $[a, b)$—the endpoints don't matter.)

Density function:

$$P(X \in dx)/dx = 1/(b - a) \quad a \leq x \leq b$$

The probability of any subinterval of (a, b) is proportional to the length of the subinterval.

Cumulative distribution function:

$$P(X \leq x) = \begin{cases} 0 & \text{if } x < a \\ (x - a)/(b - a) & \text{if } a \leq x \leq b \\ 1 & \text{if } x > b \end{cases}$$

Mean and standard deviation:

$$E(X) = (a + b)/2 \qquad SD(X) = \sqrt{\frac{(b - a)^2}{12}}$$

Transformations:

Linear: (4.4)

If X has uniform (a, b) distribution, then for constants $c > 0$ and d, the random variable $Y = cX + d$ has uniform distribution on $(ca + d, cb + d)$.

Cumulative distribution functions: (4.5)

Let X be a random variable with continuous c.d.f. F. Then the random variable $F(X)$ has uniform $(0, 1)$ distribution.

Inverse cumulative distribution functions:

Let U have uniform distribution on $(0, 1)$, and let F be any cumulative distribution function. Then $F^{-1}(U)$ is a random variable whose c.d.f. is F.

This allows random numbers with any given distribution to be generated from uniform $(0, 1)$ random numbers.

Order statistics: (4.6)

The kth order statistic of n independent uniform $(0, 1)$ random variables has beta distribution with parameters k and $n - k + 1$.

Sums: (5.4)

The density of the sum of n independent uniform $(0, 1)$ random variables is defined by polynomials of degree $n - 1$ on each of the intervals $[0, 1)$, $[1, 2)$, $\ldots$, $(n - 1, n)$.

For $n \geq 5$, the distribution of the standardized sum is very well approximated by the standard normal distribution.

Uniform distribution on a region in the plane (5.1)

A random point (X, Y) has uniform distribution on a region D of the plane, where D has finite area, if:

(i) (X, Y) is certain to lie in D

(ii) the chance that (X, Y) falls into a subregion C of D is proportional to the area of C:

$$P[(X, Y) \in C] = \text{area}(C)/\text{area}(D) \quad C \subset D$$

If X and Y are independent random variables, each uniformly distributed on an interval, then (X, Y) is uniformly distributed on the rectangle (range of X) $\times$ (range of Y).

Suppose (X, Y) is uniformly distributed on a region D in the plane. Then given $X = x$, Y has uniform distribution on the values of Y which are possible when $X = x$.

Uniform distributions on regions in three or higher dimensions have similar properties, with volumes replacing lengths and areas.

Examinations

Midterm Examination 1 (1 hour)

1. Ten dice are rolled. Five dice are red and five are green. Write down numerical expressions for:

a) The probability of the event that exactly four of the ten dice are sixes.

b) The probability of the event that exactly two of the red dice are sixes and exactly three of the green dice show even numbers.

c) The probability that there are the same number of sixes among the red dice as among the green dice.

d) The probability that there are strictly more sixes among the red dice than among the green dice.

2. Five cards are dealt from a standard deck of 52. Write down numerical expressions for

a) The probability that the third card is an ace.

b) The probability that the third card is an ace given that the last two cards are not aces.

c) The probability that all cards are of the same suit.

d) The probability of two or more aces.

3. A student takes a multiple choice examination where each question has 5 possible answers. He works a question correctly if he knows the answer, otherwise he guesses at random. Suppose he knows the answer to 70% of the questions.

a) What is the probability that on a question chosen at random the student gets the correct answer?

b) Given that the student gets the correct answer to this question chosen at random, what is the probability that he actually knew the answer?

Suppose there are 20 questions on the examination. Let N be the number of questions that the student gets correct.

c) Find $E(N)$. d) Find $SD(N)$.

4. Let A, B, and C be events which are mutually independent, with probabilities a, b, and c. Let N be the random number of events which occur.

a) What is the event $(N = 2)$ in terms of A, B and C?

b) What is the probability of this event in terms of a, b, and c?

c) What is $E(N)$ in terms of a, b, and c?

d) What is $SD(N)$ in terms of a, b, and c?

5. Let X_2 and X_3 be indicators of independent events with probabilities $\frac{1}{2}$ and $\frac{1}{3}$, respectively.

a) Display the joint distribution table of $X_2 + X_3$ and $X_2 - X_3$.

b) Calculate $E(X_2 - X_3)$.

c) Calculate $SD(X_2 - X_3)$.

Midterm Examination 2 (1 hour)

1. **Coin spinning.** I have two coins. One shows heads with probability $1/10$ when spun. The other shows heads with probability $1/2$. Suppose you pick one of my two coins at random and spin it twice. Find:

   a) $P(\text{heads on first spin})$;

   b) $P(\text{heads on second spin})$;

   c) $P(\text{heads on both spins})$;

   d) the probability that the coin is the $1/2$ coin given heads on both spins.

2. **True or false.** A student answers a set of 100 true/false questions by answering 36 questions correctly, and guessing the other 64 at random.

   a) If the pass mark is 70 questions correct, what is the student's chance of passing? Give your answer as a decimal correct to two places.

   b) Another student also knows 36 correct answers and guesses the rest at random. What is the chance that just one of these two students passes?

3. **Rare white balls**. A box contains 998 black and 2 white balls. Let $X =$ the number of whites in 500 random draws *with* replacement from this box. Calculate:

   a) $P(X = 1)/P(X = 2)$;

   b) $P(X = 1 \text{ given } X = 1 \text{ or } 2)$;

   c) repeat b) assuming draws without replacement.

4. **Reliability**. A system consists of four components which work independently with probabilities 0.9, 0.8, 0.7, and 0.6. Let $X =$ the number of components that work. Find:

   a) $E(X)$;

   b) $SD(X)$;

   c) $P(X > 0)$;

   d) $P(X = 2)$.

Final Examination 1 (3 hours)

1. A random variable N is uniformly distributed on $\{1, 2, \ldots, 10\}$. Let X be the indicator of the event $(N \leq 5)$ and Y the indicator of the event (N is even).

   a) Find $E(X)$ and $E(Y)$.

   b) Are X and Y independent?

   c) Find $Cov(X, Y)$.

   d) Find $E\big[(X+Y)^2\big]$.

2. A box contains 5 tickets. An unknown number of them are red, the rest are green. Suppose that to start off with you think there are equally likely to be 0, 1, 2, 3, 4, or 5 red tickets in the box.

   a) Three tickets are drawn from the box with replacement between draws. The tickets drawn are red, green, and red. Given this information, what is the chance that there are actually 3 red tickets in the box?

   b) What would your answer to (a) be if you knew the draws were made without replacement?

3. In the World Series, two teams play a series of games, and the first team to win four games wins the series. Suppose that each game ends in either a win or a loss for your team, and that for each game that is played the chance of a win for your team is p, independently of what happens in other games. What is the probability that your team wins the series?

4. Let X, Y, and Z be three independent normal $(0, 1)$ random variables. Calculate:

   a) $P(|X| \leq 1, |Y| \leq 2, |Z| \leq 3)$;

   b) $E\left[(X+Y+Z)^2\right]$;

   c) $P(X + Y \leq 2Z)$.

5. Suppose that T is a random variable such that $P(T > t) = e^{-t}, \quad t \geq 0$.

   a) Find a formula for the probability density function f_X of the random variable $X = 1/T$.

   b) What is the value of $E(X)$?

6. A fair coin is tossed 100 times. The probability of getting *exactly* 50 heads is close to one of the following numbers.

   $$0.001, \quad 0.01, \quad 0.1, \quad 0.5, \quad 0.9, \quad 0.99, \quad 0.999$$

   a) Circle which number you think is closest and explain your choice.

   b) How many times do you have to toss the coin to make the probability of getting exactly as many heads as tails very close to one tenth of this probability of getting 50 heads in 100 tosses?

7. A pair of dice is rolled n times, where n is chosen so that the chance of getting at least one double six in the n rolls is very close to $1/2$.

a) The number of rolls n must be very close to one of the following numbers:

$$6, \quad 12, \quad 18, \quad 20, \quad 25, \quad 30, \quad 36, \quad 50, \quad 72, \quad 100.$$

Circle which number you think n must be close to, and explain your choice.

b) What, approximately, is the chance that you actually get two or more double sixes in this many rolls? Give your answer as a decimal.

8. Let U_1 and U_2 be two independent uniform $[0,1]$ random variables. Let

$$X = \min(U_1, U_2)$$

$$Y = \max(U_1, U_2)$$

where $\min(u_1, u_2)$ is the smaller and $\max(u_1, u_2)$ the larger of two numbers u_1 and u_2. Find:

a) the probability density function f_X of X;

b) the joint density function $f_{X,Y}$ of (X, Y);

c) $P(X \leq 1/2 | Y \geq 1/2)$.

9. Suppose that on average one person in a hundred has a particular genetic defect, which can be detected only by a laboratory test.

a) Suppose fifty people chosen at random are tested. What is the probability that at least one of them will have the defect? [Answer as a decimal.]

b) About how many people have to be tested in order for the probability to be at least 99% that at least one person has the defect?

c) If this number of people are tested, what is the expected number of individuals with the defect?

10. Let $U_1, U_2, \ldots, U_n$ be independent uniform $[0,1]$ random variables. If n is large the geometric mean G_n of $U_1, U_2, \ldots, U_n$, defined by $G_n = (U_1 U_2 \ldots U_n)^{1/n}$, is most likely to be very close to a certain number g. Explain why, and find g.
[*Hint:* Use logarithms.]

Final Examination 2 (3 hours)

1. Suppose one morning you pick two eggs for lunch at random from a dozen eggs in your refrigerator, thinking that they are all hard-boiled. You then learn that in fact four of the eggs have not been hard-boiled.

a) What is the probability that your two lunch eggs are both hard-boiled?

b) Given that you crack one of your lunch eggs and find it is hard-boiled, what is the probability that the second egg is hard boiled?

2. A hat contains n coins, f of which are fair, and b of which are biased to land with heads with probability $2/3$, with $f+b=n$. A coin is drawn at random from the hat and tossed once. It lands heads. What is the probability that it is a biased coin?

3. A die has one spot painted on one face, two spots painted on each of two faces, and three spots painted on each of three faces. The die is rolled twice.

a) Calculate the distribution of the sum S_2 of the numbers on the two rolls. Display your answer in a table.

b) Calculate the numerical value of $E(S_2)$ in two different ways to check your answer to a).

c) Calculate the standard deviation of S_2.

4. Suppose the average family income in a particular area is $\$10,000$.

a) Find an upper bound for the fraction of families in the area with incomes over $\$50,000$.

b) Find a smaller upper bound than in a), given that the standard deviation is $\$8000$.

c) Do you think the normal approximation would give a good estimate for the fraction in question?

5. A random variable X has probability density function of the form

$$f_X(x) = \begin{cases} cx^2, & 0 \leq x \leq 1 \\ 0 & \text{otherwise.} \end{cases}$$

a) Find the constant c.

b) Find $P(X \leq a)$ for $0 \leq a \leq 1$.

c) Calculate $E(X)$.

d) Calculate $SD(X)$.

6. Telephone calls arrive at an exchange at an average rate of one every second. Find the probabilities of the following events, explaining briefly your assumptions.

a) No calls arriving in a given five-second period.

b) Between four and six calls arriving in the five-second period.

c) Between 90 and 110 calls arriving in a 100-second period. (Give answer as a decimal.)

7. Let T be the number of times you have to roll a die before each face has appeared at least once. Let N be the number of different faces appearing in the first six rolls. Calculate:

a) $E(T)$;

b) $E(N)$;

c) $E(T|N = 3)$.

8. Let X, Y, and Z be independent standard normal random variables. Find the probability density functions of each of the following random variables:

a) X^2;

b) $X^2 + Y^2$;

c) $X + Y + Z$.

9. A floor is ruled with equally spaced parallel lines. A needle is such that if its two ends are placed on adjacent lines the angle between the needle and the lines is α, where $0 \leq \alpha \leq \pi/2$. Calculate the probability that the needle crosses at least one of the lines when tossed at random on the floor:

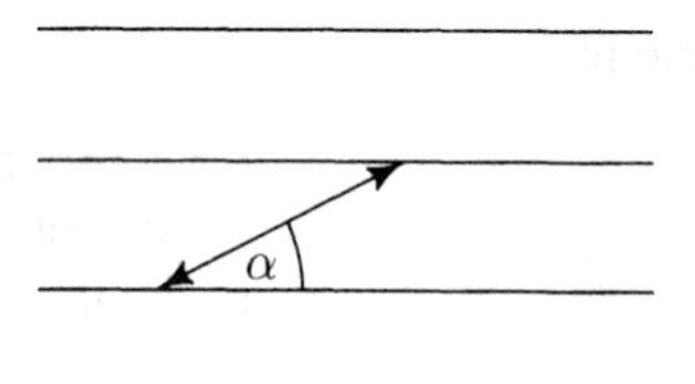

a) for $\alpha = 45°$;

b) for a general α between 0 and $\pi/2$.

10. A fair coin is tossed $2n$ times. Let p_{2n} be the probability of getting the same number of heads as tails.

a) Find constants a and b such that

$$p_{2n} \sim \frac{1}{an^b} \quad \text{as } n \to \infty.$$

b) Show that $p_{2n} \to 0$ as $n \to \infty$.

c) Why does this not contradict the law of large numbers?

Final Examination 3 (3 hours)

1. Suppose you try 5 times to hit the bull's eye. The first time you have a 0.2 chance of a hit, and each time you try your chance of hitting increases by 0.1. Let H be the number of hits in the five attempts. Assuming your attempts are independent, calculate the following quantities. Answers should be decimals.

a) $E(H)$; b) $Var(H)$.

2. Suppose that in a network of 3 computers, at a given time the event that the kth computer is down has overall probability p_k, $k = 1, 2, 3$. Calculate the probability that at this time there is at least one computer up:

a) assuming the computers are up or down independently of each other;

b) assuming that there is probability p of power failure, in which case all the computers are down, but given that there is no power failure the computers are up or down independently of each other.

3. A fair six-sided die has:

the 1 spot face opposite the 6 spot face;
the 2 spot face opposite the 5 spot face;
the 3 spot face opposite the 4 spot face.

Suppose the die is rolled once. Let X be the number of spots showing on top, Y the number of spots showing on one of its side faces, say the leftmost face from a particular point of view.

a) Display the joint probability distribution of X and Y in a suitable table.

b) Are X and Y independent?

c) Find $Cov(X, Y)$.

d) Find $Var(X + Y)$.

4. Suppose there are 50 married couples. After some years, 20 of these 100 people have died. Regard the 20 deaths as striking the 100 people at random. Find numerical expressions for:

a) the probability that a particular couple has survived;

b) the expected number of couples surviving;

c) the probability that two particular couples have survived;

d) the variance of the number of couples surviving.

5. Suppose X and Y are independent random variables, each uniformly distributed on $[0, 1]$. Calculate:

a) $P(X^2 + Y^2 \leq 1)$; b) $P(Y^2 > 3X^2)$; c) $P(X^2 + Y^2 \leq 1 \text{ given } Y^2 > 3X^2)$.

6. Suppose a particle has velocity V which is normally distributed with mean 0 and variance σ^2. Let $X = mV^2/2$ where $m > 0$ is a positive constant. Find formulae in terms of m and σ for;

a) $E(X)$; b) the probability density function of X; c) $Var(X)$.

7. A particle counter records two types of particles, Types 1 and 2. Type 1 particles arrive at an average rate of 1 per minute, Type 2's at an average rate of 2 per minute. Assume these are two independent Poisson processes. Give numerical expressions for the following probabilities:

a) Three Type 1 particles and four of Type 2's arrive in a two-minute period;

b) the total number of particles of either type in a two-minute period is 5;

c) the fourth particle arrives in the first 5 minutes;

d) the first particle to arrive is of Type 1;

e) the second particle of Type 1 turns up before the third of Type 2.

8. Consider the average $\bar{X}_n = (X_1 + X_2 + \cdots + X_n)/n$ of n independent random variables, each uniformly distributed on $[0, 1]$. Find n so that $P(\bar{X}_n < 0.51)$ is approximately 90%.

9. Two statisticians are watching a sequence of independent Bernoulli trials with probability p of success on each trial. The first statistician estimates p by the proportion of successes in the first 100 trials. The second statistician estimates p by the proportion of successes in the next 300 trials. Consider the probability that the second estimate is closer to p than the first.

a) Explain why this probability hardly depends at all on p, provided p is fairly close to $1/2$.

b) Assuming p is fairly close to $1/2$, this probability is very close to one of the following numbers:

$$0, \quad 1/10, \quad 1/5, \quad 1/4, \quad 1/3, \quad 1/2, \quad 2/3, \quad 3/4, \quad 4/5, \quad 9/10, \quad 1.$$

Which one, and why?

10. Suppose 10 dice are shaken together and rolled. Any that turn up six are set aside. The remaining dice are shaken and rolled again. Any of these that turn up six are set aside. And so on, until all the dice show six. Let N be the number of times the dice are shaken and rolled. To illustrate, if after the first roll of 10 dice, 7 non-sixes remain, and after the second roll of these 7 dice 2 non-sixes remain, and after the third roll of these 2 dice no non-sixes remain, then $N = 3$.

a) Describe the distribution of N.

[*Hint:* Consider the number of times each die is rolled.]

b) Let T be the total number of individual die rolls. To illustrate, $T = 10 + 7 + 2 = 19$ for the outcome described above. Describe the distribution of T.

c) Let L be the number of dice shaken on the last roll. To illustrate, $L = 2$ for the outcome described above. Describe the distribution of L.

Midterm Examination 1—Solutions

1. a) $\binom{10}{4}(1/6)^4(5/6)^6$

b) $\binom{5}{2}(1/6)^2(5/6)^3\binom{5}{3}(1/2)^5$

c) $\binom{5}{0}^2(5/6)^{10}+\binom{5}{1}^2(1/6)^2(5/6)^8+\binom{5}{2}^2(1/6)^4(5/6)^6+\cdots+\binom{5}{5}^2(1/6)^{10}$

d) $\frac{1}{2}$ (1 – answer to c))

2. a) 1/13 b) 4/50

c) $\dfrac{4\times\binom{13}{5}}{\binom{52}{5}}$ d) $1-\dfrac{\binom{4}{0}\binom{48}{5}+\binom{4}{1}\binom{48}{4}}{\binom{52}{5}}$

3. a) $0.7+(0.3)(0.2)=0.76$ b) $\dfrac{0.7}{0.76}$

c) 20×0.76 d) $\sqrt{6\times\frac{1}{5}\times\frac{4}{5}}$

4. a) $ABC^c\cup AB^cC\cup A^cBC$

b) $ab(1-c)+a(1-b)c+(1-a)bc$

c) $a+b+c$

d) $\sqrt{a(1-a)+b(1-b)+c(1-c)}$

5. a)

X_2-X_3	X_2+X_3 = 0	1	2
−1	0	1/6	0
0	1/3	0	1/6
1	0	1/3	0

b) $\dfrac{1}{6}$

c) $\sqrt{\frac{1}{2}\cdot\frac{1}{2}+\frac{1}{3}\cdot\frac{2}{3}}$

Midterm Examination 2—Solutions

1. a) $P(H_1) = P(1/10 \text{ coin})P(H_1 \mid 1/10 \text{ coin}) + P(1/2 \text{ coin})P(H_1 \mid 1/2 \text{ coin})$
$= \frac{1}{2} \times \frac{1}{10} + \frac{1}{2} \times \frac{1}{2} = \frac{3}{10}$

b) $P(H_2) = P(H_1) = \frac{3}{10}$

c) $P(H_1H_2) = P(1/10 \text{ coin})P(H_1H_2 \mid 1/10 \text{ coin}) + P(1/2 \text{ coin})P(H_1H_2 \mid 1/2 \text{ coin})$
$= \frac{1}{2} \times \left(\frac{1}{10}\right)^2 + \frac{1}{2} \times \left(\frac{1}{2}\right)^2 = \frac{13}{100}.$

d) $P(1/2 \text{ coin} \mid H_1H_2) = \dfrac{P(1/2 \text{ coin})P(H_1H_2 \mid 1/2 \text{ coin})}{P(H_1H_2)} = \dfrac{\frac{1}{2} \times \left(\frac{1}{2}\right)^2}{\frac{13}{100}} = \dfrac{25}{26}$

2. a) Let X be the number of correct answers in 64 questions; then X has binomial ($n = 64$, $p = 1/2$) distribution, so $EX = 32$, $SD(X) = 4$.

$$P(\text{passing}) = P(X \geq 34) = P(X \geq 33.5) = P\left(\frac{X - 32}{4} \geq \frac{33.5 - 32}{4}\right)$$
$$\approx 1 - \Phi(0.375) = 1 - 0.646 = 0.354$$

b) This is just $2pq$ for $p = 0.354$, $q = 0.646$, i.e.,

$$2pq = 2 \times 0.354 \times 0.646 = 0.457$$

3. a) X has binomial ($n = 500, p = 2/1000$) distribution, so

$$\frac{P(X = 1)}{P(X = 2)} = \frac{2}{n - 2 + 1} \cdot \frac{1 - p}{p} = \frac{2}{499} \cdot \frac{\frac{998}{1000}}{\frac{2}{1000}} = 2$$

b) $P(1 \mid 1 \text{ or } 2) = \dfrac{P(X = 1)}{P(X = 1) + P(X = 2)} = \dfrac{P(X = 1)/P(X = 2)}{P(X = 1)/P(X = 2) + 1} = \dfrac{2}{3}$

c) Now $\dfrac{P(X = 1)}{P(X = 2)} = \dfrac{\binom{2}{1}\binom{998}{499}}{\binom{2}{2}\binom{998}{498}} = 2 \times \dfrac{498!500!}{499!499!} = 2 \times \dfrac{500}{499} = \dfrac{1000}{499}$

Continue as before: $P(X = 1 \mid 1 \text{ or } 2) = \dfrac{(1000/499)}{(1000/499) + 1} = \dfrac{1000}{1499}$

4. a) $X = X_1 + X_2 + X_3 + X_4$, where X_i is the indicator that the ith component works.
So $E(X) = \sum_{i=1}^4 P(X_i = 1) = 0.9 + 0.8 + 0.7 + 0.6 = 3.0$

b) $Var(X) = \sum_{i=1}^4 Var(X_i) = 0.9 \times 0.1 + 0.8 \times 0.2 + 0.7 \times 0.3 + 0.6 \times 0.4 = 0.7$
$SD(X) = 0.8367$

c) $1 - P(X = 0) = 1 - 0.1 \times 0.2 \times 0.3 \times 0.4 = 0.9976$

d) $0.9 \times 0.8 \times 0.3 \times 0.4 + 0.9 \times 0.2 \times 0.7 \times 0.4 + 0.9 \times 0.2 \times 0.3 \times 0.6 + 0.1 \times 0.8 \times 0.7 \times 0.4 + 0.1 \times 0.8 \times 0.3 \times 0.6 + 0.1 \times 0.2 \times 0.7 \times 0.6 = 0.2144$

Final Examination 1—Solutions

1. a) $E(X) = E(Y) = 1/2$.

b) No.

c) $-1/20$.

d) 1.4.

2. a) 0.36 b) 0.4

3.

$$\binom{7}{4}p^4q^3 + \binom{7}{5}p^5q^2 + \binom{7}{6}p^6q + p^7 = p^4 + 4p^4q + \binom{5}{2}p^4q^2 + \binom{6}{3}p^4q^3,$$

where $q = 1 - p$.

4. a) 0.65 b) 3 c) 1/2

5. a) $f_X(x) = \dfrac{e^{-1/x}}{x^2}, \quad x > 0, \quad 0$ otherwise.

b) $E(X) = \infty$.

6. a) 0.1, by normal approximation or Stirling's formula.

b) 10^4 tosses.

7. a) 25.

b) 0.152 (Poisson approximation)

8. a) $f_X(x) = 2 - 2x, \quad 0 \le x \le 1, \quad 0$ otherwise.

b) $f_{X,Y}(x, y) = 2, \quad 0 \le x \le y \le 1, \quad 0$ otherwise.

c) 2/3

9. a) $1 - e^{-1/2}$.

b) $100 \log 100$.

c) $\log 100$.

10. e^{-1}, by law of large numbers.

Final Examination 2—Solutions

1. a) $\dfrac{\binom{8}{2}\binom{4}{0}}{\binom{12}{2}} = \dfrac{56}{132} = \dfrac{14}{33} = 0.424$ (sampling without replacement) b) 7/11

2. $\dfrac{4b}{4b+3f} = \dfrac{\frac{2}{3}b}{\frac{2}{3}b+\frac{1}{2}f}$ (Bayes' rule)

3. a)

2	3	4	5	6
1/36	4/36	10/36	12/36	9/36

b) $\frac{14}{3}$ c) $\frac{\sqrt{10}}{3}$

4. a) $\frac{10{,}000}{50{,}000} = \frac{1}{5}$ (Markov's inequality)

b) $\left(\frac{8{,}000}{40{,}000}\right)^2 = \frac{1}{25}$ (Chebychev's inequality)

c) No, because income ≥ 0.

5. a) 3 b) a^3 c) $\frac{3}{4}$ d) $\sqrt{\frac{3}{80}} \approx 0.194$

6. a) e^{-5} (Poisson process) b) $e^{-5}\left(\dfrac{5^4}{4!}+\dfrac{5^5}{5!}+\dfrac{5^6}{6!}\right)$ c) 0.68 (normal approximation)

7. a) $1+\frac{6}{5}+\frac{6}{4}+\cdots+6 \approx 14.7$

b) $6\left(1-\left(\frac{5}{6}\right)^6\right)$

c) $6+\frac{6}{3}+\frac{6}{2}+6 = 17$

8. a) gamma $(1/2, 1/2)$ b) gamma $(1, 1/2)$ c) normal $(0, 3)$

9. a) $\dfrac{2}{\pi}\left(\sqrt{2}-1\right)+\dfrac{1}{2}$ b) $\dfrac{2}{\pi \sin\alpha}(1-\cos\alpha)+1-\dfrac{2\alpha}{\pi}$

10. a) $a = \sqrt{\pi}$, $b = 1/2$

b) Follows from $\frac{1}{\sqrt{n}} \to 0$ as $n \to \infty$

c) The law of large numbers says that the proportion of heads is very likely to be very close to 1/2, not that it is very likely to be *exactly* 1/2.

Final Examination 3—Solutions

1. a) $0.2 + 0.3 + 0.4 + 0.5 + 0.6 = 2.0$ b) $0.2 \times 0.8 + 0.3 \times 0.7 + 0.4 \times 0.6 + 0.5 \times 0.5 + 0.6 \times 0.4 = 1.1$

2. a) $1 - P(\text{all down}) = 1 - p_1p_2p_3$

b) Answer is $(1-p)(1-r_1r_2r_3)$, where r_i is the conditional probability that computer i is down given no power failure. But $p_i = p + (1-p)r_i$, so $r_i = (p_i - p)/(1-p)$, and the answer is

$$(1-p)\left[1 - \frac{(p_1-p)(p_2-p)(p_3-p)}{(1-p)^3}\right]$$

3. a)

Values of X

Values of Y	1	2	3	4	5	6
1	0	1/24	1/24	1/24	1/24	0
2	1/24	0	1/24	1/24	0	1/24
3	1/24	1/24	0	0	1/24	1/24
4	1/24	1/24	0	0	1/24	1/24
5	1/24	0	1/24	1/24	0	1/24
6	0	1/24	1/24	1/24	1/24	0

b) No.

c) $Cov(X, Y) = 0$.

d) $Var(X+Y) = Var(X) + Var(Y) = \frac{35}{12} + \frac{35}{12} = \frac{35}{6} = 5.833$

4. a) $\frac{80}{100} \times \frac{79}{99} = 0.638$

b) Let I_i = indicator that couple i has survived. Then

$$E(\#\text{ couples surviving}) = E\sum_{i=1}^{50} I_i = \sum_{i=1}^{50} E(I_i) = 50 \times 0.638 = 31.9$$

c) $\frac{80}{100} \times \frac{79}{99} \times \frac{78}{98} \times \frac{77}{97} = 0.4033 = E(I_1I_2)$

d) $Var(S_{50}) = \sum_{i=1}^{50} Var(I_i) + 2\sum_{j<k} Cov(I_j, I_k)$; here

$$Var(I_i) = E(I_1^2) - [E(I_1)]^2 = 0.638 - 0.638^2 = 0.23085,$$
$$Cov(I_j, I_k) = Cov(I_1, I_2) = E(I_1I_2) - E(I_1)E(I_2) = 0.4033 - 0.638^2 = -0.004196,$$

so $Var(S_{50}) = 50Var(I_1) + 50 \cdot 49 Cov(I_1, I_2) = 50 \cdot (0.23085) + 2450(-0.004196)$.

5. a) $P(X^2 + Y^2 \leq 1)$

$$= \frac{\text{area of shaded region}}{\text{area of square}} = \frac{\frac{1}{4}\pi(1)^2}{1^2} = \frac{\pi}{4} = 0.785$$

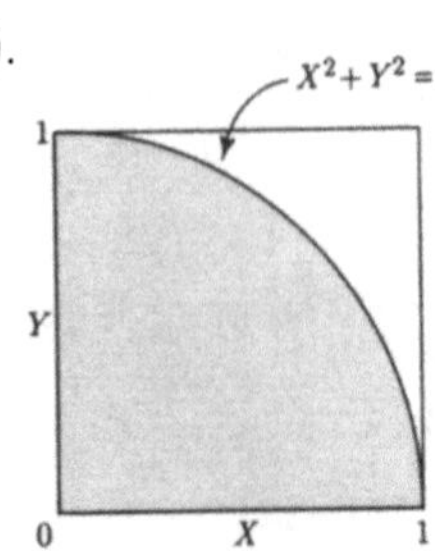

b) $P(Y^2 > 3X^2)$

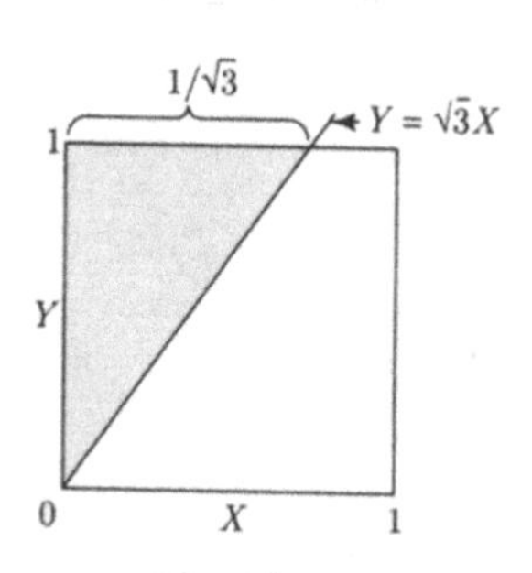

$$= \frac{\text{area of shaded region}}{\text{area of square}}$$

$$= \frac{\frac{1}{2}(1)\left(\frac{1}{\sqrt{3}}\right)}{1} = \frac{1}{2\sqrt{3}} = 0.2887$$

c) $\tan\theta = \frac{1}{\sqrt{3}}$, so $\theta = 30°$, and

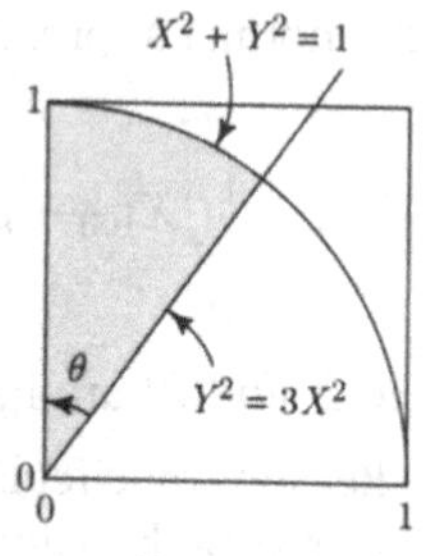

$$P(X^2 + Y^2 \leq 1 \text{ given } Y^2 > 3X^2)$$

$$= \frac{\text{area of shaded region}}{\text{area of shaded region in (b)}}$$

$$= \frac{\frac{1}{4}\pi\left(\frac{1}{3}\right)}{\frac{1}{2\sqrt{3}}} = \frac{\sqrt{3}\pi}{6} = \frac{\pi}{2\sqrt{3}} = 0.91$$

6. a) V^2 has gamma $\left(\frac{1}{2}, \frac{1}{2\sigma^2}\right)$ distribution, so

$$E(V^2) = \frac{\frac{1}{2}}{\frac{1}{2\sigma^2}} = \sigma^2 \text{ and } Var(V^2) = \frac{\frac{1}{2}}{\left(\frac{1}{2\sigma^2}\right)^2} = 2\sigma^4.$$

So $E(X) = E\left(\frac{1}{2}mV^2\right) = \frac{1}{2}m\sigma^2$.

b) $X = \frac{1}{2}mV^2$; $V = \sqrt{\frac{2X}{m}}$;

$$\frac{dV}{dX} = \sqrt{\frac{2}{m}}\left(\frac{1}{2\sqrt{X}}\right) = \frac{1}{\sqrt{2mX}}$$

$$f_V(v) = \frac{1}{\sqrt{2\pi}\sigma}e^{-\frac{v^2}{2\sigma^2}}$$

$$\Rightarrow \quad f_X(x) = 2f_V(v)\left|\frac{dv}{dx}\right| = 2 \cdot \frac{1}{\sqrt{2\pi}\sigma}e^{\frac{-\left(\frac{2x}{m}\right)}{2\sigma^2}}\left(\frac{1}{\sqrt{2mx}}\right) = \frac{1}{\sqrt{\pi m x}\sigma}e^{-x/m\sigma^2}$$

c) By part a), $Var(X) = Var\left(\frac{1}{2}mV^2\right) = \frac{m^2}{4}2\sigma^4 = m^2\sigma^4/2$

7. a) $P(3 \text{ of Type } 1)P(4 \text{ of Type } 2) = e^{-2}\frac{2^3}{3!} \cdot e^{-4}\frac{4^4}{4!}$

b) $e^{-6}6^5/5!$, since the total number of particles of either type in a two-minute period has Poisson $[2(1+2)]$ distribution.

c) $P(T_4 \leq 5) = P(N_5 \geq 4) = 1 - P(N_5 < 4) = 1 - e^{-15}\left(1 + 15 + \frac{15^2}{2!} + \frac{15^3}{3!}\right)$.

d) 1/3

e) $P(\text{3rd of Type 2 after 2nd of Type 1})$
$= P(\text{3rd of Type 2 at or after 5th of either type})$
$= P(\text{0 or 1 or 2 Type 2's in first 4 particles})$
$= (1/3)^4 + 4(1/3)^3 \cdot (2/3) + 6(1/3)^2 \cdot (2/3)^2 = \frac{1+8+24}{81} = \frac{11}{27}$

8. $\bar{X}_n$ has approximately normal $\left(0.5, \frac{1}{12n}\right)$ distribution. Want n so that

$$0.9 = P(\bar{X}_n < 0.51) \approx \Phi\left(\frac{0.01}{\sqrt{1/12n}}\right)$$

$$\iff \frac{0.01}{\sqrt{1/12n}} \approx 1.29$$

$$\iff \sqrt{12n} \approx \frac{1.29}{0.01} \iff n \approx \left(\frac{1.29}{0.01}\right)^2 \cdot \frac{1}{12} \iff n \approx 1387$$

9. a) By the normal approximation

$$P\left(|\bar{X}_{100} - p| > |\bar{Y}_{300} - p|\right) = P\left(\frac{|\bar{X}_{100} - p|}{\frac{\sqrt{pq}}{10}} > \frac{|\bar{Y}_{300} - p|}{\frac{\sqrt{pq}}{\sqrt{3}\cdot 10}} \frac{1}{\sqrt{3}}\right) \approx P\left(|Z| > \frac{|Z'|}{\sqrt{3}}\right)$$

where Z and Z' are independent normal $(0,1)$.

b) 2/3. Reason: by circular symmetry of (Z, Z'), the desired probability is

$$\frac{\arctan(\sqrt{3})}{\pi/2} = \frac{\pi/3}{\pi/2} = \frac{2}{3}$$

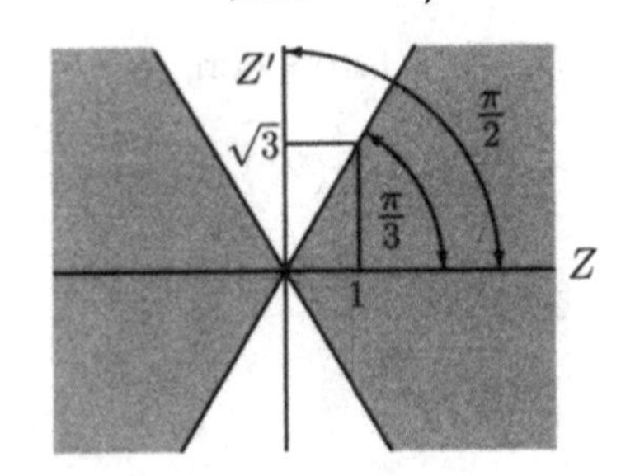

10. a) Let W_i be the number of times die i is rolled. Then $N = \max_i W_i$, and the W_i are independent with geometric $(1/6)$ distribution, so

$$P(N \le n) = P(W_i \le n \text{ for } i = 1 \text{ to } 10) = P(W \le n)^{10} = \left[1 - \left(\frac{5}{6}\right)^n\right]^{10}$$

$$P(N = n) = P(N \le n) - P(N \le n - 1).$$

b) $T = W_1 + \cdots + W_{10}$, so $T - 10$ has negative binomial $(10, 1/6)$ distribution

$$P(T = t) = \binom{t-1}{9}\left(\frac{1}{6}\right)^{10}\left(\frac{5}{6}\right)^{t-10}$$

c) Use $P(L = l) = \sum_{k=1}^{\infty} P(N = k, L = l)$ with

$$P(N = k, L = l) = P(10 - l \text{ dice have fallen 6 by roll } k - 1, l \text{ dice fall 6 on roll } k)$$

$$= \binom{10}{l}[P(W \le k-1)]^{10-l}[P(W = k)]^l$$

$$= \binom{10}{l}\left[1 - \left(\frac{5}{6}\right)^{k-1}\right]^{10-l}\left[\frac{1}{6}\left(\frac{5}{6}\right)^{k-1}\right]^l$$

Appendices

Appendix 1
Counting

Basic Rules

Let $\#(B)$ denote the number of elements in a finite set B. There are three basic rules to help evaluate $\#(B)$, the *correspondence rule*, the *addition rule*, and the *multiplication rule*. The first of these is the basis of counting on your fingers:

The Correspondence Rule

If the elements of B can be put in one-to-one correspondence with the elements of another set C, then $\#(B) = \#(C)$.

The trick to using this rule is to find a one-to-one correspondence between a set you are trying to count, and some other set you already know how to count. See examples below.

The Addition Rule

If B can be split into disjoint sets $B_1, B_2, \ldots, B_n$, then

$$\#(B) = \#(B_1) + \#(B_2) + \cdots + \#(B_n).$$

The multiplication rule is generally less applicable, but nonetheless extremely useful. The number of elements of a set B is the number of different ways in which an element B may be chosen. In many problems it is possible to regard the choice of an element in B as being made by stages. For example, if B is a set of sequences, the choice of a sequence in B can be made by choosing the first element of the sequence, then the second element, and so on. The number of elements in B is then equal to the number of ways of making the successive choices. If there are k choices to be made, and at each stage $j \leq k$ there are n_j possible choices available, where n_j does not depend on what choices were made previously, then $\#(B)$ is equal to the product $n_1 n_2 \cdots n_k$, by the following rule:

The Multiplication Rule

Suppose that k successive choices are to be made, with exactly n_j choices available at each stage $j \leq k$, no matter what choices have been made at previous stages. Then the total number of successive choices which can be made is $n_1 n_2 \cdots n_k$.

The same rule can be expressed in other words. For example, *choices* can be replaced by *decisions* or *selections*. Notice that exactly which choices are available at stage j may depend on what choices have been made earlier, provided that the *number* of these choices available, that is n_j, does not. The multiplication rule can be proved by mathematical induction, using the addition rule. A good way to visualize the setup for the multiplication rule is to think in terms of a *decision tree*, where at each stage the decision is which branch of the tree to follow. Here is a decision tree representing $k = 3$ successive choices, with $n_1 = 4$ choices available at stage 1, $n_2 = 3$ choices at stage 2, and $n_3 = 2$ choices at stage 3. The number of possible successive choices is the total number of paths through the tree. In accordance with the multiplication rule, there are $4 \times 3 \times 2 = 24$ paths.

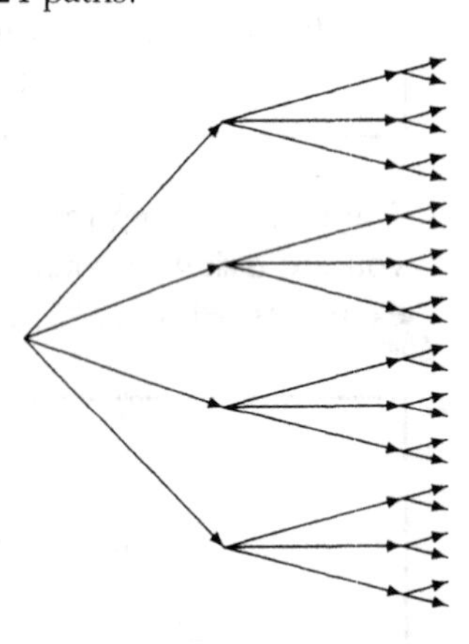

Sequences, Orderings, and Combinations

Let S be a finite set. A *sequence of length k of elements of S* is an ordered k-tuple $(s_1, s_2, \ldots, s_k)$ with $s_j \in S$ for each j. If S has n elements, the first element of the sequence can be chosen in n ways. However the first is chosen, the second can be chosen in n ways, making $n \times n = n^2$ ways to choose the first two elements of the sequence. Whichever one of those n^2 choices is made, there are n ways to choose a third element, so $n^2 \times n = n^3$ ways to choose the first three elements. This is the multiplication rule in action. Continuing in this way gives the following:

Formula for Number of Sequences

The number of sequences of length k from a set of n elements is n^k.

Example 1.

Let S be the alphabet, that is, the set of 26 letters {a, b, ..., z}. Call a sequence of five letters, such as "aargh", a *five-letter word*, no matter whether it is meaningful or not. Define a *k-letter word* similarly. There are $n = 26$ letters in the alphabet S. Hence there are

26 one-letter words,

$26^2 = 26 \times 26$ two-letter words,

$26^3 = 26 \times 26 \times 26$ three-letter words, and so on. In general, there are

$26^k = \underbrace{26 \times 26 \times \cdots \times 26}_{k \text{ factors}}$ k-letter words.

An *ordering* or *permutation of k elements of S* is a sequence of length k of elements of S with no duplications. That is to say, an arrangement of k distinct elements of S. If S has n elements, the first element in an ordering can be chosen in n ways. However this choice is made, the second element can be any one of the $n - 1$ remaining elements. So there are $n(n - 1)$ ways to choose the first two elements in an ordering. Whichever one of these choices is made, there are $n - 2$ remaining elements from which to choose a third element for the permutation, so $n(n-1)(n-2)$ ways to choose the first three elements in an ordering. Continue in this way, choosing one element of the ordering at a time from among the remaining possibilities, and use the multiplication rule to obtain the following:

Formula for Number of Orderings

The number of orderings of k out of n elements is

$$n(n-1)(n-2)\cdots(n-k+1).$$

The product of k decreasing factors $n(n-1)(n-2)\cdots(n-k+1)$ is denoted $(n)_k$, a symbol which may be read "n order k". It is the number of ways of ordering k out of n elements. Alternative notations for $(n)_k$, found in some other texts, are ${}_nP_k$ and P_n^k. Compare with the larger number n^k, the number of sequences of length k, without the restriction that there be no repetitions.

Example 2.

Let S be the alphabet as in Example 1. A permutation of length k of the 26 letters of the alphabet is a word of length k with no repetitions of letters. For example, for $k = 5$, "aorgh" is such a permutation, but "aargh" and "gargh" are not. There are

26 one-letter permutations,

$(26)_2 = 26 \times 25$ two-letter permutations,

$(26)_3 = 26 \times 25 \times 24$ three-letter permutations, and so on.

In general, there are

$$(26)_k = \underbrace{26 \times 25 \times 24 \times \cdots \times (26-k+1)}_{k \text{ factors}} \quad k\text{-letter permutations.}$$

Example 3.

In the birthday problem (Section 1.6), the probability that a group of k people all have different birthdays, assuming all possible sequences of k birthdays are equally likely, is

$$\frac{(365)_k}{365^k} = \frac{365 \times 364 \times \cdots \times (365-k+1)}{365 \times 365 \times \cdots \times 365}$$

because the denominator is the number of all possible sequences of birthdays of length k, while the numerator is the number of possible sequences with no duplication, that is, the number of possible permutations of k birthdays.

Factorials

The notation $n!$ is used for

$$(n)_n = n(n-1)\cdots 2 \cdot 1$$

and by convention $0! = 1$. The symbol $n!$ is read "n factorial". By the formula for the number of permutations in the special case $k = n$,

the number of ways of ordering a set of n elements is $n!$.

Put another way,

$n!$ is the number of different ways to arrange n objects in a row

and from above

$$(n)_k = \frac{n!}{(n-k)!}$$

is the number of different ways of arranging k of these n objects in a row.

The above expression for $(n)_k$ is correct because the factor $(n-k)!$ in the denominator cancels the last $n-k$ factors in the numerator, leaving just

$$(n)_k = n \times (n-1) \times \cdots \times (n-k+1)$$

The formula works even when $k = n$, because of the convention that $0! = 1$.

A permutation is a particular kind of sequence, namely one with no repetitions. But *combination* is just another name for "subset". A *combination of k elements from a set of n elements* is a subset consisting of k of the n elements. A combination may also be called an *unordered sample*. The number of combinations of k elements from a set of n elements is denoted $\binom{n}{k}$, a symbol which is read "n choose k". This is the number of ways of choosing k out of n elements. An ordering of k of a set of n elements can be made by the following two-stage procedure:

(i) choose a combination of k elements;

(ii) order the combination.

The number of ways of making the first choice is $\binom{n}{k}$. And no matter what combination is chosen, the number of ways of ordering it is $(k)_k = k!$. Thus by the multiplication rule,

$$(n)_k = \binom{n}{k} k!$$

Dividing both sides by $k!$ yields the following basic formula:

Formula for Number of Combinations (Subsets)

The number of ways of choosing k out of n elements is

$$\binom{n}{k} = \frac{(n)_k}{k!} = \frac{n!}{k!(n-k)!} = \frac{n(n-1)\cdots(n-k+1)}{k(k-1)\cdots 1}$$

By the convention $0! = 1$, $\binom{n}{0} = 1$, every set has just one subset with no elements, the empty set. To make the second formula in terms of $(n)_k$ work in this case, make the convention that $(n)_0 = 1$.

As well as being the number of subsets of size k of a set of n elements, one-to-one correspondences show that $\binom{n}{k}$ is:

- the number of different ways to choose k places out of n places in a row;
- the number of different ways to arrange k symbols p and $n-k$ symbols q in a row.

The numbers $\binom{n}{k}$ are also called *binomial coefficients*, as they appear in the *binomial theorem*

$$(x+y)^n = \sum_{k=0}^{n} \binom{n}{k} x^k y^{n-k}$$

Number of Subsets of a Set of *n* Elements

The number of subsets of a set of n elements is 2^n.

Note that the subsets include both the empty set and the whole set. A subset may be chosen by deciding for each of the n elements whether that element should belong to the subset, or not. There are n successive choices to be made, with two possible choices at each stage. The product rule applies once more, to show that there are 2^n subsets in all. Since each subset may be classified according to its size, the number of subsets may also be expressed using the addition rule as

$$\sum_{k=0}^{n} \binom{n}{k}$$

The equality of this expression with 2^n is the binomial theorem for $x = y = 1$.

Exercises: Appendix 1

(i) Prove that

$$\binom{n}{k} = \binom{n}{n-k}$$

(a) by using the formula for $\binom{n}{k}$;

(b) by exhibiting a one-to-one correspondence between subsets of size k and subsets of size $n - k$.

(ii) Prove that

$$\binom{n}{k} = \binom{n-1}{k-1} + \binom{n-1}{k}$$

(a) by using the formula for $\binom{n}{k}$;

(b) by breaking subsets of size k into two mutually exclusive classes, one class comprising all those subsets which contain a given element, and the other all those which don't.

(iii) Use (i) and (ii) to generate the next two rows in the following table (called Pascal's triangle), where $\binom{n}{k}$ appears in the kth column of the nth row.

n \ k	0	1	2	3	4	5
0	1					
1	1	1				
2	1	2	1			
3	1	3	3	1		
4	?	?	?	?	?	
5	?	?	?	?	?	?

(iv) Check that the formula

$$\sum_{k=0}^{n} \binom{n}{k} = 2^n$$

holds for rows $n = 0$ to 5 in Pascal's triangle. (If it doesn't work for $n = 4$ or 5, go back and redo (iii)!)

(v) Prove the formula of (iv) using (ii).

(a) using (ii);

(b) by proving that both sides of the formula represent the ***number of subsets of a set of n elements.*** For the left side use the ***addition rule*** for counting after partitioning the collection of all subsets according to size. And for the right side use the ***product rule*** for counting after identifying a subset $A \subset \{1, 2, \ldots, n\}$ with the sequence of zeros and ones which is the ***indicator*** of A.

(vi) Prove that

$$\binom{2n}{n} = \sum_{k=0}^{n} \binom{n}{k}\binom{n}{n-k} = \sum_{k=0}^{n} \binom{n}{k}^2$$

(vii) Find a formula for the number of sequences of 0's and 1's of length n such that the sum of the 0's and 1's in the sequence is k. [*Hint*: choose the places for the 1's.]

(viii) a) Prove that for $k_0 + k_1 + k_2 = n$, the number of sequences of 0's, 1's, and 2's of length n which contain exactly k_0 0's, k_1 1's and k_2 2's is $\dfrac{n!}{k_0!k_1!k_2!}$

(b) Generalize your formula to find the number of sequences of the numbers 0, 1, 2, ..., m of length n in which the number j appears k_j times. These numbers are called ***multinomial coefficients.***

(ix) Prove the binomial theorem by counting the number of terms of the form $x^k y^{n-k}$ in the expansion of $(x + y)^n$.

(x) How many different eleven-letter words (not necessarily pronounceable or meaningful!) can be made from the letters in the word MISSISSIPPI?

(xi) How many different 5-card poker hands can be dealt from a regular 52-card deck?

(xii) How many of these hands contain no aces?

(xiii) How many contain a aces, for $a = 0$ to 4?

(xiv) How many contain all cards of the same suit?

Appendix 2
Sums

The symbol $\sum_{i=1}^{n} a_i$ stands for the sum of the terms a_i from $i = 1$ to n, also denoted

$$a_1 + a_2 + \cdots + a_n$$

Note that the symbol i is an *index* or *dummy variable.* It can be replaced by any other symbol without changing the value of the sum. So

$$\sum_{i=1}^{n} a_i = \sum_{j=1}^{n} a_j$$

Sums are often made over other index sets than the first n integers. For example,

$$\sum_{i=3}^{5} a_i = a_3 + a_4 + a_5$$

If the range of i is clear from the context, a sum may be written simply $\sum_i a_i$.

General Properties of Sums

All sums are assumed to be over the same range of i:

Constants: If $x_i = c$ for every i, then

$$\sum_i x_i = (\text{number of terms}) \times c$$

Indicators: If $x_i = 0$ or 1 for every i, then

$$\sum_i x_i = (\text{number of } i \text{ such that } x_i = 1)$$

Constant factors:

$$\sum_i cx_i = c\sum_i x_i$$

Addition:

$$\sum_i (x_i + y_i) = \sum_i x_i + \sum_i y_i$$

Inequalities: If $x_i \le y_i$ for every i, then

$$\sum x_i \le \sum y_i$$

Particular Sums

$$1 + 2 + \cdots + n = \sum_{i=1}^{n} i = n(n+1)/2$$

Provided $R \neq 1$, $$1 + R + R^2 + \cdots + R^n = \sum_{i=0}^{n} R^i = \frac{1 - R^{n+1}}{1 - R}$$

Appendix 3
Calculus

Infinite Series

Let $a_1, a_2, \ldots$ be a sequence of numbers. The infinite sum

$$\sum_{i=1}^{\infty} a_i = a_1 + a_2 + \cdots$$

is called an *infinite series.*

The finite sum

$$\sum_{i=1}^{n} a_i = a_1 + a_2 + \cdots + a_n$$

is called the nth *partial sum* of the a's.

Convergence of Infinite Series

The series $\sum_{i=1}^{\infty} a_i$ *converges* if the sequence of partial sums converges to a finite limit, that is, if $\lim_{n\to\infty} \sum_{i=1}^{n} a_i$ exists and is finite.

The series $\sum_{i=1}^{\infty} a_i$ *diverges* if the sequence of partial sums does not converge to a finite limit, that is, if $\lim_{n\to\infty} \sum_{i=1}^{n} a_i$ either does not exist or is infinite.

If $a_1, a_2, \ldots$ are all positive, then the sequence of partial sums is increasing, and thus has a limit, though the limit may be $+\infty$. So the series $\sum_{i=1}^{\infty} a_i$ either converges, or diverges to $+\infty$.

Some Common Infinite Series

$$\sum_{n=1}^{\infty} \frac{1}{n} = +\infty$$

$$\sum_{n=1}^{\infty} \frac{1}{n^2} = \frac{\pi^2}{6}$$

Geometric series: If $|r| < 1$, $\quad \sum_{i=i_0}^{\infty} r^i = \frac{r^{i_0}}{1-r} = \frac{\text{first term}}{1 - \text{common ratio}}$

Exponential series: $\sum_{n=0}^{\infty} \frac{x^n}{n!} = e^x \quad$ (see Appendix 4)

Derivatives

The function $f(x)$ is said to be *differentiable at* x_0 if

$$\lim_{\Delta x \to 0} \frac{f(x_0 + \Delta x) - f(x_0)}{\Delta x}$$

exists. In that case, the *derivative of $f(x)$ at x_0* is defined as the limit

$$f'(x_0) = \lim_{\Delta x \to 0} \frac{f(x_0 + \Delta x) - f(x_0)}{\Delta x}$$

If $f(x)$ is differentiable at every x_0 in its domain, then $f(x)$ is called *differentiable.*

Interpretations of the derivative

The derivative $f'(x_0)$ may be interpreted as

the *rate of change* of $f(x)$ at x_0,

or

the *slope* of the graph of $f(x)$ at x_0.

If $y = f(x)$, the derivative function $f'(x)$ is often written as

$$\frac{dy}{dx} \quad \text{or} \quad \frac{d}{dx}f(x)$$

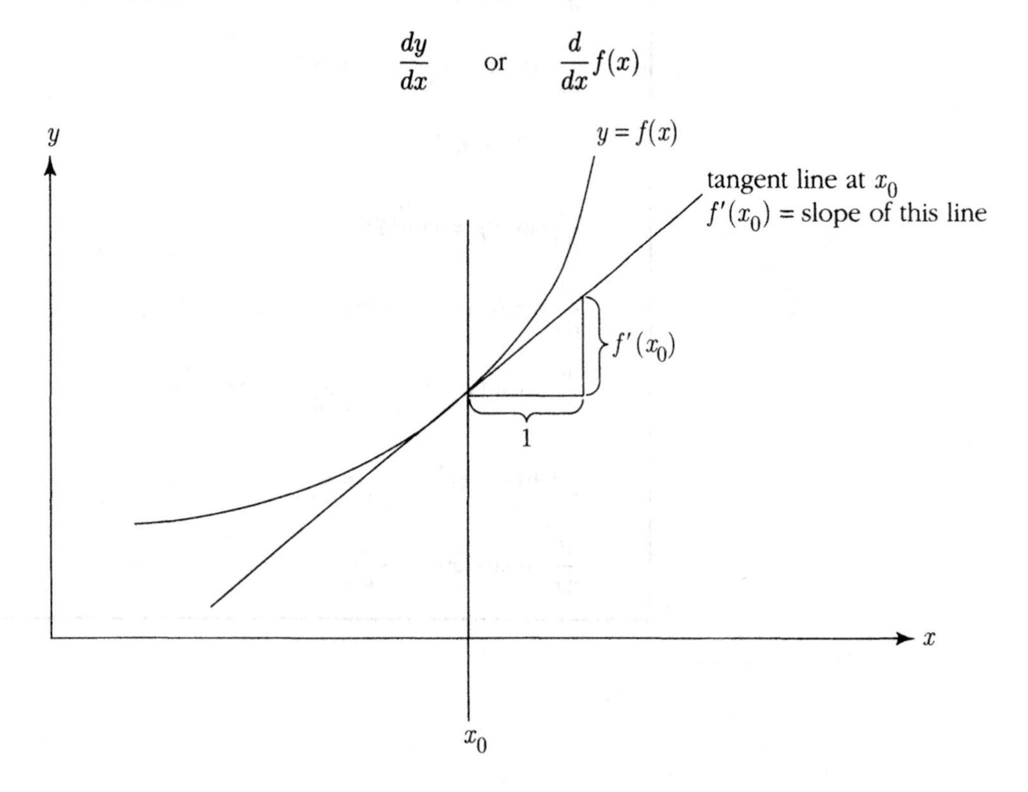

Properties of Derivatives

Constants: If $f(x) = c$ for all x, then $f'(x) = 0$ for all x.

Constant factors: $(cf)'(x) = c(f'(x))$

Addition: $(f + g)'(x) = f'(x) + g'(x)$

Multiplication: $(fg)'(x) = f'(x)g(x) + f(x)g'(x)$

Chain rule: $\frac{d}{dx}f(g(x)) = f'(g(x))g'(x)$

Some Common Derivatives

$$\frac{d}{dx}x^n = nx^{n-1}, \qquad n = 1, 2, \ldots$$

$$\frac{d}{dx}\log(x) = \frac{1}{x}, \qquad x > 0$$

$$\frac{d}{dx}e^{\beta x} = \beta e^{\beta x}$$

$$\frac{d}{d\theta}\sin(\theta) = \cos(\theta)$$

$$\frac{d}{d\theta}\cos(\theta) = -\sin(\theta)$$

$$\frac{d}{dx}\arcsin(x) = \frac{1}{\sqrt{1 - x^2}}, \qquad |x| < 1$$

$$\frac{d}{dx}\arccos(x) = \frac{-1}{\sqrt{1 - x^2}}, \qquad |x| < 1$$

$$\frac{d}{dx}\arctan(x) = \frac{1}{1 + x^2}$$

Integrals

Consider a non-negative function $h(x)$ defined for x on the line $(-\infty, \infty)$. For example, $h(x)$ might be the following curve.

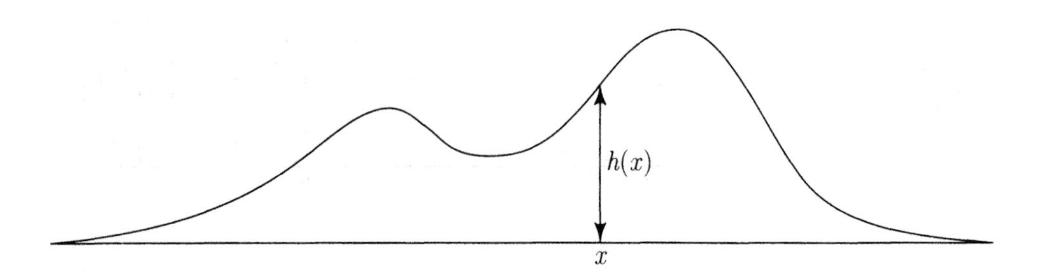

The area under the graph of $h(x)$ between two points a and b on the line is by definition the integral of $h(x)$ from a to b:

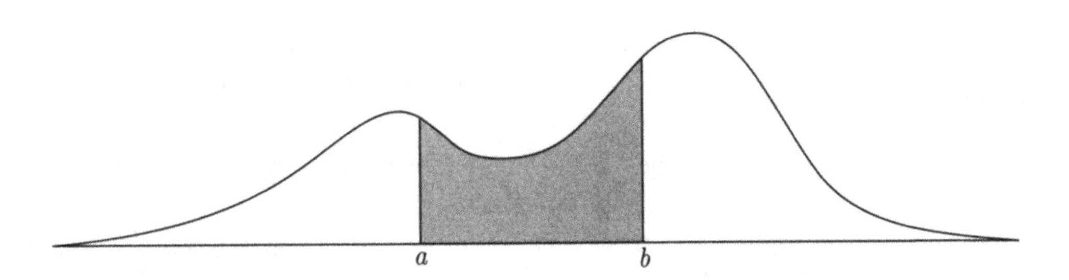

$$\text{Area } (a \text{ to } b) = \int_a^b h(x)dx.$$

This area integral is a limit of areas obtained by approximating $h(x)$ with step functions which take a finite number of different values on a finite number of disjoint intervals, as in the following diagram:

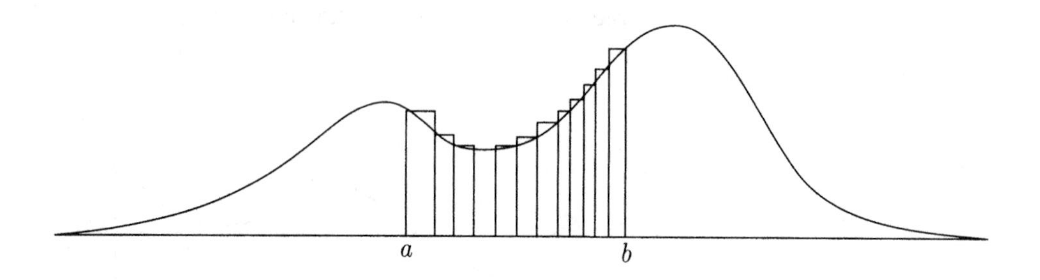

It is shown by calculus that such area integrals exist, and that they can be evaluated by finding a function H which is an *anti-derivative* or *indefinite integral* of h,

$$H'(x) = h(x), \qquad \frac{d}{dx}H(x) = h(x), \qquad \text{or} \qquad \int h(x)dx = H(x),$$

to express the same relation with three different standard notations. Such an indefinite integral H is unique apart from the addition of arbitrary constants, and

$$\text{Area } (a \text{ to } b) = \int_a^b h(x)dx = H(b) - H(a) \stackrel{def}{=} H(x)\Big|_a^b$$

The total area under the graph of h

$$\text{Area } (-\infty \text{ to } \infty) = \int_{-\infty}^{\infty} h(x)dx$$

is defined as the limit of Area (a to b) as $a \to -\infty$ and $b \to \infty$.

Fundamental Theorem of Calculus

$$\int_a^b F'(x)dx = F(b) - F(a) \stackrel{def}{=} F(x)\Big|_a^b$$

Some Indefinite Integrals

$$\int x^n dx = \frac{x^{n+1}}{n+1} \qquad (n \neq -1, \quad x > 0)$$

$$\int \frac{1}{x} dx = \log(x) \qquad (x > 0)$$

$$\int e^{\beta x} dx = \frac{1}{\beta} e^{\beta x}$$

$$\int \log(x) dx = x \log(x) - x \qquad (x > 0)$$

$$\int \sin(\theta) d\theta = -\cos(\theta)$$

$$\int \cos(\theta) d\theta = \sin(\theta)$$

$$\int \frac{1}{\sqrt{1-x^2}} dx = \arcsin(x) \qquad (|x| < 1)$$

$$\int \frac{1}{1+x^2} dx = \arctan(x)$$

Some Definite Integrals

$$\int_{-\infty}^{\infty} e^{-\frac{1}{2}x^2} \, dx = \sqrt{2\pi}$$

$$\int_0^{\infty} x^n e^{-x} \, dx = n! \qquad (n \; integer, n \geq 0)$$

$$\int_0^1 x^m (1-x)^n \, dx = \frac{m! \, n!}{(m+n+1)!}$$

The first four properties of integrals in the box below should be compared to the corresponding properties of sums, listed in Appendix 2.

Properties of Integrals

Assume $a < b$.

Constants: If $f(x) = c$ for all x, then

$$\int_a^b f(x)dx = \int_a^b c\,dx = (b-a)c = (\text{length of interval}) \times c$$

Constant factors:

$$\int_a^b cf(x)dx = c\int_a^b f(x)dx$$

Addition:

$$\int_a^b (f(x)+g(x))dx = \int_a^b f(x)dx + \int_a^b g(x)dx$$

Inequalities: If $f(x) \leq g(x)$ for all x, then

$$\int_a^b f(x)dx \leq \int_a^b g(x)dx$$

Splitting the range of integration: If $a < b < c$,

$$\int_a^c f(x)dx = \int_a^b f(x)dx + \int_b^c f(x)dx$$

Integration by parts:

$$\int_a^b f(x)g'(x)dx = [f(x)g(x)]_a^b - \int_a^b f'(x)g(x)dx$$

Appendix 4
Exponents and Logarithms

Suppose that $b > 0$. For each positive integer x, a number b^x, called *b to the power x*, *b to the exponent x*, or just *b to the x*, is defined by $b^1 = b$, $b^2 = b.b$, and so on. So b^x is the product of x factors of b. This implies the first two rules stated in the box for positive integer exponents x and y. The definition of b^x is extended to $x = 0$, negative integers x, and rational numbers x, by requiring these two rules to hold for all these values of x as well. This implies the rest of the laws stated for rational x and y. The definition of b^x is further extended to all real x by assuming b^x is a continuous function of x.

Laws of Exponents

For $b, c > 0$, and all real numbers x and y:

(i) $b^{x+y} = b^x b^y$

(ii) $b^{xy} = (b^x)^y$

(iii) $b^0 = 1$

(iv) $b^{-x} = 1/b^x$

(v) $b^{x-y} = b^x/b^y$

(vi) $(bc)^x = b^x c^x$

To illustrate, for a positive integer n, $b^{1/n}$ is the positive nth root of b, also denoted $\sqrt[n]{b}$. This comes from rule (ii) for $x = 1/n$ and $y = n$. For positive rational $x = m/n$, (ii) gives

$$b^x = (b^{1/n})^m = (b^m)^{1/n}$$

Negative exponents are defined by rule (iv). The idea of multiplying together x factors of b does not make sense if x is not a positive integer. But the extended definition of exponents is very useful for algebraic manipulations with powers.

For $y > 0$ and $b > 0$ the equation $y = b^x$ is solved by a unique number $x = \log_b(y)$, called the logarithm of x to base b. In other words, the function $y \mapsto \log_b(y)$ is the *inverse function* of $x \mapsto b^x$. The laws of exponents imply the following:

Laws of Logarithms

For $b > 0$, $x > 0$, $y > 0$,

$$\log_b(xy) = \log_b(x) + \log_b(y)$$

$$\log_b(x^y) = y \log_b(x) \qquad \text{(true also for } y \leq 0\text{)}$$

$$\log_b(1) = 0$$

$$\log_b(1/x) = -\log_b(x)$$

$$\log_b(y/x) = \log_b(y) - \log_b(x)$$

$$\log_a(x) = \log_a(b) \log_b(x) \qquad \text{(change of base)}$$

As the graphs suggest, b^x is a differentiable function of x for every $b > 0$. This involves the constant

$$e = 2.71828\ldots$$

defined precisely by any of the formulae in the next box. While the function $x \to b^x$ may be called an exponential function for any $b > 0$, *the* exponential function is

$$\exp(x) = e^x$$

FIGURE 1. Graphs of $y = b^x$

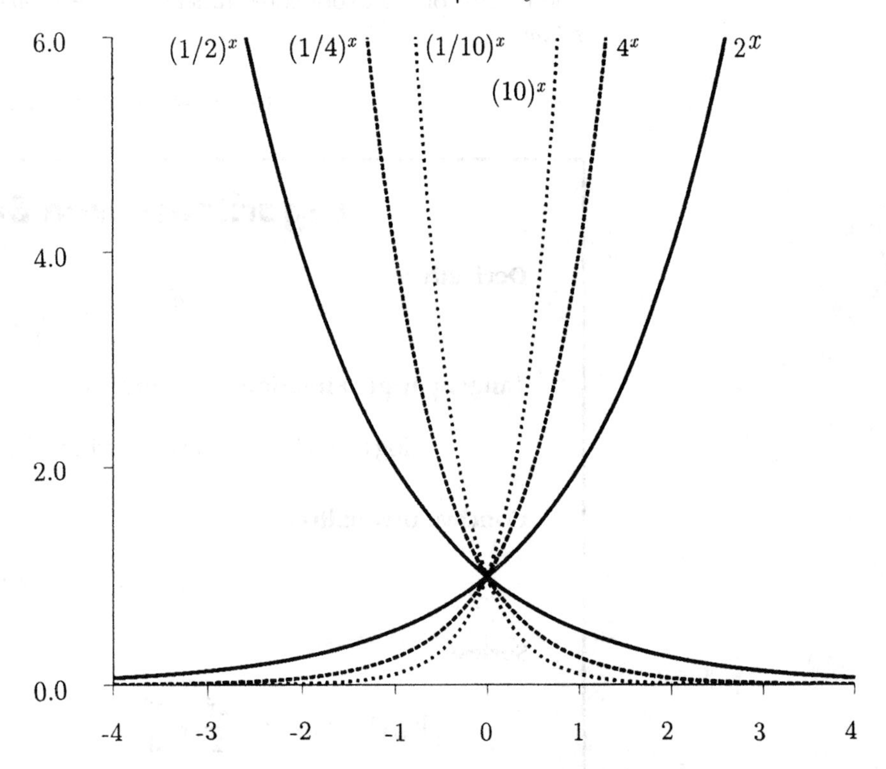

Exponentials with Base e

Derivative:

$$\frac{d}{dx}(e^x) = e^x.$$

Tangent approximation near zero:

$$e^x = 1 + x + \epsilon(x) \qquad \text{where } \epsilon(x)/x \to 0 \text{ as } x \to 0$$

Convex inequality:

$$e^x \geq 1 + x \qquad \text{for all } x$$

Series:

$$e^x = 1 + x + \frac{x^2}{2!} + \frac{x^3}{3!} + \cdots = \sum_{k=0}^{\infty} \frac{x^k}{k!}$$

Product limit:

$$e^x = \lim_{n \to \infty} (1 + \frac{x}{n})^n$$

The inverse of the exponential function is the *logarithm to base e*, or *natural logarithm*,

$$\log(x) = \log_e(x), \quad x > 0.$$

Logarithms with Base e

Derivative:

$$\frac{d}{dx}\log(x) = \frac{1}{x}$$

Tangent approximation near one:

$$\log(1+z) = z - \delta(z) \qquad \text{where } \delta(z)/z \to 0 \text{ as } z \to 0$$

Concave inequality:

$$\log(1+z) \le z \qquad \text{for all } z$$

Series:

$$\log(1+z) = z - \frac{z^2}{2} + \frac{z^3}{3} - \frac{z^4}{4} + \cdots \qquad \text{for } -1 < z \le 1$$

FIGURE 2. **Graphs of e^x and log x.** The graph of log x, the inverse function of e^x, is obtained by reflection of the graph of e^x about the 45° line $y = x$. Just as the slope of $y = e^x$ is $e^0 = 1$ as this curve passes through the point $(0, 1)$, the slope of the curve $y = \log x$ is also 1 as it passes through the point $(1, 0)$. So
the 45° line $y = x + 1$ is tangent below the curve $y = e^x$ at $x = 0$,
the 45° line $y = x - 1$ is tangent above the curve $y = \log x$ at $x = 1$.
This gives the tangent approximations and inequalities for exp and log.

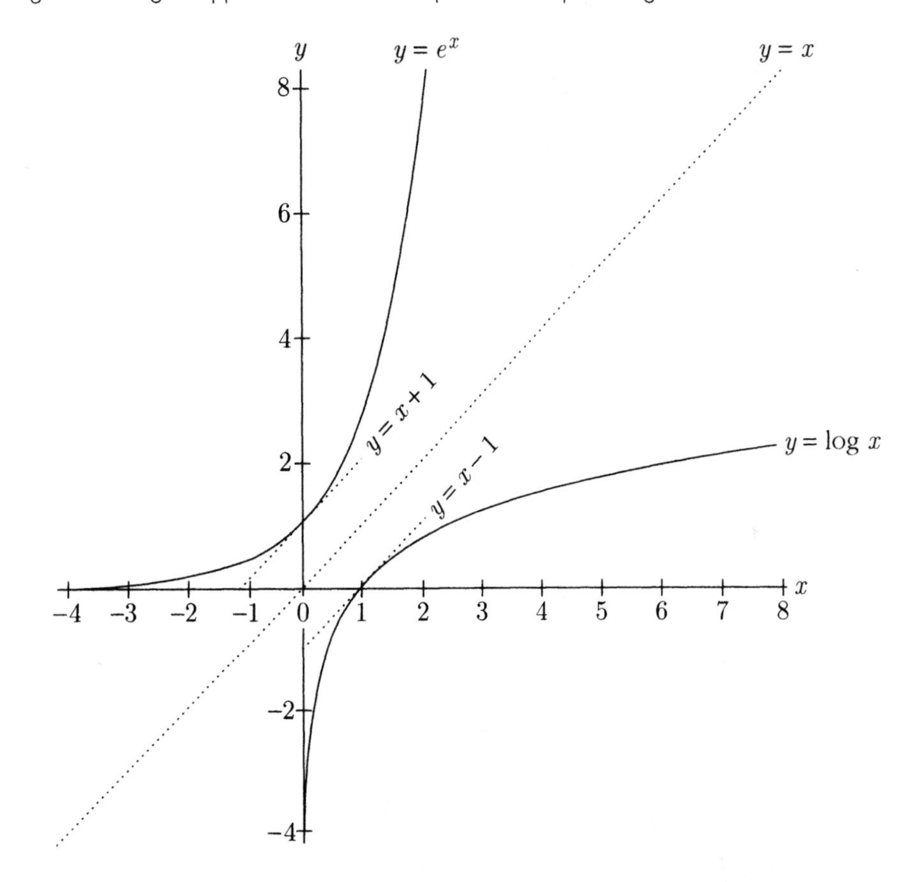

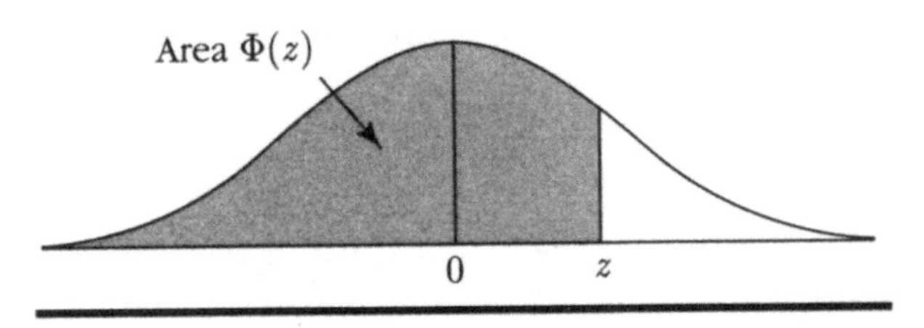

Appendix 5 Normal Table

Table shows values of $\Phi(z)$ for z from 0 to 3.59 by steps of .01. Example: to find $\Phi(1.23)$, look in row 1.2 and column .03 to find $\Phi(1.2 + .03) = \Phi(1.23) = .8907$. Use $\Phi(z) = 1 - \Phi(-z)$ for negative z.

	.0	.01	.02	.03	.04	.05	.06	.07	.08	.09
0.0	.5000	.5040	.5080	.5120	.5160	.5199	.5239	.5279	.5319	.5359
0.1	.5398	.5438	.5478	.5517	.5557	.5596	.5636	.5675	.5714	.5753
0.2	.5793	.5832	.5871	.5910	.5948	.5987	.6026	.6064	.6103	.6141
0.3	.6179	.6217	.6255	.6293	.6331	.6368	.6406	.6443	.6480	.6517
0.4	.6554	.6591	.6628	.6664	.6700	.6736	.6772	.6808	.6844	.6879
0.5	.6915	.6950	.6985	.7019	.7054	.7088	.7123	.7157	.7190	.7224
0.6	.7257	.7291	.7324	.7357	.7389	.7422	.7454	.7486	.7517	.7549
0.7	.7580	.7611	.7642	.7673	.7703	.7734	.7764	.7794	.7823	.7852
0.8	.7881	.7910	.7939	.7967	.7995	.8023	.8051	.8078	.8106	.8133
0.9	.8159	.8186	.8212	.8238	.8264	.8289	.8315	.8340	.8365	.8389
1.0	.8413	.8438	.8461	.8485	.8508	.8531	.8554	.8577	.8599	.8621
1.1	.8643	.8665	.8686	.8708	.8729	.8749	.8770	.8790	.8810	.8830
1.2	.8849	.8869	.8888	.8907	.8925	.8944	.8962	.8980	.8997	.9015
1.3	.9032	.9049	.9066	.9082	.9099	.9115	.9131	.9147	.9162	.9177
1.4	.9192	.9207	.9222	.9236	.9251	.9265	.9279	.9292	.9306	.9319
1.5	.9332	.9345	.9357	.9370	.9382	.9394	.9406	.9418	.9429	.9441
1.6	.9452	.9463	.9474	.9484	.9495	.9505	.9515	.9525	.9535	.9545
1.7	.9554	.9564	.9573	.9582	.9591	.9599	.9608	.9616	.9625	.9633
1.8	.9641	.9649	.9656	.9664	.9671	.9678	.9686	.9693	.9699	.9706
1.9	.9713	.9719	.9726	.9732	.9738	.9744	.9750	.9756	.9761	.9767
2.0	.9772	.9778	.9783	.9788	.9793	.9798	.9803	.9808	.9812	.9817
2.1	.9821	.9826	.9830	.9834	.9838	.9842	.9846	.9850	.9854	.9857
2.2	.9861	.9864	.9868	.9871	.9875	.9878	.9881	.9884	.9887	.9890
2.3	.9893	.9896	.9898	.9901	.9904	.9906	.9909	.9911	.9913	.9916
2.4	.9918	.9920	.9922	.9925	.9927	.9929	.9931	.9932	.9934	.9936
2.5	.9938	.9940	.9941	.9943	.9945	.9946	.9948	.9949	.9951	.9952
2.6	.9953	.9955	.9956	.9957	.9959	.9960	.9961	.9962	.9963	.9964
2.7	.9965	.9966	.9967	.9968	.9969	.9970	.9971	.9972	.9973	.9974
2.8	.9974	.9975	.9976	.9977	.9977	.9978	.9979	.9979	.9980	.9981
2.9	.9981	.9982	.9982	.9983	.9984	.9984	.9985	.9985	.9986	.9986
3.0	.9987	.9987	.9987	.9988	.9988	.9989	.9989	.9989	.9990	.9990
3.1	.9990	.9991	.9991	.9991	.9992	.9992	.9992	.9992	.9993	.9993
3.2	.9993	.9993	.9994	.9994	.9994	.9994	.9994	.9995	.9995	.9995
3.3	.9995	.9995	.9995	.9996	.9996	.9996	.9996	.9996	.9996	.9997
3.4	.9997	.9997	.9997	.9997	.9997	.9997	.9997	.9997	.9997	.9998
3.5	.9998	.9998	.9998	.9998	.9998	.9998	.9998	.9998	.9998	.9998

Brief Solutions to Odd-Numbered Exercises

1.1.1. a) $2/3$ b) 66.67% c) 0.6667 d) $4/7$ e) 57.14% f) 0.5714

1.1.3. a) $1/n^2$ b) $(n-1)/n^2$ c) $(1-1/n)/2$ d) $1/n(n-1)$, $1/n$, $1/2$

1.1.5. a) 2652 b) $1/13$ c) $1/13$ d) $1/221$ e) $33/221$

1.1.7. a) $1/9$ b) $1/4$ c) $5/36$ d) $x^2/36, (2x-1)/36$ e) 1

1.1.9. $1/11$, $1/6$

1.1.11. Use the definition of fair odds, substitute in the formula for the house percentage.

1.2.1. The opinion of the judge.

1.2.3. a) $\Sigma > 1$ b) Yes. In this situation, you can get back more than you bet.

1.3.1. $4/7$ of the cake

1.3.3. $\Omega = \{1, 2, \ldots, 500\}$ a) $\{17, 93, 202\}$
b) $\{17, 93, 202, 4, 101, 102, 398\}^c$ c) $\{16, 18, 92, 94, 201, 203\}$

1.3.5. a) first coin lands heads b) second coin lands tails c) first coin lands heads d) at least two heads e) exactly two tails f) first two coins land the same way

1.3.7. a) $P(1) = P(6) = p/2$ and $P(2) = P(3) = P(4) = P(5) = (1-p)/4$
b) $(3-p)/4$

1.3.9. a) 0.9 b) 1 c) 0.1

1.3.13. Use inclusion–exclusion for two sets.

1.3.15. *Hint*: Let $A_i = B_i^c$.

1.4.1. a) can't be decided; the rest are true.

1.4.3. 75%

1.4.5. c) 17/35

1.4.7. a) 0.3 b) 0.6

1.4.9. $p_1 = 0.1$, $p_2 = 0.4$, and $p_3 = 0.5$

1.4.11. a) $\frac{1+p}{4}$ b) $\frac{1-p}{4}$ c) $\frac{1-p}{2}$ d) $\frac{1+p}{2}$

1.5.1. a) 7/24 b) 8/17

1.5.3. a) 40/41 b) 1/41

1.5.5. a) 0.0575 b) 0.002 c) 0.9405 d) $16/115 \approx 0.139$ e) yes

1.5.7. a) 5/12 b) no c) You would be right 7/16 of the time.
d) Respond by always guessing box 1. Your probability of correct guessing is 1/2.

1.6.1. 5

1.6.3. a) 0.7692 b) 0.2308

1.6.5. a) $1 - (364/365)^{n-1}$ b) at least 254

1.6.7. a) $p_3p_1q_2 + p_3q_1p_2 + p_3p_1p_2$
b) $p_4 + P(\text{flows along top}) - p_4 \cdot P(\text{flows along top})$ where $P(\text{flows along top})$ was calculated in a)

1.rev.1. 6/11

1.rev.3. False

1.rev.5. The chance of passing when you use the first order is $zh(2-z)$. With the second order, it's $hz(2-h)$.

1.rev.7. a) $\frac{20}{50} \cdot \frac{19}{49} \cdot \frac{18}{48} \cdot \frac{17}{47} = .021$ b) 1 − answer to a) c) $4 \cdot \frac{30}{50} \cdot \frac{20}{49} \cdot \frac{19}{48} \cdot \frac{18}{47}$
d) $4 \cdot \frac{30}{50} \cdot \frac{29}{49} \cdot \frac{28}{48} \cdot \frac{20}{47} + \frac{30}{50} \cdot \frac{29}{49} \cdot \frac{28}{48} \cdot \frac{27}{47}$

1.rev.9. a) 1/60 b) 3/5 c) 13/30

1.rev.11. $\frac{9f}{9f+8b}$

1.rev.13. *Hint*: Write $P(A|B) = \sum_{i=1}^{n} P(AB_i|B)$.

1.rev.15. 2/3

1.rev.17. False.

2.1.1. a) $\binom{7}{4}$; b) $\binom{7}{4}(1/6)^4(5/6)^3$

2.1.3. a) 0.1608 b) 0.1962 c) 0.9645 d) 0.3125 e) 0.5

2.1.5. a) $\frac{\binom{19}{11}}{\binom{20}{12}}$ b) $\frac{\binom{18}{10}}{\binom{20}{12}}$ c) $1 - \left\{ \frac{\binom{15}{12}}{\binom{20}{12}} + 5 \times \frac{\binom{15}{11}}{\binom{20}{12}} \right\}$

2.1.7. 0.1005

2.1.9. a) 8, with probability 0.1387 b) 0.1128 c) 0.1133

2.1.11. a) 11 b) 0.2186

2.1.13. a) no b) 0.5 c) 0.75 d) 0.5

2.1.15. b) Note that $np = int\ (np) + [np - int\ (np)]$.

2.2.1. a) 0.7062 b) 0.1509 c) 0.0398 d) 0.0242

2.2.3. a) the first one b) $0.1841 > 0.0256$

2.2.5. 0.3974

2.2.7. a) City B has better accuracy. b) Both have same accuracy. c) City B has better accuracy.

2.2.9. a) 0.0495 b) Increase c) 0.1093

2.2.11. a) 0.4562 b) 0.2929 c) 0.2929 d) Increase e) Could be due to chance

2.2.13. Sample 9604 people

2.3.1. *Hint*: $P(k) = R(k) \cdot R(k-1) \cdots R(1)P(0)$.

2.3.3. a) Use odds ratios. b) Condition. e) Use the inequality $1 - x \le e^{-x}$.

2.4.1. a) Approximately Poisson(1) b) Approximately Poisson(2) c) Approximately Poisson(0.3284)

2.4.3. a) 0.999674. b) 0.997060.

2.4.5. 0.5945, 0.3092, 0.0804.

2.4.7. a) 2 b) 0.2659 c) 0.2475 d) 0.2565 e) $m = 250$; Normal approx: 0.0266 f) $m = 2$; Poisson approx: 0.2565

2.4.9. Use Poisson approximation: 0.9828

2.5.1. a) $\frac{\binom{20}{4}\binom{30}{6}}{\binom{50}{10}}$ b) $\binom{10}{4}(2/5)^4(3/5)^6$

2.5.3. a) $\frac{\binom{4}{4}\binom{48}{9}}{\binom{52}{13}}$ b) $\frac{\binom{3}{3}\binom{48}{9}}{\binom{51}{12}}$ c) $\frac{\binom{4}{4}\binom{48}{9}}{\binom{52}{13}-\binom{48}{13}}$ d) 0

2.5.5. $n \geq 537$ will do.

2.5.7. a) 0.1456 b) 0.3716 c) 0.0929

2.5.9. a) 0.282409 b) 0.459491

2.5.11. $\max\{0, n - N + G\}$ to $\min\{n, G\}$

2.5.13. 0.0028

2.rev.1. a) $\binom{10}{4}(1/6)^4(5/6)^6$ b) $\binom{10}{4}(1/5)^4(4/5)^6$ c) $\frac{10!}{4!3!2!}/6^{10}$ d) $\frac{\binom{7}{4}}{\binom{10}{4}}$

2.rev.3. a) $1/6$ b) $1/4$

2.rev.5. $\frac{\binom{97}{57}}{\binom{100}{60}}$

2.rev.7. $k \approx 1025$

2.rev.9. a) 0.8 b) guess 3 c) 0.4375

2.rev.11. 0.0102

2.rev.13. 0.99; the chance that any particular packet needs to be replaced is about 0.0144.

2.rev.15. a) $\binom{20}{5}(0.4)^5(0.6)^15$ b) $\frac{20!}{2!4!6!8!}(0.1)^2(0.2)^4(0.3)^6(0.4)^8$ c) $\binom{24}{2}(0.1)^3(0.9)^{22}$

2.rev.17. a) $\frac{\binom{6}{1}}{6^4}$ b) $\binom{6}{1} \times \binom{5}{1} \times \frac{\binom{4}{1}}{6^4}$ c) $\binom{6}{2} \times \frac{\binom{4}{2}}{6^4}$

2.rev.19. a) $(2/3)^4$ b) $\binom{4}{1}(2/3)^4(1/3) + (2/3)^4$

2.rev.21. $\sum_{x=0}^{(n-1)/2} \binom{n}{x} q^x p^{n-x}$.

2.rev.23. $\frac{(.4 \times 1/2)+(.2 \times 6/8)+(.1 \times 14/16)}{(.2 \times 1/2)+(.4 \times 3/4)+(.2 \times 7/8)+(.1 \times 15/16)}$

2.rev.25. a) p^3, $3p^3q$, $6p^3q^2$ b) $p^3 + 3p^3q + 6p^3q^2$ c) $\frac{1}{1+3q+6q^2}$ d) 0.375 e) no.

2.rev.27. 0.3971

2.rev.29. 0.0579

2.rev.31. *Hint:* $np \geq npq$

2.rev.33. *Hint*: Think about conditional probabilities.

2.rev.35. a) $\sum_{k=20}^{35}\binom{1000}{k}(1/38)^k(37/38)^{1000-k}$ b) Use normal: 0.876

3.1.1. a) $P(X=0)=1/8$, $P(X=1)=3/8$, $P(X=2)=3/8$, $P(X=3)=1/8$
b) $P(|X-1|=0)=3/8$, $P(|X-1|=1)=4/8$, $P(|X-1|=2)=1/8$

3.1.3. a) All integers from 2 to 12 inclusive. b) Partial answer: $P(S=2)=1/36$, $P(S=3)=2/36$, $P(S=4)=3/36$, $P(S=5)=4/36$

3.1.5. Partial answer: $P(X_1X_2=1)=1/36$, $P(X_1X_2=2)=2/36$, $P(X_1X_2=3)=2/36$, $P(X_1X_2=4)=3/36$, $P(X_1X_2=5)=2/36$, $P(X_1X_2=6)=4/36$

3.1.7. a) $(ABC^c)\cup(AB^cC)\cup(A^cBC)$ b) $ab(1-c)+a(1-b)c+(1-a)bc$

3.1.9. Partial answer: $P(X=2)=5/35$, $P(X=3)=10/35$

3.1.11. a) binomial $(n+m,p)$ e) $\binom{2n}{n}$

3.1.13. a) $\frac{2n-2}{2n}\cdot\frac{2n-4}{2n}\cdot\dots\cdot\frac{2n-2(k-1)}{2n}$ b) $\sqrt{n\log 4}$

3.1.15. a) $1/n$ b) $(n-1)/2n$ c) $(n-1)/2n$ d) $(2k-1)/n^2$ e) $[2(n+1-k)-1]/n^2$
f) $(k-1)/n^2$ for $k=2$ to $n+1$; $(2n-k+1)/n^2$ for $k=n+2$ to $2n$

3.1.17. a) $P(Z=k)=(k/21)\binom{20}{k}(1/2)^{20}+(1/21)\sum_{i=0}^{k}\binom{20}{i}(1/2)^{20}$

3.1.19. a) Partial answer: $P(S=7)=p_1r_6+p_2r_5+p_3r_4+p_4r_3+p_5r_2+p_6r_1$
d) yes

3.1.21. yes

3.1.23. $P(X\le T)\le P(Y\le T)$

3.2.1. 41.5

3.2.3. The expected number of sixes is 1/2, the expected number of odds is 3/2.

3.2.5. Expect to lose about 8 cents per game.

3.2.7. $\sum_{i=1}^{n}p_i$

3.2.9. $p-2pr+r$

3.2.11. Simple upper bound: 0.3 Actual probability: 0.271

3.2.13. a) 35 b) 8.458 c) 5.43 d) 10/3 e) 0.9690 f) 5.0310

3.2.15. Show that $E[L(Y,b)]=(\lambda+\pi)\sum_{y\le b}(b-y)p(y)-\pi b$.

3.2.17. a) $\frac{\binom{10}{9-3}}{\binom{13}{9}}$ b) $\frac{\binom{10}{9-3}}{\binom{13}{9}}-\frac{\binom{10}{8-3}}{\binom{13}{8}}$ c) 10.5

3.2.19. a) $\frac{5!}{1!1!1!2!}[2(1/7)^3(2/7)(3/7)+(1/7)^2(2/7)^2(3/7)+(1/7)^2(2/7)(3/7)^2]$
b) $2[1-(6/7)^5]+1-(5/7)^5+1-(4/7)^5$

3.3.1. a) $E(X)=30.42$, $SD(X)=0.86$ b) $E(X)=30.44$, $SD(X)=0.86$

3.3.3. a) 5 b) 26 c) 1 d) 26

3.3.5. *Hint*: Use the computational formula for the variance.

3.3.7. a) no b) $E(X)=\sum_{i=1}^{3} n_i p_i$, $Var(X)=\sum_{i=1}^{3} n_i p_i q_i$

3.3.9. a) $r(1-p_1)+(n-r)p_2$ b) $r(1-p_1)p_1+(n-r)p_2(1-p_2)$

3.3.11. $E(Y)=a+b\left(\frac{n-1}{2}\right)$, $Var(Y)=b^2(n^2-1)/12$

3.3.13. a) 111112 b) 55556 c) 1300

3.3.15. b) $10\sqrt{8}$

3.3.17. a) 0.05 b) 0.03 c) 0.92

3.3.19. Approximately $1-\Phi(1.66)=1-0.9515=0.0485$

3.3.21. a) 0.0876 b) 0.0489

3.3.23. Approximately $\Phi(-0.77)\approx 0.22$

3.3.27. For b), reduce to a). For c): Half the list are zeros, the rest are nines.

3.3.29. a) Guess 4. b) $(n=1)$ 1/10; $(n=2)$ 19/100; $(n=33)$ 0.6826; $(n=66)$ 0.8414; $(n=132)$ 0.9544. c) $n\geq 220$ will do.

3.3.31. a) $9/2, \sqrt{33}/2$ d) $2\Phi(2b/\sqrt{33})-1$

3.4.1. a) $\binom{9}{5}p^5(1-p)^4$ b) $(1-p)^6\cdot p$ c) $\binom{11}{4}p^4(1-p)^7\cdot p$
d) $\sum_{k=0}^{5}\binom{8}{k}p^k(1-p)^{8-k}\cdot\binom{5}{k}p^k(1-p)^{5-k}$

3.4.3. 12

3.4.5. Let $q_i=1-p_i$. a) q_2^n b) $(q_1q_2q_3)^n$ c) $(q_1q_2q_3)^{n-1}-(q_1q_2q_3)^n$
d) $p_2/(1-q_1q_2q_3)$

3.4.7. a) $\frac{1}{1+q}$ b) $\frac{q}{1+q}$ c) $\frac{p}{1-q^3}, \frac{q-q^3}{1-q^3}$ d) $p=\frac{3-\sqrt{5}}{2}$ e) 2/3

3.4.9. Expect to lose \$4 per game.

3.4.11. a) $\frac{p_Aq_B}{1-q_Aq_B}$ b) $\frac{q_Ap_B}{1-q_Aq_B}$ c) $\frac{p_Ap_B}{1-q_Aq_B}$ d) $P(N=k)=(q_Aq_B)^{k-1}(1-q_Aq_B)$ for $k=1,2,3,\ldots$

3.4.13. a) $P(\text{Black wins})=\frac{p}{1-qp}$ b) $(3-\sqrt{5})/2$ c) no d) 13

3.4.15. a) Use Exercise 3.4.6 b) *Hint*: Look at the tail probabilities $P(F\geq k)$

3.4.17. $\frac{2(1-p)p^k}{(2-p)^{k+1}}$ $(k \geq 0)$

3.4.19. *Hints:* a) Negative binomial b) Symmetry c) Use the result of b)

3.4.21. a) μ b) μ

3.4.23. a) $p/(1-qz)$ b) $f_1 = \frac{q}{p}$, $f_2 = 2(\frac{q}{p})^2$, $f_3 = 6(\frac{q}{p})^3$

3.5.1. 0.1428

3.5.3. a) 0.222 b) About 44

3.5.5. 0.39

3.5.7. a) Poisson(3), Poisson(2), Poisson(5) b) 0.3679

3.5.9. a) 0.0996 b) 0.8008 c) 0.3951

3.5.11. a) $e^{-2}2^4/4!$ b) 6 c) $e^{-3}3^4/4!$

3.5.13. a) $2.69 \times 10^{19}x^3$, $5.19 \times 10^9 x^{3/2}$ b) 7.19×10^{-6} cm

3.5.15. a) 198.01, 1.97 b) 1.79 c) 0.59

3.5.17. a) 0.0067 b) 0.0037

3.5.19. b) μ, μ^2, μ^3 c) μ, $\mu^2 + \mu$, $\mu^3 + 3\mu^2 + \mu$

3.5.21. c) 0.58304 d) 0.5628 e) 0.58306

3.6.1. a) 1/13 b) 4/50 c) $4 \times \frac{\binom{13}{5}}{\binom{52}{5}}$ d) $1 - \frac{\binom{4}{0}\binom{48}{5}}{\binom{52}{5}} - \frac{\binom{4}{1}\binom{48}{4}}{\binom{52}{5}}$

3.6.3. a) 8/47 b) $(12 \times 11 \times 10 \times 9 \times 8)/(51 \times 50 \times 49 \times 48 \times 47)$ c) 1/4
d) 1/13 e) 1/13 f) 1/4

3.6.5. a) $b\left(\frac{b-1}{b}\right)^n$ b) $b\left(\frac{b-1}{b}\right)^n + b(b-1)\left(\frac{b-2}{b}\right)^n - b^2\left(\frac{b-1}{b}\right)^{2n}$

3.6.7. a) $n \cdot \frac{26}{52}$ b) $\left(\frac{52-n}{52-1}\right) \cdot n \cdot \frac{26}{52} \cdot \frac{26}{52}$

3.6.9. a) $\frac{N+1}{G+1}$ b) $\sqrt{\frac{BG(N+1)}{(G+1)^2(G+2)}}$

3.6.11. a) $P(x_1, \ldots, x_n) = 1/\binom{n}{g}$ if $x_1 + \cdots + x_n = g$ and 0 otherwise
b) no c) yes

3.6.13. a) Uniform on all ordered $(n+1)$-tuples of non-negative integers with sum $N-n$ c) $(N-n)_w n/(N)_{w+1}$ d) $E(W_i) = (N-n)/(n+1)$,
$E(T_i) = i(N+1)/(n+1)$, 9.6, 10.6, 21.2, 31.8, 42.4 e) $\frac{\binom{n}{1}\binom{N-n}{t}}{\binom{N}{t+1}} \frac{(n-1)}{(N-t-1)}$
f) $P(D_n = d) = P(W_1 + W_{n+1} = N - 2 - d)$. Now use e).
$E(D_n) = \frac{(n-1)(N+1)}{(n+1)} - 1$

3.6.15. c) $t_1 = 0 = t_{n+1}$; $t_2 = t_3 = \cdots = t_n = 1$; so $t = n - 1$.

3.rev.1. a) $1-(5/6)^{10}$ b) $10/6$ c) 35 d) $\frac{\binom{5}{2}\binom{5}{2}}{\binom{10}{4}}$ e) $\frac{1}{2}\left(1-\sum_{k=0}^{5}\left[\binom{5}{k}(1/6)^k(5/6)^{5-k}\right]^2\right)$

3.rev.3. a) $(2x-1)/36$ b) $2/5$ for $y = 1, 2$ and $1/5$ for $y = 3$.
c) $2/36$ for $1 \leq y < x \leq 6$ and $1/36$ for $y = x$ d) 7

3.rev.5. a) -18.4 cents b) 2.111 c) 12.667

3.rev.7. a) 0.1875 b) 0.5 c) 0.219

3.rev.9. a) $5/12$ b) $7/12$ c) 441 d) Approximately 796

3.rev.11. $P(X < 2)$ is largest, $P(X > 2)$ is smallest.

3.rev.15. a) Binomial$(100, 1/38)$ b) Poisson$(100/38)$ c) Negative binomial $(3, 1/38)$ shifted to $\{3, 4, ...\}$ d) 3×38

3.rev.17. a) $N/6$ b) 0.3604

3.rev.19. a) $e^{-p\mu}$ b) 0.6065

3.rev.21. Negative binomial distribution on $\{0, 1, ...\}$ with parameters $r = 3$ and p

3.rev.23. a) $\frac{2^k (n)_k}{(2n)_k}$ b) $H/\sqrt{2n}$ tends to the Rayleigh distribution (See section 6.3).
c) $\sqrt{\pi n}$ d) 17 or so.

3.rev.25. a) Partial answer: $P(Y_1 + Y_2 = 0) = 9/36$, $P(Y_1 + Y_2 = 1) = 12/36$, $P(Y_1 + Y_2 = 2) = 10/36$. b) $10/3$

3.rev.27. 343.047

3.rev.29. c) uniform on $\{0, 1, \ldots, n\}$ d) no, yes e) $\frac{b}{b+w}$ f) $\frac{b+d}{b+w+d}$

3.rev.33. b) $\frac{1}{2^{n-1}}, \frac{1}{2^{n-2}\left(1+\frac{1}{n}\right)}$ d) $\frac{1}{2^{n-3}\left(1+\frac{3}{n}\right)}$

3.rev.37. a) $\binom{n}{k}\frac{(G)_k}{(N)_k}$

3.rev.41. a) 2350 b) 70 c) 9400 d) 8700 e) 730

4.1.1. a) 0.000399 b) 0.000242

4.1.3. a) 6 b) $1/2$ c) $7/27$ d) $13/54$ e) $1/2$, $1/20$

4.1.5. b) $7/12$ c) $1/2$ d) no

4.1.7. 0.096

4.1.9. 0.0418

4.1.11. a) 0.2325 b) 0.6102 c) 0.84

4.1.13. a) $1/16$ b) $n \geq 134$

4.1.15. a) $(0, 1/2)$ b) $\text{erf}(x) = 2\Phi(\sqrt{2}x) - 1$ c) $\Phi(z) = (\text{erf}(z/\sqrt{2}) + 1)/2$

4.2.1. a) $1/32$ b) 3.32 years c) 10 years d) 0.3679

4.2.3. a) 0.6321 b) 0.3935 c) 0.8647 d) 0.99995

4.2.5. a) 0.86 b) 0.73 c) 4 seconds

4.2.7. $-\frac{1}{\lambda}\log(1-p)$

4.2.9. c) $E(T^n) = \Gamma(n+1)$.

4.2.11. b) $e^{-\lambda} - e^{-2\lambda}$

4.2.13. a) 5% per day
b) $(d = 10)$ 6065, 49; $(d = 20)$ 3679, 48; $(d = 30)$ 2231, 42.

4.2.15. a) 80 days b) 40 days c) 0.6472

4.2.17. a) $E(T_{\text{total}}) = 80$ days, $SD(T_{\text{total}}) = 20\sqrt{2}$ days, $P(T_{\text{total}} \geq 60) \approx 0.744$.
b) four spares will do.

4.3.1. a) $1 - G(b)$ b) $G(a) - G(b)$

4.3.3. a) $\left(\frac{b}{b+t}\right)^a$ if $t > 0$. b) $\left(\frac{a}{b+t}\right)\left(\frac{b}{b+t}\right)^a = \frac{ab^a}{(b+t)^{a+1}}$ if $t > 0$.

4.3.5. b) Mean: $\lambda^{-1/\alpha}\Gamma\left(\frac{1}{\alpha}+1\right)$. Variance: $\lambda^{-2/\alpha}\left\{\Gamma\left(\frac{2}{\alpha}+1\right) - \left[\Gamma\left(\frac{1}{\alpha}+1\right)\right]^2\right\}$

4.3.7. b) 9.265 c) About $1 - \Phi(2.456) = 0.007$

4.4.1. Exponential (λ/c)

4.4.3. $f_Y(y) = \frac{1}{2\sqrt{y}}$ if $0 < y < 1$.

4.4.5. If $0 < y < 1$ then $f_Y(y) = \frac{1}{3\sqrt{y}}$; if $1 < y < 4$ then $f_Y(y) = \frac{1}{6\sqrt{y}}$.

4.4.7. Apply Exercise 4.4.6

4.4.9. One to one change of variable formula

4.5.3. a) Y has the same distribution as X.
b) If $0 < r < 1$ then $F_R(r) = r^2$ and $f_R(r) = 2r$.

4.5.5. If $x \leq 0$, then $F_X(x) = \frac{1}{2}e^x$; if $x \geq 0$, then $F_X(x) = 1 - \frac{1}{2}e^{-x}$.

4.5.7. a) $f_Y(y) = 2\lambda y e^{-\lambda y^2}$ $(y > 0)$ b) 0.51 c) Let $Y = \sqrt{-\log(1-U)/\lambda}$

4.6.1. a) 0.0881 b) 0.0056 c) 0.0399

4.6.3. a) $(y-x)^n$ b) $(1-x)^n-(y-x)^n$ c) $y^n-(y-x)^n$
d) $1-(1-x)^n-y^n+(y-x)^n$ e) $\binom{n}{k}x^k(1-y)^{n-k}$
f) $\binom{n}{k}x^k(1-y)^{n-k}+\binom{n}{k+1}x^{k+1}(1-y)^{n-k-1}+\frac{n!}{k!1!(n-k-1)!}x^k(y-x)(1-y)^{n-k-1}$

4.6.5. a) $P(X_{(k)}\le x)=\sum_{i=k}^{n}\binom{n}{i}[F(x)]^i[1-F(x)]^{n-i}$

4.rev.1. a) $ne^{-\lambda t}$ b) $ne^{-\lambda t}(1-e^{-\lambda t})$

4.rev.3. Density: $3x^2$ if $0<x<1$. Expectation: $3/4$

4.rev.5. a) 0.4 b) If $0<t\le 30$, then $P(T>t)=\frac{100-2t}{100}$; If $30<t\le 70$, then $P(T>t)=\frac{70-t}{100}$. c) $f_T(t)=2/100$ if $0<t\le 30$, $=1/100$ if $30<t\le 70$.
d) mean 29, SD 19.8 e) Locate the station at the midpoint of the road.

4.rev.7. a) $\alpha=\frac{\beta}{2}$ b) $E(X)=0$, $Var(X)=2/\beta^2$ c) $e^{-\beta y}$ if $y>0$
d) $1-(1/2)e^{-\beta x}$ if $x>0$; $(1/2)e^{\beta x}$ if $x<0$

4.rev.11. a) $1/2$ b) $\frac{7!}{2^8}$ c) $\frac{100^5}{30}$

4.rev.13. a) $\frac{e^{-\lambda_{\text{loc}}}(\lambda_{\text{loc}})^5}{5!}\frac{e^{-\lambda_{\text{dis}}}(\lambda_{\text{dis}})^3}{3!}$ b) $e^{-3(\lambda_{\text{loc}}+\lambda_{\text{dis}})}\frac{[3(\lambda_{\text{loc}}+\lambda_{\text{dis}})]^{50}}{50!}$ c) $\left(\frac{\lambda_{\text{loc}}}{\lambda_{\text{dis}}+\lambda_{\text{loc}}}\right)^{10}$

4.rev.15. 0.2518

4.rev.19. a) $(20-2)\log_2 10$ b) $20\log_2 10-\log_e 2$

4.rev.21. a) $f_Y(y)=2ye^{-y^2}$ $(y>0)$ b) exponential (1) c) 1

4.rev.23. a) 5, 4 b) $f_M(m)=0.5e^{-0.5(m-3)}$ $(m>3)$ c) 0.3679

4.rev.25. a) uniform $(0,1/2)$ b) uniform $(0,1)$ c) $1/4$, $1/48$

4.rev.27. a) Use the fact that all the $n!$ orderings of $U_1,\ldots,U_n$ are equally likely.

4.rev.29. a) When $c<\sqrt{\frac{2}{\pi}}$
b) Expected net gain is maximized at b satisfying $e^{-b^2/2}=\sqrt{\frac{\pi}{2}}c$.

5.1.1. a) $7/12$ b) $5/36$

5.1.3. $7/12$

5.1.5. a) 0.1 b) 0.81

5.1.7. a) $(1-x)^2$ b) If $0<x<1$ then $P(M\le x)=1-(1-x)^2$ and $f_M(x)=2(1-x)$

5.1.9. $1/4$

5.2.1. a) If $0<|y|<x<1$ then $f_{X,Y}(x,y)=1$ b) If $0<x<1$ then $f_X(x)=2x$; if $0<y<1$ then $f_Y(y)=1-|y|$ c) no d) $E(X)=2/3$, $E(Y)=0$

5.2.3. a) $3/4$ b) $\frac{3}{4}\left(\frac{a^3}{3}+a^2\right)$ c) $\frac{3}{4}\left(\frac{b}{3}+b^2\right)$

5.2.5. $\frac{\mu}{3\lambda+\mu}$

5.2.7. 1/8

5.2.9. a) $2\lambda^2 e^{-\lambda(x+y)}$ $(0 < x < y)$, no; b) $2\lambda^2 e^{-2\lambda x - \lambda z}$ $(x > 0, z > 0)$, yes; c) X is exponential (2λ) and Z is exponential (λ).

5.2.11. a) 3/2 b) 1/2 c) 4/3 d) ∞

5.2.13. The distributions are all the same, with density $2(1-x)$ for $0 < x < 1$.

5.2.15. a) $F(b,d) - F(a,d) - F(b,c) + F(a,c)$ b) $F(x,y) = \int_{-\infty}^{x} \int_{-\infty}^{y} f(u,v)dudv$
c) $f(x,y) = \frac{\partial}{\partial x}\frac{\partial}{\partial y}F(x,y)$ d) $F(x,y) = F_X(x)F_Y(y)$
e) $F(x,y) = y^n - (y-x)^n$ for $0 < x < y < 1$;
$f(x,y) = n(n-1)(y-x)^{n-2}$ for $0 < x < y < 1$

5.2.17. a) $f(x,r) = \frac{2}{\pi}\frac{r}{\sqrt{r^2-x^2}}$ for $0 \le r \le 1$ and $-r \le x \le r$
b) $f(x,r) = \frac{3}{2}r$ for $0 \le r \le 1$ and $-r \le x \le r$

5.2.19. a) $f_{\text{Lon}}(x) = 1/360$ if $-180 < x < 180$
b) $f_{\text{Lat}}(y) = \frac{\pi}{360}\cos\left(\frac{\pi}{180}y\right)$ if $-90 < y < 90$
c) $f(x,y) = \frac{1}{360} \cdot \frac{\pi}{360}\cos\left(\frac{\pi}{180}y\right)$ if $-180 < x < 180$ and $-90 < y < 90$
d) yes

5.2.21. a) 0.3825, 0.765 b) 1/3 c) 0.577 d) 0.5197 ± 0.0048

5.3.1. a) 0.1175 b) 0.1178 c) $\sqrt{\frac{2}{\pi}}$ d) 0.762 e) 0.58 f) 0.3521 g) 0.29

5.3.3. a) $1 - \Phi(0.5)$ b) 1/2 c) 5 d) $\sqrt{14}$

5.3.5. About 2.1

5.3.7. a) 97.72% b) 88.49% c) 0.9795

5.3.9. a) 0.1307 b) 0.0062 c) The answer to b) will be approximately the same.

5.3.11. a) normal with mean 0 and variance $t\sigma^2$ b) $\frac{R_t}{\sigma\sqrt{t}}$ has Rayleigh distribution so R_t has expectation $\sigma\sqrt{\frac{t\pi}{2}}$ and SD $\sigma\sqrt{t\left(\frac{4-\pi}{2}\right)}$ c) 0.1353

5.3.13. c) Try $h(u) = \sqrt{-2\log(1-u)}$ and $k(v) = 2\pi v$

5.3.15. *Hints:* a) Example 4.4.5 b) induction c) linear change of variable

5.3.17. a) Skew-normal approximations: 0.1377, 0.5940, 0.9196, 0.9998, 1.0000
Compare to the exact values: 0.0902, 0.5940, 0.9389, 0.9970, 1.0000
b) 0.441, 0.499. Skew-normal is better.

5.4.1. a) 3/4 b) $f_{X_1+X_2}(z) = z/2$ if $0 \le z \le 1$; $= 1/2$ if $1 \le z \le 2$; $= (3-z)/2$ if $2 \le z \le 3$ c) $F_{X_1+X_2}(z) = z^2/4$ if $0 \le z \le 1$; $= (2z-1)/4$ if $1 \le z \le 2$; $= 1 - (3-z)^2/4$ if $2 \le z \le 3$.

5.4.3. a) If $\alpha \neq \beta$, $f_{X+Y}(z) = \frac{\alpha\beta}{\alpha-\beta}(e^{-\beta z} - e^{-\alpha z})$ b) $\frac{1}{\alpha} + \frac{1}{\beta}$ c) $\frac{\sqrt{\alpha^2+\beta^2}}{\alpha\beta}$

5.4.5. a) Uniform over $(10, 70)$ b) 0.483

5.4.7. a) $f_{XY}(z) = \int_{-\infty}^{\infty} \frac{1}{|x|} f_{X,Y}(z, \frac{z}{x})dx$. b) $f_{X-Y}(z) = \int_{-\infty}^{\infty} f_{X,Y}(x, x-z)dx$.
c) $f_{X+2Y}(z) = \int_{-\infty}^{\infty} \frac{1}{2} f_{X,Y}(x, \frac{z-x}{2})dx$.

5.4.9. $f_X(x) = -\log(x) \qquad (0 < x < 1)$

5.4.11. uniform $(0, 1)$

5.4.13. $f_Z(z) = \frac{1}{2}\lambda e^{-\lambda|z|}$

5.4.15. a) reduce to the case $\lambda = 1$ by scaling. b) $P(Z \leq z) = 2z/(1+z)$
c) $f_Z(z) = 2/(1+z)^2$ for $0 < z < 1$.

5.4.17. a) $\frac{\sqrt{3}}{2}t^2$ if $0 < t < 1$; $\sqrt{3}(-t^2 + 3t - \frac{3}{2})$ if $1 < t < 2$; $\frac{\sqrt{3}}{2}(3-t)^2$ if $2 < t < 3$
b) If $t \leq 1$ then the cross section is an equilateral triangle having side length $t\sqrt{2}$; if $t = 3/2$ then the cross section is a regular hexagon having side length $1/\sqrt{2}$.

5.rev.1. $1 - \frac{\sqrt{2}}{4}$

5.rev.3. a) 0.04 b) 0.039 c) 0.29

5.rev.5. a) $1 - \pi/8$ b) $5/12$

5.rev.7. 0.0124

5.rev.9. a) $f_{X+U}(x) = 1/4$ if $0 < x < 1$; $= 1/2$ if $1 < x < 2$; $= 1/4$ if $2 < x < 3$.
b) Uniform$(-1/2, 1/2)$

5.rev.11. a)$P(X > x) = 1 - (1/2)x$ for $0 < x < 1$. b)$F_X(x) = (1/2)x$ for $0 < x < 1$ and $F_X(x) = 1 - \frac{1}{2x}$ for $x > 1$. c)$f_X(x) = 1/2$ for $0 < x < 1$ and $f_X(x) = \frac{1}{2x^2}$ for $x > 1$

5.rev.13. a) $F_X(x) = 1 - \frac{1}{\pi}\arccos(x)$ for $|x| \leq 1$ b) Y has the same distribution function as X. c) $F_{X+Y}(z) = 1 - \frac{1}{\pi}\arccos\frac{z}{\sqrt{2}}$ for $|z| \leq \sqrt{2}$

5.rev.15. a) 1/6 b) 0 c) 1 d) 1/2 e) 2/3 f) $e^{-\frac{1}{2}}$ g) 0.8759 h) 3/4
i) 0.5737

5.rev.17. a) 0.92 b) About 27.7

5.rev.19. a) $(K_1 = k) = (W_k < \min_{i \neq k} W_i)$; b) $p_k = \lambda_k/(\lambda_1 + \cdots \lambda_d)$; c) use the memoryless property of the exponential waiting times; d) the answer to g) must be p_k by the law of large numbers; e) $\lambda_k T$; f) $(\lambda_1 + \cdots \lambda_d)T$;
g) $\lambda_k/(\lambda_1 + \cdots \lambda_d)$.

5.rev.21. a)$F_R(r) = 1 - e^{-\lambda\pi r^2}$ and $f_R(r) = 2\lambda\pi r e^{-\lambda\pi r^2}$ for $r > 0$ c) $E(R) = \frac{1}{2\sqrt{\lambda}}$; $SD(R) = \frac{1}{2\sqrt{\lambda}}\sqrt{\frac{4-\pi}{\pi}}$ d) mode: $\frac{1}{\sqrt{2\lambda\pi}}$; median: $\sqrt{\frac{\log 2}{\lambda\pi}}$

5.rev.23. a) 1/8 b) 7/19

5.rev.25. a) $\frac{n!x^{k-1}(y-x)^{m-k-1}(1-y)^{n-m-1}}{(k-1)!(m-k-1)!(n-m-1)!}$ $(0 < x < y < 1)$
b) beta $(m-k, n-m+k+1)$ c) beta $(k, m-k+1)$

5.rev.29. Let the spacing between the parallel lines be $2a$. a) If $0 < x \leq a$ then $P(X \leq x) = \frac{2}{\pi a}x$; if $x \geq a$ then $P(X \leq x) = \frac{2}{\pi a}\left[a \arccos\frac{a}{x} + x - \sqrt{x^2 - a^2}\right]$
b) If $0 < x < a$ then $f_X(x) = \frac{2}{\pi a}$; if $x \geq a$ then $f_X(x) = \frac{2}{\pi a}\left[1 - \sqrt{\frac{x^2-a^2}{x}}\right]$

5.rev.31. a) $r = \sqrt{4/3}$ b) no, X_1 and Y_1 are not independent d) $e^{-1/2}$

5.rev.33. a)$r = \sqrt{1/2}$ b) no, X_1 and Y_1 are not independent d) $e^{-1/2}$ g) 2^{-n}

6.1.1. a) binomial $(3, 1/2)$ b) binomial $(3-x, 1/2)$ distribution shifted to $\{x, x+1, \ldots 3\}$ c) Partial answer: $P(X=0, Y=1) = 3/64$, $P(X=1, Y=1) = 3/32$, $P(X=2, Y=1) = 0$, $P(X=3, Y=1) = 0$. d) $P(Y=y) = 1/64, 9/64, 27/64, 27/64$ for $y = 0, 1, 2, 3$. e) Partial answer: The conditional distribution of X given $Y=0$ is given by $P(X=0|Y=0) = 1/3$, $P(X=1|Y=0) = 2/3$. f) For $y = 0, 1, 2, 3$, guess $x = 0, 1, 1$ (or 2), 2 respectively. g) 31/64

6.1.3. a) $P(U=u) = 0, 0.1, 0.4, 0.3, 0.2$ for $u = 0, 1, 2, 3, 4$ b) 0.1125 c) 0.075

6.1.5. b) By the normal approximation to the binomial, $1 - \Phi\left(\frac{89.5-75}{6.124}\right) = 0.0089$

6.1.7. a) Write $P(X=k) = \sum_{n=k}^{\infty} P(X=k, N=n)$. b) 0.0000016

6.1.9. Further equivalent condition: $P(X=x|Y=y, Z=z) = P(X=x|Z=z)$.

6.2.1. a) $E(Y|X=x) = 41/11, 38/9, 33/7, 26/5, 17/3, 6/1$ for $x = 1, 2, 3, 4, 5, 6$
b) $E(X|Y=y) = 1, 4/3, 9/5, 16/7, 25/9, 36/11$ for $y = 1, 2, 3, 4, 5, 6$

6.2.3. a) $E(Y|X=x) = \frac{n+x+1}{2}$ for $x = 1$ to $n-1$ b) $E(X|Y=y) = \frac{y}{2}$ for $y = 2$ to n.

6.2.5. a) $F_1(x)p + F_2(x)(1-p)$ b) $E(X_1)p + E(X_2)(1-p)$
c) $Var(X_1)p + Var(X_2)(1-p) + p(1-p)(E(X_1) - E(X_2))^2$

6.2.7. Condition on the value of N.

6.2.9. a) $j/2$ b) $j + \frac{N-j+1}{k}$ c) $h\frac{j}{i}$ if $h < i$; $j + (h-i)\left(\frac{N-j+1}{k-i+1}\right)$ if $h > i$.

6.2.11. 25/78

6.2.13. c) $\frac{(n-m)}{(n-1)} m \frac{k}{n} \frac{(n-k)}{n}$

6.2.15. a) $E(S) = nE(\Pi)$ b) $Var(S) = nE(\Pi)(1 - E(\Pi)) + n(n-1)Var(\Pi)$ c) As large as possible: Π with values only 0 and 1. As small as possible: Π constant.

6.3.1. $1/3$

6.3.3. $P(Y \le y | X = x) = y/(2 - x)$ for $0 < x < 2$ and $0 < y < 2 - x$

6.3.5. a) If $|x| < 1/2$ then $P(Y \ge 1/2 | X = x) = \frac{1/2 - |x|}{1 - |x|}$. b) One minus the answer in a). c) If $|x| < 1$ then $E(Y|X = x) = (1 - |x|)/2$ d) If $|x| < 1$ then $Var(Y|X = x) = (1 - |x|)^2/12$

6.3.7. a) $f_Y(y) = 3(1 - y)^2$ for $0 < y < 1$ b) $1/9$

6.3.9. a) $2/3$ b) $P(Y \in dp | AB^c) = 6p(1 - p)dp$ for $0 < p < 1$

6.3.11. no

6.3.13. a) $1 - \frac{1}{3}e^{-\lambda}(3 + 2\lambda + \lambda^2/2)$ b) Partial answer: $P(X \in dx | X < Y) = \frac{1 - e^{-\lambda}}{3 - e^{-\lambda}(3 + 2\lambda + \frac{\lambda^2}{2})} dx$ for $0 \le x < 1$ c) $\frac{9 - e^{-\lambda}(9 + 8\lambda + \frac{5}{2}\lambda^2)}{6 - 2e^{-\lambda}(3 + 2\lambda + \frac{1}{2}\lambda^2)}$

6.3.15. b) $\frac{\Gamma(r+s)}{\Gamma(r)\Gamma(s)} \binom{n}{k} \frac{\Gamma(r+k)\Gamma(s+n-k)}{\Gamma(r+s+n)}$
d) mean: $\frac{r+k}{r+s+n}$, variance: $\frac{(r+k)(s+n+k)}{(r+s+n)^2(r+s+n+1)}$

6.3.17. a)Independent negative binomial (r_i, p) $(i = 1, 2)$ b) negative binomial $(r_1 + r_2, p)$ c) negative binomial $(\sum_i r_i, p)$

6.4.1. a) 0.5 b) positively dependent c) 0.2 d) 0.356

6.4.3. less likely; more likely.

6.4.5. Uncorrelated, not independent.

6.4.7. a) Partial answer: $P(X_2 + X_3 = 0, X_2 - X_3 = 0) = 1/3$,
$P(X_2 + X_3 = 1, X_2 - X_3 = 0) = 0$, $P(X_2 + X_3 = 2, X_2 - X_3 = 0) = 1/6$
b) $1/6$
c) uncorrelated

6.4.9. a) $k(n + 1)/2$ b) $\frac{k(n^2 - 1)(n - k)}{12(n - 1)}$

6.4.11. $\sqrt{1/3}$

6.4.13. True: note that $E(N_A N_B) = nP(AB) + n(n - 1)P(A)P(B)$.

6.4.15. d) Write $N_R = \sum_{i=1}^{n} X_i$ and $N_B = \sum_{j=1}^{n} Y_j$, where $X_i = 1$ if the ith spin is red, $= 0$ otherwise; and $Y_j = 1$ if the jth spin is black, $= 0$ otherwise.

6.4.17. Apply Exercise 6.4.16.

6.4.19. a) 375 b) 26.25 c) 0.0021 d) higher; lower.

5.2.9. a) $2\lambda^2 e^{-\lambda(x+y)}$ $(0 < x < y)$, no; b) $2\lambda^2 e^{-2\lambda x - \lambda z}$ $(x > 0, z > 0)$, yes; c) X is exponential (2λ) and Z is exponential (λ).

5.2.11. a) 3/2 b) 1/2 c) 4/3 d) ∞

5.2.13. The distributions are all the same, with density $2(1-x)$ for $0 < x < 1$.

5.2.15. a) $F(b,d) - F(a,d) - F(b,c) + F(a,c)$ b) $F(x,y) = \int_{-\infty}^{x} \int_{-\infty}^{y} f(u,v)dudv$
c) $f(x,y) = \frac{\partial}{\partial x}\frac{\partial}{\partial y} F(x,y)$ d) $F(x,y) = F_X(x)F_Y(y)$
e) $F(x,y) = y^n - (y-x)^n$ for $0 < x < y < 1$;
$f(x,y) = n(n-1)(y-x)^{n-2}$ for $0 < x < y < 1$

5.2.17. a) $f(x,r) = \frac{2}{\pi}\frac{r}{\sqrt{r^2-x^2}}$ for $0 \le r \le 1$ and $-r \le x \le r$
b) $f(x,r) = \frac{3}{2}r$ for $0 \le r \le 1$ and $-r \le x \le r$

5.2.19. a) $f_{\text{Lon}}(x) = 1/360$ if $-180 < x < 180$
b) $f_{\text{Lat}}(y) = \frac{\pi}{360}\cos\left(\frac{\pi}{180}y\right)$ if $-90 < y < 90$
c) $f(x,y) = \frac{1}{360}\cdot\frac{\pi}{360}\cos\left(\frac{\pi}{180}y\right)$ if $-180 < x < 180$ and $-90 < y < 90$
d) yes

5.2.21. a) 0.3825, 0.765 b) 1/3 c) 0.577 d) 0.5197 ± 0.0048

5.3.1. a) 0.1175 b) 0.1178 c) $\sqrt{\frac{2}{\pi}}$ d) 0.762 e) 0.58 f) 0.3521 g) 0.29

5.3.3. a) $1 - \Phi(0.5)$ b) 1/2 c) 5 d) $\sqrt{14}$

5.3.5. About 2.1

5.3.7. a) 97.72% b) 88.49% c) 0.9795

5.3.9. a) 0.1307 b) 0.0062 c) The answer to b) will be approximately the same.

5.3.11. a) normal with mean 0 and variance $t\sigma^2$ b) $\frac{R_t}{\sigma\sqrt{t}}$ has Rayleigh distribution so R_t has expectation $\sigma\sqrt{\frac{t\pi}{2}}$ and SD $\sigma\sqrt{t\left(\frac{4-\pi}{2}\right)}$ c) 0.1353

5.3.13. c) Try $h(u) = \sqrt{-2\log(1-u)}$ and $k(v) = 2\pi v$

5.3.15. *Hints:* a) Example 4.4.5 b) induction c) linear change of variable

5.3.17. a) Skew-normal approximations: 0.1377, 0.5940, 0.9196, 0.9998, 1.0000
Compare to the exact values: 0.0902, 0.5940, 0.9389, 0.9970, 1.0000
b) 0.441, 0.499. Skew-normal is better.

5.4.1. a) 3/4 b) $f_{X_1+X_2}(z) = z/2$ if $0 \le z \le 1$; $= 1/2$ if $1 \le z \le 2$; $= (3-z)/2$ if $2 \le z \le 3$ c) $F_{X_1+X_2}(z) = z^2/4$ if $0 \le z \le 1$; $= (2z-1)/4$ if $1 \le z \le 2$; $= 1 - (3-z)^2/4$ if $2 \le z \le 3$.

5.4.3. a) If $\alpha \ne \beta$, $f_{X+Y}(z) = \frac{\alpha\beta}{\alpha-\beta}(e^{-\beta z} - e^{-\alpha z})$ b) $\frac{1}{\alpha} + \frac{1}{\beta}$ c) $\frac{\sqrt{\alpha^2+\beta^2}}{\alpha\beta}$

Index

S

Santner and Duffy: The Statistical Analysis of Discrete Data
Saville and Wood: Statistical Methods: The Geometric Approach
Sen and Srivastava: Regression Analysis: Theory, Methods, and Applications
Whittle: Probability via Expectation, Third Edition
Zacks: Introduction to Reliability Analysis: Probability Models and Statistical Methods